Environmental and Microbial Biotechnology

Series Editor

Ram Prasad, Department of Botany, Mahatma Gandhi Central University, Motihari, Bihar, India

Innovative and novel advances in microbial biotechnology are providing great understandings in to the machineries of nature, presenting fascinating prospects to apply principles of biology to different arenas of science. Sustainable elucidations are emerging to address the concerns on improving crop productivity through microbes, depleting natural resources, environmental pollution, microbial degradation of pollutants, nanomaterials, nanotoxicity & safety issues, safety of food & agricultural products etc. Simultaneously, there is an increasing demand for natural bio-products of therapeutic and industrial significance (in the areas of healthcare, environmental remediation, microbial biotechnology). Growing awareness and an increased attention on environmental issues such as climate change, energy use, and loss of non-renewable resources have carried out a superior quality for research that provides potential solutions to these problems. Emerging microbiome approaches potentially can significantly increase agriculture productivity & human healthcare and henceforth can contribute to meet several sustainable development goals.

The main objectives have provided an impetus for research on plants and microorganisms that produce novel bio-products with variable properties and understanding their mechanisms of action at cellular and molecular level. Hence, research activities of the environmental and microbial Biotechnology are comprehensively focused up on major sectors viz., bioresources, biorefining, bioremediation of organic and inorganic pollutants, environmental risk analysis of microorganisms, environmental assessment using microbiological indicators, enzymes for environment, food & industrial applications, nanomaterials & nanotoxicity, sustainable ecobiotechnology, biofertilizer, biocontrol agents for agriculture improvement and natural products for healthcare applications.

This book series is a state-of-the-art for a wide range of scientists, researchers, students, policy makers and academician involve in understanding and implementing the knowledge on environmental and microbial biotechnology to develop biologics for proper health care to continue life in smooth and sustainable strategy without any adverse effect.

More information about this series at http://www.springer.com/series/16324

Ram Prasad
Editor

Environmental Pollution and Remediation

Springer

Editor
Ram Prasad
Department of Botany
Mahatma Gandhi Central University
Motihari, Bihar, India

ISSN 2662-1681 ISSN 2662-169X (electronic)
Environmental and Microbial Biotechnology
ISBN 978-981-15-5498-8 ISBN 978-981-15-5499-5 (eBook)
https://doi.org/10.1007/978-981-15-5499-5

This Springer imprint is published by the registered company Springer Nature Singapore Pte Ltd.
The registered company address is: 152 Beach Road, #21-01/04 Gateway East, Singapore 189721, Singapore

Contents

Editor and Contributors

About the Editor

Ram Prasad is presently working as an Associate Professor in the Department of Botany at Mahatma Gandhi Central University, Motihari, Bihar, India. His research interest includes applied and environmental microbiology, plant-microbe-interactions, sustainable agriculture and nanobiotechnology. Dr. Prasad has more than one 175 publications to his credit, including research papers, review articles and book chapters as well as five patents issued or pending, and has edited or authored several books. Dr. Prasad has 12 years of teaching experience and has been awarded the Young Scientist Award and Prof. J.S. Datta Munshi Gold Medal by the International Socicty for Ecological Communications; FSAB fellowship by the Society for Applied Biotechnology; the American Cancer Society UICC International Fellowship for Beginning Investigators, USA; Outstanding Scientist Award in the field of Microbiology by Venus International Foundation; and BRICPL Science Investigator Award and Research Excellence Award. He has been serving as editorial board member of *Frontiers in Microbiology*, *Frontiers in Nutrition*, *Academia Journal of Biotechnology*, *Phyton*, *International Journal of Experimental Botany*, *Archives of Phytopathology and Plant Protection*, *Journal of Renewable Materials*, *Biocell*, *IET Nanobiotechnology*, including Series editor of *Nanotechnology in the Life Sciences*, Springer Nature, USA. Previously, Dr. Prasad served as Assistant Professor, Amity University, Uttar Pradesh, India;

Visiting Assistant Professor, Whiting School of Engineering, Department of Mechanical Engineering at Johns Hopkins University, USA; and Research Associate Professor at School of Environmental Science and Engineering, Sun Yat-sen University, Guangzhou, China.

Contributors

Madhoolika Agrawal Laboratory of Air Pollution and Global Climate Change, Department of Botany, Institute of Science, Banaras Hindu University, Varanasi, India

Ricardo Aguilar-López Departamento de Biotecnología y Bioingeniería, CINVESTAV-IPN, Ciudad de México, México

Amtul Bari Tabinda Akhtar Sustainable Development Study Centre, Government College University, Lahore, Pakistan

M. M. Balamurali Chemistry Division, School of Advanced Sciences, Vellore Institute of Technology, Chennai, Tamilnadu, India

J. Brindha Chemistry Division, School of Advanced Sciences, Vellore Institute of Technology, Chennai, Tamilnadu, India

Monica Butnariu Banat's University of Agricultural Sciences and Veterinary Medicine "King Michael I of Romania" from Timisoara, Timis, Romania

Marian Butu National Institute of Research and Development for Biological Sciences, Splaiul Independentei, Bucharest, Romania

Mihaela Corneanu Banat's University of Agricultural Sciences and Veterinary Medicine "King Michael I of Romania" from Timisoara, Timis, Romania

Milind Shrinivas Dangate Chemistry Division, School of Advanced Sciences, Vellore, Institute of Technology, Chennai, Tamilnadu, India

Alok Prasad Das Department of Life Science, Rama Devi Women's University, Bhubaneswar, Odisha, India

Meenu Gautam Laboratory of Air Pollution and Global Climate Change, Department of Botany, Institute of Science, Banaras Hindu University, Varanasi, India

Anna Gorczyca Department of Microbiology and Biomonitoring, Faculty of Agriculture and Economics, University of Agriculture in Krakow, Krakow, Poland

Anand S. Gupta Amity Institute of Biotechnology, Amity University Mumbai, Mumbai, Maharashtra, India

Pankaj Kumar Gupta Faculty of Environment, University of Waterloo, Waterloo, ON, Canada

Sanjay Kumar Gupta Environmental Engineering, Department of Civil Engineering, Indian Institute of Technology Delhi, New Delhi, India

Saurabh Gupta Department of Microbiology, Mata Gujri College, Fatehgarh Sahib, Punjab, India

Ingrid Hagarová Institute of Laboratory Research on Geomaterials, Faculty of Natural Sciences, Comenius University in Bratislava, Bratislava, Slovakia

Jose Gnanaleela Aswin Jeno Department of Biotechnology, Sri Venkateswara College of Engineering (Autonomous), Sriperumbudur Tk, Tamil Nadu, India

Damanjeet Kaur Department of Microbiology, Mata Gujri College, Fatehgarh Sahib, Punjab, India

Gaganpreet Kaur Department of Biotechnology & Medical Sciences, Baba Farid Group of Institutions, Baba Farid College, Bathinda, Punjab, India

Suphiya Khan Department of Bioscience and Biotechnology, Banasthali Vidyapith, Banasthali Tonk, Rajasthan, India

Ajay Kumar Indian Institute of Technology Roorkee, Roorkee, Uttrakhand, India

Piyush Kumar Amity Institute of Biotechnology, Amity University Mumbai, Mumbai, Maharashtra, India

Ewelina Matras Department of Microbiology and Biomonitoring, Faculty of Agriculture and Economics, University of Agriculture in Krakow, Krakow, Poland

Srishti Mishra Laboratory of Air Pollution and Global Climate Change, Department of Botany, Institute of Science, Banaras Hindu University, Varanasi, India

Sunanda Mishra Department of Life Science, Rama Devi Women's University, Bhubaneswar, Odisha, India

Ekambaram Nakkeeran Department of Biotechnology, Sri Venkateswara College of Engineering (Autonomous), Sriperumbudur Tk, Tamil Nadu, India

Samra Naseem Sustainable Development Study Centre, Government College University, Lahore, Pakistan

Zunaira Naseem Department of Civil Engineering, Monash University, Clayton, VIC, Australia

Lucia Nemček Institute of Laboratory Research on Geomaterials, Faculty of Natural Sciences, Comenius University in Bratislava, Bratislava, Slovakia

María Isabel Neria-González División de Ingeniería Química y Bioquímica, Tecnológico de Estudios Superiores de Ecatepec, Ecatepec de Morelos, EdoMéxico, México

Soumya Pandit Department of Life Sciences, School of Basic Sciences and Research, Sharda University, Greater Noida, Uttar Pradesh, India

Ewa Pociecha Department of Plant Breeding, Physiology and Seed Science, Faculty of Agriculture and Economics, University of Agriculture in Krakow, Krakow, Poland

Ram Prasad Department of Botany, Mahatma Gandhi Central University, Motihari, Bihar, India

Sivashankari M. Ramamoorthi Department of Materials Science and Engineering, Tokyo Institute of Technology, Yokohama, Japan

Ravichandran Rathna Department of Biotechnology, Sri Venkateswara College of Engineering (Autonomous), Sriperumbudur Tk, Tamil Nadu, India

Prasant Kumar Rout Department of Material Science and Engineering, Tripura Central University, Agartala, Tripura, India

Ioan Sarac Banat's University of Agricultural Sciences and Veterinary Medicine "King Michael I of Romania" from Timisoara, Timis, Romania

Neerja Srivastava Department of Biochemistry, IBSBT, CSJM University, Kanpur, India

Alexandros I. Stefanakis School of Environmental Engineering, Technical University of Crete, Crete, Greece
Bauer Nimr LLC, Muscat, Sultanate of Oman

Takeharu Tsuge Department of Materials Science and Engineering, Tokyo Institute of Technology, Yokohama, Japan

Vijaylaxmi Department of Bioscience and Biotechnology, Banasthali Vidyapith, Banasthali Tonk, Rajasthan, India

Abdullah Yasar Sustainable Development Study Centre, Government College University, Lahore, Pakistan

Nanotechnology in Agriculture, the Food Sector, and Remediation: Prospects, Relations, and Constraints

1

Anna Gorczyca, Ewa Pociecha, and Ewelina Matras

Abstract

The applications and benefits of nanotechnology in the agricultural and food sector are now established as a matter of fact, but scientists still have a problem to properly acknowledge the influence of nanocompounds on the environment. We are starting to use nanopesticides and nanofertilizers, but the effects they have on organisms in the environment can be diverse in nature – ranging from the beneficial to the harmful. Environmental microbiome can be strongly modified by nanoparticles. Soil microbiomes are the basis for healthy plants and crops. At the same time, the unique properties of nanomaterials make them particularly suitable for environmental remediation due to their high surface area-to-volume ratio, which often results in higher reactivity. Nanomaterials are starting to be used for the remediation of problematic contaminants (organic and organophosphorus compounds, volatile organic compounds, some heavy metals, or herbicides), and such use offers some very promising prospects. This chapter provides an overview of nanotechnology in general, including a description of the main categories of nanomaterials, the utilization of nanotechnology in the agricultural and food sector, as well as the use of nanomaterials in environmental remediation. Lastly, some toxicological aspects of nanomaterials present in the environment are presented on the basis of the scientific literature in the field.

A. Gorczyca (✉) · E. Matras
Department of Microbiology and Biomonitoring, Faculty of Agriculture and Economics, University of Agriculture in Krakow, Krakow, Poland
e-mail: rrgorczy@cyf-kr.edu.pl; ewelina.matras@urk.edu.pl

E. Pociecha
Department of Plant Breeding, Physiology and Seed Science, Faculty of Agriculture and Economics, University of Agriculture in Krakow, Krakow, Poland
e-mail: rrchilmo@cyf-kr.edu.pl

R. Prasad (ed.), *Environmental Pollution and Remediation*, Environmental and Microbial Biotechnology, https://doi.org/10.1007/978-981-15-5499-5_1

Keywords

Nanotechnology · Nanoremediation · Ecotoxicity · Agriculture · The food industry

1.1 Introduction

Arable land is a limited and, to a large extent, nonrenewable natural resource of productive and cultural importance. The main component of arable land is the soil, the characteristics, and use values of which can be sustained and even enriched through appropriate treatments. Land is a multifunctional asset, which meets many needs and can be used for different purposes. Currently, resources of cultivatable land are estimated to be 1530 mln ha globally. A safe operating space is considered to be an area of 1640 mln ha. The need for arable land is set to grow very quickly and significantly exceed that safe level. It is estimated that by 2030 an additional 81–140 mln ha of arable land will need to be cultivated. This is above all a result of dynamic population growth. United Nations data indicates that, by 2100, the world population will grow by 32% compared to 2017 to reach 11.2 bn (UN 2017). According to the Food and Agriculture Organization of the United Nations (FAO 2009), the global production of cereals will then have to rise by over 70% to meet nutritional needs. In the face of such data, the search for technologies in the field of intensification of agriculture or the recovery of land use value through remediation is a global priority. The green revolution, which in the twentieth century was tasked with eliminating world hunger through the intensification of agriculture, caused many environmental changes, including a significant decrease in biodiversity and the fertility of soil, as well as the pollution of groundwater and the eutrophication of water reservoirs as a result of the use of nitrogen and phosphorus fertilizers (Conley et al. 2009; Grillo et al. 2016; Duhan et al. 2017). Changes in climate, which are contributing to a decrease in agricultural production, are now a global problem (Kang et al. 2009; Venkatramanan et al. 2020). As a result of the decrease in efficiency, the basic prices of food products are rising constantly, which has an adverse impact on the wealth of the population and even on the health of many people (Green et al. 2013). Malnutrition accounts for one in three child deaths worldwide (Bain et al. 2013). In 2016, 815 million people suffered from chronic malnutrition (FAO 2017). According to the World Food Programme, shortages can lead to higher rates of mortality than those of diseases of civilization, such as AIDS, tuberculosis, malaria, and others (Grillo et al. 2016). In addition to guaranteeing sufficient amounts of foods, a modern, aware society also expects food to be healthy and to be produced with care for the environment.

The development of a technologically advanced, sustainable system of land cultivation is necessary, and the use of nanotechnology in order to increase production and allow plants to make better use of nutrients is possible (Chen and Yada 2011; Liu and Lal 2015). The application of nanotechnology may also be a strategy in limiting or eliminating the adverse impact of modern agriculture on the

ape and size at

s Institution has
connected with

that deals with
d compares the
ated differences.
scale dimension
cal applications.
on the nanoscale

ieral nanoscale

sions. The terms
(NP), when the
ent.
d a third larger
fiber.
n the nanoscale

n the nanoscale

e nanostructure.

potential lies in
to develop at a
t logical step in
ie, and physics
its way into an
ials are making
now an area of

concerned with
stances change
xhibit different
rical conductiv-
l magnetic and
s (Prasad et al.
ighter, cheaper,
ninimal amount
ials is above all
Commission's

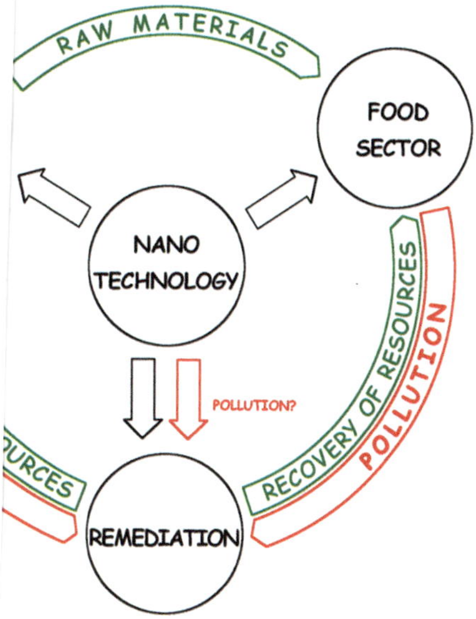

nd and recovery of its use value (Fig. 1.1)
et al. 2018).
g increasingly popular in agriculture and
A, the government is spending over 1 bn
l. The EU, Japan, China, and India are also
iller and Senjen 2008). The use of nano-
ture and remediation is a new and very
al. 2011; Khot et al. 2012; Liu and Lal

?

ogy related to the nanoscale. According to
e's definition, "Nanotechnology is science,
t the nanoscale, which is about 1–100 nm.
e study and application of extremely small
r science fields, such as chemistry, biology,
ing" (https://www.nano.gov/nanotech-101/

oyal Society and the Royal Academy of
on of dimensions at all: *Nanoscience is the*
of materials at atomic, molecular, and
s differ significantly from those at a larger
design, characterization, production, and

application of structures, devices, and systems by controlling s nanometre scale (http://www.nanotec.org.uk/report/chapter2.pdf).

According to Jeevanandam et al. (2018), the British Standar proposed the following basic definitions for some of the terms nanotechnology:

- *Nanoscale*: Approximately 1–1000 nm size range.
- *Nanoscience*: The science and study of matter at the nanoscal understanding their size and structure-dependent properties a emergence of individual atoms or molecules or bulk material–re
- *Nanotechnology*: Manipulation and control of matter on the nan by using scientific knowledge of various industrial and biomed
- *Nanomaterial*: Material with any internal or external structures dimension.
- *Nano-object*: Material that possesses one or more perip dimensions.
- *Nanoparticle*: Nano-object with three external nanoscale dimen nanorod or nanoplate are employed, instead of nanoparticle longest and the shortest axes lengths of a nano-object are differ
- *Nanofiber*: When two similar exterior nanoscale dimensions a dimension are present in a nanomaterial, it is referred to as nan
- *Nanocomposite*: Multiphase structure with at least one phase dimension.
- *Nanostructure*: Composition of interconnected constituent parts region.
- *Nanostructured materials*: Materials containing internal or surfa

Just half a century ago, people did not yet understand how much miniaturization, and it was only in the 1990s that this field starte faster pace. Many scientists considered nanotechnology to be the ne science, integrating engineering with biology, chemistry, medic (Ray et al. 2009). Currently, nanotechnology is slowly making increasingly wide range of fields of human activity, and nanomate numerous innovations possible. Nanoscience and nanotechnology a priority for technological research worldwide.

According to the aforementioned definition, nanotechnology is structures of dimensions of 1–100 nm in size. The properties of su significantly after their reduction to nanosize, and they then characteristics to classic materials. Below 100 nm, thermal and elec ity, or the tensile strength of materials, undergo modification, a optical properties start to be subject to the laws of quantum phys 2016). Nanotechnology makes it possible to provide tools which are and smaller, built from a minimal amount of materials and using a of energy. The fundamental and unique characteristic of nanomate the high surface area-to-volume ratio (according to the Europea definition – greater than 60 $m^2 \cdot cm^{-3}$), which becomes greater as the particle size becomes smaller. This has an effect on the reactivity, toughness, and adsorption properties of nanomaterials and antimicrobial activity. Among nanomaterials, nanoparticles (NPs) of metals are particularly noteworthy – forming a group which has already been commercialized on numerous occasions in many fields, including in medicine, agriculture, the automotive sector, electronics, dentistry, the construction industry, and photography (Bandyopadhyay et al. 2013; Jeevanandam et al. 2018; Thangadurai et al. 2020a, b). The number of patent applications filed in the nanotechnology sector has risen around tenfold over the last two decades, which is indicative of the very major potential of nanomaterials in commercial products (Dang et al. 2010; Kim et al. 2018).

Nanotechnology is something that we consider to be a technology of the future, but we already know that in some areas it will lead to breakthroughs, while in others it will not prove to be any use (Kim et al. 2018). Nanotechnology is inspiring many people to come up with futuristic visions, which are widely covered in the mass media, but these visions are not always confirmed by the results of scientific studies. The information that commercial companies which manufacture products involving different nanomaterials provide on the subject is often incomplete. This means that it is difficult to estimate the actual significance of the nanotechnologies deployed in products currently on the market. The excitement surrounding nanotechnology shows no sign of relenting, but there has now been a breakthrough regarding the uncritical assessment of its significance and safety. More care is starting to be taken in this area. Gross et al. (2018) estimated that every new technology needs at least 20 years to be effectively and safely commercialized. A situation cannot be allowed to occur, where materials and products which have not been thoroughly tested and described find their way onto the market, as the consequences of such an oversight could be very dangerous in many ways. Everyone still remembers the example of asbestos – which was quickly and widely commercialized – and the problems which had to be dealt with later as a result.

1.3 Nanotechnology in Agriculture and the Food Industry

Nanotechnology has the potential to transform agriculture and food production, and the leading examples in this field are innovative tools for the rapid diagnosis of pests, mycotoxins, residues, the nutrient needs of plants, nanopesticides or their nanoformulations, as well as tools for monitoring the quality of agricultural products or new functional materials (Prasad et al. 2014, 2017; Marzbani et al. 2015; Sertova 2015; Anderson et al. 2016; Nuruzzaman et al. 2016; Iavicoli et al. 2017). Nanotechnology may be used to increase yields by improving the capacity of plants to absorb nutrients (Gruère 2012; Tarafdar et al. 2013; Mukhopadhyay 2014). Numerous attempts are being made to develop new nanoagrochemicals (Kim et al. 2018). In addition to this, nanotechnology has also made it possible to find new directions in the development of genetically modified crops and the means of production used in animal breeding, as well as smart systems for the processing, packaging, and monitoring of foodstuffs (Raliya et al. 2013; Luvisi 2016; Iavicoli et al. 2017).

Fertilizers play a key role in increasing agricultural production, whereby their overuse can be the cause of damage to the environment. In comparison to conventional mineral fertilizers, nanofertilizers may significantly improve plant growth and yield, provide plants with necessary nutrients (macroelements or microelements), increase the efficiency of use of fertilizers, and minimize the adverse effect of fertilizers on the environment (Liu and Lal 2015; Giri et al. 2019). Nano-formulation protects against premature degradation and allows release of nutrients in a controlled manner (Nair et al. 2010). Oancea et al. (2009) have confirmed that the controlled release of nutrients enclosed in nanocomposites may be a viable technology for organic farming. The use of a nanocomposite consisting of nitrogen, phosphorus, potassium, microelements, mannose, and amino acids had a positive effect on the cultivation of cereal crops through the increased availability of nutrients (Liu et al. 2006). Zheng et al. (2005) have observed that titanium dioxide TiO_2NPs help with the absorption of water and improve the germination of seeds. They showed that a 30-day treatment of spinach seeds with TiO_2NPs caused an improvement in the dry matter content of seeds, a threefold improvement in photosynthesis efficiency, and a higher chlorophyll content compared to the control. It was found that TiO_2NPs in certain concentrations may increase the absorption of mineral nutrients, accelerate the decomposition of organic substances, and also neutralize the production of reactive oxygen species (ROS). It increases the availability of water and oxygen for seeds and the rate of germination as a result. Yang et al. (2006) have also shown the positive role of TiO_2NPs in plant growth by increasing nitrogen metabolism and photosynthesis. Composites containing calcium phosphate in the form of a nanogel improved the germination of plants *Oryza sativa*, *Arachis hypogea*, and *Amaranthus spinosus* (Umarani and Mala 2013). The foliar application of potassium- and calcium-chelated nanofertilizers had a positive effect on the growth of *Ocimum basilicum* (Ghahremani et al. 2014). Similar results were obtained by Tarafdar et al. (2014) after treatment of *Pennisetum americanum* with zinc oxide (ZnO)NPs. Preparations containing chitosan NPs have been successfully used for controlled release of NPK fertilizers, while hydroxyapatite NPs have been used to similar effect in providing plants with nutrients, especially phosphorus (Corradini et al. 2010; Hasaneen et al. 2014; Sarkar et al. 2015). Nanosized zeolites enriched with nutrients have proved to be equally effective. Malekian et al. (2011) have shown that corn grain yield and dry matter were increased by 4.9% and 10.3%, respectively, in relation to the control after the application of nanozeolites. In addition, the rate of uptake of fertilizer by plants increased from 77.4% to 85.8%. The positive effect of zeolite on wheat has also been shown by Zwingmann et al. (2011), who observed a 64% increase in dry matter compared to the control. Zeolites, including nanometric zeolites, proved to be an ideal substrate for improving the efficiency of uptake of nutrients by plants, better plant growth, as well as improved infiltration and retention of water. It may also contribute to decreasing the negative effects of the use of conventional fertilizers and thus to the reduction of environmental risk.

One of the main problems of modern agricultural production is to limit losses of crop yield caused by plant diseases, pests, and weeds. It is estimated that such losses amount to 20–40% and continue to remain high despite a clear increase in the use of pesticides (Savary et al. 2012). Total losses caused by pests amounted to 26–28% for soybean and wheat and 31%, 37%, and 40% for maize, rice, and potatoes, respectively (Oerke 2006). The use of nanotechnology in agriculture to protect crops against pests appears to be a new, attractive option (Khot et al. 2012; Ragaei and Sabry Al-Kazafy 2014; Abd-Elsalam and Prasad 2018, 2019). The two main directions of use of nanotechnology in the production of pesticides are nanoformulations constituting nanocarriers of active ingredients (AIs) or nanocompounds with biocidal effect.

Nanocarriers protect pesticides against unfavorable environmental conditions (high temperature, radiation) while at the same time increasing their chemical stability. They can also improve the dispersion and wettability of pesticides and reduce the risk of pesticides being washed off plants. Nanoformulations increase surface area, thermal stability, biodegradability, and affinity with regard to target species (Prasad et al. 2019). Smart systems for the delivery of pesticides may be characterized by the controlled release of AIs over time or spatially conditioned release or also result in better penetration through biological barriers, which is unachievable for conventional pesticides. The main function of a nanocarrier or nanoencapsulation is to protect AIs against their release and to improve solubility and tissue penetration. This fits perfectly with practices of sustainable agriculture and the principles of integrated plant protection – small, targeted doses and the absence of any residues in the environment – and as a result, the impact on organisms and people is positive (Xu et al. 2010; Kah and Hofman 2014; Vijayalakshmi et al. 2015; Nuruzzaman et al. 2016; Bhattacharyya et al. 2016; Kumar et al. 2019).

Nanoencapsulation of AIs protects against degradation and improves the effectiveness of pesticides (Nuruzzaman et al. 2016). Similarly, nanoemulsions, in other words colloidal systems with very fine internal phase particles, increase the solubility of AIs in biocidal products while maintaining the concentration of surfactants at a lower level than in microemulsions (5–10% of surfactant required in nanoemulsions, compared to 20% in microemulsions) (Song et al. 2009; Yang et al. 2009; Anjali et al. 2010; Xu et al. 2010). Nanoencapsulated pesticides also display greater resistance to changing atmospheric conditions compared to their conventional counterparts (Choi et al. 2011).

The insecticidal effectiveness of liposome-based agents has been described by Hwang et al. (2011). The initial effectiveness of liposomes in controlling pests was similar to that of conventional pesticides (100% mortality observed). However, after 13 days, the effectiveness of the commercial pesticide decreased significantly, while the pesticide containing liposome retained 50% of its effectiveness. The longer effectiveness can be explained by the slower rate of release of AIs and the protection of unstable AIs resulting therefrom. Teodoro et al. (2010) studied the insecticidal activity of Al_2O_3NPs against selected pests in stocks of food in storage. Three days after the treatment of wheat with Al_2O_3NPs, mortality of the species *Sarocladium oryzae* and *Rhyzopertha dominica* was observed. In turn, the effectiveness of polyethylene glycol-coated nanoparticles loaded with garlic oil against *Tribolium castaneum* was 80%, while the effectiveness of garlic oil alone was 11% (Yang et al. 2009). Adak et al. (2012) investigated the effect of a nanoencapsulated formulation of imidacloprid against the soybean pests *Melanagromyza sojae* and *Bemisia tabaci*. The results of studies showed that nanocapsules are more effective than conventional formulations already at a lower dose and do not form any residues in crops or in the soil. The high potential that nanocapsules also have against *Myzus persicae* has been shown Kang et al. (2012).

Nanotechnology is also used to improve the effectiveness of herbicides (Prasad et al. 2014). The main direction being investigated here is the method of formulation that can be used for conventional and natural herbicides. Nanoencapsulated or NP-embedded herbicides, like the other pesticides, show greater effectiveness over time and at low doses. The secured or embedded AI is introduced into the environment at a lower dose and, working more effectively than conventional herbicides, has a lesser impact on the environment. Kumar et al. (2017) using nanoparticles of pectin (polysaccharides) loaded with herbicide (metsulfuron-methyl) under laboratory and field conditions showed that they are highly effective against *Chenopodium* and found that they can be used as a substitute for conventional herbicides with an effect on reducing the consumption of AIs and thus on environmental safety. At the same time, nanoemulsions of the controversial substance glyphosate showed slightly higher effectiveness against tested weeds than the commercial preparation Roundup (Jiang et al. 2012; Lim et al. 2013). Chi et al. (2017), studying nanocomposites containing glyphosate, found that that formulation allows not only for the controlled release of AIs but also creates a profile for the release of herbicide at a controlled temperature, which reduces losses of AI in the event of precipitation. A small dose, controlled release, and a totally natural herbicide is an excellent solution for weed control, which can be achieved by using nanoformulations of a substance of plant origin, e.g., essential oils. Apart from those mentioned above, other additional benefits resulting from the use of such herbicides is the absence of any effect on pollinators and a repellent effect on harmful insects (Kumar et al. 2017). Hazrati et al. (2017) have developed a nanoemulsion of the essential oil *Satureja hortensis* (carvacrol), which proved to be effective against *Chenopodium album* and *Amaranthus retroflexus*. The activity of nanoemulsions of essential oils of plants may consist of disrupting the germination, the physiological processes, and the growth of weeds.

Searching for effective means of eliminating phytopathogens is another direction of studies in the field of the application of nanotechnology in plant protection. It has been shown that NPs of metal, especially of silver and copper, can also be used as fungicides (Lee et al. 2013; Gopinath and Velusamy 2013; Mishra et al. 2014; Adil et al. 2015; Gupta et al. 2018). Already a decade ago, Bergeson (2010) found over 100 pesticides which contain Ag to have been commercialized due to its strong antimicrobial properties. The biological activity of NPs against pathogens may outweigh the effects exhibited by classic bacterio- and fungicides (Morones et al. 2005). Silver is able to kill 650 pathogens, which are the cause of diseases (Jeong et al. 2005; Holt and Bard 2005; Shrivastava et al. 2007). Silver has an oligodynamic effect, in other words, it works even in minute concentrations (Percival et al. 2005). In turn, copper is the oldest fungicide there is and is one of the few whose use is permitted in organic agriculture. There are numerous examples of the biocidal activity of NPs with regard to phytopathogenic microorganisms under controlled conditions. Laboratory tests have shown the potential of AgNPs produced using a physical and chemical method in controlling *Fusarium* (Kasprowicz et al. 2010, Gorczyca et al. 2015). AgNPs change the metabolism of fungi and can stimulate the production of pigments (Kasprowicz et al. 2013). Aguilar-Mendez et al. (2011) showed that AgNPs significantly inhibit the growth of *Colletotrichum gloeosporioides*. AgNPs also showed an inhibitory effect on the germination of spores of *Bipolaris* and *Magnaporthe*, which are pathogens of cereals (Jo et al. 2009). Lamsal et al. (2011) observed a decrease in the incidence of powdery mildew on cucumber and pumpkin caused by the fungal species *Golovinomyces cichoracearum* and *Sphaerotheca fusca*, after the application of AgNPs. Boxi et al. (2016) combined TiO_2NPs with pure and Ag doped (solid and hollow) to provide an effective means of controlling *Fusarium solani* and *Venturia inaequalis*. AgNPs have also been shown to have a broad spectrum of action against wood-degrading fungi, i.e., against *Gloeophyllum abietinum*, *Gloeophyllum trabeum*, *Chaetomium globosum*, and *Phanerochaete sordida* (Narayanan and Park 2014). Kanhed et al. (2014) have demonstrated the effectiveness of CuNPs against *Curvularia lunata*, *Phoma destructiva*, and *Alternaria alternata*. Saharan et al. (2013), using Cu-chitosan NPs, were able to inhibit the growth and development of *Alternaria alternate*, *Macrophomina phaseolina*, and *Rhizoctonia solani*.

Even NPs, which are produced by the biological method with the use of fungi, bacteria, algae, or plants, show biocidal action against phytopathogens (Pantidos and Horsfall 2014; Elbeshehy et al. 2015; Sabri et al. 2016; Aziz et al. 2016; Prasad et al. 2016; Srivastava et al. 2021). AgNPs have the capacity to form ROS, which cause irreversible damage to bacteria, and also have a strong affinity to bind with DNA or RNA, which interferes with the replication of microorganisms (Aziz et al. 2015, 2019). Like AgNPs, CuNPs can also be used as antibacterial agents (Yadav et al. 2017). It has been found that bimetal NPs (Ag and Cu) display better antibacterial activity than NPs of a single metal (Zain et al. 2014). In turn, a difference in the surface charge of NPs may have an influence on the total community of bacteria in the rhizoplane of the plants. NPs with a positive charge may reduce the total amount of bacteria in contrast to negatively charged NPs, which lead to an increase in the amount of these microorganisms (Gorczyca et al. 2018).

In summary, it is worth underlining that many studies have shown that nanopesticides (nanoinsecticides, nanoherbicides, nanofungicides, nanobacteriocides) pose less of a threat to people and the environment (de Oliveira et al. 2014; Kah and Hofmann 2014; Kumar et al. 2015; Bhattacharyya et al. 2016; Grillo et al. 2016; Nuruzzaman et al. 2016; Wais et al. 2016).

Without a doubt though, from the point of view of protection of the environment, it is most useful to deploy nanotechnology in the detection of undesirable compounds or in precision agriculture. Sensors are used in both these fields. These devices are usually not in themselves nanostructures but make use of the unique properties of nanocompounds, which are elements of their structure and can be parts or coatings.

Precision agriculture is a technique which uses information technology to manage crops in order to maximize production while limiting inputs (fertilizers, pesticides, water, etc.). This is achieved by precise monitoring of the environment and a localized response to needs. Precision agriculture is already successfully making use of remote sensing, the Global Positioning System (GPS), and Geographic Information System (GIS). Precision systems take detailed measurements of local environmental conditions in order to determine the type and area of problems which may affect crops. This data is collected and processed. This management approach also leads to a decrease in the amount of precipitation and in expenditure on labor. Precision agriculture is already making use of nanosensors, nanobiosensors, and smart systems for the transport of different compounds based on nanotechnology. Biosensors are compact analytic devices, which are made up of biological elements such as proteins, oligonucleotides, cells, or tissues. Biosensors consist of a probe (a biologically sensitized element, such as a receptor, enzyme, antibody, nucleic acid, microorganism), a transducer (converts the measurement into computable electrical power), and a detector (collects and transmits data to the microprocessor that it amplifies and records the data). NPs increase the efficiency of biosensors by enabling them to cover a larger area of response. In this technology, carbon nanotubes, nanowires, nanoparticles, or nanocrystals are used for signal transduction. Nanosensors can be used in precision agriculture for the monitoring of soil and of the condition of cultivated crops, including the incidence of disease and physiological state. In addition, they can detect residues of pesticides and other polluting substances in plants or the soil (Shin et al. 2006; Sekhon 2014; Kaushal and Wani 2017; Kim et al. 2018).

As tools of precision agriculture, nanosensors may in the future be decisive in the sustainable development of agricultural intensification. In this field, they will above all be used to counter the excessive consumption of water, fertilizers, and pesticides while maintaining appropriate crop yields (Aktar et al. 2009; Das et al. 2015). Nanosensors are an effective alternative to traditional, time-consuming analytic techniques, such as gas spectroscopy or liquid chromatography. Through multiplexing and real-time detection allowing for detection based on electrochemical, colorimetric, fluorescence, chemiluminescence, surface plasmon resonance, voltammetry, electronic nose, and electronic tongue, using nanobarcode and wireless technologies, nanosensors have revolutionized sensors in the food and agricultural sectors, and this is just the beginning of their application in practice (Srivastava et al. 2018).

In the field of plant protection, Singh et al. (2010) have developed a technique for the detection of *Tilletia indica* in wheat using immunosensors based on nanogold. Yao et al. (2009) used silica nanoparticles to detect *Xanthomonas axonopodis* pv. *vesicatoria* in *Solanaceae* plants. Wang et al. (2012a) used electrode-modified NPs to detect *Sclerotonia sclerotiorum* by monitoring the level of salicylic acid in oil seeds.

Nanosensors utilizing electrochemically functionalized single-walled carbon nanotubes can be used to detect gases typical of agricultural pollutants, such as ammonia, nitrogen oxides, sulfur dioxide, and volatile organic compounds (Wanekaya et al. 2006).

Nanotechnology may also have a useful impact on the food and animal feed industry, by changing the method of production, processing, packaging, and transport. Some achievements relating to the introduction of nanotechnology into the food sector are changes in food quality (improvement in taste, color, and texture; longer shelf life; and reduction in the use of preservatives, salts, fats) (Duncan 2011; Bhushani and Anandharamakrishnan 2014; Ghaani et al. 2016). Thanks to nanotechnology, it is possible to increase the availability of nutrients, dietary supplements, and fodder for animals (Prasad et al. 2017).

Another direction being focused on by nanotechnology is the creation of a new generation of packaging (Prasad et al. 2017). Packaging materials with the addition of nanocomponents are being used to improve protective and mechanical properties and extend product shelf life and as a factor in preventing the development of microorganisms (Berekaa 2015). It is possible to obtain a material with antibacterial properties by coating the surface of packaging with metal NPs, mainly those of silver (Bumbudsanpharoke and Ko 2015). A decrease in numbers of bacteria was observed in packages containing AgNPs for fruit and vegetables and in the case of fruit juices and meat products (Llorens et al. 2012). Positive effects have also been obtained (Cárdenas et al. 2009; Llorens et al. 2012) in the area of the antibacterial properties (inhibition of the growth of bacteria by 3–4 orders or magnitude) of packaging material based on chitosan with the addition of nanocopper with regard to *Staphylococcus aureus* and *Salmonella enterica*.

In turn, food products themselves originating from nanotechnology are something of a novelty and may arouse concern among customers. A lack of knowledge about the potential effects of new technology is widespread, but the awareness of consumers is growing (Donaldson et al. 2004; Peralta-Videa et al. 2011).

In the food industry, like in agriculture, the breakthrough technology is nanosensors used in food safety and quality control. Nanosensors are used for monitoring food ingredients (e.g., sugars, amino acids, alcohol, vitamins, and minerals), detecting harmful pathogens and pollutants, e.g., pesticides, heavy metals, and mycotoxins (Sertova 2015; Fraceto et al. 2016). Nanosensors are employed in the monitoring of the quality of food products during transport and storage (Vanderroost et al. 2014). Nanosensors embedded into food packaging may prove to be an excellent system for monitoring physical product parameters (humidity, pH, temperature, exposure to light), as well as for detecting pathogens and toxins or checking freshness (Fraceto et al. 2016). Nanosensors may also be used to warn consumers, if the food product is not safe for consumption (Sorrentino et al. 2007). Sensors based on NPs can be used to check for food oxidation. This technology has already been applied successfully in milk and meat packaging (Bumbudsanpharoke and Ko 2015). Currently, the only restriction on the deployment of nanosensors in the food sector is the cost of such technology (Fogel and Limson 2016).

1.4 Nanoremediation

The soil environment is a complex system, and its quality depends on biological and biochemical processes and chemical composition. Water guarantees that reactions and the development of organisms take place correctly, forms part of the composition of organisms, transports compounds, dissolves them, and thus is necessary for all processes occurring in ecosystems. Water also forms a separate ecosystem that provides an environment for numerous living organisms. The amount and quality of water and soil are basic factors in determining sustainable socioeconomic development. The technological and economic progress which is currently being made disrupts the balance of ecosystems on land and in water. One of the most important factors contributing to this phenomenon is the large amount of toxic substances emitted into the environment. Toxic substances lead to degradation of the water and soil. Water and soil are limited resources, and it is estimated that they will be insufficient in the near future. There is a clear and urgent need to develop new technologies to improve the quality of or recover these resources. Nanotechnology may have an impact on reducing the production of waste. As already indicated, in agriculture, nanofertilizers and nanopesticides limit the contamination of soil and water caused by agrochemicals (Guerra et al. 2018). Nanotechnology may also provide cheaper technologies for desalination and water treatment and enable the development of technology using renewable energy sources, e.g., solar energy conversion (Wallentin et al. 2013; Wang et al. 2013). Sensors based on nanomaterials may be used for monitoring water quality by detecting various pollutants. This is an important contribution to protection of the environment, but nanotechnology also has a role to play in soil and water remediation. Nanocompounds are considered to be more effective in remediation processes due to their aforementioned unique thermal, optical, mechanical, electromagnetic, structural, and morphological properties. NPs have the capacity to adsorb or break down pollutants by means of ion exchange, oxidation, reduction, surface processes, adsorption, precipitation, complexation, and electrostatic interactions. Nanocompounds are used in remediation as nanoadsorbents, nanomembranes, nanosensors, and disinfectants (Das et al. 2015; Ibrahim et al. 2016; Fernández-Luqueño et al. 2017).

It has been shown that NPs can effectively break down organic halogenated hydrocarbons, polycyclic aromatic hydrocarbons, nitrates, heavy metals, and dyes (Medina-Pérez et al. 2019). One of the nanomaterials most commonly used in environmental remediation is nano zerovalent iron (nZVI), which is a highly efficient and environmentally friendly adsorbent, thanks to its excellent mobility, high specific surface area, high reactivity, high rate of reaction, and low toxicity (Zou et al. 2016a; Pang et al. 2018). Qiu et al. (2011) demonstrated that decabromodiphenyl ether, which easily accumulates in soil, can be broken down using nZVI. nZVI was also found to be effective to 78% in the process of removing decabromodiphenyl ether from soil by Xie et al. (2016). Similarly, Kanel et al. (2006) proved that nZVIs successfully remove As(V) from groundwater. Zhang (2003) showed that nZVIs are effective in the transformation and removal of chlorinated organic solvents, heavy metals, polychlorinated biphenyls (PCBs), and organochlorine pesticides commonly found in polluted soil and groundwater. In field tests, Henn and Waddill (2006) noted a reduction in trichloroethene (TCE) with the use of nZVI. In their tests, Macé et al. (2006) showed there to be a reduction in volatile organic compounds (VOCs) found in groundwater treated with nZVI. The authors suggested that nZVIs may be effective over large areas, as they are transported in the groundwater over large distances from the point of introduction. In recent times, the number of real applications of nZVIs is rapidly growing. PARS Environmental Inc. estimated the potential costs of remediation of land polluted with TCE and perchloroethylene (PCE) in a production plant in New Jersey (Karn et al. 2009). Application of the pump and treat method was estimated to cost around 4,160,000 USD, and the permeable reactive barrier (PRB) method at 2,200,000 USD, while the estimated cost of nanotechnology with nZVI was 450,000 USD, a saving of around 80–90% compared to the pump and treat method. The use of nanotechnology not only reduces operating costs but also reduces the time for which employees are exposed to the contamination occurring on the site of the land being reclaimed. Zhang (2003) noted a 99% reduction in the level of TCE within a few days after use of nZVI.

The sorption of pollutants, mainly heavy metal ions, by nanomaterials depends to a large extent on environmental conditions (Wu et al. 2019). Lv et al. (2012) demonstrated that sorption of Cr(VI) by nZVI-Fe_3O_4 rises to 96.4% within 2 h. Temperature may have an influence on the rate of removal of pollutants by changing the energy in the reaction system. Xiong et al. (2014) proved that sorption of Cr (VI) by the composite γ-Fe_2O_3/C@HKUST-1 was more favorable at higher temperatures. Under neutral pH conditions, nanomaterials are more effective at removing heavy metal ions, thanks to surface coordination, electrostatic sorption, and precipitation. According to Zou et al. (2016b), sorption of Pb(II) by g-C_3N_4 and g-C_3N_4/β-CD was more efficient in the range of pH 4.0–7.0 than in that of pH 2.0–4.0. Under extreme pH conditions, the detoxication of heavy metal ions by nanomaterials decreases significantly, due to the strong electrostatic repulsion between the adsorbate and the adsorbent (Bhowmick et al. 2014). The dose of adsorbent is also a key factor influencing the capacity to remove pollutants. Fu et al. (2015) showed, that together with an increase in the dose of sepiolite-supported nanoscale zerovalent iron (S-nZVI), the efficiency of removing Cr(VI) rose from 45.1% to 99.2%.

With a view to increasing the stability of solutions of NPs and make them more effective, nanomaterials can be deliberately improved with different stabilizers, surfactants, specific chemical substances, or functional groups (Campbell et al. 2015; Guerra et al. 2017). The adjustment of the physical and chemical properties of nanomaterials, i.e., their size, morphology, porosity, and chemical composition, may give them additional positive characteristics that may be useful in environmental remediation (Pandey and Fulekar 2012; Tao 2012). nZVI produced by the borohydride method (surface of 20–40 $m^2 \cdot g^{-1}$) shows 10–1000 times higher reactivity than Fe (surface of $<1\ m^2 \cdot g^{-1}$) (Wang and Zhang 1997). In their studies, Shi et al. (2011) used nZVI and bentonite-supported nanoscale zerovalent iron (B-nZVI) to remove pollutants caused by Cr(VI) in solution in water and soil. B-nZVI was very effective at reducing aggregation and increasing specific surface area compared to nZVI. Liu et al. (2015) demonstrated that the composite nZVI@$Mg(OH)_2$ can remove 94% of Pb(II) within 15 min. The maximum sorption capacity was 1986 $mg \cdot g^{-1}$. The stabilizer $Mg(OH)_2$ prevents the aggregation and oxidation of nZVI particles by high specific surface area and the characteristic pore structure. Li et al. (2015) used nZVIs with the addition of bentonite (nZVI/Al-bent) for the adsorption of Se(VI) from aqueous solutions. Efficiency of removal of Se(VI) was reported to be 95.7%. Wang et al. (2009) demonstrated increased dehalogenation of chlorinated organic compounds under the influence of nZVIs with the addition of Pd. Su et al. (2012) showed there to be an 86% decrease in the total weight of chlorinated volatile organic compounds (CVOCs) found in groundwater caused by nanoparticles of emulsified zerovalent iron (EZVI) over a period of 2.5 years. Mackenzie et al. (2016) used a composite made up of colloidal active carbon and embedded nanoiron (carbo-iron) to remove tetrachloroethene (PCE) from polluted land in Germany. The authors recorded a successful outcome, as the initial concentration of PCE was determined to be 20 $mg \cdot L^{-1}$, while after 1000 days from the application of nanoiron, the concentration had decreased to 8–9 $mg \cdot L^{-1}$. Poguberovic et al. (2016) successfully used nZVIs synthesized using extracts from the leaves of oak, mulberry, and cherry to remove As(III), Cr(VI), Ni(II), and Cu (II) from aqueous solutions. pH was to a large extent decisive in determining the effectiveness of the process. The "green" synthesis of nZVIs demonstrated by the authors makes them safer and contributes to sustainable development – the use of extracts from leaves, which are considered to be waste, makes it an inexpensive adsorbent of toxic metals. Such results are very promising, but further studies are necessary in order for this technology to be deployed in sewage treatment.

Titanium compounds, which are used in photocatalytic processes, also have some very interesting properties from the point of view of applications in protection of the environment. Photocatalysis is an effective and advanced technology for the removal of organic pollutants from water and air (Nakata and Fujishima 2012). TiO_2NPs are the most frequently studied photocatalysts due to the strong oxidation capacity, superhydrophilicity, chemical stability, long-term stability, nontoxicity, and low cost (Fujishima and Zhang, 2006). Rasalingam et al. (2014) demonstrated the high activity TiO_2-SiO_2 mixed oxide nanomaterials in degrading a chemical organic compound – the dye methylene blue. This composite may have a potential application in industrial sewage treatment systems. The photodecomposition of harmful perfluorooctanoic acid (PFOAs) from sewers also proved to be efficient, thanks to the application of photocatalysts based on TiO_2NPs, Ga_2O_3NPs, and In_2O_3NPs under the conditions of UV radiation (da Silva et al. 2017). Compared to other nano-photocatalysts, In_2O_3NPs showed the best potential for degrading PFOA while using less energy over the same period of time (Gao and Chorover 2012, Chularueangaksorn et al. 2014). TiO_2NPs are also capable of producing reactive oxidants, i.e., hydroxyl radicals, which can act as a disinfectant against various pathogens (Zan et al. 2007).

In addition to metals, silica materials, carbon nanotubes, graphene, lime powder, hydroxyapatite, and polymers can also be deployed to good effect in the process of environmental remediation. Nano-silica have been used effectively for the removal of aldehydes, ketones, heavy metals, CO_2, H_2S, and cationic dyes (Guerra et al. 2018). Nanocomposites of attapulgite proved to be repairing agent used in mining areas and on agricultural land polluted with heavy metals (Medina-Pérez et al. 2019). Carbon-based nanomaterials have been shown to be capable of adsorbing and highly effective in the removal of fluorides and gases, i.e., SO_X, H_2, NH_3, heavy metals, benzene, pesticides, pharmaceuticals, and organic dyes (Ibrahim et al. 2016). Polymer-based nanomaterials have also been the subject of many studies. The properties of those materials were particularly useful in the process of removing polynuclear aromatic hydrocarbons, heavy metals, volatile organic compounds (VOCs), metal ions, dyes, and microorganisms (Tungittiplakorn et al. 2004; Campbell et al. 2015).

The results of studies to date indicate that nanoremediation is a good alternative to traditional technologies. The choice of the best nanomaterial to remove or reduce pollution in a specific environmental medium requires a full analysis of the type and concentration of the pollution removed, the suitable characteristics of the remediated land, the amount of material necessary for the deployment of effective remediation, and the possibility of recycling (Karn et al. 2009; Ram and Aranda 2018). Studies are also necessary to make it clear what happens to nanomaterials used to capture or degrade pollutants to avoid the possibility that these materials might prove to be a source of environmental pollution. Nonetheless, nanotechnology provides many effective strategies which can be used for the removal and prevention of environmental contamination (Guerra et al. 2018).

1.5 Ecotoxicity of Nanomaterials

Nanotechnology has been accepted by the whole of the scientific community as an innovative and rapidly developing advanced technology. It has embraced nearly all scientific disciplines, and its development is resulting in numerous applications. At the same time, the uncontrolled presence of nanomaterials in the environment may pose a threat to human health and cause changes in ecosystems (Drake and Hazelwood 2005). Literature is available, many aspects of which point to the toxic activity nanomaterials on living organisms (Yang and Watts 2005; Lin and Xing 2007; Kasemets et al. 2009; Ge et al. 2011; Rana and Kalaichelvan 2013; Du et al. 2017; Tripathi et al. 2017a, b; Prasad 2019).

The growing production and use of nanomaterials inevitably lead to their migration through trophic chains to all components of the environment (Klaine et al. 2008). One implication of nanomaterials may be their accumulation in the environment, and the main place where they will be deposited is the soil (Ruffini and Cremonini 2009). The negative effects that nanomaterials can cause in the environment have already been shown, including the disturbance of biodiversity of soil microorganisms, the influence on the circulation of elements, and the transfer of nanomaterials accumulated by microorganisms to higher trophic levels (biomagnification) (Werlin et al. 2011). The properties of NPs, i.e., size, chemical composition, surface structure, catalytic processes, form, or solubility in water and fats, may be modified through their contact with the soil environment, which may subsequently have an effect on their transport and toxicity (Sayes et al. 2004; Cornelis et al. 2010; Tolaymat et al. 2010; Shoults-Wilson et al. 2011; Sagee et al.

2012; VandeVoort and Arai 2012; Benoit et al. 2013; Rana and Kalaichelvan 2013; Jampilek and Kralova 2015).

Plants are an important component of most ecosystems on which many environmental pollutants including nanomaterials have an effect. The effect on plants depends on the properties of nanoparticles (NPs), i.e., size, shape, concentration, surface coating, reactivity, and most importantly on the dose. The major sources of contaminants are the soil and air. Soil is also the final sink of NPs, which can pose a threat to soil microorganisms and plants (Cornelis et al. 2014).

Exposure to NPs has many physiological and biochemical consequences in plants. Dietz and Herth (2011) determined five models for interaction between NPs and living organisms: (a) NPs release metal ions in solution, which causes a chemical effect; (b) hard spherical particles and their aggregates produce mechanical effects; (c) catalytic effects occur on the surface of NPs; (d) the surface of the NPs binds with proteins non-covalently or covalently or produces oxidative effects; (e) NPs change the chemical environment, in particular the pH. Plants uptake NPs and transport them to aerial parts of the plant where they cause beneficial or adverse effects (Siddiqui et al. 2015). A number of studies have demonstrated both the positive and phytotoxic effects of NPs in plants. However, due to the inability to decipher the mechanism of how NPs work, contradictory results, for instance, a reduction or increase in chlorophyll content or the inhibition or stimulation of growth, look set to remain commonplace.

After the uptake of NPs across the cell wall and cell membrane of the root epidermis, they enter the plant vascular bundle (xylem) and move to the leaves by long-distance transport. The cell wall, as a porous network of polysaccharide fibers, acts as a semipermeable barrier which facilitates only the selected particles to pass through the pores (Lin et al. 2009; Kurepa et al. 2010; Miralles et al. 2012). The uptake of NPs via pores on the cell membrane is also size-specific (Ma et al. 2010). As a consequence, regulation of the transport and accumulation of nanoparticles depend on the differences in hydraulic conductivity and the pore size of the cell wall. The mechanical presence of AgNPs in intracellular spaces affects the function of transporter proteins and blocks cell connections. Aggregated AgNPs (Geisler-Lee et al. 2013) and ZnONPs (Ma et al. 2010) have been found in the cell wall, indicating blockage of intercellular communication through obstruction in the plasmodesmata. Thus, mechanical disruption may affect the transport of nutrients and other materials when they are present in the plasmodesmata and vascular tract (Zhai et al. 2014). Aquaporins may be also involved in the regulation of uptake of NPs (Abu-Hamdah et al. 2004). The most important way via which NPs are distributed and translocated is the xylem, but transport in the reverse direction to the root through the phloem is also possible, as for example, in the case of the transport of nCuO (20–40 nm) in *Zea mays* (Wang et al. 2012b). Foliar uptake of NPs is possible via cuticular and stomatal routes (Buick et al. 1993). Sizes of $\geq$40 nm enable the transfer of hydrophilic particles through the pores of stomata (Eichert and Goldbach 2008). Moreover, leaves exposed to NPs accumulated them in the stomata instead of in the xylem and translocated them to other parts of the plant via the phloem.

The reported data from various studies suggests that effect of NPs on plant growth and seed germination depends on their properties and doses. In most cases, the view is that NPs cause degradation in the quality of crops because, after penetration into the plant, they decrease the germination rate of seeds, decrease fresh and dry biomass as well as the length of roots and shoots, and affect the functioning of many physiological processes. Nanoparticles can affect the time of flowering, fruiting, senescence, and dormancy (Vernay et al. 2008; Thul and Sarangi 2015) and thereby impact the overall growth and development of plants. Upon interaction with plants, NPs can lower the rate of transpiration in *Cucurbita pepo* treated with AgNPs (Musante and White 2012) and in *Zea mays* treated with TiO_2NP at concentrations of 30 and 1000 $mg{\cdot}L^{-1}$, respectively (Asli and Neumann 2009). AgNPs and ZnONPs in *Z. mays* and *Brassica oleracea* var. *capitata* were able to reduce the size of vacuoles, which consequently led to a reduction in the cell turgidity and size of the cell (Pokhrel and Dubey 2013). Exposure to ZnONPs resulted in a shrinkage of root tips due to disintegration of cortical, and epidermis cells (Ma et al. 2010). Zhao et al. (2013) in turn observed the collapsed structure of vacuoles, cortical cells, epidermis, and roots after treatment with ZnONPs at a concentration of 1000 $mg{\cdot}L^{-1}$. Similarly, $C_{70}(C(COOH)_2)_4 - 8$NPs have an adverse effect on root elongation, whereas they stimulate negative geotropism in *A. thaliana* roots (Liu et al. 2010). The effect of silver nanoparticles at a concentration of 0.5 $mg{\cdot}L^{-1}$ was phytotoxic in *O. sativa* for root and shoot biomass, while the increase in concentration also negatively affected the leaf surface, chlorophyll, and carotenoid content (Nair and Chung 2014).

In the literature, not only the adverse effects of AgNPs on plant vegetative growth were observed but also the phytostimulatory ones. For instance, in *Crocus sativus*, the application of AgNPs, under conditions of flooding, may promote root growth by blocking ethylene signaling (Rezvani et al. 2012). Some NPs, such as AgNPs, in adequate doses activate the gene expression involved in cellular events in *Arabidopsis* and promote the accumulation of proteins that are related to the cell cycle, chloroplast biogenesis, and carbohydrate metabolism (Syu et al. 2014). AgNPs can also enhance root growth of *Arabidopsis* and *Brassica juncea* (Syu et al. 2014; Sharma et al. 2012).

The detrimental impacts of NPs on photosynthesis have been manifested in the form of net decreases in the rate of photosynthesis, impediment to seedling growth and enhancement of chlorosis, and defoliation after exposure to TiO_2NPs (Siddiqui et al. 2015). In *O. sativa* seedlings grown in a hydroponic system with CuONPs at concentration 1000 $mg{\cdot}L^{-1}$, photosynthetic rate, transpiration rate, stomatal conductance, maximal quantum yield of PSII photochemistry, and photosynthetic pigment contents declined, with a complete loss of PSII photochemical quenching (Da Costa and Sharma 2016). In turn, ZnONPs inhibited the expression of genes involved in chlorophyll synthesis and photosystem structure (Wang et al. 2016). In another study on spinach leaves, TiO_2NPs increase the Hill reaction through enhancing light absorption in chlorophyll a, electron transfer, and oxygen evolution rate (Hong et al. 2005; Su et al. 2007; Zheng et al. 2007a, b; Wang et al. 2008; Monica and Cremonini 2009). Interactions between gold nanoparticles and plants have been related to the increase in chlorophyll a fluorescence quenching due to enhanced electron transfer (Falco et al. 2011). NPs also affect the amount of photosynthetic pigments. CuONPs and AgNPs reduced carotenoids and chlorophylls content in soybean, mustard, and rice (Nair and Chung 2014), while in mung bean plants only reduction in chlorophyll content occurred (Nair et al. 2014). Similarly, ZnONPs at concentrations of 125, 250, and 500 mg·L^{-1} decreased chlorophyll content in *Pisum sativum* (Mukherjee et al. 2014). SiO_2NPs enhance the activity of carbonic anhydrase, the enzyme supplying CO_2 to the Rubisco and synthesis of photosynthetic pigments, which may improve photosynthesis (Siddiqui et al. 2014; Xie et al. 2012). In turn, AgNPs induced a sharp decrease in photosynthetic performance due to a change in the maximum quantum yield for primary photochemistry and electron transport activity in the green algae *Chlorella vulgaris* and *Dunaliella tertiolecta*. The chlorophyll a fluorescence yield decreased strongly with elevated concentrations of AgNPs, which may be explained by the inhibition of electron transport at the donor side of the PS II (Oukarroum et al. 2012).

It has also been shown that TiO_2NPs have an influence on light absorption and electron transfer due to the photocatalytic and thermal conductivity properties of this nanomaterial (Rezaei et al. 2015; Rico et al. 2015). Some studies indicate that electron transfer between metal NPs can increase photosynthetic performance by inducing the efficiency of chemical energy production in photosynthetic systems. The chlorophyll molecule in the reaction center binds to the AuNPs and Ag nanocrystals, thus creating a new hybrid system that can produce ten times more excited electrons, thanks to plasmon resonance and fast electron-hole separation (Govorov and Carmeli 2007). Such mechanisms can thus help in the design of artificial light-harvesting systems.

A number of reports are available which show phytotoxicity of NPs caused by the generation of reactive oxygen species (ROS), which result in oxidative stress, lipid peroxidation, damage to DNA and proteins, and apoptosis-related changes (Arruda et al. 2015; Kim and Ryu 2013; Panda et al. 2011). Interaction of NPs with the cellular components – namely, chloroplasts, mitochondria, peroxisomes, and plasma membranes – may result in the formation of ROS, such as singlet oxygen ($^1O_2$), superoxide radical ($O_2^{\bullet -}$), hydroperoxy radical ($HO_2^{\bullet}$), hydrogen peroxide (H_2O_2), and hydroxyl radical ($OH^{\bullet}$). ROS molecules are continually removed nonenzymatically or enzymatically by a complex antioxidant system including the following enzymes: catalase (CAT), ascorbate peroxidase (APX), superoxide dismutase (SOD), guaiacol peroxidase (GPX), glutathione reductase (GR), dehydroascorbate reductase (DHAR), and low-molecular-weight ascorbate and reduced (GSH) and oxidized (GSSG) glutathione. The literature indicates that NPs modulate both enzymatic and nonenzymatic antioxidants. However, it still remains unclear how the chemical properties of NPs induce ROS formation in plants. The effects of interactions of NPs with antioxidant enzymes reported by numerous studies seem to be irregular and in some cases contradictory. Oxidative stress in plants exposed to metal NPs was caused by 10-nm-sized ZnONPs at a concentration of 500 ppm in *P. sativum* (Mukherjee et al. 2014) and by (PVP-coated) AgNPs of 20 nm in size in *A. thaliana* (Kaveh et al. 2013). Exposure to NiONPs caused higher activity of SOD and CAT, as well as higher GSH content and lipid peroxidation in

Lycopersicum esculentum (Faisal et al. 2013). The presence of $nCeO_2NPs$ in concentrations of 62.5, 125, 250, and 500 mg·L^{-1} induced generation of ROS and oxidative damage in *O. sativa* (Rico et al. 2013a, b; Siddiqui et al. 2015). TiO_2NPs at concentrations of 10 and 30 ppm modified the activities of enzymatic antioxidants in *Phaseolus vulgaris* (Jacob et al. 2013). Further, reduced GSH was found under quantum dot exposure at a concentration of 5.8 μM in *A. thaliana* (Navarro et al. 2012), while higher ascorbic acid levels were induced in *Asparagus officinalis* under treatment with AgNPs (An et al. 2008).

Moreover, according to Wei and Wang (2013), nanoparticles exhibit antioxidant capacity in their own right and mimic the activity of natural antioxidative enzymes. The following nanoparticles exhibit enzyme-like activities: CeO_2NPs, PtNPs, and fullerenes demonstrate properties similar to superoxide dismutase; CeO_2NPs, Fe_3O_4NPs, and Co_3O_4NPs mimic catalase; CeO_2NPs, Fe_3O_4NPs, Co_3O_4NPs, MnO_2NPs, CuONPs, and AuNPs exhibit peroxidase-like activity.

The presence of numerous NPs at some concentrations demonstrates detrimental effects on plants altering not only their morphological, physiological, and biochemical traits but also their genotoxic traits (Panda et al. 2011). NPs enter plants by various pathways, by targeting the genes or DNA of particular organelles and through the transporters (Sahebi et al. 2015; Siddiqui et al. 2015). When NPs enter the cell, a whole range of macromolecular interactions begins. It has been found that AgNPs change the expression of the matrix metalloproteinases, tumor necrosis factor (TNF), and interleukin (IL) -12 and IL-1 genes (Bhol and Schechter 2005). In addition to this, Kaveh et al. (2013) observed that, in *Arabidopsis*, AgNPs upregulated the expression of genes that are related to the response to metals and oxidative stress but downregulated the expression of genes involved in the response to pathogens, i.e., systemic acquired resistance (SAR) and hormonal signals. Furthermore, another study showed that AgNPs induce the expression of genes involved in ABA signaling and auxin-responsive genes as IAA8 (Arase et al. 2012). Genes downregulated in response to NPs were primarily related to cell organization and biogenesis (Atha et al. 2012; Landa et al. 2012). It was shown that carbon nanotubes modify the DNA structure in plant tissues. Multiwalled carbon nanotubes (MWCNTs) caused chromosomal aberrations in the roots of *Allium cepa*, which affected the expression of genes involved in cell division and apoptosis (Ghosh et al. 2011). In turn, the proximity of single-walled carbon nanotubes (SWCNT) to DNA led to the unzipping of the strands, impairing the normal matching of the nucleobases in rice DNA (Katti et al. 2015).

NPs showing biocidal properties against pathogens may also be toxic to positive microorganisms. The activity of NPs depends on many parameters: shape, size, the limiting or stabilizing agent, and reduction method used (type of reducing agent used in production) and surface charge (Pal et al. 2007; Carlson et al. 2008; Choi et al. 2009; Jiang et al. 2009; Kvitek et al. 2009; Badawy et al. 2010, 2011; Abbaszadegan et al. 2015; Kujda et al. 2015; Rana and Kalaichelvan 2011). The mechanisms most frequently involved in the toxic effects shown by NPs are damage to cell membranes, generation of ROSs, and genotoxicity (Jampilek and Kralova 2015). In the case of bacteria, AuNPs can modify their cell membrane; interfere with the binding of tRNA to ribosomes, which are necessary for the synthesis of proteins; and, through binding with DNA, inhibit transcription (Cui et al. 2012). ZnONPs are highly toxic to bacteria, as they may lead to disintegration of the cell membrane and cause oxidative stress, and as a result cause cell damage and inhibit growth (Xie et al. 2011). In vitro tests with TiO_2NPs and fullerenes have confirmed the formation of ROSs in microorganisms (Sayes et al. 2004). SiO_2NPs may limit the growth and development of bacteria and reduce bacterial adhesion, which disrupts the proliferation of bacterial cells (Lv et al. 2010; Besinis et al. 2014).

The toxicity of specific NPs has not yet been clearly determined, because it differs widely, and it is not possible to establish a common criterion. Every NP, depending on the type of material from which it was produced, is characterized by its own different structure, shape, surface, physical and chemical properties, solubility, and, as a result, cytotoxicity (Navarro et al. 2008; Kookana et al. 2014; Prasad et al. 2017). This is also why it is necessary to devote more attention to toxicological tests on nanomaterials concerning the risk related to their accumulation and retention in soil and their direct impact on organisms at different trophic levels, which, as a result, may lead to the dysfunction of agrocenosis components (Rana and Kalaichelvan 2013; Tripathi et al. 2017b). Ecotoxicological studies should pay particular attention to the environmental consequences of nanomaterials (Cox et al. 2017; Singh et al. 2017). The results obtained will be important for further research concerning the potential of NPs.

1.6 Conclusion

Nanotechnology may contribute to sustainable development of the bioeconomy, environmental protection, and remediation. Many benefits have been diagnosed (new technologies, improvements, modifications, less waste, less expenditure), resulting from nanotechnology. However, the success of nanotechnology in practice will depend not only on effective implementations but also on a comprehensive assessment of the safety of the use of nanocompounds. Innovations in nanotechnology are of paramount importance in meeting contemporary global challenges – population growth, climate change, depletion of natural resources, and sustainable development. At the same time, nanomaterials pose a certain threat to the natural environment and organisms, including people. It is for this reason that the envisaged intensified application of nanomaterials will require a thorough environmental assessment at the outset, so that it does not prove to be the cause of adverse disturbances in ecosystems. In understanding of the nature, structure, and properties of nanomaterials, the development of standards in the assessment of their impact on environmental components must be put before the deployment of nanotechnology in practice in all fields of human life and activity. The connection of nanotechnology with such important areas as agriculture, food industry, and remediation is very promising, but the implementation of technology in these and other fields cannot result in further need for remediation because the applied nanotechnology has contributed to contamination of the environment.

Acknowledgments The authors would like to thank Jerzy Wieczorek for his support in preparing the graphical abstract.

References

Abbaszadegan A, Ghahramani Y, Gholami A, Hemmateenejad B, Dorostkar S, Nabavizadeh M, Sharghi H (2015) The effect of charge at the surface of silver nanoparticles on antimicrobial activity against gram-positive and gram-negative bacteria: a preliminary study. J Nanomater: 8 p

Abu-Hamdah R, Cho WJ, Cho SJ, Jeremic A, Kelly M, Ilie AE, Jena BP (2004) Regulation of the water channel aquaporin-1: isolation and reconstitution of the regulatory complex. Cell Biol Int 28:7–17

Abd-Elsalam K, Prasad R (2018) Nanobiotechnology applications in plant protection, vol 2. Springer International Publishing. ISBN 978–3–030-13295-8. https://www.springer.com/gp/book/9783030132958

Abd-Elsalam KA, Prasad R (2019) Nanobiotechnology applications in plant protection. Springer International Publishing. ISBN 978–3–319-91161-8. https://www.springer.com/us/book/9783319911601

Adak T, Kumar J, Dey D, Shakil NA, Walia S (2012) Residue and bio-efficacy evaluation of controlled release formulations of imidacloprid against pests in soybean (*Glycine max*). J Environ Sci Health 47:226–231

Adil SF, Assal ME, Khan M (2015) Biogenic synthesis of metallic nanoparticles and prospects toward green chemistry. Dalton Trans 44:9709–9717

Aguilar-Mendez MA, San Martin-Martinez E, Ortega-Arroyo L, Cobian-Portillo G, Sanchez-Espindola E (2011) Synthesis and characterization of silver nanoparticles: effect on phytopathogen *Colletotrichum gloeosporioides*. J Nanopart Res 13:2525–2532

Aktar MW, Sengupta D, Chowdhury A (2009) Impact of pesticides use in agriculture: their benefits and hazards. Interdiscip Toxicol 2:1–12

An J, Zhang M, Wang S, Tang J (2008) Physical, chemical and microbiological changes in stored green asparagus spears as affected by coating of silver nanoparticles-PVP. LWT Food Sci Technol 41:1100–1107

Anderson JA, Gipmans M, Hurst S, Layton R, Nehra N, Pickett J, Shah DM, Souza TL, Tripathi L (2016) Emerging agricultural biotechnologies for sustainable agriculture and food security. J Agric Food Chem 64:383–393

Anjali CH, Khan SS, Margulis-Goshen K, Magdassi S, Mukherjee A, Chandrasekaran N (2010) Formulation of water-dispersible nanopermethrin for larvicidal applications. Ecotoxicol Environ Saf 73:1932–1936

Arase F, Nishitani H, Egusa M, Nishimoto N, Sakurai S, Sakamoto N, Kaminaka H (2012) IAA8 involved in lateral root formation interacts with the TIR1 auxin receptor and ARF transcription factors in *Arabidopsis*. PLoS One 7:43414

Arruda SCC, Silva ALD, Galazzi RM, Azevedo RA, Arruda MAZ (2015) Nanoparticles applied to plant science: a review. Talanta 131:693–705

Asli S, Neumann PM (2009) Colloidal suspensions of clay or titanium dioxide nanoparticles can inhibit leaf growth and transpiration via physical effects on root water transport. Plant Cell Environ 32:577–584

Atha DH, Wang H, Petersen EJ, Cleveland D, Holbrook RD, Jaruga P, Dizdaroglu M, Xing B, Nelson BC (2012) Copper oxide nanoparticle mediated DNA damage in terrestrial plant models. Environ Sci Technol 46:1819–1827

Aziz N, Faraz M, Pandey R, Sakir M, Fatma T, Varma A, Barman I, Prasad R (2015) Facile algae-derived route to biogenic silver nanoparticles: synthesis, antibacterial and photocatalytic properties. Langmuir 31:11605–11612

Aziz N, Pandey R, Barman I, Prasad R (2016) Leveraging the attributes of Mucor hiemalis-derived silver nanoparticles for a synergistic broad-spectrum antimicrobial platform. Front Microbiol 7:1984. https://doi.org/10.3389/fmicb.2016.01984

Aziz N, Faraz M, Sherwani MA, Fatma T, Prasad R (2019) Illuminating the anticancerous efficacy of a new fungal chassis for silver nanoparticle synthesis. Front Chem 7:65. https://doi.org/10.3389/fchem.2019.00065

Badawy AM, Luxton TP, Silva RG, Scheckel KG, Suidan MT, Tolaymat TM (2010) Impact of environmental conditions (pH, ionic strength, and electrolyte type) on the surface charge and aggregation of silver nanoparticles suspensions. Environ Sci Technol 44:1260–1266

Badawy AM, Silva RG, Morris B, Scheckel KG, Makram TS, Tolaymat TM (2011) Surface charge-dependent toxicity of silver nanoparticles. Environ Sci Technol 45:283–287

Bain LE, Awah PK, Geraldine N, Kindong NP, Sigal Y, Bernard N, Tanjeko AT (2013) Malnutrition in sub-Saharan Africa: burden, causes and prospects. Pan Afr Med J 15:120

Bandyopadhyay S, Peralta-Videa JR, Gardea-Torresdey JL (2013) Advanced analytical techniques for the measurement of nanomaterials in food and agricultural samples: a review. Environ Eng Sci 30:118–125

Benoit R, Wilkinson KJ, Sauve´ S (2013) Partitioning of silver and chemical speciation of free Ag in soils amended with nanoparticles. Chem Cent J 7:75

Berekaa MM (2015) Nanotechnology in food industry; advances in food processing, packaging and food safety. Int J Curr Microbiol App Sci 4:345–357

Bergeson LL (2010) Nanosilver: US EPA's pesticide office considers how best to proceed. Environ Qual Manag 19:79–85

Besinis A, De Peralta T, Handy RD (2014) The antibacterial effects of silver, titanium dioxide and silica dioxide nanoparticles compared to the dental disinfectant chlorhexidine on *Streptococcus* mutans using a suite of bioassays. Nanotoxicology 8:1–16

Bhattacharyya A, Duraisamy P, Govindarajan M, Buhroo AA, Prasad R (2016) Nano-biofungicides: emerging trend in insect pest control. In: Prasad R (ed) Advances and applications through fungal nanobiotechnology. Springer, pp 307–319

Bhol KC, Schechter PJ (2005) Topical nanocrystalline silver cream suppresses inflammatory cytokines and induces apoptosis of inflammatory cells in a murine model of allergic contact dermatitis. Br J Dermatol 152:1235–1242

Bhowmick S, Chakraborty S, Mondal P, Renterghem WV, Berghe SVD, Roman-Ross G, Chatterjee D, Iglesias M (2014) Montmorillonite-supported nanoscale zero-valent iron for removal of arsenic from aqueous solution: kinetics and mechanism. Chem Eng J 243:14–23

Bhushani JA, Anandharamakrishnan C (2014) Electrospinning and electrospraying techniques: potential food based applications. Trends Food Sci Technol 38:21–33

Boxi SS, Mukherjee K, Paria S (2016) Ag doped hollow TiO_2 nanoparticles as an effective green fungicide against *Fusarium solani* and *Venturia inaequalis* phytopathogens. Nanotechnology 27(8):085103

Buick RD, Buchan GD, Field RJ (1993) The role of surface tension of spreading droplets in absorption of a herbicide formulation via leaf stomata. Pestic Sci 38:227–235

Bumbudsanpharoke N, Ko S (2015) Nano-food packaging: an overview of market, migration research, and safety regulations. J Food Sci 80:910–923

Campbell ML, Guerra FD, Dhulekar J, Alexis F, Whitehead DC (2015) Target-specific capture of environmentally relevant gaseous aldehydes and carboxylic acids with functional nanoparticles. Chem Eur J 21:14834–14842

Cárdenas G, Días J, Meléndrez MF, Cruzat C, Garcia Cancino A (2009) Colloidal Cu nanoparticles/chitosan composite film obtained by microwave heating for food package applications. Polym Bull 62:511–524

Carlson C, Hussain SM, Schrand AM, Braydich-Stolle LK, Hess KL, Jones RL, Schlager JJ (2008) Unique cellular interaction of silver nanoparticles: size-dependent generation of reactive oxygen species. J Phys Chem B 112:13608–13619

Chen H, Yada R (2011) Nanotechnologies in agriculture: new tools for sustainable development. Trends Food Sci Technol 22:585–594

Chi Y, Zhang G, Xiang Y, Cai D, Wu Z (2017) Fabrication of a temperature-controlled-release herbicide using a nanocomposite. ACS Sustain Chem Eng 5:4969–4975

Choi O, Clevengera TE, Deng B, Surampalli RY, Ross L, Hu Z (2009) Role of sulfide and ligand strength in controlling nanosilver toxicity. Water Res 43:1879–1886

Choi AJ, Kim CJ, Cho YJ, Hwang JK, Kim CT (2011) Characterization of capsaicin-loaded nanoemulsions stabilized with alginate and chitosan by self-assembly. Food Bioprocess Technol 4:1119–1126

Chularueangaksorn P, Tanaka S, Fujii S, Kunacheva C (2014) Adsorption of perfluorooctanoic acid (PFOA) onto anion exchange resin, non-ion exchange resin, and granular-activated carbon by batch and column. Desalin Water Treat 52:6542–6548

Conley DJ, Paerl HW, Howarth RW, Boesch DF, Seitzinger SP, Havens KE, Lancelot C, Likens GE (2009) Controlling eutrophication: nitrogen and phosphorus. Ecology 323:1014–1015

Cornelis G, Kirby JK, Beak D, Chittleborough D, McLaughlin MJ (2010) A method for determination of retention of silver and cerium oxide manufactured nanoparticles in soils. Environ Chem 7:298–308

Cornelis G, Hund-Rinke K, Kuhlbusch T, Van den Brink N, Nickel C (2014) Fate and bioavailability of engineered nanoparticles in soils: a review. Crit Rev Environ Sci Technol 44:2720–2764

Corradini E, De Moura M, Mattoso L (2010) A preliminary study of the incorporation of NPK fertilizer into chitosan nanoparticles. Express Polym Lett 4:509–515

Cox A, Venkatachalam P, Sahi S, Sharma N (2017) Reprint of: silver and titanium dioxide nanoparticle toxicity in plants: a review of current research. Plant Physiol Biochem 110:33–49

Cui Y, Zhao Y, Tian Y, Zhang W, Lü X, Jiang X (2012) The molecular mechanism of action of bactericidal gold nanoparticles on *Escherichia coli*. Biomaterials 33:2327–2333

Da Costa MVJ, Sharma PK (2016) Effect of copper oxide nanoparticles on growth, morphology, photosynthesis, and antioxidant response in *Oryza sativa*. Photosynthetica 54:110–119

da Silva FL, Laitinen T, Pirilla M, Keiski RL, Ojala S (2017) Photocatalytic degradation of perfluorooctanoic acid (PFOA) from wastewaters by TiO_2, In_2O_3 and Ga_2O_3 catalysts. Top Catal 60:1345–1358

Dang Y, Zhang Y, Fan L, Chen H, Roco MC (2010) Trends in worldwide nanotechnology patent applications: 1991 to 2008. J Nanopart Res 12:687–706

Das S, Sen B, Debnath N (2015) Recent trends in nanomaterials applications in environmental monitoring and remediation. Environ Sci Pollut Res 22:18333–18344

de Oliveira JL, Campos EV, Bakshi M, Abhilash PC, Fraceto LF (2014) Application of nanotechnology for the encapsulation of botanical insecticides for sustainable agriculture: prospects and promises. Biotechnol Adv 32:1550–1561

Dietz KJ, Herth S (2011) Plant nanotoxicology. Trends Plant Sci 16:582–589

Donaldson K, Stone V, Tran CL, Kreyling W, Borm PJ (2004) Nanotoxicology. Occup Environ Med 619:727–728

Drake PL, Hazelwood KJ (2005) Exposure-related health effects of silver and silver compounds: a review. Ann Occup Hyg 49:575–585

Du W, Tan W, Peralta-Videa JR, Gardea-Torresdey JL, Ji R, Yin Y, Guo H (2017) Interaction of metal oxide nanoparticles with higher terrestrial plants: physiological and biochemical aspects. Plant Physiol Biochem 110:210–225

Duhan JS, Kumar R, Kumar N, Kaur P, Nehra K, Duhan S (2017) Nanotechnology: the new perspective in precision agriculture. Biotechnol Rep 15:11–23

Duncan TV (2011) Applications of nanotechnology in food packaging and food safety: barrier materials, antimicrobials and sensors. J Colloid Interface Sci 363:1–24

Eichert T, Goldbach HE (2008) Equivalent pore radii of hydrophilic foliar uptake routes in stomatous and astomatous leaf surfaces-further evidence for a stomatal pathway. Physiol Plant 132:491–502

Elbeshehy EKF, Elazzazy AM, Aggelis G (2015) Silver nanoparticles synthesis mediated by new isolates of *Bacillus* spp., nanoparticle characterization and their activity against bean yellow mosaic virus and human pathogens. Front Microbiol 6:453

Faisal M, Saquib Q, Alatar AA, Al-Khedhairy AA, Hegazy AK, Musarrat J (2013) Phytotoxic hazards of NiO-nanoparticles in tomato: a study on mechanism of cell death. J Hazard Mater 250:318–332

Falco WF, Botero ER, Falcao EA, Santiago EF, Bagnato VS, Caires ARL (2011) In vivo observation of chlorophyll fluorescence quenching induced by gold nanoparticles. J Photochem Photobiol A 225:65–71

FAO (2009) How to feed the world in 2050. In: Proceedings of the expert meeting on how to feed the world in 2050. 24–26 June 2009. FAO Headquarters, Rome

FAO (2017) The state of food security and nutrition in the world. Rome

Fernández-Luqueño F, Lopez-Valdez F, Sarabia-Castillo CR, García-Mayagoitia S, Perez-Rios SR (2017) Bioremediation of polycyclic aromatic hydrocarbons-polluted soils at laboratory and field scale: a review of the literature on plants and microorganisms. In: Anjum NA, Gill SS, Tuteja N (eds) Enhancing cleanup of environmental pollutants, vol 1: biological approaches. Springer, Cham

Fogel R, Limson J (2016) Developing biosensors in developing countries: South Africa as a case study. Biosensors 6:5

Fraceto LF, Grillo R, de Medeiros GA, Scognamiglio V, Rea G, Bartolucci C (2016) Nanotechnology in agriculture: which innovation potential does it have? Front Environ Sci 4:20

Fu R, Yang Y, Xu Z, Zhang X, Guo X, Bi D (2015) The removal of chromium (VI) and lead (II) from groundwater using sepiolite-supported nanoscale zero-valent iron (S-NZVI). Chemosphere 138:726–734

Fujishima A, Zhang X (2006) Titanium dioxide photocatalysis: present situation and future approaches. C R Chim 9:750–760

Gao X, Chorover J (2012) Adsorption of perfluorooctanoic acid and perfluorooctanesulfonic acid to iron oxide surfaces as studied by flow-through ATR-FTIR spectroscopy. Environ Chem 9:148–157

Ge Y, Schimel JP, Holden PA (2011) Evidence for negative effects of TiO_2 and ZnO nanoparticles on soil bacterial communities. Environ Sci Technol 45:1659–1664

Geisler-Lee J, Wang Q, Yao Y, Zhang W, Geisler M, Li K, Huang Y, Chen Y, Kolmakov A, Ma X (2013) Phytotoxicity, accumulation and transport of silver nanoparticles by *Arabidopsis thaliana*. Nanotoxicology 7:323–337

Ghaani M, Cozzolino CA, Castelli G, Farris S (2016) An overview of the intelligent packaging technologies in the food sector. Trends Food Sci Technol 51:1–11

Ghahremani A, Akbari K, Yousefpour M, Ardalani H (2014) Effects of nano-potassium and nano-calcium chetaled fertilizers on qualitative and quantitative characteristics of *Ocimum basilicum*. Int J Pharm Res Sch 3:235–241

Ghormade V, Deshpande MV, Paknikar KM (2011) Perspectives for nano-biotechnology enabled protection and nutrition of plants. Biotechnol Adv 29:792–803

Ghosh M, Chakraborty A, Bandyopadhyay M, Mukherjee A (2011) Multi-walled carbon nanotubes (MWCNT): induction of DNA damage in plant and mammalian cells. J Hazard Mater 197:327–336

Giri B, Prasad R, Wu Q-S, Varma A (2019) Biofertilizers for sustainable agriculture and environment. Springer International Publishing. ISBN 978–3–030-18932-7. https://www.springer.com/gp/book/9783030189327

Gopinath V, Velusamy P (2013) Extracellular biosynthesis of silver nanoparticles using *Bacillus* sp. GP-23 and evaluation of their antifungal activity towards *Fusarium oxysporum*. Spectrochim Acta Mol Biomol Spectrosc 106:170–174

Gorczyca A, Pociecha E, Kasprowicz MJ, Niemiec M (2015) Effect of nanosilver in wheat seedlings and *Fusarium culmorum* culture systems. Eur J Plant Pathol 142:251–261

Gorczyca A, Przemieniecki S, Kurowski T, Oćwieja M (2018) Early plant growth and bacterial community in rhizoplane of wheat and flax exposed to silver and titanium dioxide nanoparticles. Environ Sci Pollut R 25:33820–33826

Govorov AO, Carmeli I (2007) Hybrid structures composed of photosynthetic system and metal nanoparticles: plasmon enhancement effect. Nano Lett 7:620–625

Green R, Cornelsen L, Dangour AD, Turner R, Shankar B, Mazzocchi M, Smith RD (2013) The effect of rising food prices on food consumption: systematic review with meta-regression. BMJ 346:1–9

Grillo R, Abhilash PC, Fraceto LF (2016) Nanotechnology applied to bio-encapsulation of pesticides. J Nanosci Nanotechnol 16:1231–1234

Gross R, Hanna R, Gambhir A, Heptonstall P, Speirs J (2018) How long does innovation and commercialisation in the energy sectors take? Historical case studies of the timescale from invention to widespread commercialisation in energy supply and end use technology. Energ Policy 123:682–699

Gruère GP (2012) Implications of nanotechnology growth in food and agriculture in OECD countries. Food Policy 37:191–198

Guerra FD, Campbell ML, Whitehead DC, Alexis F (2017) Tunable properties of functional nanoparticles for efficient capture of VOCs. Chem Sel 2:9889–9894

Guerra FD, Attia MF, Whitehead DC, Alexis F (2018) Nanotechnology for environmental remediation: materials and applications. Molecules 23:1760

Gupta N, Upadhyaya CP, Singh A, Abd-Elsalam KA, Prasad R (2018) Applications of silver nanoparticles in plant protection. In: Abd-Elsalam K, Prasad R (eds) Nanobiotechnology applications in plant protection. Springer International Publishing AG, Cham, pp 247–266

Hasaneen MNA, Abdel-Aziz HMM, El-Bialy DMA, Omer AM (2014) Preparation of chitosan nanoparticles for loading with NPK fertilizer. Afr J Biotechnol 13:3158–3164

Hazrati H, Saharkhiz MJ, Niakousari M, Moein M (2017) Natural herbicide activity of *Satureja hortensis* L. essential oil nanoemulsion on the seed germination and morphophysiological features of two important weed species. Ecotoxicol Environ Saf 142:423–430

Henn KW, Waddill DW (2006) Utilization of nanoscale zero-valent iron for source remediation-a case study. Remediation 16:57–77

Holt KB, Bard AJ (2005) Interaction of silver(I) ions with the respiratory chain of *Escherichia coli*: an electrochemical and scanning electrochemical microscopy study of the antimicrobial mechanism of micromolar Ag+. Biochemistry 44:13214–13223

Hong FS, Yang P, Gao FQ, Liu C, Zheng L, Zhou J (2005) Effect of nano-anatase TiO_2 on spectral characterization of photosystem II particles from spinach. Chem Res Chin Univ 21:196–200

http://www.nanotec.org.uk/report/chapter2.pdf

https://www.nano.gov/nanotech-101/what/definition

Hwang IC, Kim TH, Bang SH, Kim KS, Kwon HR, Seo MJ, Youn YN, Park HJ, Yasunaga-Aokt C, Yo YM (2011) Insecticidal effect of controlled release formulations of etofenprox based on nano-bio technique. J Fac Agric Kyushu Univ 56:33–40

Iavicoli I, Leso V, Beezhold DH, Shvedova AA (2017) Nanotechnology in agriculture: opportunities, toxicological implications, and occupational risks. Toxicol Appl Pharmacol 329:96–111

Ibrahim RK, Hayyan M, AlSaadi MA, Hayyan A, Ibrahim S (2016) Environmental application of nanotechnology: air, soil, and water. Environ Sci Pollut Res 23:13754–13788

Jacob DL, Borchardt JD, Navaratnam L, Otte ML, Bezbaruah AN (2013) Uptake and translocation of Ti from nanoparticles in crops and wetland plants. Int J Phytoremediation 15:142–153

Jampilek J, Kralova K (2015) Application of nanotechnology in agriculture and food industry, its prospects and risks. Ecol Chem Eng S 22:321–361

Jeevanandam J, Barhoum A, Chan YS, Dufresne A, Danquah MK (2018) Review on nanoparticles and nanostructured materials: history, sources, toxicity and regulations. Beilstein J Nanotechnol 9:1050–1074

Jeong SH, Hwnag YH, Yi SC (2005) Antibacterial properties of padded PP/PE nonwovens incorporating nano-sized silver colloids. J Mater Sci 40:5413–5418

Jiang J, Oberdorster G, Biswas P (2009) Characterization of size, surface charge, and agglomeration state of nanoparticle dispersions for toxicological studies. J Nanopart Res 11:77–89

Jiang LC, Basri M, Omar D, Rahman MBA, Salleh AB, Rahman RNZRA, Salleh AB, Rahman RNZRA, Selamat A (2012) Green nano-emulsion intervention for water-soluble glyphosate isopropylamine (IPA) formulations in controlling *Eleusine indica* (E-indica). Pestic Biochem Physiol 102:19–29

Jo Y, Kim BH, Jung G (2009) Antifungal activity of silver ions and nanoparticles on phytopathogenic fungi. Plant Dis 93:1037–1043

Kah M, Hofmann T (2014) Nanopesticides research: current trends and future priorities. Environ Int 63:224–235

Kanel S, Greneche J, Choi H (2006) Arsenic(V) removal from groundwater using nano scale zero-valent iron as a colloidal reactive barrier material. Environ Sci Technol 40:2045–2050

Kang Y, Khan S, Ma X (2009) Climate change impacts on crop yield, crop water productivity and food security – a review. Prog Nat Sci 19:1665–1674

Kang MA, Seo MJ, Hwang IC, Jang C, Park HJ, Yu YM, Youn YN (2012) Insecticidal activity and feeding behavior of the green peach aphid, *Myzus persicae*, after treatment with nano types of pyrifluquinazon. J Asia Pac Entomol 15:533–541

Kanhed P, Birla S, Gaikwad S, Gade A, Seabra AB, Rubilar O, Duran N, Rai M (2014) In vitro antifungal efficacy of copper nanoparticles against selected crop pathogenic fungi. Mater Lett 115:13–17

Karn B, Kuiken T, Otto M (2009) Nanotechnology and in situ remediation: a review of the benefits and potential risks. Environ Health Perspect 117:1813–1831

Kasemets K, Ivask A, Dubourguier H, Kahru A (2009) Toxicity of nanoparticles of ZnO, CuO and TiO_2 to yeast *Saccharomyces cerevisiae*. Toxicol In Vitro 23:1116–1121

Kasprowicz MJ, Kozioł M, Gorczyca A (2010) The effect of silver nanoparticles on phytopathogenic spores of *Fusarium culmorum*. Can J Microbiol 56:247–253

Kasprowicz MJ, Gorczyca A, Frandsen RJN (2013) The effect of nanosilver on pigments production by *Fusarium culmorum* (W. G. Sm.) Sacc. Pol J Microbiol 62:365–372

Katti DR, Sharma A, Pradham SM, Katti KS (2015) Carbon nanotube proximity influences rice DNA. Chem Phys 455:17–22

Kaushal M, Wani SP (2017) Nanosensors: frontiers in precision agriculture. In: Prasad R, Kumar M, Kumar V (eds) Nanotechnology. Springer, Singapore

Kaveh R, Li YS, Ranjbar S, Tehrani R, Brueck CL, Van Aken B (2013) Changes in *Arabidopsis thaliana* gene expression in response to silver nanoparticles and silver ions. Environ Sci Technol 47:10637–10644

Khot LR, Sankarana S, Majaa JM, Ehsania R, Schuster EW (2012) Applications of nanomaterials in agricultural production and crop protection: a review. Crop Prot 35:64–70

Kim S, Ryu DY (2013) Silver nanoparticle-induced oxidative stress, genotoxicity and apoptosis in cultured cells and animal tissues. J Appl Toxicol 33:78–89

Kim DY, Kadam A, Shinde S, Saratale RG, Patra J, Ghodake G (2018) Recent developments in nanotechnology transforming the agricultural sector: a transition replete with opportunities. J Sci Food Agric 98:849–864

Klaine SJ, Alvarez PJ, Batley GE, Fernandes TF, Handy RD, Lyon DY, Mahendra S, McLaughlin MJ, Lead JR (2008) Nanomaterials in the environment: behavior, fate, bioavailability, and effects. Environ Toxicol Chem 27:1825–1851

Kookana RS, Boxall ABA, Reeves PT, Ashauer R, Beulke S, Chaudhry Q, Corneslis G, Fernandes TF, Gan J, Kah M, Lynch I, Ranville J, Sinclair C, Spurgeon D, Tiede K, Brink PJV (2014) Nanopesticides: guiding principles for regulatory evaluation of environmental risks. J Agric Food Chem 62:4227–4240

Kujda M, Oćwieja M, Adamczyk Z, Bocheńska O, Kozik A, Bielańska E, Barabasz J (2015) Charge stabilized silver nanoparticles applied as antibacterial agents. J Nanosci Nanotechnol 15:3574–3583
Kumar S, Bhanjana G, Sharma A, Sarita SMC, Dilbaghi N (2015) Herbicide loaded carboxymethyl cellulose nanocapsules as potential carrier in agri-nanotechnology. Sci Adv Mater 7:1143–1148
Kumar S, Bhanjana G, Sharma A, Dilbaghi N, Sidhu MC, Kim KH (2017) Development of nanoformulation approaches for the control of weeds. Sci Total Environ 586:1272–1278
Kumar S, Nehra M, Dilbaghi N, Marrazza G, Hassan AA, Kim KH (2019) Nano-based smart pesticide formulations: emerging opportunities for agriculture. J Control Release 294:131–153
Kurepa J, Paunesku T, Vogt S, Arora H, Rabatic BM, Lu J, Wanzer MB, Woloschak GE, Smalle JA (2010) Uptake and distribution of ultrasmall anatase TiO_2 Alizarin red S nanoconjugates in *Arabidopsis thaliana*. Nano Lett 10:2296–2302
Kvitek L, Vanickova M, Panacek A, Soukupova J, Dittrich M, Valentova E, Prucek R, Bancirova M, Milde D, Zboril R (2009) Initial study on the toxicity of silver nanoparticles (NPs) against *Paramecium caudatum*. J Phys Chem C 113:4296–4300
Lamsal K, Kim SW, Jung JH, Kim YS, Kim KS, Lee YS (2011) Inhibition effects of silver nanoparticles against powdery mildews on cucumber and pumpkin. Mycobiology 39:26–232
Landa P, Vankova R, Andrlova J, Hodek J, Marsik P, Storchova H, White JC, Vanek T (2012) Nanoparticle-specific changes in *Arabidopsis thaliana* gene expression after exposure to ZnO, TiO_2, and fullerene soot. J Hazard Mater 241–242:55–62
Lee KJ, Park SH, Govarthanan M (2013) Synthesis of silver nanoparticles using cow milk and their antifungal activity against phytopathogens. Mater Lett 105:128–131
Li YM, Cheng W, Sheng GD, Li JF, Dong HP, Chen Y, Zhu LZ (2015) Synergetic effect of a pillared bentonite support on SE(VI) removal by nanoscale zero valent iron. Appl Catal B 174–175:329–335
Lim CJ, Basri M, Omar D, Rahman MBA, Salleh AB, Rahman RNZRA (2013) Green nanoemulsion-laden glyphosate isopropylamine formulation in suppressing creeping foxglove (*A. gangetica*), slender button weed (*D. ocimifolia*) and buffalo grass (*P. conjugatum*). Pest Manag Sci 69:104–111
Lin D, Xing B (2007) Phytotoxicity of nanoparticles: inhibition of seed germination and root growth. Environ Pollut 150:243–250
Lin S, Reppert J, Hu Q, Hudson JS, Reid ML, Ratnikova TA, Rao AM, Luo H, Ke PC (2009) Uptake, translocation, and transmission of carbon nanomaterials in rice plants. Small 5:1128–1132
Liu R, Lal R (2015) Potentials of engineered nanoparticles as fertilizers for increasing agronomic productions. Sci Total Environ 514:131–139
Liu X, Feng Z, Zhang S, Zhang J, Xiao Q, Wang Y (2006) Preparation and testing of cementing nano-subnano composites of slow or controlled release of fertilizers. Agric Sci China 5:700–706
Liu Q, Zhao Y, Wan Y, Zheng J, Zhang X, Wang C, Fang X, Lin J (2010) Study of the inhibitory effect of water-soluble fullerenes on plant growth at the cellular level. ACS Nano 4:5743–5748
Liu M, Wang Y, Chen L, Yan Z, Zhang L (2015) $Mg(OH)_2$ supported nanoscale zero valent iron enhancing the removal of Pb(II) from aqueous solution. ACS Appl Mater Interfaces 7:7961–7969
Llorens A, Lloret E, Picouet PA, Trbojevich R, Fernandez A (2012) Metallic-based micro and nanocomposites in food contact materials and active food packaging. Trends Food Sci Technol 24:19–29
Luvisi A (2016) Electronic identification technology for agriculture, plant, and food. A review. Agron Sustain Dev 36:13
Lv M, Su S, He Y, Huang Q, Hu W, Li D, Lee ST (2010) Long-term antimicrobial effect of silicon nanowires decorated with silver nanoparticles. Adv Mater 22:5463–5467
Lv X, Xu J, Jiang G, Tang J, Xu X (2012) Highly active nanoscale zero-valent iron (nZVI)-Fe3O4 nanocomposites for the removal of chromium(VI) from aqueous solutions. J Colloid Interface Sci 369:460–469

Ma X, Geiser-Lee J, Deng Y, Kolmakov A (2010) Interactions between engineered nanoparticles (ENPs) and plants: phytotoxicity, uptake and accumulation. Sci Total Environ 408:3053–3061

Macé C, Desrocher S, Gheorghiu F, Kane A, Pupeza M, Cernik M, Kvapil P, Venkatakrishnan R, Zhang W (2006) Nanotechnology and groundwater remediation: a step forward in technology understanding. Remediation 16:23–33

Mackenzie K, Bleyl S, Kopinke FD, Doose H, Bruns J (2016) Carbo-Iron as improvement of the nanoiron technology: from laboratory design to the field test. Sci Total Environ 563-564:641–648

Malekian R, Abedi-Koupai J, Eslamian SS (2011) Influences of clinoptilolite and surfactant-modified clinoptilolite zeolite on nitrate leaching and plant growth. J Hazard Mater 185:970–976

Marzbani P, Afrouzi YM, Omidvar A (2015) The effect of nano-zinc oxide on particleboard decay resistance. Maderas Cienc Tecnol 17:63–68

Medina-Pérez G, Fernández-Luqueño F, Vazquez-Nuñez E, López-Valdez F, Prieto-Mendez J, Madariaga-Navarrete A, Miranda-Arámbula M (2019) Remediating polluted soils using nanotechnologies: environmental benefits and risks. Pol J Environ Stud 28:1013–1030

Miller G, Senjen R (2008) Out of the laboratory and on to our plates. Nanotechnology in food & agriculture. In: Friends of the Earth, Australia, Europe & U.S.A (ed). Washington

Miralles P, Church TL, Harris AT (2012) Toxicity, uptake, and translocation of engineered nanoparticles in vascular plants. Environ Sci Technol 46:9224–9239

Mishra S, Singh BR, Singh A (2014) Biofabricated silver nanoparticles act as a strong fungicide against *Bipolaris sorokiniana* causing spot blotch disease in wheat. PLoS One 9:97881

Monica RC, Cremonini R (2009) Nanoparticles and higher plants. Caryologia 62:161–165

Morones JR, Elechiguerra JL, Camacho A, Holt K, Kouri JB, Ramirez JT, Yacaman MJ (2005) The bactericidal effect of silver nanoparticles. J Nanotechnol 16:2346–2353

Mukherjee A, Peralta-Videa JR, Bandyopadhyay S, Rico CM, Zhao L, Gardea- Torresdey JL (2014) Physiological effects of nanoparticulate ZnO in green peas (*Pisum sativum* L.) cultivated in soil. Metallomics 6:132–138

Mukhopadhyay SS (2014) Nanotechnology in agriculture: prospects and constraints. Nanotechnol Sci Appl 7:63–71

Musante C, White JC (2012) Toxicity of silver and copper to *Cucurbita pepo*: differential effects of nano and bulk- size particles. Environ Toxicol 27:510–517

Nair PMG, Chung IM (2014) Physiological and molecular level effects of silver nanoparticles exposure in rice (*Oryza sativa* L.) seedlings. Chemosphere 112:105–113

Nair R, Varghese SH, Nair BG, Maekawa T, Yoshida Y, Kumar DS (2010) Nanoparticulate material delivery to plants. Plant Sci 179:154–163

Nair PMG, Kim SH, Chung IM (2014) Copper oxide nanoparticle toxicity in mung bean (*Vigna radiata* L.) seedlings: physiological and molecular level responses of in vitro grown plants. Acta Physiol Plant 36:2947–2958

Nakata K, Fujishima A (2012) TiO_2 photocatalysis: design and applications. J Photochem Photobiol 13:169–189

Narayanan KB, Park HH (2014) Antifungal activity of silver nanoparticles synthesized using turnip leaf extract (*Brassica rapa* L.) against wood rotting pathogens. Eur J Plant Pathol 140:185–192

Navarro E, Baun A, Behra R, Hartmann NB, Filser J, Miao A, Quigg A, Santschi PH, Sigg L (2008) Environmental behavior and ecotoxicity of engineered nanoparticles to algae, plants, and fungi. Ecotoxicology 17:372–386

Navarro DA, Bisson MA, Aga DS (2012) Investigating uptake of water-dispersible CdSe/ZnS quantum dot nanoparticles by *Arabidopsis thaliana* plants. J Hazard Mater 211:427–435

Nuruzzaman M, Rahman MM, Liu Y, Naidu R (2016) Nanoencapsulation, nano-guard for pesticides: a new window for safe application. J Agric Food Chem 64:1447–1483

Oancea S, Padureanu S, Oancea AV (2009) Growth dynamics of corn plants during anionic clays action. Lucrari Stiintifice Seria Agronomie 52:212–217

Oerke EC (2006) Crop losses to pests. J Agric Sci 144:31–43

Oukarroum A, Bras S, Perreault F, Popovic R (2012) Inhibitory effects of silver nanoparticles in two green algae, *Chlorella vulgaris* and *Dunaliella tertiolecta*. Ecotoxicol Environ Saf 78:80–85

Pal S, Tak YK, Song JM (2007) Does the antimicrobial activity of silver nanoparticles depend on the shape of the nanoparticle? A study of the gram-negative bacterium *Escherichia coli*. Appl Environ Microbiol 73:1712–1720

Panda KK, Achary VMM, Krishnaveni R, Padhi BK, Sarangi SN, Sahu SN, Panda BB (2011) In vitro biosynthesis and genotoxicity bioassay of silver nanoparticles using plants. Toxicol In Vitro 25:1097–1105

Pandey B, Fulekar MH (2012) Nanotechnology: remediation technologies to clean up the environmental pollutants. Res J Chem Sci 2:90–96

Pang H, Wu Y, Huang S, Ding C, Li S, Wang X, Yu S, Chen Z, Song G, Wang X (2018) Macroscopic and microscopic investigation of uranium elimination by Ca-Mg-Al-layered double hydroxide supported nanoscale zero valent iron. Inorg Chem Front 5:2657–2665

Pantidos N, Horsfall LE (2014) Biological synthesis of metallic nanoparticles by bacteria, fungi and plants. J Nanomed Nanotechnol 5:233

Peralta-Videa JR, Zhao L, Lopez-Moreno ML, de la Rosa G, Hong J, Gardea-Torresdey JL (2011) Nanomaterials and the environment: a review for the biennium 2008–2010. J Hazard Mater 186:1–15

Percival SL, Bowler PG, Russell D (2005) Bacterial resistance to silver in wound care. J Hosp Infect 60:1–7

Poguberović SS, Krčmar DM, Dalmacija BD, Maletić SP, Tomašević-Pilipović DD, Kerkez DV, Rončević SD (2016) Removal of Ni(II) and Cu(II) from aqueous solutions using 'green' zero-valent iron nanoparticles produced by oak and mulberry leaf extracts. Water Sci Technol 74:2115–2123

Pokhrel LR, Dubey B (2013) Evaluation of developmental responses of two crop plants exposed to silver and zinc oxide nanoparticles. Sci Total Environ 452:321–332

Prasad R, Kumar V, Prasad KS (2014) Nanotechnology in sustainable agriculture: present concerns and future aspects. Afr J Biotechnol 13:705–713

Prasad R, Pandey R, Barman I (2016) Engineering tailored nanoparticles with microbes: quo vadis. WIREs Nanomed Nanobiotechnol 8:316–330. https://doi.org/10.1002/wnan.1363

Prasad R, Bhattacharyya A, Nguyen Quang D (2017) Nanotechnology in sustainable agriculture: recent developments, challenges, and perspectives. Front Microbiol 8:1–13

Prasad R, Aranda E (2018) Approaches in bioremediation. Springer International Publishing. https://www.springer.com/de/book/9783030023683

Prasad R (2019) Plant nanobionics: approaches in nanoparticles biosynthesis and toxicity. Springer International Publishing. ISBN 978–3–030-16379-2. https://www.springer.com/gp/book/9783030163785

Prasad R, Kumar V, Kumar M, Choudhary D (2019) Nanobiotechnology in bioformulations. Springer International Publishing. ISBN 978–3–030-17061-5. https://www.springer.com/gp/book/9783030170608

Qiu X, Fang Z, Liang B, Gu F, Xu Z (2011) Degradation of decabromodiphenyl ether by nano zero-valent iron immobilized in mesoporous silica microspheres. J Hazard Mater 193:70

Ragaei M, Sabry Al-Kazafy H (2014) Nanotechnology for insect pest control. Int J Sci Environ Technol 3:528–545

Raliya R, Tarafdar JC, Gulecha K, Choudhary K, Rameshwar R, Prakash M, Saran PR (2013) Review article; scope of nanoscience and nanotechnology in agriculture. J Appl Biol Biotechnol 1:41–44

Rana S, Kalaichelvan PT (2011) Antibacterial effects of metal nanoparticles. Adv Biotechnol 2:21–23

Rana S, Kalaichelvan PT (2013) Ecotoxicity of nanoparticles. ISRN Toxicol:1–11

Rasalingam S, Peng R, Koodali RT (2014) Removal of hazardous pollutants from wastewaters: applications of TiO_2-SiO_2 mixed oxide materials. J Nanomater 617405:1–42

Ray PC, Yu H, Fu PP (2009) Toxicity and environmental risks of nanomaterials: challenges and future needs. J Environ Sci Health C Environ Carcinog Ecotoxicol Rev 27(1):1–35

Rezaei F, Moaveni P, Mozafari H (2015) Effect of different concentrations and time of nano TiO_2 spraying on quantitative and qualitative yield of soybean (*Glycine max* L.) at Shahr-e-Qods. Iran Biol Forum 7:957–964

Rezvani N, Sorooshzadeh A, Farhadi N (2012) Effect of nano-silver on growth of saffron in flooding stress. World Acad Sci Eng Technol 61:606–611

Rico CM, Hong J, Morales MI, Zhao L, Barrios AC, Zhang JY, Peralta-Videa JR, Gardea-Torresdey JL (2013a) Effect of cerium oxide nanoparticles on rice: a study involving the antioxidant defense system and in vivo fluorescence imaging. Environ Sci Technol 47:5635–5642

Rico CM, Morales MI, McCreary R, Castillo-Michel H, Barrios AC, Hong J, Tafoya A, Lee WY, Varela-Ramirez A, Peralta-Videa JR, Gardea-Torresdey JL (2013b) Cerium oxide nanoparticles modify the antioxidative stress enzyme activities and macromolecule composition in rice seedlings. Environ Sci Technol 47:14110–14118

Rico CM, Peralta-Videa JR, Gardea-Torresdey JL (2015) Chemistry, biochemistry of nanoparticles, and their role in antioxidant defense system in plants. In: Nanotechnology and plant sciences. Springer, Cham, pp 1–17

Ruffini CM, Cremonini R (2009) Nanoparticles and higher plants. Caryologia 62:161–165

Sabri MA, Umer A, Awan GH, Hassan MF, Hasnain A (2016) Selection of suitable biological method for the synthesis of silver nanoparticles. Nanomater Nanotechnol 6:29

Sagee O, Dror I, Berkowitz B (2012) Transport of silver nanoparticles (AgNPs) in soil. Chemosphere 88:670–675

Saharan V, Mehrotra A, Khatik R, Rawal P, Sharma SS, Pal A (2013) Synthesis of chitosan based nanoparticles and their *in vitro* evaluation against phytopathogenic fungi. Int J Biol Macromol 62:677–683

Sahebi M, Hanafi MM, Siti Nor Akmar A, Rafii MY, Azizi P, Tengoua FF, Nurul Mayzaitul Azwa J, Shabanimofrad M (2015) Importance of silicon and mechanisms of biosilica formation in plants. Biomed Res Int 2015:396010

Sarkar S, Datta SC, Biswas DR (2015) Effect of fertilizer loaded nanoclay/superabsorbent polymer composites on nitrogen and phosphorus release in soil. Proc Natl Acad Sci B Biol Sci 85:415–421

Savary S, Ficke A, Aubertot JN, Hollier C (2012) Crop losses due to diseases and their implications for global food production losses and food security. Food Sec 4:519–537

Sayes CM, Fortner JD, Guo W, Lyon D, Boyd AM, Ausman KD, Tao YJ, Sitharaman B, Wilson LJ, Hughes JB, West JL, Colvin VL (2004) The differential cytotoxicity of water-soluble fullerenes. Nano Lett 4:1881–1887

Sekhon BS (2014) Nanotechnology in Agri-food production: an overview. Nanotechnol Sci Appl 7:31–53

Sertova NM (2015) Application of nanotechnology in detection of mycotoxins and in agricultural sector. J Cent Eur Agric 16:117–130

Sharma P, Bhatt D, Zaidi MG, Saradhi PP, Khanna PK, Arora S (2012) Silver nanoparticle-mediated enhancement in growth and antioxidant status of *Brassica juncea*. Appl Biochem Biotechnol 167:2225–2233

Shi LN, Zhang X, Chen ZL (2011) Removal of chromium (VI) from wastewater using bentonite-supported nanoscale zero-valent iron. Water Res 45:886

Shin KS, Kim YH, Min JA, Kwak SM, Kim SK, Yang EG, Park JH, Ju BK, Kim TS, Kang JY (2006) Miniaturized fluorescence detection chip for capillary electrophoresis immunoassay of agricultural herbicide atrazine. Anal Chem Acta 573-574:164–171

Shoults-Wilson WA, Reinsch BC, Tsyusko OV, Bertsch PM, Lowry GV, Unrine JM (2011) Role of particle size and soil type in toxicity of silver nanoparticles to earthworms. Soil Sci Soc Am J 75:365–377

Shrivastava S, Bera T, Roy A, Singh G, Ramachandrarao P, Dash D (2007) Characterization of enhanced antibacterial effects of novel silver nanoparticles. Nanotechnol 18:225103

Siddiqui MH, Al-Whaibi MH, Faisal M, Al Sahli AA (2014) Nano-silicon dioxide mitigates the adverse effects of salt stress on *Cucurbita pepo* L. Environ Toxicol Chem 33:2429–2437

Siddiqui MH, Al-Whaibi MH, Mohammad F (eds) (2015) Nanotechnology and plant sciences: nanoparticles and their impact on plants. Springer, Cham

Singh S, Singh M, Agrawal VV, Kumar A (2010) An attempt to develop surface plasmon resonance based immunosensor for Karnal bunt (*Tilletia indica*) diagnosis based on the experience of nano-gold based lateral flow immunodipstick test. Thin Solid Films 519:1156–1159

Singh S, Vishwakarma K, Singh S, Sharma S, Dubey NK, Singh VK, Liu S, Tripathi DK, Chauhan DK (2017) Understanding the plant and nanoparticle interface at transcriptomic and proteomic level: a concentric overview. Plant Gene 11:265–272

Song SL, Liu XH, Jiang JH, Qian YH, Zhang N, Wu QH (2009) Stability of triazophos in self-nanoemulsifying pesticide delivery system. Colloid Surf A 350:57–62

Sorrentino A, Gorrasi G, Vittoria V (2007) Potential perspectives of bio-nanocomposites for food packaging applications. Trends Food Sci Technol 18:84–95

Srivastava AK, Dev A, Karmakar S (2018) Nanosensors and nanobiosensors in food and agriculture. Environ Chem Lett 16:161

Srivastava S, Usmani Z, Atanasov AG, Singh VK, Singh NP, Abdel-Azeem AM, Prasad R, Gupta G, Sharma M, Bhargava A (2021) Biological nanofactories: using living forms for metal nanoparticle synthesis. Mini-Rev Med Chem 21(2):245–265. https://doi.org/10.2174/1389557520999201116163012

Su MY, Hong FS, Liu C, Wu X, Liu XQ, Chen L, Gao FQ, Yang F, Li Z (2007) Effects of nano-anatase TiO_2 on absorption, distribution of light, and photoreduction activities of chloroplast membrane of spinach. Biol Trace Elem Res 131:101–101

Su C, Puls RW, Krug TA, Watling MT, O'Hara SK, Quinn JW, Ruiz NE (2012) A two and half-year-performance evaluation of a field test on treatment of source zone tetrachloroethene and its chlorinated daughter products using emulsified zero valent iron nanoparticles. Water Res 46:5071–5084

Syu YY, Hung JH, Chen JC, Chuang HW (2014) Impacts of size and shape of silver nanoparticles on *Arabidopsis* plant growth and gene expression. Plant Physiol Biochem 83:57–64

Tao F (2012) Synthesis, catalysis, surface chemistry and structure of bimetallic nanocatalysts. Chem Soc Rev 41:7977–7979

Tarafdar JC, Sharma S, Raliya R (2013) Nanotechnology: interdisciplinary science of applications. Afr J Biotechnol 12:219–226

Tarafdar JC, Raliya R, Mahawar H, Rathore I (2014) Development of zinc nanofertilizer to enhance crop production in pearl millet (Pennisetum americanum). Agric Res 3:257–262

Teodoro S, Micaela B, David KW (2010) Novel use of nano-structured alumina as an insecticide. Pest Manag Sci 66:577–579

Thangadurai D, Sangeetha J, Prasad R (2020a) Functional bionanomaterials. Springer International Publishing. ISBN 978–3–030-41464-1. https://www.springer.com/gp/book/9783030414634

Thangadurai D, Sangeetha J, Prasad R (2020b) Nanotechnology for food, agriculture, and environment. Springer International Publishing. ISBN 978–3–030-31938-0. https://www.springer.com/gp/book/9783030319373

Thul ST, Sarangi BK (2015) Implications of nanotechnology on plant productivity and its rhizospheric environment. In: Nanotechnology and plant sciences. Springer, Cham, pp 37–53

Tolaymat TM, El Badawy AM, Genaidy A, Scheckel KG, Luxton TP, Suidan M (2010) An evidence-based environmental perspective of manufactured silver nanoparticle in syntheses and applications: a systematic review and critical appraisal of peer-reviewed scientific papers. Sci Total Environ 408:999–1006

Tripathi DK, Singh S, Singh S, Srivastava PK, Singh VP, Singh S, Prasad SM, Singh PK, Dubey NK, Pandey AC, Chauhan DK (2017a) Nitric oxide alleviates silver nanoparticles (AgNPs)-induced phytotoxicity in *Pisum sativum* seedlings. Plant Physiol Biochem 110:167–177

Tripathi DK, Tripathi A, Shweta SS, Singh Y, Vishwakarma K, Yadav G, Sharms S, Singh VK, Mishra RK, Upadhvav RG, Dubey NK, Lee Y, Chauhan DK (2017b) Uptake, accumulation and toxicity of silver nanoparticle in autotrophic plants, and heterotrophic microbes: a concentric review. Front Microbiol 8:07

Tungittiplakorn W, Lion LW, Cohen C, Kim JY (2004) Engineered polymeric nanoparticles for soil remediation. Environ Sci Technol 38:1605–1610

Umarani R, Mala R (2013) Influence of calcium phosphate nano gel fertilizer composites on germination of crop plants. Asian J Microbiol Biotechnol Environ Sci 15:327–331

UN (2017) Department of Economic and Social Affairs, Population Division. World population prospects: the 2017 revision, vol I: comprehensive tables. ST/ESA/SER.A/399

Vanderroost M, Ragaert P, Devlieghere F, DeMeulenaer B (2014) Intelligent food packaging: the next generation. Trends Food Sci Technol 39:47–62

VandeVoort AR, Arai Y (2012) Effect of silver nanoparticles on soil denitrification kinetics. Ind Biotechnol 8:358–364

Venkatramanan V, Shah S, Prasad R (2020) Global climate change and environmental policy: resilient and smart agriculture. Springer Singapore. ISBN 978–981–329-855-2. https://www.springer.com/gp/book/9789813298552

Vernay P, Gauthier-Moussard C, Jean L, Bordas F, Faure O, Ledoigt G, Hitmi A (2008) Effect of chromium species on phytochemical and physiological parameters in *Datura innoxia*. Chemosphere 72:763–771

Vijayalakshmi C, Chellaram C, Kumar SL (2015) Modern approaches of nanotechnology in agriculture-a review. Biosci Biotechnol Res Asia 12:327–331

Wais U, Jackson AW, He T, Zhang H (2016) Nanoformulation and encapsulation approaches for poorly water-soluble drug nanoparticles. Nanoscale 8:1746–1769

Wallentin J, Anttu N, Asoli D, Huffman M, Åberg I, Magnusson MH, Siefer G, Fuss-Kailuweit P, Dimroth F, Witzigmann B, Xu HQ, Samuelson L, Deppert K, Borgström MT (2013) InP nanowire array solar cells achieving 13.8% efficiency by exceeding the ray optics limit. Science 339:1057–1060

Wanekaya AK, Chen W, Myung NV, Mulchandani A (2006) Nanowire-based electrochemical biosensors. Electroanalysis 18:533–550

Wang CB, Zhang WX (1997) Synthesizing nanoscale iron particles for rapid and complete dechlorination of TCE and PCBs. Environ Sci Technol 31:2154–2156

Wang XM, Gao FQ, Ma LL, Liu J, Yin ST, Yang P, Hong FS (2008) Effects of nano-anatase on ribulose-1, 5-bisphosphate carboxylase/oxygenase mRNA expression in spinach. Biol Trace Elem Res 126(1–3):280–289

Wang X, Chen C, Chang Y, Liu H (2009) Dechlorination of chlorinated methanes by Pd/Fe bimetallic nanoparticles. J Hazard Mater 161:815–823

Wang Z, Tan X, Zhang Z, Gu S, Li G, Shi H (2012a) Defense to *Sclerotinia sclerotiorum* in oilseed rape is associated with the sequential activations of salicylic acid signaling and jasmonic acid signaling. Plant Sci 184:75–82

Wang Z, Xie X, Zhao J, Liu X, Feng W, White JC, Xing B (2012b) Xylem-and phloem-based transport of CuO nanoparticles in maize (*Zea mays* L.). Environ Sci Technol 46:4434–4441

Wang H, Sun K, Tao F, Stacchiola DJ, Hu YH (2013) 3D honeycomb-like structured graphene and its high efficiency as a counter-electrode catalyst for dye-sensitized solar cells. Angew Chem Int 52:9210–9214

Wang X, Yang X, Chen S, Li Q, Wang W, Hou C, Gao X, Wang L, Wang S (2016) Zinc oxide nanoparticles affect biomass accumulation and photosynthesis in *Arabidopsis*. Front Plant Sci 6:1243

Wei H, Wang E (2013) Nanoparticles with enzyme-like characteristics (nanozymes): next- 1023 generation artificial enzymes. Chem Soc Rev 42:6060–6093

Werlin R, Priester JH, Mielke RE, Kramer S, Jackson SP, Stoimenov K, Stucky GD, Cherr GN, Orias E, Holden PA (2011) Biomagnification of cadmium selenide quantum dots in a simple experimental microbial food chain. Nat Nanotechnol 6:65–71

Wu Y, Pang H, Liu Y, Wang X, Yu S, Fu D, Chen J, Wang X (2019) Environmental remediation of heavy metal ions by novel-nanomaterials: a review. Environ Pollut 246:608–620

Xie Y, He Y, Irwin PL, Jin T, Shi X (2011) Antibacterial activity and mechanism of action of zinc oxide nanoparticles against *Campylobacter jejuni*. Appl Environ Microbiol 77:2325–2331

Xie Y, Li B, Zhang Q, Zhang C (2012) Effects of nano-silicon dioxide on photosynthetic fluorescence characteristics of *Indocalamus barbatus* McClure. J Nanjing For Univ (Nat Sci Edn) 2:59–63

Xie Y, Cheng W, Tsang PE, Fang Z (2016) Remediation and phytotoxicity of decabromodiphenyl ether contaminated soil by zero valent iron nanoparticles immobilized in mesoporous silica microspheres. J Environ Manag 166:478

Xiong Y, Ye F, Zhang C, Shen S, Su L, Zhao S (2014) Synthesis of magnetic porous γ-Fe2O3/C@HKUST-1 composites for efficient removal of dyes and heavy metal ions from aqueous solution. RSC Adv 5:5164–5172

Xu L, Liu Y, Bai R, Chen C (2010) Applications and toxicological issues surrounding nanotechnology in the food industry. Pure Appl Chem 82:349–372

Yadav L, Tripathi RM, Prasad R, Pudake RN, Mittal J (2017) Antibacterial activity of Cu nanoparticles against E. coli, Staphylococcus aureus and Pseudomonas aeruginosa. Nano Biomed Eng 9(1):9–14. https://doi.org/10.5101/nbe.v9i1.p9-14

Yang L, Watts DJ (2005) Particle surface characteristics may play an important role in phytotoxicity of alumina nanoparticles. Toxicol Lett 158:122–132

Yang F, Hong F, You W, Liu C, Gao F, Wu C, Yang P (2006) Influences of nano-anatase TiO2 on the nitrogen metabolism of growing spinach. Biol Trace Elem Res 110:179–190

Yang FL, Li XG, Zhu F, Lei CL (2009) Structural characterization of nanoparticles loaded with garlic essential oil and their insecticidal activity against *Tribolium castaneum* (Herbst) (Coleoptera: Tenebrionidae). J Agric Food Chem 57:10156–10162

Yao KS, Li SJ, Tzeng KC, Cheng TC, Chang CY, Chiu CY, Liao CY, Hsu JJ, Lin ZP (2009) Fluorescence silica nanoprobe as a biomarker for rapid detection of plant pathogens. Adv Mater Res 79-82:513–516

Zain NM, Stapley AG, Shama G (2014) Green synthesis of silver and copper nanoparticles using ascorbic acid and chitosan for antimicrobial applications. Carbohydr Polym 112:195–202

Zan L, Fa W, Peng T, Gong Z (2007) Photocatalysis effect of nanometer TiO_2 and TiO_2-coated ceramic plate on Hepatitis B virus. J Photochem Photobiol B Biol 86:165–169

Zhai G, Walters KS, Peate DW, Alvarez PJ, Schnoor JL (2014) Transport of gold nanoparticles through plasmodesmata and precipitation of gold ions in woody poplar. Environ Sci Technol Lett 1:146–151

Zhang WX (2003) Nanoscale iron particles for environmental remediation: an overview. J Nanopart Res 5:323–332

Zhao L, Sun Y, Hernandez-Viezcas JA, Servin AD, Hong J, Niu G, Peralta-Videa JR, Duarte-Gardea M, Gardea-Torresdey JL (2013) Influence of CeO_2 and ZnO nanoparticles on cucumber physiological markers and bioaccumulation of Ce and Zn: a life cycle study. J Agric Food Chem 61:11945–11951

Zheng L, Hong F, Lu S, Liu C (2005) Effect of nano-TiO_2 on strength of naturally aged seeds and growth of spinach. Biol Trace Elem Res 104:83–91

Zheng L, Su MY, Chao L, Liang C, Hao H, Xiao W, Liu XQ, Yang F, Gao FQ, Hong FS (2007a) Effects of nanoanatase TiO_2 on photosynthesis of spinach chloroplasts under different light illumination. Biol Trace Elem Res 119:68–76

Zheng L, Su MY, Xiao W, Chao L, Qu CX, Chen L, Hao H, Liu XQ, Hong FS (2007b) Effects of nano-anatase on spectral characteristics and distribution of LCHII on the thylakoid membranes of spinach. Biol Trace Elem Res 120:273–283

Zou Y, Wang X, Khan A, Wang P, Liu YH, Alsaedi A, Hayat T, Wang X (2016a) Environmental remediation and application of nanoscale zero-valent iron and its composites for the removal of heavy metal ions: a review. Environ Sci Technol 50:7290–7304

Zou Y, Wang X, Ai Y, Liu Y, Ji Y, Wang H, Hayat T, Alsaedi A, Hu W, Wang X (2016b) β-Cyclodextrin modified graphitic carbon nitride for the removal of pollutants from aqueous solution: experimental and theoretical calculation study. J Mater Chem 4:14170–14179
Zwingmann N, Mackinnon IDR, Gilkes RJ (2011) Use of a zeolite synthesised from alkali treated kaolin as a K fertiliser: glasshouse experiments on leaching and uptake of K by wheat plants in sandy soil. Appl Clay Sci 53:684–690

Nanotechnology-Based Treatment Systems for Arsenic Sequestration in Groundwater: Contamination, Challenges and Future Scope of Studies

2

Ajay Kumar, Pankaj Kumar Gupta, and Sanjay Kumar Gupta

Abstract

Arsenic has a global concern as a groundwater contaminant due to its potential for fatal health consequences. The long-term consumption of arsenic-contaminated water and food results in damage to the liver, kidney and gall bladder. Both developed and developing countries are under a potential threat of groundwater arsenic contamination. To combat the problems related to its contamination, the WHO and the US EPA have set the standard limit of 10 μg L^{-1} for drinking water since the year of 2006. Its removal employing nanoadsorbents has gained considerable attention through both ex situ and in situ techniques among the scientific groups since the last two and half decades. Moreover, nanotechnology-based water treatment systems in general are also finding favour due to better resource and energy efficiency.

Among the arsenic removal technologies (ART), oxidation and filtration lead to addition of chemicals in the solution and individually is kinetically slow process. Co-precipitation leads to the generation of large amount of sludge, and extraction of removed ions is difficult. The ion-exchange phenomena are pH dependent. Whereas membrane technology is not considered to be cost-effective,

A. Kumar
Environmental Hydrology Laboratory, Department of Hydrology, Indian Institute of Technology Roorkee, Roorkee, Uttrakhand, India
e-mail: akumar1@hy.iitr.ac.in

P. K. Gupta (✉)
Faculty of Environment, University of Waterloo, Waterloo, ON, Canada
e-mail: pk3gupta@uwaterloo.ca

S. K. Gupta
Environmental Engineering, Department of Civil Engineering, Indian Institute of Technology Delhi, New Delhi, India

R. Prasad (ed.), *Environmental Pollution and Remediation*, Environmental and Microbial Biotechnology, https://doi.org/10.1007/978-981-15-5499-5_2

in literature, removal through adsorption has been considered as a cost-effective and promising technology.

Keywords

Arsenic contamination · Nanoparticles · Remediation · Health risk

2.1 Introduction

Arsenic is a toxic, bio-accumulating, redox and pH-sensitive element. It is a life-threatening contaminant, to which millions are exposed through contaminated groundwater from the beginning of the twentieth century (World Health Organization 2010). The high mobility of arsenic species in sub-surface systems is of concern among the scientific communities (Gupta et al. 2020). Worldwide, about 105 countries and approximately 226 million people are under the potential threat of arsenic contamination (Naujokas et al. 2013). Around 90,000 km^2 area with 50 million people are critically affected by its contamination in Bengal Delta basin comprising Bangladesh (Pal et al. 2009). Its contamination in groundwater broadly reported in flood plains of the Ganga River is a subject of concern among the scientific communities across the globe since the last two and half decades due to its high mobility in ecosystem, high toxicity, carcinogenicity to human beings and its biomagnification characteristics. The menace among the communities across the globe is mainly due to its contamination in groundwater as compared to that of surface water. To combat the problems related to its contamination, the WHO and the US EPA have set its standard limit as 10 $\mu g\ L^{-1}$ for drinking water since 2006 (World health organisation 2017; Office of Water U.S. Environmental Protection Agency 2018).

2.2 Arsenic Pollution in Groundwater and Its Health Consequences

Both developed and developing countries are under a potential threat of problems related to arsenic-contaminated groundwater. The countries such as Argentina (Smedley et al. 2005), Bangladesh (Ahmed 2001a), Brazil (Ciminelli et al. 2017), Canada (Bondu et al. 2017), Cambodia (Berg et al. 2007), Chile (Pino et al. 2015), Ghana (Adomako et al. 2011), Hungary (Sugár et al. 2014), Mexico (Navarro et al. 2017), Pakistan (Berg et al. 2007; Richards et al. 2015), Republic of China, Taiwan, United States of America and Vietnam (Berg et al. 2007) are affected with arsenic-contaminated groundwater. It may be interesting to note that one of the first countries to report the arsenic contamination in the groundwater in the beginning of the twentieth century (1917) was Argentina, whereas in India, its potential health hazards were reported as late as in early 1990s. In India, the states of Assam (Das et al. 2017), Bihar (Chakraborti et al. 2016), Chhattisgarh (Patel et al. 2017; Rana

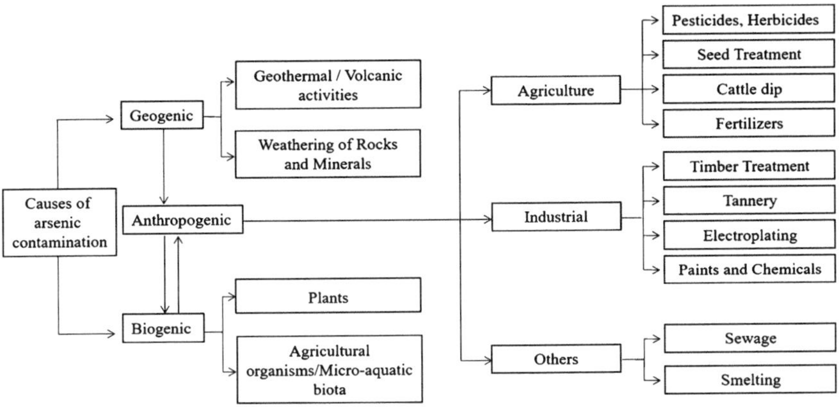

Fig. 2.1 Representation of possible causes of arsenic pollution in groundwater. (Nordstrom 2002; Ng et al. 2003; Ali et al. 2006)

et al. 2016), Jharkhand (Alam et al. 2016), Punjab (Sharma et al. 2016), Manipur (Chandrashekhar et al. 2016), Uttar Pradesh (Olea et al. 2018) and West Bengal (Bhowmick et al. 2018a) are reported to have arsenic in groundwater with the concentration exceeding the permissible limit set for drinking water (Prasad et al. 2013).

Most reported arsenic problems (As^{III} species) have been found in groundwater supply systems and are caused primarily by natural processes, such as mineral weathering and dissolution resulting from a change in the geochemical environment to a reductive condition. Millions of people in West Bengal and Bangladesh had been drinking groundwater from wells contained arsenic concentration ranging from 100 to 3900 μg L^{-1}, with many of them succumbing to diseases caused by it (Chatterjee et al. 2010; Bhowmick et al. 2018b). A pictorial representation of the possible causes of arsenic pollution is shown in Fig. 2.1. It may exist as oxyanions in different types of species due to its sensitivity towards redox and pH conditions. At redox potential (pE > 600 mV), $HAsO_4^{2-}$ is a dominant species of As^{V} in groundwater between the pH range of 6.5 and 11.5. Below pH < 6.5, it exists as $H_2AsO_4^{-}$ (Ahmed et al. 2010). However, zero-valent arsenic (As°) and As^{3-} occur rarely in aquatic environments.

Early clinical symptoms in acute poisoning are severe nausea and vomiting, weakness, muscular pain and flushing skin and colicky abdominal pain (Uede and Furukawa 2003). Chronic exposure is much more insidious in nature, and sometimes symptoms appear between 6 months and 2 years, leading to anaemia and leukopenia (Rahman et al. 2009). The major dermatological diseases, such as melano-keratosis, melanosis, spotted melanosis and spotted and diffuse keratosis, were reported to affect the human beings. Further, it has been widely reported that the long-term consumption of arsenic-contaminated water and food results to damage the liver, kidney and gall bladder (Smith et al. 1992; Tokar et al. 2010). Some potential health hazards of arsenic to human beings are highlighted in Fig. 2.2.

Fig. 2.2 An overview of arsenic health consequences: (**a**). skin lesions (**b**). skin cancer (**c**). blackfoot disease. (Source: http://www.sos-arsenic.net/english/contamin/index.html)

2.3 Sources of Arsenic in Subsurface Environment

Arsenic from Earth's crust and bedrocks, enriched with its ores, leaches moderately into groundwater (Vahter 2008). The mineral iron arsenate ($FeAsO_4$) is considered to be a common source of arsenic in the subsurface environment. The compounds scorodite ($FeAs_4$. $2H_2O$) and pittcite [$Fe_x^{3+}(AsO_4)_y$ $(SO_4)_z$]. nH_2O are reported as direct and instantaneous source of arsenic in subterranean conditions (Rimstidt et al. 1987; Langmuir et al. 1999; Singh 2006). Both of these are alteration products of mineral arsenopyrite (FeAsS), which constitutes a widespread source of arsenic in natural water (Chowdhury et al. 1999).

2.3.1 Causes of Its Release in Aquifer Systems

Arsenic mobilization in subsurface environment and aquifer systems are controlled by two major factors, viz. (1) adsorption and desorption and (2) precipitation and dissolution of solid phase. The possible mechanisms that cause its release into groundwater are discussed below:

2.3.1.1 Reductive Dissolution

Under reducing environment, the release of arsenic attached to iron has been considered as an important reason for the presence of elevated levels of arsenic in the sedimentary aquifer systems (Kumar et al. 2018). The generation of reducing conditions in subsurface environment primarily depends on microbial activities in the sediments along with the abundance of organic matter. The major factors responsible for its mobilization in surrounding environment are diffusion of gases, sedimentation rate and microbial reactions. Nickson et al. (Nickson et al. 2000) have described that the lowering of pH accelerates the dissolution of iron(oxy) hydroxides and subsequent release of associated arsenic through laboratory scale studies. In another study, the generation of highly reducing environment at near neutral pH has been suggested due to elevated concentration of carbonates and is considered to be the possible reason of its release into the aquifer media (Richards et al. 2017; Appelo et al. 2002). Recently, oxidation of organics along with microbial triggered reductive processes has been reported to be crucial in the mobilization of arsenic in alluvial aquifers of Terai Region, Nepal (Diwakar et al. 2015). However, high concentrations of Fe, Mn, HCO_3^- ions, NH_4^+ ions and CH_4 gas and an absence of oxidized species, such as NO_3^- and SO_4^{2-} ions, are the indicators of strongly reducing conditions (Lockwood et al. 2014). Moreover, the microbial mechanisms of arsenic release require the presence of abundant of organic matter in subsurface environment.

2.3.1.2 Alkali Desorption

Arsenic attachment in the subterranean environment onto the surface of iron oxides is an example of adsorption, and its detachment occurs through desorption. Its adsorption and release in the aquifer systems mainly occur through solid-phase precipitation and solid-phase dissolution, respectively. In alluvium plains, arsenic is present as FeAsS (arsenopyrite) along with its transformed phases, such as $FeAsO_4$ (ferric arsenate) and $FeAsO_3$ (ferric arsenite) (Bhattacharya et al. 1997). It has been reported that the presence of an oxidizing agent, most commonly atmospheric oxygen (as O_2), controls the oxidation rate of arsenic containing sulphide minerals (Chakraborty et al. 2015; Poreda et al. 2002). Thus, increment in the discharge rate of groundwater makes the arsenic-bearing sulphide mineral exposed to oxygen, which causes their oxidation and further release of arsenic to aquifer system.

2.3.1.3 Geothermal Trigger

The elevated levels of arsenic have been reported in the hot springs on Qinghai-Tibet Plateau that caused thorough leaching of rocks under high temperature conditions (Rodríguez-Lado et al. 2013). The generation of these conditions occurred due to either circulation of groundwater (deep and rapid) or shallow volcanism. Incidentally, the rivers and groundwater of Chile have also been reported with severe arsenic pollution caused due to seepage triggered by geothermal forces from the mountain, Andes, located hundreds of kilometres from the point of groundwater abstraction areas (Bundschuh et al. 2012).

2.4 Conventional Treatment Methods and Their Limitations

Several technologies have been developed so far to remove the arsenic from groundwater. However, the most common removal methods which have been widely explored include oxidation and adsorption (Luo et al. 2012; Luo et al. 2013), chemical precipitation/sedimentation and filtration (Meng et al. 2001; Sheoran and Sheoran 2006), use of ion-exchange resins (Greenleaf et al. 2006) and membrane technology, including reverse osmosis (Ning 2002). Recently, the in situ arsenic removal technology using permeable reactive barriers (PRBs) has drawn considerable attention among the research communities and is being practiced in few countries. All these methods have been discussed briefly below:

2.4.1 Coagulation/Flocculation

It involves the mixing of chemicals such as salts of iron (ferric chloride), aluminium (alum) and manganese (manganese sulphate) into water to settle down the arsenic through precipitation, co-precipitation or combination of both these processes. Afterwards, the arsenic is separated from the solution by filtration as it gets adsorbed onto the solid matrix (Wickramasinghe et al. 2004). In general, this approach removes As^{III} as compared to As^{V}, and ferric chloride has been reported to be more efficient than alum. The effectiveness of this method depends on the pH, concentration of arsenic, arsenic speciation and presence of competing ions in the water. However, it requires pretreatment process to oxidize As^{III} to As^{V} using chlorine or permanganate ions since it causes partial removal of As^{III} from aqueous solution (Ahmed 2001b). Nowadays, solar and microbial oxidation have also been commonly utilized to convert As^{III} to As^{V} (Jain and Singh 2012).

2.4.2 Ion-Exchange Resins

It is a physio-chemical process involving exchange of ions between solid phase and resin (solid) phase of similar charge. The exchanged ions are held electrostatically on the surface of solid phase with ions of similar charge in a solution. The ion-exchange media is usually packed into a column, and then water containing arsenic is allowed to pass for removing the desired contaminant. The columns can be regenerated with the brine solution for preparation of next cycle of removal (Ghurye et al. 1999).

2.4.3 Oxidation and Adsorption

Oxidation followed by adsorption is one of the conventional methods to treat arsenic-contaminated water. Mostly, As^{III} is oxidized to As^{V} using several oxidants, such as chlorine, potassium permanganate, ozone and hydrogen peroxide. Typically, adsorption involves passing of contaminated water through a packed media to which arsenic get attached via physical or chemical bonds. The common adsorbents that have been reportedly employed in the field include granular ferric hydroxides, activated alumina, iron coated sand and activated carbon with surface modifications (Westerhoff et al. 2005; Hristovski et al. 2009).

2.4.4 Membrane Processes

In this technique, a semipermeable membrane has been utilized to separate the arsenic from contaminated water. This membrane acts as a physical barrier to some ions depending on their physio-chemical characteristics. The driving force responsible for the movement of ions across the membrane is the generation of potential difference between the two sides of membrane. The process involved for the arsenic removal depends on the mechanisms of filtration, adsorption of arsenic-bearing compounds and electric repulsion (Brandhuber and Amy 1998). The removal rate of arsenic through this process is affected by shape, size and chemical characteristics of contaminant species. Several membranes have been developed so far for the removal of arsenic, which can be broadly classified into two categories: high-pressure membrane (75–250 psi) and low-pressure membrane (10–30 psi). Low-pressure membrane includes microfiltration (MF) and ultrafiltration (UF), whereas high-pressure membrane includes nanofiltration (NF) and reverse osmosis (RO).

2.5 Adoption of Technology at Rural Scale

In order to fulfil the requirement of arsenic-free drinking water for densely populated countries such as India and Bangladesh, adoption of the technology at rural scale in a convenient, economic and effective manner is a challenging task among the scientific communities. Several conventional technologies with modifications have been reported to be accomplished in rural and remote areas to provide safe drinking water (Fan et al. 2017; Sarkar and Blaney 2008). These technologies have been generally classified as household treatment systems or community-based treatment systems. A brief description of these technologies is presented below:

Aeration, boiling of water, sand filters, ceramic filters and oxidation using solar light are among the typical methods of household treatment systems, where contaminated water is treated from a single source and utilized for livestock's purposes such as cooking and drinking (Crabbe et al. 2017; Ahsan 2010). Whereas in community-based treatment systems, many households invest money as a society and make a single treatment unit. The advanced and costly methods for the treatment of arsenic-contaminated groundwater are not affordable to most of the populations in the world (Khettab and Chabbi-Chemrouk 2017; Singh 2017; Shahraki 2017). Therefore, the focus of this research also describes the development of arsenic remediation materials by utilizing industry waste as precursor material, which might present a sustainable approach to upgrade the existing arsenic removal units by providing indigenously generated nanoadsorbents.

2.5.1 Domestic or Household Treatment Systems

This type of technology was developed to provide arsenic-free water to the poor people. Many developing countries are unable to afford centralized treatment methods for arsenic remediation; therefore, different types of economic treatments have been developed. Among these, the bucket treatment units (BTU), sono-filters and bio-sand filter are quite common.

Bucket treatment unit is based on the arsenic removal processes, such as coagulation, co-precipitation and adsorption. It consists of two buckets placed one above another. From the top, the arsenic-contaminated water is poured along with the mixing of alum (as a flocculant) and potassium permanganate (as an oxidizing agent). This mixture is allowed to settle after stirring for a couple of minutes. Then, the settled water is passed to lower bucket through the pipe attached to the lower end of upper bucket (Tahura et al. 1998; Souter et al. 2003). The arsenic-free water is collected from the lower part of the bucket after passing through sand filter. The obtained aluminium-arsenic complex can be removed by filtration process at the end of remediation process.

Sono-filter is an another simpler household technique, which involves the process of oxidation, precipitation, adsorption and filtration during removal of arsenic (Leupin et al. 2005; Hussam 2009). It consists of the top and middle pots as reactors along with the lower one as storage pot for treated water. The layers of polyester cloth, coarse sand, iron chips and polyester cloth, fine sand and charcoal are placed as removal material medium in top and middle pot of the filter system, respectively. This approach has been reported to be quite efficient for arsenic removal in the pH range of groundwater, which involves surface complexation reaction mechanisms.

Arsenic bio-sand filter (ABF) is among the one of the most effective and economical technologies adopted for arsenic removal to provide drinking water in rural areas (Shah et al. 2015). It consists of two layers, such as pathogen removal unit (PRU) and ATU (ARU), representing the lower part and upper part of the filter system, respectively. These parts are equipped with iron nails, polyester cloth, metal diffuser box and fine sand, coarse sand and gravels as components of ARU and PRU, respectively.

This filter works on the principal of arsenic removal through adsorption onto iron hydroxides and slow sand filtration mechanism (Ahammed and Davra 2011). The formation of iron hydroxide particles responsible for arsenic adsorption is produced after the oxidation of iron nails in the presence of air. Then, arsenic-loaded water is allowed to pass through underlined sand filter to acquire potable water. Further, the PRU unit is considered to remove the pathogens by steps, such as mechanical trapping (occurs due to entrapment of pathogens along with sediments in the space between the consolidated materials), adsorption (attachment of pathogens to sediments and each other), predation (consumption due to already existing microbes in biofilm layer) and natural death due to scarcity of food.

2.5.2 Community-Based Treatment Systems

In rural areas, many people consume water from common tube wells and handpumps. To provide safe drinking water, arsenic treatment unit (ATU) is generally attached to water source, which is operated with an intermittent flow of water. Generally, the conventional treatment is done through the processes, such as mixing of chemicals, flocculation, sedimentation and filtration (Duarte et al. 2009; Chen et al. 1999). In the first stage, sodium hypochlorite and alum are added to the arsenic infested water for the oxidation and coagulation, respectively. In the second stage, the mixing of these chemicals leads to the formation of flocks containing arsenic, which is followed by sedimentation and filtration to get arsenic-free water. Recently, the importance and possible scope of upgrading the conventional arsenic treatment systems with newly developed nanoadsorbets are reviewed, and might be considered for future application in real-world scenario (Kumar et al. 2020).

2.6 Scope of Nanotechnology-Based Treatment Systems

The above-mentioned removal methods have been reported to have the disadvantages in terms of generating highly toxic waste that causes its easy mobilization in the living ecosystems. Arsenic removal through coagulation/flocculation requires addition of chemicals and thereafter removal in the form of precipitates after the completion of process. This serves as one of the drawbacks of this technology. Also, there are the possibilities of generation of secondary pollutants in treated water, which may cause diseases related to these pollutants. Moreover, the technologies such as ion-exchange and membrane processes have limitations in terms of their higher energy consumption, higher costs and complex removal mechanisms.

Among the above-mentioned removal methods, arsenic removal by adsorption has still gained considerable attention even considering its limitation in handling toxic waste after the exhaustion of adsorbed material. It is widely acceptable technology as on today due to its easy operation, less maintenance cost and cost-effectiveness (Grassi et al. 2012). The literature is widely replete in cases of development of different nanoadsorbents, which have been explored for arsenic removal (Lata and Samadder 2016). From the past two decades, several metallic nanoadsorbents, including oxides of iron, aluminium, cerium, copper, zirconium and titanium, have been reported to developed for arsenic removal (Ali 2012). Metallic iron-based nanoadsorbents have been widely explored due to their strong affinity towards arsenic and ecological-friendly nature. Therefore, the development of its polymorphs, such as nZVI, iron^{III} oxides (α-Fe_2O_3, β- Fe_2O_3, γ- Fe_2O_3), iron^{III} (oxy) hydroxides (α-FeOOH, β-FeOOH, γ-FeOOH) and iron^{II,III} oxides (magnetite), are being reported in literature for arsenic removal. Among these, nZVI (nanoscale zero-valent iron) has been extensively explored in the literature both at laboratory- and pilot-scale studies. These nanoparticles are unstable in natural environmental conditions and prone to oxidize into iron oxides/hydroxides after oxidation which limit its application for pilot-scale studies significantly (Nurmi et al. 2005).

The literature is also replete in providing a number of adsorbents for arsenic removal in groundwater, which involve the use of naturally originated, synthetic adsorbent materials and low cost materials (Zhang and Selim 2005; Genc-Fuhrman et al. 2004; Gu et al. 2005; Elizalde-Gonzalez et al. 2001; Gillman 2006; Malik et al. 2009; Sullivan et al. 1997). In recent years, a number of nanoadsorbents, such as nZVI alone (Yin et al. 2012) supported on clay, tea waste, polyaniline and starch (Tandon et al. 2013; Bhaumik et al. 2014; Mosaferi et al. 2014); oxides of iron (De et al. 2009; Wang et al. 2011; Lin et al. 2014); titania as anatase (Özlem Kocabaş-Atakli and Yürüm 2013; Danish et al. 2013); cerium (IV) oxide (Sun et al. 2012); copper (Reddy et al. 2013); and polymer-based adsorbents (Davodi and Jahangiri 2014) including the sulphides of iron (Han et al. 2011) and zinc (Piquette et al. 2012), have been explored for arsenic removal.

2.7 Nanoadsorbents for Arsenic Remediation

2.7.1 Iron-Based Nanomaterials

NZVI is a widely explored phase among the iron-based nanomaterials utilized for arsenic removal through both ex situ and in situ treatment processes. As compared to bulk, its nanosize does not only enhance the surface area (Kanel et al. 2006), but the high intrinsic activity of these nanoparticles delivers the high density of reactive sites to the contaminant for electrostatic interactions (Morgada et al. 2009; Robalds et al. 2016). Generally, the arsenic removal mechanism of bare nZVI involves the precipitation, co-precipitation, oxidation and reduction reactions.

Magnetite is a predominant constituent of the subsurface environment (N. Compounds 2001). This phase has been found to be more efficient for the arsenic removal as compared to extensively explored bare zero-valent iron nanoparticles (Hunt and Amrhein 2002). Υ-FeOOH is a predominant and highly reactive phase which occurs as ore deposits, sediments and terrestrial soil (Cornell and Schwertmann 2003; van der Zee et al. 2003). The vacant sites are located between the Fe occupied double chain rows and considered as false tunnel-like structures (Cornell and Schwertmann 2004). The nanostructured β-FeOOH has been considered as having a true tunnel-like structure along with the single row of anions missing in the unit cell (Scheck et al. 2015). The presence of tunnel-like formation has been considered to be responsible for its efficient adsorption capabilities, providing large surface area and reactive sites. The nanostructured iron oxides containing iron in trivalent state, such as hematite (α-Fe_2O_3), beta phase (β-Fe_2O_3), maghemite (Υ-Fe_2O_3), have been investigated for various environmental applications (Kaloti et al. 2015; Kaloti and Kumar 2016).

2.7.2 Ti-, Al-, Zr-, Ce-, Zn- and Cu-Based Nanomaterials

Anatase, brookite and rutile are common polymorphs of titanium dioxide, which have been reported to be explored for arsenic. These have been considered as significant nanomaterials for the environmental applications due to their non-corrosive, non-toxic and chemically stable properties (Hung et al. 2007). Alumina is an important adsorbent as it acquires excellent physical and textural properties comparable to those of other transitional metal oxides (Takanashi et al. 2004). It has been found in several industrial wastes, such as blast furnace slag, coal combustion ash and waterworks sludge. Nanoscale zirconia- and ceria-based nanomaterials have demonstrated an adequate affinity for As^V due to which it has been explored for arsenic treatment (Hristovski et al. 2007; Haron et al. 2008). These have been considered as suitable adsorbents for field-scale applications due to their stable characteristic features (Ahmed 2001a). The polymorphs of metallic zinc nanoparticles, such as ZnO and ZnS have, been explored for the removal of arsenic. The structural stability of ZnS at nanoscale is sensitive to natural environmental conditions due to their ability to adsorb water and aggregation characteristics. Copper oxide(s) are other class of nanomaterials which do not require pretreatment and post-pH adjustments during adsorption. It has two types of polymorphs: Cu^{I} oxide and Cu^{II} oxide. Lastly, the bulk Cu_2O is known to get oxidized into CuO at ambient conditions; however, nanostructured Cu_2O has been observed to be fairly stable (Ram and Mitra 2001).

2.8 Scope of Future Nanoadsorbent Development

The stable characteristics of nanoadsorbents in a wide range of environmental conditions are expected to be extremely helpful during field-scale applications. Also, the upgradation of conventional treatment systems by utilizing the nanoscale-based absorbents as remediation material is required in future scenario in providing potable water to affected areas. Further studies might be conducted for synthesizing its nanohybrids using bio-templates and organics to support them on the different matrices so as to overcome its limitation of dispersion and stability in the aqueous medium.

2.9 Subsurface Sequestration of Arsenic

Generally, removal of arsenic from the contaminated systems has been categorized into: ex situ (pump and treat) and in situ (underground treatment) processes. Till now, pump-and-treat remediation has been practiced on large scale in providing arsenic-free water to the affected population (Jain and Singh 2012). However, such technologies have several disadvantages like high operational cost and generation of toxic sludge (Sarkar et al. 2010). The advantages of in situ removal of contaminants compared to ex situ technologies include natural treatment of groundwater and low operational cost. Moreover, the sequestration of groundwater contaminants using in situ technologies has been considered among sustainable approaches (Kanel 2006). Therefore, the development of technologies to sequester the arsenic in the subsurface is gaining a considerable attention. From the last two decades, in situ technologies such as biosorption, permeable reactive barriers, chemical oxidation and natural attenuation are being developed. Among these, PRB (permeable reactive barrier), because of its convenient operation and compactness, makes it easy to manage for remediation and has been considered as a promising technology (Jain and Singh 2012; Lee et al. 2009).

2.9.1 Technology Advantages

Subsurface remediation of contaminant of organic volatile pollutants has been considered an environmental-friendly approach in terms of its attenuation and release to the atmosphere (Azubuike et al. 2016; Limmer and Burken 2016). Similarly, the treatment of arsenic-contaminated groundwater in subterranean environment (below surface) is a logical choice in respect to reduce the chance of its mobilization in living environment. In situ treatment has been considered more environmental friendly, cost-effective and reliable as compared to ex situ mode of treatment (Spira et al. 2006). Developing a better understanding about in situ remediation methods and further improvement appears to be more rational and relevant choice now rather than to improve the already developed conventional techniques. Some recent methods explored for the removal of arsenic in subsurface environment are discussed further.

2.9.2 Methods of Treatment

2.9.2.1 Permeable Reactive Barriers (PRBs)

In the early 1990s, permeable reactive barriers (PRBs) made use of zero-valent iron, iron work slag materials. An overview is shown in the Fig. 2.3. Initially, ZVI has been applied in permeable reactive barrier (PRB) system for the remediation of groundwater (Fiúza et al. 2015). In this approach, multiple functional barriers were employed for the remediation of arsenic from groundwater (Beak and Wilkin 2009). The system was widely explored for the remediation of both As^{III} and As^{V} due to its easy availability and environmental-friendly characteristics.

In recent years, the nanoscale zero-valent iron (nZVI) has been explored in PRB system and is considered as an extension of conventional ZVI technology (Beak and Wilkin 2009; Waybrant et al. 1998). In subsurface environment, nZVI has proved as one of the most popular adsorbents for the removal of dominant arsenic species as compared to μm size iron (Kanel 2006).

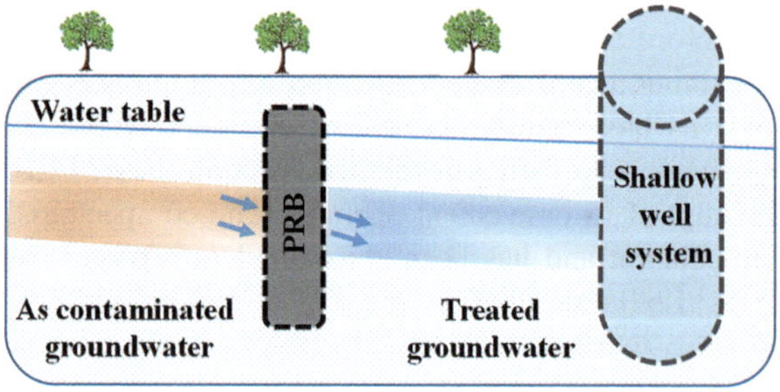

Fig. 2.3 An illustration of PRB-based technique representing remediation of arsenic in shallow groundwater system. (Zhou et al. 2014; Wilkin et al. 2005)

2.9.2.2 In Situ Chemical Treatment

Generally, it involves the direct injection of oxidants to the subsurface such as $KMnO_4$ or oxygen, which promotes the oxidation of As^{III} to As^{V} and then the co-precipitation of As^{V} from iron oxides. In this process, the chemical oxidant is injected in an aquifer upstream of the contaminated site, and a closed loop can be generated by pumping of water in the downstream site to utilize it for reinjection. The reaction between injected chemical and the contaminant inhibits the mobilization of arsenic in aquifer systems. The most efficient chemicals injected into aquifer are hydrogen peroxide and ferric chloride (Alp et al. 2016). Both of these chemicals have to be injected sequentially because the oxidation by hydrogen peroxide is a rapid process. The oxidation and precipitation reactions included in this process are:

$$FeAsO_2 + H_2O_2 \rightarrow H_3AsO_4 + 2H^+ + 2e^- \left(E^0 = 0.56\ V\right) \tag{2.1}$$

$$Fe^{3+} + AsO_4^{3-} \rightarrow FeAsO_4 \left(K_{sp} = 5.7\ X\ 10^{-21}\right) \tag{2.2}$$

where E° is standard electrode potential and K_{sp} is solubility product constant. Ipsen et al. (Ipsen et al. 2005) have utilized this technology for the remediation of a contaminated site in Tacoma, Washington, USA, where contamination of arsenic in groundwater was caused by sodium arsenite.

2.9.2.3 Concern Associated to Technology

Although, permeable reactive barriers (PRBs) have been proved to offer a promising technology as compared to the conventional mode of treatment methods for arsenic removal, but there are still significant limitations, which occur due to geochemical and physical characteristics of contaminated sites (Oikawa et al. 1994). The major limitation is the lack of information on the long-term effectiveness of large-scale remediation systems. Therefore, it has not been yet approved by the US EPA (United States Environmental Protection Agency) till now.

The PRB-based treatment requires to study the effects of aging of the reagent, decrease in permeability due to precipitation, microbiological growth and accumulation of gas and evaluating the long-term performance which cannot be predicted in short-term laboratory experiments (Lo et al. 2007; Morris et al. 2007). Also, the literature has not been updated about the possible issues, which may take place during the elimination process of system media after its exhaustion. Further, the injection of the dissolved phase of remediation materials (iron salts) in aquifer systems may cause the generation of secondary pollutants due to their possible chemical reactions with other ions. The possibility of recontamination due to dissolution of the compounds is also a concern in this application (Yin et al. 2012).

2.9.2.4 Scope for Future Studies

The installation of PRBs requires the process of excavation for the remediation of arsenic, which is viable to treat the shallow aquifer systems (such as wells) effectively, where the source of arsenic is well defined. Figure 2.4 presents a pictorial

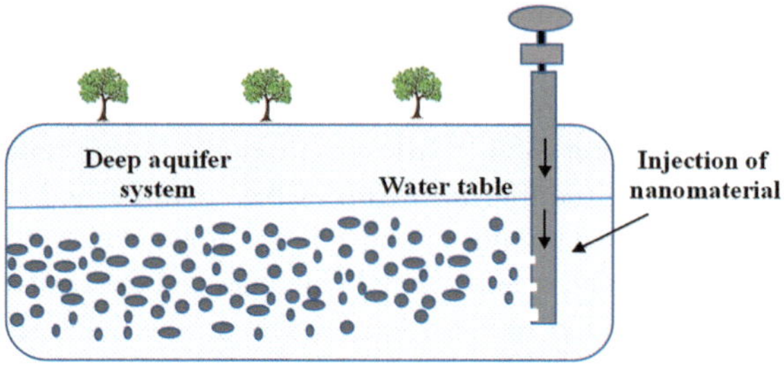

Fig. 2.4 An overview of injection-based technique representing remediation of arsenic in deep aquifer system. (He et al. 2010)

representation demonstrating an approach of nanomaterial injection in deep aquifer systems displaying a distributed source of arsenic.

Future research could be useful in exploring the utilization of colloidal or sub-colloidal iron or iron nanoparticles for injection as substitutes for using iron salts (Kaplan et al. 1996). In this method, the injection of nanomaterials creates a reactive which further sequester the arsenic through different removal (Baxter et al. 2008). Recently, the possible scope of maghemite NPs as in-situ remediation material was explored through 3D experiments simulating groundwater elemental concentration of Ballia district, UP, India (Kumar et al. 2019). It also requires in-depth analysis of fate and transport of nanomaterials in porous media as well as possibilities of secondary pollution. Moreover, injection of materials may be done through exiting deep wells or abandoned wells, in which the reactive material may be introduced.

2.10 Lack of Studies at Laboratory Scale

2.10.1 Studies Representing Real-World Conditions

Generally, the water resources are sensitive to physical, chemical and thermal variations. There are several ions in groundwater which might affect arsenic removal during adsorption processes. The batch experiments are not feasible using actual groundwater at laboratory scale due to a large volume requirement of samples and fluctuating water quality issues. In order to evaluate the arsenic removal potential of different nanoadsorbents, using the groundwater concentration representing the real-world conditions is the most appropriate approach. Therefore, proper calculation of the required quantity of ingredients for the synthesis of artificial water is a necessity (Kaloti et al. 2015). Lin et al. (Lin et al. 2012) have investigated the effects of Cl^-, SO_4^{2-}, NO_3^- and PO_4^{3-} ions on arsenic removal using γ-Fe_2O_3 nanoparticles. Recently, the arsenic sequestration capacity of maghemite NPs was explored in formulated synthetic water representing the elemental concentrations equivalent to those of groundwater arsenic contaminated prone areas of Ballia District, UP, India (Kumar et al. 2019).

2.10.2 Cost Evaluation of Nanomaterials Production

In developing a cost-effective arsenic treatment system, it is required to examine the production cost of different nanoadsorbents during laboratory-scale studies. Needless to say, the chemical (co-precipitation) approach of synthesis has been reported to be a common mode of generation of nanoadsorbents at laboratory scale. A new approach of calculating the cost of produced NPs using chemical method covering the maximum possible aspects is also suggested by Kumar et al. (2020). This methodology might help and motivates the future researcher to evalaute the cost of synthesized NPs and subsequently the estimation of treatment cost of arsenic contaminated groundwater.

2.10.3 Fabrication of Columns

The adsorption data employing laboratory columns have provided a base for the application of nanoadsorbents at pilot- and full-scale remediation projects (Su and Puls 2003; Kundu and Gupta 2005). In most of the studies, the laboratory columns have been reported to be conventionally designed, which are unambiguously based on the batch removal experiments (Wang et al. 2009). In the reviewed literature, the ratio of $d_{column}/d_{particles}$ (d-diameter) has been used for the laboratory scale column experiments, which depends on the ignorance of wall effect on the mass transfer (Benenati and Brosilow 1962).

Further, suitable binder is necessary, when the nanoadsorbents have to be used as packing materials because the loss of particles might occur under the non-uniformity of flow (Sharma et al. 2010). At the pilot scale in the household treatment units of arsenic, the nanoadsorbents have been allowed to mix with easily available porous materials like soil particles. However, the pursuance of an absolute methodology for the development of laboratory columns is a need among the research community for the better comparability and interpretation of arsenic removal data at a common platform. Therefore, for designing of laboratory columns, the factors which need to be considered are height of reactive zone and column with constant porosity and their correlation with the required mass of adsorbents, as established by Noubactep (Noubactep and Caré 2011). The mathematical equation to calculate the height of reactive zone is shown below:

$$h_{rz} = \frac{4V_{rz}}{\pi D^2} = \frac{4V_{solid}}{C\pi D^2} \tag{2.3}$$

where D, C, h_{rz}, V_{rz} and V_{solid} represent internal diameter of the column, packing density of the nanoadsorbents, height of reactive zone, volume of reactive zone and volume of solids (for the usage of the two types of nanoadsorbents), respectively. The h_{rz} is the fraction of L (length of column), and length of column should be more than the magnitude of height of reactive zone ($L > h_{rz}$).

2.11 Conclusions

Although arsenic removal through adsorption is a widely acceptable technology, however, fewgaps and challenges still exist for the scientific communities to address. The followingefforts can be addressed for further development of nanoadsorbents along with the improvement in laboratory-based experiments. For deployment of nanoparticles at a larger and wider scale, the green approach of synthesis is required for upscaling the bulk production of nanoadsorbents. To assess feasibility of nanoadsorbents for pilot-scale application, comprehensive investigations related to the removal capabilities of nanoadsorbents under varied experimental conditions are recommended for such remediation studies, along with the use of representative test water samples in both batch and column (appropriately designed) experimental studies, are required. Assessment of the durability of system through time series analyses under different environmental conditions using laboratory-scale experiments is also a prerequisite of literature for future investigations. Besides, an attempting of techno-financial analysis of the laboratory process to get an insight into further possible upscaling can be explored further.

References

Adomako EE, Williams PN, Deacon C, Meharg AA (2011) Inorganic arsenic and trace elements in Ghanaian grain staples. Environ Pollut 159(10):2435–2442

Ahammed MM, Davra K (2011) Performance evaluation of biosand filter modified with iron oxide-coated sand for household treatment of drinking water. Desalination 276(1–3):287–293

Ahmed MF (2001a) An overview of arsenic removal Technologies in Bangladesh and India. Civ Eng:251–269

Ahmed MF (2001b) An overview of arsenic removal technologies in Bangladesh and India. Technologies for arsenic removal from drinking water. A compilation. Papers presented at the international workshop technologies for removing arsenic from drinking water, pp 251–269

Ahmed J et al (2010) Ecotoxicology and environmental safety speciation and evaluation of arsenic in surface water and groundwater samples: a multivariate case study. Ecotoxicol Environ Saf 73 (5):914–923

Ahsan T (2010) Technologies for arsenic removal from groundwater Tanveer Ahsan. SamSam Water

Alam MO, Shaikh WA, Chakraborty S, Avishek K, Bhattacharya T (2016) Groundwater arsenic contamination and potential health risk assessment of Gangetic Plains of Jharkhand, India. Expo Health 8(1):125–142

Ali I (2012) New generation adsorbents for water treatment. Chem Rev 112(10):5073–5091

Ali MA et al (2006) Groundwater dynamics and arsenic contamination in Bangladesh. Chem Geol 228(1–3):112–136

Alp S et al (2016) Chemical composition and antioxidant activity Ziziphora clinopodioides ecotypes from Turkey. Rom Biotechnol Lett 21(2):11298–11303

Appelo CAJ, Van Der Weiden MJJ, Tournassat C, Charlet L (2002) Surface complexation of ferrous iron and carbonate on ferrihydrite and the mobilization of arsenic. Environ Sci Technol 36(14):3096–3103

Azubuike CC, Chikere CB, Okpokwasili GC (2016) Bioremediation techniques–classification based on site of application: principles, advantages, limitations and prospects. World J Microbiol Biotechnol 32(11)

Baxter R, Hastings N, Law A, Glass EJ (2008) A rapid and robust sequence-based genotyping method for BoLA-DRB3 alleles in large numbers of heterozygous cattle. Anim Genet 39 (5):561–563

Beak DG, Wilkin RT (2009) Performance of a zerovalent iron reactive barrier for the treatment of arsenic in groundwater: part 2. Geochemical modeling and solid phase studies. J Contam Hydrol 106(1–2):15–28

Benenati RF, Brosilow CB (1962) Void fraction distribution in beds of spheres. AICHE J 8 (3):359–361

Berg M et al (2007) Magnitude of arsenic pollution in the Mekong and Red River Deltas — Cambodia and Vietnam. Sci Total Environ 372(2–3):413–425

Bhattacharya P, Chatterjee D, Jacks G (Mar. 1997) Occurrence of arsenic-contaminatedGroundwater in alluvial aquifers from Delta Plains, eastern India: options for safe drinking water supply. Int J Water Resour Dev 13(1):79–92

Bhaumik M, Choi HJ, McCrindle RI, Maity A (2014) Composite nanofibers prepared from metallic iron nanoparticles and polyaniline: high performance for water treatment applications. J Colloid Interface Sci 425:75–82

Bhowmick S, Pramanik S, Singh P, Mondal P, Chatterjee D, Nriagu J (2018a) Arsenic in groundwater of West Bengal, India: a review of human health risks and assessment of possible intervention options. Sci Total Environ 612:148–169

Bhowmick S, Pramanik S, Singh P, Mondal P, Chatterjee D, Nriagu J (2018b) Science of the total environment arsenic in groundwater of West Bengal, India: a review of human health risks and assessment of possible intervention options. Sci Total Environ 612:148–169

Bondu R, Cloutier V, Rosa E, Benzaazoua M (2017) Mobility and speciation of geogenic arsenic in bedrock groundwater from the Canadian Shield in western Quebec, Canada. Sci Total Environ 574:509–519

Brandhuber P, Amy G (1998) Alternative methods for membrane filtration of arsenic from drinking water. Desalination 117(1):1–10

Bundschuh J et al (2012) One century of arsenic exposure in Latin America: a review of history and occurrence from 14 countries. Sci Total Environ 429:2–35

Chakraborti D, Rahman MM, Ahamed S, Dutta RN, Pati S, Mukherjee SC (2016) Arsenic groundwater contamination and its health effects in Patna district (capital of Bihar) in the middle Ganga plain, India. Chemosphere 152:520–529

Chakraborty M, Mukherjee A, Ahmed KM (2015) A review of groundwater arsenic in the Bengal Basin, Bangladesh and India: from source to sink. Curr Pollut Rep 1(4):220–247

Chandrashekhar AK, Chandrasekharam D, Farooq SH (2016) Contamination and mobilization of arsenic in the soil and groundwater and its influence on the irrigated crops, Manipur Valley, India. Environ Earth Sci 75(2):1–15

Chatterjee D, Halder D, Majumder S, Biswas A (2010) Assessment of arsenic exposure from groundwater and rice in Bengal Delta Region, West Bengal, India. Water Res 44 (19):5803–5812

Chen H, Frey MM, Clifford D, McNeill LS, Edwards M (1999) Arsenic treatment considerations. J Am Water Works Assoc 91(3):74–85

Chowdhury TR, Basu GK, Mandal BK, Biswas BK, Samanta G, Chowdhury UK, Chanda CR, Lodh D, Roy SL, Saha KC, Roy S, Kabir S, Quamruzzaman Q, Chakraborti D (1999) Arsenic poisoning in the Ganges delta. Nature 40:1049

Ciminelli VST, Gasparon M, Ng JC, Silva GC, Caldeira CL (2017) Dietary arsenic exposure in Brazil: the contribution of rice and beans. Chemosphere 168:996–1003

Cornell RM, Schwertmann U (2003) The Iron Oxides 39(8)

Cornell RM, Schwertmann U (2004) Introduction to the iron oxides. In: The iron oxides. Wiley-VCH Verlag GmbH & Co. KGaA, Weinheim, pp 1–7

Crabbe H et al (2017) Hazard ranking method for populations exposed to arsenic in privatewater supplies: relation to bedrock geology. Int J Environ Res Public Health 14(12)

Danish MI, Qazi IA, Zeb A, Habib A, Awan MA, Khan Z (2013) Arsenic removal from aqueous solution using pure and metal-doped titania nanoparticles coated on glass beads: adsorption and column studies. J Nanomater 2013:1–17

Das S, Bora SS, Yadav RNS, Barooah M (2017) A metagenomic approach to decipher the indigenous microbial communities of arsenic contaminated groundwater of Assam. Genomics Data 12:89–96

Davodi B, Jahangiri M (2014) Determination of optimum conditions for removal of As (III) and As (V) by polyaniline/polystyrene nanocomposite. Synth Met 194:97–101

De D, Mandal SM, Bhattacharya J, Ram S, Roy SK (2009) Iron oxide nanoparticle-assisted arsenic removal from aqueous system. J Environ Sci Health A Tox Hazard Subst Environ Eng 44 (2):155–162

Diwakar J, Johnston SG, Burton ED, Das Shrestha S (2015) Arsenic mobilization in an alluvial aquifer of the Terai region, Nepal. J Hydrol Reg Stud 4:59–79

Duarte AALS, Cardoso SJA, Alçada AJ (2009) Emerging and innovative techniques for arsenic removal applied to a small water supply system. Sustainability 1(4):1288–1304

Elizalde-Gonzalez MP, Mattusch J, Einicke W-D, Wennrich R (2001) Sorption on natural solids for arsenic removal. Chem Eng J 81:187–195

Fan D et al (2017) Sulfidation of iron-based materials: a review of processes and implications for water treatment and remediation. Environ Sci Technol 51(22):13070–13085

Fiúza A, Futuro A, Silva A, Ferreira A, Guimarães F (2015) In-situ removal of arsenic from groundwater using permeable reactive barriers with iron based sorbents

Genc-Fuhrman H, Tjell JC, McConchie D (2004) Adsorption of arsenic from water using activated neutralized red mud. Environ Sci Technol 38(8):2428–2434

Ghurye GL, Clifford DA, Tripp AR (1999) Combined arsenic and nitrate removal by ion exchange. J Am Water Works Assoc 91(10):85–96

Gillman GP (2006) A simple technology for arsenic removal from drinking water using hydrotalcite. Sci Total Environ 366(2–3):926–931

Grassi M, Kaykioglu G, Belgiorno V (2012) Emerging compounds removal from wastewater. Springer, Dordrecht, pp 15–38

Greenleaf JE, Lin JC, Sengupta AK (2006) Two novel applications of ion exchange fibers: arsenic removal and chemical-free softening of hard water. Environ Prog 25(4):300–311

Gu Z, Fang J, Deng B (2005) Preparation and evaluation of adsorbents for arsenic removal. Environ Sci Technol 39(10):3833–3843

Gupta PK, Yadav B, Kumar A, Singh RP (2020) India's Major subsurface pollutants under future climatic scenarios: challenges and remedial solutions. In: Singh P, Singh R, Srivastava V (eds) contemporary environmental issues and challenges in era of climate change. Springer, Singapore. https://doi.org/10.1007/978-981-32-9595-7_6

Han YS, Gallegos TJ, Demond AH, Hayes KF (2011) FeS-coated sand for removal of arsenic(III) under anaerobic conditions in permeable reactive barriers. Water Res 45(2):593–604

Haron MJ, Ab Rahim F, Abdullah AH, Hussein MZ, Kassim A (2008) Sorption removal of arsenic by cerium-exchanged zeolite P. Mater Sci Eng B Solid-State Mater Adv Technol 149 (2):204–208

He F, Zhao D, Paul C (2010) Field assessment of carboxymethyl cellulose stabilized iron nanoparticles for in situ destruction of chlorinated solvents in source zones. Water Res 44 (7):2360–2370

Hristovski K, Baumgardner A, Westerhoff P (2007) Selecting metal oxide nanomaterials for arsenic removal in fixed bed columns: from nanopowders to aggregated nanoparticle media. J Hazard Mater 147(1–2):265–274

Hristovski KD, Westerhoff PK, Möller T, Sylvester P (2009) Effect of synthesis conditions on nano-iron (hydr)oxide impregnated granulated activated carbon. Chem Eng J 146(2):237–243

Hung WC, Fu SH, Tseng JJ, Chu H, Ko TH (2007) Study on photocatalytic degradation of gaseous dichloromethane using pure and iron ion-doped TiO2 prepared by the sol-gel method. Chemosphere 66(11):2142–2151

Hunt ML, Amrhein C (2002) Arsenic (III) and Arsenic (V) reactions with zerovalent iron corrosion products. Environ Sci Technol 36(24):5455–5461

Hussam A (2009) Contending with a development disaster: SONO filters remove arsenic from well water in Bangladesh (innovations case discussion: SONO filters). Innov Technol Gov Glob 4 (3):89–102

Ipsen SO, Gerth J, Förstner U (2005) Identifying and testing materials for arsenic removal by permeable reactive barriers

Jain CK, Singh RD (2012) Technological options for the removal of arsenic with special reference to South East Asia. J Environ Manag 107:1–18

Kaloti M, Kumar A (2016) Synthesis of chitosan-mediated silver coated γ-Fe2O3 (Ag-γ-Fe2O3@Cs) superparamagnetic binary nanohybrids for multifunctional applications. J Phys Chem C 120(31):17627–17644

Kaloti M, Kumar A, Navani NK (2015) Synthesis of glucose-mediated Ag–γ-Fe_2O_3 multifunctional nanocomposites in aqueous medium – a kinetic analysis of their catalytic activity for 4-nitrophenol reduction. Green Chem 17(10):4786–4799

Kanel SRAJ (2006) Arsenic (V) removal from groundwater using nano scale zero-valent Iron as a colloidal reactive barrier material. Environ Sci Technol 40(6):2045–2050

Kanel SR, Grenèche J-M, Choi H (2006) Arsenic(V) removal from groundwater using nano scale zero-valent Iron as a colloidal reactive barrier material. Environ Sci Technol 40(6):2045–2050

Kaplan DI, Cantrell KJ, Wietsma TW, Potter MA (1996) Retention of zero-valent iron colloids by sand columns: application to chemical barrier formation. J Environ Qual 25:1086–1094

Khettab S, Chabbi-Chemrouk N (2017) Sense of place in the coastal town of Tipaza in Algeria: local-community's socio-cognitive representations. Int J Sustain Built Environ 6(2):544–554

Kumar M, Ramanathan AL, Mukherjee A, Verma S, Rahman MM, Naidu R (2018) Hydrogeomorphological influences for arsenic release and fate in the central Gangetic Basin, India. Environ Technol Innov 12:243–260

Kumar A, Joshi H, Kumar A (2019) An approach of Multi-variate statistical design (Taguchi) and numerical tool (COMSOL) in exploring the arsenic sequestration potential of γ-Fe2O3 NPs in groundwater of Ballia district, Uttar-Pradesh, India. In AGU fall meeting abstracts (Vol. 2019, pp. GH23B-1227). https://ui.adsabs.harvard.edu/abs/2019AGUFMGH23B1227K

Kumar A, Joshi H, Kumar A (2020) Remediation of Arsenic by Metal/ Metal Oxide Based Nanocomposites/ Nanohybrids: Contamination Scenario in Groundwater, Practical challenges, and future perspectives. Sep Purif Rev 00(00):1–32. https://doi.org/10.1080/15422119.2020.1744649

Kundu S, Gupta AK (2005) Analysis and modeling of fixed bed column operations on As (V) removal by adsorption onto iron oxide-coated cement (IOCC). J Colloid Interface Sci 290 (1):52–60

Langmuir D, Mahoney J, MacDonald A, Rowson J (1999) Predicting arsenic concentrations in the porewaters of buried uranium mill tailings. Geochim Cosmochim Acta 63(19–20):3379–3394

Lata S, Samadder SR (2016) Removal of arsenic from water using nano adsorbents and challenges: a review. J Environ Manag 166:387–406

Lee K, Lee Y, Yoon J, Kamala-Kannan S, Park S, Oh B (2009) Assessment of zero-valent iron as a permeable reactive barrier for long-term removal of arsenic compounds from synthetic water. Environ Technol 30(13):1425–1434

Leupin OX, Hug SJ, Badruzzaman a BM (2005) Arsenic removal from Bangladesh tube well water with filter columns containing zerovalent Iron filings and sand. Environ Sci Technol 39 (20):8032–8037

Limmer M, Burken J (2016) Phytovolatilization of organic contaminants. Environ Sci Technol 50 (13):6632–6643

Lin S, Lu D, Liu Z (2012) Removal of arsenic contaminants with magnetic γ-Fe2O3 nanoparticles. Chem Eng J 211–212:46–52

Lin YF, Chen JL, Xu CY, Chung TW (2014) One-pot synthesis of paramagnetic iron(III) hydroxide nanoplates and ferrimagnetic magnetite nanoparticles for the removal of arsenic ions. Chem Eng J 250:409–415
Lo IMC, Surampalli R, Lai KCK (2007) Zero-valentiron reactive materials for hazardous waste and inorganics removal. Zero-Valent Iron React Mater Hazard Waste Inorg Removal
Lockwood CL et al (2014) Mobilisation of arsenic from bauxite residue (red mud) affected soils: effect of pH and redox conditions. Appl Geochem 51:268–277
Luo X, Wang C, Luo S, Dong R, Tu X, Zeng G (2012) Adsorption of As (III) and As (V) from water using magnetite Fe3O4-reduced graphite oxide–MnO2 nanocomposites. Chem Eng J 187:45–52
Luo X et al (2013) Nanocomposites of graphene oxide-hydrated zirconium oxide for simultaneous removal of As(III) and As(V) from water. Chem Eng J 220:98–106
Malik AH, Khan ZM, Mahmood Q, Nasreen S, Bhatti ZA (2009) Perspectives of low cost arsenic remediation of drinking water in Pakistan and other countries. J Hazard Mater 168(1):1–12
Meng X, Korfiatis GP, Christodoulatos C, Bang S (2001) Treatment of arsenic in Bangladesh well water using a household co-precipitation and filtration system. Water Res 35(12):2805–2810
Morgada ME, Levy IK, Salomone V, Farías SS, López G, Litter MI (2009) Arsenic (V) removal with nanoparticulate zerovalent iron: effect of UV light and humic acids. Catal Today 143 (3–4):261–268
Morris J et al (2007) Nanotechnology white paper. Environmental Protection Agency (EPA), U.S., 2007. Sr. Policy Council Nanotechnology White Paper USEPA, Washington, DC, p 136
Mosaferi M, Nemati S, Khataee A, Nasseri S, Hashemi AA (2014) Removal of Arsenic (III, V) from aqueous solution by nanoscale zero-valent iron stabilized with starch and carboxymethyl cellulose. J Environ Health Sci Eng 12(1):74
N. Compounds (2001) Longevity of granular iron in groundwater treatment processes: solution composition, pp 1–9
Naujokas MF et al (2013) The broad scope of health effects from chronic arsenic exposure: update on a worldwide public health problem. Environ Health Perspect 121(3):295–302
Navarro O, González J, Júnez-Ferreira HE, Bautista CF, Cardona A (2017) Correlation of Arsenic and Fluoride in the groundwater for human consumption in a semiarid region of Mexico. Proc Eng 186:333–340
Ng JC, Wang J, Shraim A (2003) A global health problem caused by arsenic from natural sources. Chemosphere 52(9):1353–1359
Nickson RT, McArthur JM, Ravenscroft P, Burgess WG, Ahmed KM (2000) Mechanism of arsenic release to groundwater, Bangladesh and West Bengal. Appl Geochem 15(4):403–413
Ning R (2002) Arsenic removal by reverse osmosis. Desalination 143(3):237–241
Nordstrom DK (2002) Worldwide occurrences of arsenic in ground water. Science 296 (5576):2143–2145
Noubactep C, Caré S (2011) Designing laboratory metallic iron columns for better result comparability. J Hazard Mater 189(3):809–813
Nurmi JT et al (2005) Characterization and properties of metallic Iron nanoparticles: spectroscopy, electrochemistry, and kinetics. Environ Sci Technol 39(5):1221–1230
Office of Water U.S. Environmental Protection Agency 2018 edition of the drinking water standards and health advisories tables, *EPA 822-F-18-001 Off.*, no. March, 2018
Oikawa N, Nakagawa Y, Nishimura K, Ueno T, Fujita T (1994) Quantitative structure-activity studies of insect growth regulators X. Substituent effects on larvicidal activity of 1-tert-butyl-1-(2-chlorobenzoyl)-2-(substituted benzoyl)hydrazines against Chilo suppressalis and design synthesis of potent derivatives. Pestic Biochem Physiol 48(2):135–144
Olea RA, Raju NJ, Egozcue JJ, Pawlowsky-Glahn V, Singh S (2018) Advancements in hydrochemistry mapping: methods and application to groundwater arsenic and iron concentrations in Varanasi, Uttar Pradesh, India. Stoch Environ Res Risk Assess 32(1):241–259
Özlem Kocabaş-Atakli Z, Yürüm Y (2013) Synthesis and characterization of anatase nanoadsorbent and application in removal of lead, copper and arsenic from water. Chem Eng J 225:625–635

Pal P et al (2009) Contamination of groundwater by arsenic: a review of occurrence, causes, impacts, remedies and membrane-based purification. J Integr Environ Sci 6(4):295–316
Patel KS et al (2017) Groundwater arsenic and fluoride in Rajnandgaon District, Chhattisgarh, northeastern India. Appl Water Sci 7(4):1817–1826
Pino P et al (2015) Chile confronts its environmental health future after 25 years of accelerated growth. Ann Glob Health 81(3):354–367
Piquette A, Cannon C, Apblett AW (2012) Remediation of arsenic and lead with nanocrystalline zinc sulfide. Nanotechnology 23:294014
Poreda RJ, Dowling CB, Basu AR, Aggarwal PK, Peters SL (2002) Geochemical study of arsenic release mechanisms in the Bengal Basin groundwater. Water Resour Res 38(9):12–1-12–18
Prasad KS, Ramanathan AL, Paul J, Subramanian V, Prasad R (2013) Biosorption of arsenite (As^{+3}) and arsenate (As^{+5}) from aqueous solution by Arthrobacter sp. biomass. Environ Technol 34:2701–2708
Rahman MM, Ng JC, Naidu R (2009) Chronic exposure of arsenic via drinking water and its adverse health impacts on humans. Environ Geochem Health 31(1):189–200
Ram S, Mitra C (2001) Formation of stable Cu 2 O nanocrystals in a new orthorhombic crystal structure. Mater Sci Eng 306:805–809
Rana A, Bhardwaj SK, Thakur M, Verma S (2016) Assessment of heavy metals in surface and ground water sources under different land uses in mid-hills of Himachal Pradesh. Int J Bio-Resour Stress Manag 7:461–465
Reddy KJ, McDonald KJ, King H (2013) A novel arsenic removal process for water using cupric oxide nanoparticles. J Colloid Interface Sci 397:96–102
Richards LA, Magnone D, van Dongen BE, Ballentine CJ, Polya DA (2015) Use of lithium tracers to quantify drilling fluid contamination for groundwater monitoring in Southeast Asia. Appl Geochem 63:190–202
Richards LA et al (2017) High resolution profile of inorganic aqueous geochemistry and key redox zones in an arsenic bearing aquifer in Cambodia. Sci Total Environ 590–591:540–553
Rimstidt JD, Dove PM, Nordstrom DK, Parks GA, Robins RG (1987) Solubility and stability of scorodite, FeAsO 4 *2H 2 O: discussion. Am Mineral 72:852
Robalds A, Naja GM, Klavins M (2016) Highlighting inconsistencies regarding metal biosorption. J Hazard Mater 304:553–556
Rodríguez-Lado L et al (2013) Groundwater arsenic contamination throughout China. Science (80-.) 341(6148):866–868
Sarkar S, Blaney LEEM (2008) Arsenic removal from groundwater and its safe containment in a rural environment: validation of a sustainable approach. Environ Sci Technol 42 (12):4268–4273
Sarkar S et al (2010) Evolution of community-based arsenic removal systems in remote villages in West Bengal, India: assessment of decade-long operation. Water Res 44(19):5813–5822
Scheck J, Lemke T, Gebauer D (2015) The role of chloride ions during the formation of akaganéite revisited. Fortschr Mineral 5(4):778–787
Shah SH, Mahmood Q, Raja IA, Pervez A, Kalsoom A (2015) Bio-Sand filter to treat arsenic contaminated drinking water. Desalin Water Treat 53(11):2999–3006
Shahraki AA (2017) Urban planning and design in unauthorized neighborhoods using case studies. Int J Sustain Built Environ 6(2):273–284
Sharma A, Verma N, Sharma A, Deva D, Sankararamakrishnan N (2010) Iron doped phenolic resin based activated carbon micro and nanoparticles by milling: synthesis, characterization and application in arsenic removal. Chem Eng Sci 65(11):3591–3601
Sharma S, Kaur J, Nagpal AK, Kaur I (2016) Quantitative assessment of possible human health risk associated with consumption of arsenic contaminated groundwater and wheat grains from Ropar Wetand and its environs. Environ Monit Assess 188(9)
Sheoran AS, Sheoran V (2006) Heavy metal removal mechanism of acid mine drainage in wetlands: a critical review. Miner Eng 19(2):105–116

Singh AK (2006) Chemistry of arsenic in groundwater of Ganges-Brahmaputra river basin. Curr Sci 91(5):599–606
Singh SK (2017) An analysis of the cost-effectiveness of arsenic mitigation technologies: implications for public policy. Int J Sustain Built Environ 6(2):522–535
Smedley PL et al (2005) Arsenic associations in sediments from the loess aquifer of La Pampa, Argentina. Appl Geochem 20(5):989–1016
Smith AH et al (1992) Cancer risks from arsenic in drinking water. Environ Health Perspect 97:259–267
Souter PF et al (2003) Evaluation of a new water treatment for point-of-use household applications to remove microorganisms and arsenic from drinking water. J Water Health 1(2):73–84
Spira Y, Henstock J, Nathanail P, Müller D, Edwards D (2006) A European approach to increase innovative soil and groundwater remediation technology applications. Remediat J 16:81–96
Su C, Puls RW (2003) In situ remediation of arsenic in simulated groundwater using zerovalent iron: laboratory column tests on combined effects of phosphate and silicate. Environ Sci Technol 37(11):2582–2587
Sugár É, Mihucz VG, Záray G (2014) Determination of arsenic in drinking water and several food items. Elelmiszervizsgalati Kozlemenyek 60(2):163–176
Sullivan EJ, Bowman RS, a Legiec I (1997) Sorption of arsenic from soil-washing leachate by surfactant-modified zeolite. J Environ Qual 32(1997):2387–2391
Sun W, Li Q, Gao S, Shang JK (2012) Exceptional arsenic adsorption performance of hydrous cerium oxide nanoparticles: part B. Integration with silica monoliths and dynamic treatment. Chem Eng J 185–186:136–143
Tahura S, Shahidullah SM, Rahman T, Milton AH (1998) Evaluation of an arsenic removal household device: bucket treatment unit (BTU). Technol Arsenic Removal Drink Water:158–170
Takanashi H, Tanaka A, Nakajima T, Ohki A (2004) Arsenic removal from groundwater by a newly developed adsorbent. Water Sci Technol 50(8):23–32
Tandon PK, Shukla RC, Singh SB (2013) Removal of arsenic(III) from water with clay-supported zerovalent Iron nanoparticles synthesized with the help of tea liquor. Ind Eng Chem Res 52 (30):10052–10058
Tokar EJ, Diwan BA, Ward JM, Delker DA, Waalkes MP (2010) Carcinogenic effects of 'whole-life' exposure to inorganic arsenic in CD1 mice. Toxicol Sci 119(1):73–83
Uede K, Furukawa F (2003) Skin manifestations in acute arsenic poisoning from the Wakayama curry-poisoning incident. Br J Dermatol 149(4):757–762
Vahter M (2008) Health effects of early life exposure to arsenic. Basic Clin Pharmacol Toxicol 102 (2):204–211
van der Zee C, Roberts DR, Rancourt DG, Slomp CP (2003) Nanogoethite is the dominant reactive oxyhydroxide phase in lake and marine sediments. Geology 31(11):993–996
Wang TH, Li MH, Teng SP (2009) Bridging the gap between batch and column experiments: a case study of Cs adsorption on granite. J Hazard Mater 161(1):409–415
Wang Y, Morin G, Ona-Nguema G, Juillot F, Calas G, Brown GE (2011) Distinctive arsenic (V) trapping modes by magnetite nanoparticles induced by different sorption processes. Environ Sci Technol 45(V):7258–7266
Waybrant KR, Blowes DW, Ptacek CJ (1998) Selection of reactive mixtures for use in permeable reactive walls for treatment of mine drainage. Environ Sci Technol 32(13):1972–1979
Westerhoff P, Highfield D, Badruzzaman M, Yoon Y (2005) Rapid small-scale column tests for arsenate removal in Iron oxide packed bed columns. J Environ Eng 131(February):262–271
Wickramasinghe SR, Han B, Zimbron J, Shen Z, Karim MN (2004) Arsenic removal by coagulation and filtration: comparison of groundwaters from the United States and Bangladesh. Desalination 169(3):231–244
Wilkin RT, Su C, Ford RG, Paul CJ (2005) Chromium-removal processes during groundwater remediation by a zerovalent iron permeable reactive barrier. Environ Sci Technol 39 (12):4599–4605

World health organisation (2017) Guidelines for drinking-water quality: fourth edition incorporating the first addendum
World Health Organization (2010) Exposer to arsenic: a major public health concern, Report, vol. World Heal
Yin K, Lo IMC, Dong H, Rao P, Mak MSH (2012) Lab-scale simulation of the fate and transport of nano zero-valent iron in subsurface environments: aggregation, sedimentation, and contaminant desorption. J Hazard Mater 227–228:118–125
Zhang New.pdf
Zhang H, Selim HM (2005) Kinetics of arsenate adsorption-desorption in soils. Environ Sci Technol 39(16):6101–6108
Zhou D et al (2014) Column test-based optimization of the permeable reactive barrier (PRB) technique for remediating groundwater contaminated by landfill leachates. J Contam Hydrol 168:1–16

Development of Polyhydroxyalkanoate (PHA) and Its Copolymers as a Possible "Cure" for the Plastic Pollution

3

Ramamoorthi Sivashankari and Takeharu Tsuge

Abstract

Plastic pollution is one of the growing critical global issues. The accumulation of plastics in the environment mainly in marine is causing indeed a serious environmental issue. Indisputably, it will impact the environment enormously if this issue is let to be continued with no proper solution. In order to come up with a potential solution to curb this issue from growing, it is imperative to identify the main "culprit" behind such occurrence. As per our knowledge, it is apparent that the overuse of petroleum-based plastic (PBP) is one of the main causes of plastic pollution. It is evident that PBP eases our day-to-day life due to their versatility and lightweight. It is now being used in numerous packaging purposes such as food, easy-to-carry bags, and so on. Therefore, imagining a life without it is definitely hard, or to a certain extent is impossible too. Thus, having an alternative which is similar to PBP and able to serve the same purpose as PBP is highly effective to bring down the use of PBP and eventually the damages to the environment prior to its dumping. Polyhydroxyalkanoate (PHA) is one of the best candidates to be proposed as a substitute for PBP. However, unlike PBP, PHA is a bio-based polymer and thus leaves a less negative impact on the environment. Therefore, it has garnered enormous attention worldwide to replace PBP. This chapter will review recent progress on PHA, how it is being biosynthesized, the emergence of new copolymers to improve the properties of existing PHA, a detailed comparison on mechanical properties of PHA with PBP, and other efforts being made to make it as a potential "cure" for the plastic pollution.

R. Sivashankari · T. Tsuge (✉)
Department of Materials Science and Engineering, Tokyo Institute of Technology, Yokohama, Japan
e-mail: tsuge.t.aa@m.titech.ac.jp

R. Prasad (ed.), *Environmental Pollution and Remediation*, Environmental and Microbial Biotechnology, https://doi.org/10.1007/978-981-15-5499-5_3

Keywords

Plastic pollution · Polyhydroxyalkanoate · Petroleum-based plastics · Biosynthesis · Copolymer

3.1 Introduction: Polyhydroxyalkanoate (PHA) as a Possible Substitute for Petroleum-Based Plastics (PBP)

Polyhydroxyalkanoate (PHA) is a family of biopolymer and falls into polyester class (Anderson and Dawes 1990). PHAs are typically made from biomass, unlike conventional plastics which are made from petroleum. Biomass is a renewable energy; thus, it will be present as long as this planet exists. Soybean oil (Kahar et al. 2004), rice husk (Heng et al. 2017), vegetable oil (Xia and Larock 2010), and molasses (Albuquerque et al. 2010) are few types of biomasses that have been tested for the biosynthesis of PHA. On the other hand, petroleum is a nonrenewable source of energy, so it will be completely depleted with continuous exploitation for various applications. Moreover, petroleum also contributes to climate change (Ruddiman 2010) by emitting excessive carbon to the atmosphere throughout the refinement processes. Meanwhile, processes involving in the making of biopolymers such as PHA will be indisputably nature friendly.

Despite the drawbacks, mankind is heavily dependent on petroleum-based plastics (PBP) as it is highly convenient mainly for its lightweight properties and overall versatility. It is almost impossible to imagine a life without PBP because it has been used in various items we use in daily life such as in food packaging, clothing, electronic devices, medical equipment, and many others. Therefore, it is very crucial to find an alternative material which is similitude to PBP in terms of physicochemical properties but capable to be produced in environmentally friendly ways and importantly can be disposed without leaving any damages to nature.

From previous studies, PHA has been demonstrated to fit this criterion, hence garnered an enormous research interest. PHA is a naturally occurring polymer in bacteria (Anderson and Dawes 1990); thus, it can be biosynthesized by feeding microbes with carbon sources from biomass. In this section, the biosynthesis of PHAs and its material properties are described. The basic chemical structure of PHA is shown in Fig. 3.1, where the side group (denoted as R) can contain up to

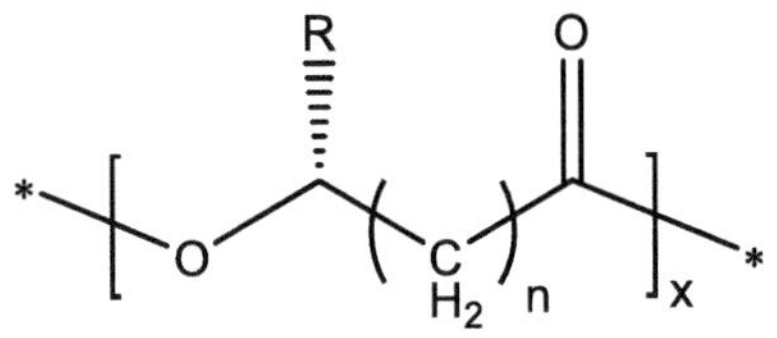

Fig. 3.1 General chemical structure of PHA, where *n* and *x* indicate the number of repeating units in the PHA chain. $n = 0$, 2-hydroxyalkanoate (2HA); $n = 1$, 3-hydroxyalkanoate (3HA); $n = 2$, 4-hydroxyalkanoate (4HA)

11 carbons. Being a polyester, PHA is formed when the carboxyl group of one monomer established an ester bond with the adjacent monomer's hydroxyl group inside the microorganisms. The term biopolymer also clarifies that PHA can be degraded with the help of microorganisms (Lee 1996). The unique part is PHA was proven to be biodegraded even in the marine system (Thellen et al. 2008). Its biodegradability in the marine system has made PHA outshine other bio-based plastics such as polylactide (PLA) and polybutylene succinate (PBS). This is because marine microbes possess PHA depolymerase (Mukai et al. 1993; Volova et al. 2011) so it can degrade PHA, whereas other types of biopolymers might take quite a long time to be degraded. Such unique property of PHA is therefore expected to solve the worrying plastic pollution in the marine system.

3.2 Biosynthesis of PHA

Poly[(*R*)-3-hydroxybutyrate)], P(3HB), was the first member of the PHA family to be discovered. It was discovered in *Bacillus megaterium* by Lemoigne in 1926 (Lemoigne 1926). Following this discovery, many other researchers also have reported the occurrence of P(3HB) in other microorganisms (Stapp 1924; Forsyth et al. 1958); hence, it hints the role of PHA as a carbon and energy source for bacteria (Doudoroff and Stanier 1959).

PHA is accumulated in a bacterium when it is exposed to a stressful condition mainly to support their survival by acting as an alternative energy source (Doudoroff and Stanier 1959). For example, under nitrogen-limiting condition, TCA cycle in a bacterium will be affected and elevates the level of acetyl-coenzyme A (acetyl-CoA), and the increased amount of acetyl-CoA will be then channeled for the synthesis of PHA monomer (Wang et al. 2009). Therefore, under unconducive environment bacteria can produce PHA to continue its survival.

In earlier days, 3HB was thought to be the only type of hydroxyalkanoate (HA) to occur in microorganisms. This well-perceived fact was proved to be wrong with Wallen and Rohwedder's finding on the occurrence of 3-hydroxyvalerate (3HV) and 3-hydroxyhexanoate (3HHx) in domestic sewage sludge in 1974 (Wallen and Rohwedder 1974). Their finding is indeed a watershed breakthrough because it has intrigued many researchers to look for other new HA units that as for to date there are nearly 160 monomer units (Choi et al. 2020) that have been reported. PHAs are grouped into three groups based on the number of carbon atoms in the monomer. Short-chain-length PHA (scl-PHA) contains 3–5 carbon atoms; medium-chain-length PHA (mcl-PHA) contains 6–14 carbon atoms; and long-chain-length PHA (lcl-PHA) contains more than 14 carbon atoms (Anderson and Dawes 1990; Steinbüchel 1992). A lot of bacteria are known to be the natural PHA producers (Taguchi et al. 2005), and via genetic approaches, many other bacteria are now able to synthesize PHA (Fidler and Dennis 1992); thus, a myriad of bacteria have the capability to produce PHA under specific cultivation condition. Among the numerous PHA producers, *Ralstonia eutropha* (also known as *Cupriavidus necator*) is the best-studied microorganism (Reinecke and Steinbüchel 2009) and is able to accumulate PHA over 90 wt% of dry cell. The literature on this versatile microbe's ability in PHA production from the past 50 years has revealed its vast biotechnological potential. The availability of the complete genome sequence of *R. eutropha* (Pohlmann et al. 2006) has given a useful insight to fine-tune the microorganism for enhanced PHA production meeting industrial requirements.

In this section, three key components for the biosynthesis of PHA which are PHA synthase, potential carbon sources, and metabolic pathways will be discussed, and recent approaches in PHA biosynthesis are summarized.

3.2.1 PHA Synthase

PHA synthase holds a unique ability to polymerize high molecular weight of water-insoluble PHA monomers in a water-based environment such as in cell cytoplasm without needing a template unlike other macromolecules (Tsuge 2016) at ambient temperature. It polymerizes monomeric substrates acyl-CoA with the release of CoA.

The best understanding of PHA biosynthesis has clarified that substrate specificity of PHA synthase is one of the salient factors in deciding the type of PHA polymer synthesized in a microorganism (Sudesh et al. 2000). Therefore, many efforts to understand PHA synthase in details are being carried out, and this has gotten easier with the availability of partial crystal structure of the catalytic subunit PhaC of PHA synthase from *R. eutropha* (Kim et al. 2017) and *Chromobacterium* sp. USM (Chek et al. 2017). The information from crystal structure aids in the process of elucidating the mechanism of PHA polymerization at molecular levels which remained as a bigger challenge in previous days. Also, many PhaC structural genes from a diverse group of gram-positive, gram-negative bacteria and cyanobacteria have been cloned (Sudesh et al. 2000). PHA can be classified into 4 distinct classes based on its substrate specificities and subunit compositions (Tsuge et al. 2015) (Table 3.1).

A few studies have reported that class I PhaC was also shown to polymerize a very small fraction of 3HHx and 3HO units (Dennis et al. 1998; Antonio et al. 2000). Class II PhaCs mainly from *Pseudomonas* have broad substrate specificity, yet they

Table 3.1 Classification of PHA synthase. (Sudesh et al. 2000; Tsuge 2016; Tsuge et al. 2015)

Class	Substrate specificity	Subunit composition	Representative microorganism	Type of PHA
I	C3–C5 monomers	Single subunit of PhaC (60–70 kDa)	*R. eutropha*	Mainly scl
II	C6–C14 monomers	Single subunit of PhaC (60–70 kDa)	*Pseudomonas oleovorans*	Mainly mcl
III	C3–C5 monomers	Two subunits of PhaC (~40 kDa) and PhaE (~40 kDa)	*Chromatium vinosum*	scl
IV	C3–C5 monomers	Two subunits of PhaC (~40 kDa) and PhaR (~20 kDa)	*Bacillus megaterium*	scl

scl short-chain length, *mcl* medium-chain length

do not polymerize C4 monomers. However, PhaC from *Pseudomonas* sp. 61-3 was shown to be able to polymerize scl-HA prior to the expression of respective genes in *R. eutropha* PHA-negative mutants (Matsusaki et al. 1998, 2000; Tsuge et al. 2005). Besides, PhaC from *Bacillus megaterium* is interestingly dissimilar to other PhaCs despite composing two subunits of PhaC and PhaR (Tomizawa et al. 2011) such as of class III synthase, and therefore, it has been classified into a new category, class IV. At present PhaC class IV category encompasses only PhaCs from *Bacillus* (Tsuge et al. 2015). Generally, with the supply of precursors, PhaCs from *Bacillus* were able to polymerize copolymers. For instance, P(3HB-*co*-3HV) could be biosynthesized using *Bacillus* with the supply of propionate and valerate precursors. However, it has shown that some *Bacillus* sp. are able to polymerize P(3HB-*co*-3HV) with up to 2 mol% 3HV fraction using glucose as a sole carbon source (Mizuno et al. 2010). A similar situation was observed in another study, where it was then known that P(3HB-*co*-3HV) with 2–12 mol% 3HV can be produced in some *Bacillus* sp. using glucose alone (Masood et al. 2013).

Pursuing the current trends, PHA copolymer compositions can be altered using mutated PHA synthase genes (Taguchi et al. 2003). Change in monomer composition in PHA copolymer using PHA synthase can be linked to the change in substrate specificities toward target substrate monomers; thus, noticeable changes happen in related metabolic pathways. Point mutagenesis of Ala510 in $PhaC_{Re}$ has changed the polymerization activity of PhaC and hinted the magnificent influence of amino acid at the position 510 in PHA biosynthesis (Tsuge et al. 2004a). In addition, PhaC from *Aeromonas caviae* ($PhaC_{Ac}$) with two mutations, N149S/D171G (NSDG), has expressed it in PHA-negative strain and obtains higher 3HHx fraction than of the wild-type PhaC (Tsuge et al. 2007). Besides, the PhaC gene from *Pseudomonas putida* KT2440 ($phaC1_{Pp}$) was randomly mutated to achieve a high yield of poly (3-hydroxydodecanoate), P(3HDD) (Hiroe et al. 2018). From the random mutant library, $PhaC1_{Pp}$ mutants with increased activity and specificity for the polymerization of 3HDD were selected for the further enzyme optimization to enhance the production of P(3HDD). Therefore, it is evident that causing a change in PhaC using molecular techniques such as point mutation or random mutation indeed affects the overall biosynthesis results.

3.2.2 Potential Carbon Sources for PHA Biosynthesis

The high production cost of PHA, unlike PBP, is one of the major factors that hinder it from getting commercialized. One of the plausible ways to bring down the production cost of PHA is by using cheaper carbon sources. A good carbon source is required for microbes to grow well before entering to PHA production phase prior to stress induced through nutrient (nitrogen/phosphorus) limitation. A variety of cheap structurally related and unrelated carbon sources were/are being evaluated to be used in large-scale PHA production using wild-type or recombinant strains. Typically, using wild-type PHA producers, structurally unrelated carbon sources such as glucose are used to produce P(3HB) homopolymer, whereas structurally related carbon sources or precursors are needed to produce copolymers. For instance, valeric acid and propionic acid are used to produce copolymers P(3HB-*co*-3HV). These precursors are very expensive, and therefore, it limits the production of such copolymer at an industrial scale. Therefore, the search for novel and cheap carbon sources is highly required.

Sugars such as glucose, sucrose, fructose, and xylose are structurally unrelated carbon sources used to produce PHA because they are available at relatively cheaper rates. As of now, the highest production of P(3HB) was achieved in a recombinant *E. coli* containing *A. latus* PHA biosynthesis genes using glucose as the sole carbon source (Choi et al. 1998). Besides, recombinant *E. coli* LS5218 was able to biosynthesize P(3HB)-based copolymer comprising of 94.6 mol% 3HB and 5.4 mol% mcl-3HA consisting of C_8, C_{10}, C_{12}, and C_{14} using glucose as the single carbon source (Hokamura et al. 2015). Interestingly, a wild-type *Bacillus cereus* fed with only glucose also has shown to have the ability to polymerize P(3HB-*co*-3HV) (Mizuno et al. 2010), suggesting that this microorganism has an intriguing metabolic versatility of channeling the second monomer using a single carbon source. An interesting PHA copolymer, P(3HB-*co*-2-hydroxyalkanoate), with different monomer molar ratio was biosynthesized using recombinant *E. coli* strain with 20 g/L of glucose or xylose as the sole carbon source in Luria-Bertani (LB) medium cultivated at 30 °C for 72 h (Mizuno et al. 2018). In LB medium with glucose as the monocarbon source, the strain could store up to 55 wt% of PHA and yield around 10.6 g/L of dry cell weight. Meanwhile, when xylose was used, about 43 wt% of PHA and 9.6 g/L of the dry cell was obtained. However, when the strain was cultivated under the same conditions in M9 medium with glucose or xylose as the sole carbon source, 2-hydroxy-3-methylbutyrate (2H3MB) monomer unit in the copolymer could not be biosynthesized (Fig. 3.2). Cultivation of the recombinant *E. coli* strain in M9 media with glucose or xylose also affected the PHA content and dry cell weight. In terms of PHA content, it was found that about 70 wt% of PHA was obtained for glucose, and for xylose it declined to 28 wt%. However, for cell dry weight, in both cases, quite a drastic reduction was noticed.

Next, plant oils and their derivatives are considered as a potential structurally unrelated carbon source for PHA biosynthesis. An estimate of 0.3–0.4 g of PHA can be produced using 1 g of glucose (Ryu et al. 1997), meanwhile 0.6–0.8 g of PHA using 1 g of plant oil (Akiyama et al. 2003). This clarifies that plant oil has more carbon per weight in comparison to sugar (Akiyama et al. 1992); thus, it is worth evaluation for PHA biosynthesis. The possibility of corn oil to be a carbon source for PHA production was demonstrated by Akiyama et al. (Akiyama et al. 1992). *Alcaligenes* sp. AK201 fed with 3 g/L of corn oil for 48 h could yield 39 wt% of PHA and 2.8 g/L of dry cell. Besides, soybean oil is also widely used for PHA production. The wild type of *R. eutropha* was shown to yield about 72–76 wt% of P(3HB) with 118–126 g/L of the dry cell when soybean oil was used as a sole carbon source (Kahar et al. 2004). Tsuge et al. have also used soybean oil to test for PHA production in *R. eutropha* mutants encompassing mutated PhaC (Tsuge et al. 2009). Twenty recombinant strains of *R. eutropha* with mutated PhaC were subjected to this evaluation, and the highest PHA content observed was 57 wt% with 1.6 g/L of dry

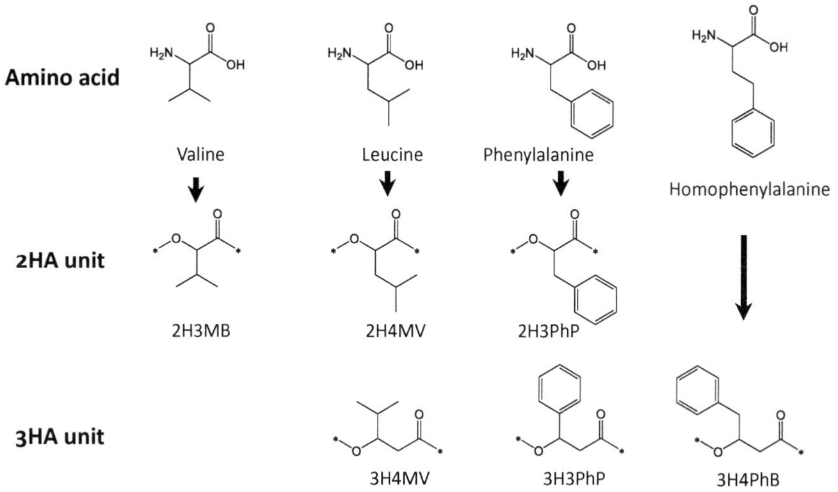

Fig. 3.2 2-Hydoxyalkanoate (2HA) and 3-hydroxyalkanoate (3HA) units that have corresponding carbon backbone with amino acids. *2H3MB* 2-hydroxy-3-methylbutyrate, *2H4MV* 2-hydroxy-4-methylvalerate, *2H3PhP* 2-hydroxy-3-phenylpropionate, *3H4MV* 3-hydroxy-4-methylvalerate, *3H3PhP* 3-hydroxy-3-phenylpropionate, *3H4PhB* 3-hydroxy-4-phenylbutyrate

cell weight. The potential of sunflower oil, olive oil, coconut oil, and palm olein for PHA production was also investigated by Lee et al. with the supply of 3HV precursors (Lee et al. 2008). From the study, it was found that around 60–80 wt% of PHA could be obtained when the wild type of *R. eutropha* strain is fed with an almost equal amount of oil and precursors. Biosynthesis of a tetrapolymer poly (94 mol% 3HB-*co*-3HV-*co*-3HHx-*co*-3HO) was attained in recombinant *R. eutropha* strain using canola oil (Valdés et al. 2018). However, in this case, fructose was also used in the cell growth phase before canola oil was used at PHA production. Nevertheless, continuous use of edible plant oils as carbon feedstocks for PHA production is not a wise idea because it will lead to food shortage and inflation problems in few underdeveloped countries. Therefore, few nonedible plant oils also were tested for the biosynthesis of PHA. Firstly, palm kernel oil is an example of oil obtained from palm seeds which are used mainly for nonedible purposes such as for manufacturing of soap, detergent, and cosmetics. It also has been demonstrated to have the possibility for PHA production. Recombinant *R. eutropha* PHB$^-$4 strain was shown to produce P(3HB-*co*-3HV-*co*-3HHx) terpolymer from mixtures of palm kernel oil and 3HV precursors for the first time (Bhubalan et al. 2008). 3HV molar ratio of 34% could be attained when higher concentrations of sodium valerate were mixed with palm kernel oil as the carbon source, whereas 29 mol% of 3HV was obtained with sodium propionate. Jatropha oil is also one of the nonedible plant oils, which holds the potential to be a carbon feedstock for PHA production. Jatropha oil has shown to be a potential renewable carbon source for the large-scale production of P(3HB-*co*-3HV) by recombinant *R. eutropha* PHB$^-$4 and wild-type strain (Chia et al. 2010). The recombinant strain fed with only 10 g/L jatropha oil as the carbon source showed to be able to synthesize P(3HB-*co*-3 mol% 3HHx). Up to 12.5 g/L of jatropha oil was found to be safe to be fed, and beyond that, it exerts toxicity to cells. On the other hand, *R. eutropha* wild-type strain fed with a mixture of jatropha oil with 3HV precursors could synthesize P(3HB-*co*-3HV) with 3HV fraction of 41 mol%. The results discussed here using promising plant oils are tabulated in Table 3.2.

A fatty acid is a structurally related carbon source, and it is pivotal in synthesizing PHA especially mcl-PHA using microorganisms. Generally, fatty acid will exert a certain level of toxicity to microbes, thus utilizing efficient feeding strategy in terms of the concentration, and feeding time will give a better output. It was shown that intermittent feeding of low concentration of sodium decanoate (C10Na) or sodium dodecanoate (C12Na) along with the low concentration of glucose, glycerol, or xylose yielded better PHA production in *E. coli* strain with defective β-oxidation (Fadzil et al. 2018). The efficient intermittent feeding strategy proposed in this study has enabled the engineered *E. coli* to have better conversion yield of C10Na and C12Na to P(3HD) and P(3HDD), respectively. A near homopolymer of 5.44 g/L P (3HD) and 3.50 g/L P(3HDD) was obtained.

The addition of amino acid also plays an interesting role in PHA biosynthesis. For instance, an obligate anaerobic bacterium *Clostridium difficile* can convert leucine to 4-methylvalerate (Saika et al. 2014). In the conversion process, 4-methyl-2-pentenoyl-CoA (4M2PE-CoA) is intermediate, and this can also be converted to 3-hydroxy-4-methylvalerate (3H4MV) and eventually channeled for the synthesis of PHA copolymer with 3H4MV unit (Table 3.3) (Saika et al. 2014). Besides, it has confirmed that amino acid is a promising precursor for the synthesis of 2HA with various side-chain structures when related enzymes are used to convert amino acids to 2HA-CoA as the substrate for the lactate-polymerizing enzyme (LPE). For supplementation of 1 g/L of valine, the fraction of 2H3MB, which has the same carbon backbone as valine does, increased to 8.3 mol%. Meanwhile, for 1 g/L of leucine supplementation, the fraction of 2-hydroxy-4-methylvalerate (2H4MV), which has the same carbon backbone as leucine does (Fig. 3.2), was detected in the amount of 23.6 mol%. On the other hand, phenylalanine supplementation at 1 g/L increased the 2-hydroxy-3-phenylpropionate (2H3PhP) fraction to 17.2 mol%. After the increase in the phenylalanine supplementation to 5 g/L, the 2H3PhP fraction slightly increased to 20 mol% (Mizuno et al. 2018). However, usually, a supply of amino acid can hamper cell growth due to the end product feedback inhibition, and thus, the concentration of amino acid has to be adjusted for desirable cell growth and PHA production.

Apart from all these carbon sources, agricultural and domestic waste materials also were tested for PHA production. Such waste materials are rich in xylose, but PHA-producing microorganisms are generally weak in assimilating them. However, via anaerobic fermentation of xylose which is easier to assimilate, lactic acid and acetic acid can be produced; thus, they can be utilized for PHA production (Tsuge 2002). Rice husk is one of the notable agricultural wastes and it is rich in cellulose and hemicellulose. The richness of rice husk has enabled it to be a potential carbon source in PHA production. *B.cepacia* USM was shown to have a dry cell weight of 7.8 g/L and PHA content of 50 wt% when it was cultivated in a 5 L fermenter supplemented

Table 3.2 Type of plant oils used as a carbon source for PHA biosynthesis by *R. eutropha*

Type of plant oil	Plant oil conc. (g/L)	Co-carbon source (g/L)	Bacteria (Class of PHA synthase)	Culture time (h)	Dry cell wt. (g/L)	PHA content (wt%)	Type of PHA
Corn oil (Fukui and Doi 1998)	10	–	*R. eutropha* Wild type (I)	72	3.6	81	P(3HB)
		–	*R. eutropha* $PHB^{-}4$ recombinant (I)	72	3.6	77	P(3HB-*co*-5 mol% 3HHx)
Soybean oil (Tsuge et al. 2009; Hyakutake et al. 2011)	20	–	*R. eutropha* $PHB^{-}4$ recombinant (IV)	72	6.0	81	P(3HB-*co*-0.5 mol% 3HHx)
	20	–	*R. eutropha* $PHB^{-}4$ recombinant (II)	72	1.6	57	P(3HB-*co*-6 mol% 3HA)
Sunflower oil (Lee et al. 2008)	5	Sodium propionate (5)	*R. eutropha* Wild type (I)	72	6.7	78	P(3HB-*co*-8 mol% 3HV)
		Sodium valerate (5)	*R. eutropha* Wild type (I)	72	5.2	80	P(3HB-*co*-14 mol% 3HV)
Olive oil (Lee et al. 2008)	5	Sodium propionate (5)	*R. eutropha* Wild type (I)	72	6.0	78	P(3HB-*co*-8 mol% 3HV)

(continued)

Table 3.2 (continued)

Type of plant oil	Plant oil conc. (g/L)	Co-carbon source (g/L)	Bacteria (Class of PHA synthase)	Culture time (h)	Dry cell wt. (g/L)	PHA content (wt%)	Type of PHA
		Sodium valerate (5)	*R. eutropha* Wild type (I)	72	6.1	89	P(3HB-*co*-14 mol% 3HV)
Coconut oil (Lee et al. 2008)	5	Sodium propionate (5)	*R. eutropha* Wild type (I)	72	5.4	75	P(3HB-*co*-2 mol% 3HV)
		Sodium valerate (5)	*R. eutropha* Wild type (I)	72	6.0	78	P(3HB-*co*-3 mol% 3HV)
Palm olein (Lee et al. 2008)	5	Sodium propionate (5)	*R. eutropha* Wild type (I)	72	6.7	88	P(3HB-*co*-4 mol% 3HV)
		Sodium valerate (5)	*R. eutropha* Wild type (I)	72	3.5	68	P(3HB-*co*-4 mol% 3HV)
Canola oil (Valdés et al. 2018)	5% v/v for PHA accumulation	Fructose (10) at growing phase	*R. eutropha* PHB$^-$4 recombinant (II)	48	6.2	96	P(3HB-*co*-1 mol% 3HV-*co*-4 mol% 3HHx-*co*-1 mol% 3HO)
Palm kernel oil (PKO) (Bhubalan et al. 2008; Chia et al. 2010; Loo et al., 2005; Murugan et al. 2016)	5	–	*R. eutropha* PHB$^-$4 recombinant (I)	72	4.3	87	P(3HB-*co*-5 mol% 3HHx)
		–	*R. eutropha* H16 recombinant (I)	48	5.0	73	P(3HB-*co*-33 mol% 3HHx)

		Sodium propionate (10)	*R. eutropha* PHB⁻4 recombinant (I)	72	2.0	26	P(3HB-*co*-34 mol% 3HV-*co*-3 mol% 3HHx)
		Sodium valerate (10)	*R. eutropha* PHB⁻4 recombinant (I)	72	0.7	6	P(3HB-*co*-29 mol% 3HV-*co*-6 mol% 3HHx)
		Sodium 4-metylvalerate (1)	*R. eutropha* PHB⁻4 recombinant (I)	48	1.7	50	P(3HB-*co*-1 mol% 3HV-*co*-4 mol% 3H4MV-*co*-16 mol% 3HHx)
Jatropha oil (Ng et al. 2011)	10	–	*R. eutropha* PHB⁻4 recombinant (I)	48	8.0	84	P(3HB-*co*-3% mol% 3HHx)
	5.19	Sodium propionate (4.32)	*R. eutropha* Wild type (I)	48	5.6	67	P(3HB-*co*-27 mol% 3HV)
	5.19	Sodium valerate (4.32)	*R. eutropha* Wild type (I)	48	6.7	76	P(3HB-*co*-41 mol% 3HV)

Table 3.3 Common types of 3HB-based copolymers synthesized using bacteria

3HB-based copolymers	Chemical structure
Poly(3-hydroxybutyrate-*co*-3-hydroxyvalerate) [P(3HB-*co*-3HV)]	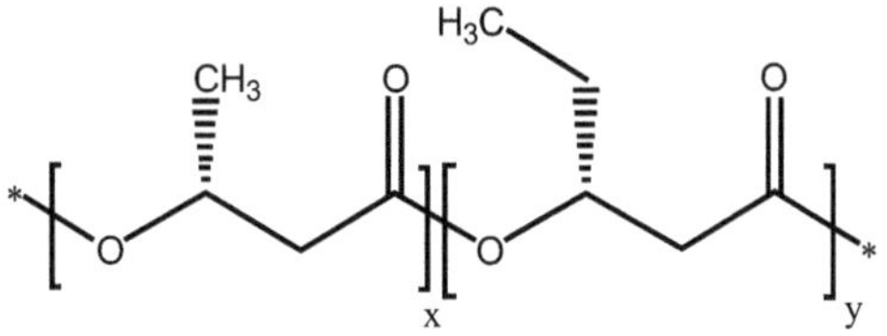
Poly(3-hydroxybutyrate-*co*-4-hydroxybutyrate) [P(3HB-*co*-4HB)]	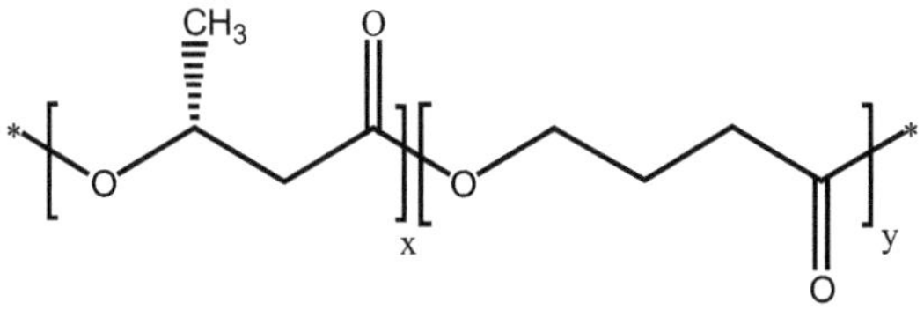
Poly(3-hydroxybutyrate-*co*-3-hydroxyhexanoate) [P(3HB-*co*-3HHx)]	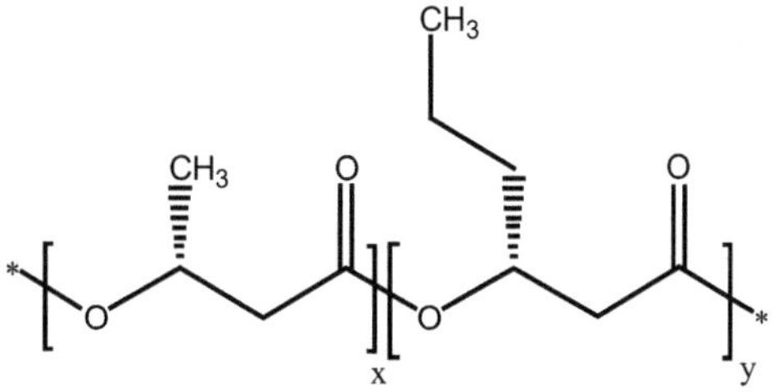
Poly(3-hydroxybutyrate-*co*-3-hydroxy-4-metylvalerate) [P(3HB-*co*-3H4MV)]	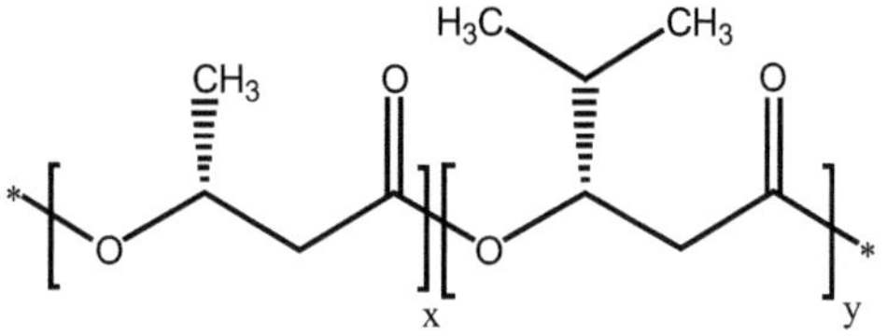

with the hydrolysate of rice husk (Heng et al. 2017). Domestic waste such as waste cooking oil was also shown to have some potential to be a carbon source for PHA production. An approximate of 1.2 g/L of P(3HB) obtained from waste frying oil was 1.2 g/L, which is almost similar to a concentration that can be obtained from glucose (Verlinden et al. 2011). In parallel to the growing research interest on PHA production, many other novel carbon sources will be added to the list.

3.2.3 Metabolic Pathways for PHA Biosynthesis

Metabolic pathways operating in a microorganism is also an essential deciding factor in PHA biosynthesis because it gives rise to the relevant precursors for desired PHA production. Earlier attempts at producing interesting PHAs include basic fermentation technology using naturally occurring PHA-producing bacteria with different types of carbon sources. However, recent advancement in biotechnology studies has led researchers to manipulate microorganisms instead to uptake the designated carbon sources, and this has set a new dimension in PHA biosynthesis. In this subtopic, naturally occurring PHA biosynthetic pathways and engineered pathways will be summarized.

3.2.3.1 Naturally Occurring PHA Biosynthetic Pathways

Dimerization pathway, fatty acid de novo synthesis, and fatty acid β-oxidation are three main metabolic pathways in microorganisms that involved in the biosynthesis of PHA (Fig. 3.3). Apart from these main metabolic pathways, PHA also can be biosynthesized by means of other pathways such as pathways utilizing amino acids. Commonly, the dimerization pathway by β-ketothiolase (PhaA) and acetoacetyl-CoA reductase (PhaB) generates 3HB monomers from the condensation of acetyl-CoA in a PHA-producing microorganism. Meanwhile, mcl-PHA is mainly biosynthesized through fatty acid de novo synthesis and fatty acid β-oxidation. These two pathways responsible for mcl-PHA production are mostly found in PHA producers coming from the *Pseudomonas* genus. Mainly mcl-(*R*)-3HA monomers will be supplied from fatty acid β-oxidation and de novo fatty acid synthesis. Many *Pseudomonas* strains such as *P. putida, P. citronellolis, P. oleovorans*, and *P. pseudoalcaligenes* were shown to biosynthesize mcl-PHA (Huang et al. 2018). However, PhaC is only active toward (*R*)-enantiomers that are linked to CoA, so it is crucial for the intermediates in both fatty acid synthesis and fatty acid degradation pathway to get it converted. In fatty acid degradation, (*S*)-3-hydroxyacyl-CoA can be converted to (*R*)-3-hydroxyacyl-CoA via enoyl-CoA with the help of two stereospecific enoyl-CoA hydratases (Sato et al. 2007). On the other hand, in fatty acid biosynthesis, the intermediate is (*R*)-enantiomer 3-hydroxyacyl-ACP.

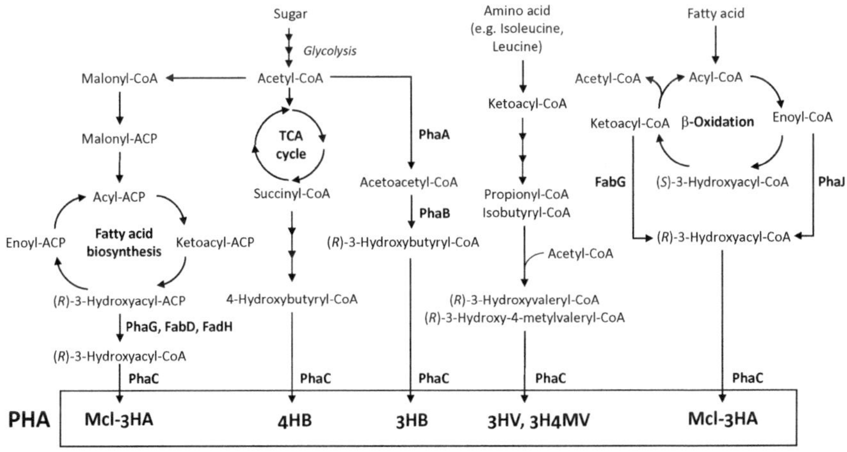

Fig. 3.3 PHA biosynthetic pathways

3.2.3.2 Genetically Engineered Pathways for PHA Biosynthesis

Genetic engineering has garnered research interest in PHA biosynthesis because it is useful to biosynthesize PHA with targeted monomer composition and desired properties and at a higher yield. Furthermore, genetic approaches also open a possibility in the production of 3HB-based copolymers using various cheap and renewable carbon substrates such as waste materials and carbon dioxide, ensuring PHA production to be more economical for replacing PBP. Concerning PHA biosynthesis, genetic or metabolic engineering can be used either to upgrade the existing PHA biosynthetic pathway in a natural PHA producer to increase the flux for PHA monomer synthesis or to modify a non-PHA producer by incorporating essential genes for PHA production. A number of essential PHA synthetic genes such as *phaC*, *phaA*, *phaB*, *phaG*, and *phaJ* from PHA producers could be cloned into *E. coli* or other nonnatural PHA producers. For example, mcl-PHA homopolymers are generally not naturally and widely occurring PHAs, unlike scl-PHA. Therefore, gene modification is required to construct an artificial biosynthetic pathway in bacteria to biosynthesize near mcl homopolymers using renewable resources. Two mcl-PHA homopolymers, poly(3-hydroxydecanoate) [P(3HD)] and poly(3-hydroxydodecanoate) [P(3HDD)], have biosynthesized from metabolically engineered *E. coli* LSBJ using pure fatty acids (Hiroe et al. 2016). The production of this mcl-PHA was further enhanced to 1.47 g/L and 0.42 g/L, respectively, when metal components in the culture medium were increased (Hiroe et al. 2016). PHAs with controlled monomer composition can be produced by recombinant bacteria encompassing a combination of the substrate specificities of PHA synthase and monomer-supplying enzymes. The use of (*R*)-specific enoyl-CoA hydratase (PhaJ) in PHA biosynthesis from fatty acid ß-oxidation cycle in recombinant bacteria has been determined (Tsuge et al. 2003). The presence of either $PhaJ1_{Pa}$ or $PhaJ4_{Pa}$ in recombinant *E. coli* led to the high levels of PHA accumulation of 36–41 wt.% in dry cells consisting of mainly scl- and mcl-3HA units, respectively. In addition, $phaJ1_{Pa}$ and $phaJ2_{Pa}$, homologous to the *A. caviae* PhaJ ($phaJ_{Ac}$), were cloned in *E. coli* DH5α strain to investigate the monomer-supplying ability for PHA synthesis from β-oxidation cycle. The recombinants harboring $phaJ1_{Pa}$ or $phaJ2_{Pa}$ showed high (*R*)-specific enoyl-CoA hydratase activity with different substrate specificities, that is, specific for scl- and mcl-enoyl-CoAs, respectively. Meanwhile, co-expression of these two hydratase genes with *phaC* in *E. coli* LS5218 resulted in the accumulation of PHA up to 14–29 wt% of cell dry weight from dodecanoate as a sole carbon source (Tsuge et al. 2000).

3.3 Material Properties of PHA

A variety of optically pure PHA has been biosynthesized using microorganisms. P (3HB) is the first type of PHA to be discovered and to be characterized. Following the earlier characterization of P(3HB) (Holmes et al. 1984), researchers were interested to compare its properties to polypropylene (PP). Interestingly, P(3HB) has a melting temperature (T_m) of about 177 °C and glass transition temperature (T_g) around 4 °C (Taguchi et al. 2012) which are nearly similar to PP. However, owing to the stereoregular nature of P(3HB) obtained from microbes, it is highly crystalline, unlike PP which is relatively flexible. Highly crystalline property has made it be a very stiff and brittle polymer. Moreover, as for the hot-drawing process of P(3HB) materials, one of the major problems is the decrease in molecular weight of polymer due to rapid thermal degradation near its melting temperature. All these cumulative drawbacks of P(3HB) have hindered it from being developed as an industrial material targeting a wide range of applications. Therefore, 3HB monomer unit has to be copolymerized with other monomers to circumvent the aforementioned limitations. Table 3.3 summarizes chemical structures of some of the biosynthesized 3HB-based copolymers for better material properties. 3HB-based copolymers with even small fraction of other monomers will help to reduce the T_m of the polymer in overall. It was shown that with just about 3.5 mol% of 3HHx fractions, the T_m of the copolymer was reduced noticeably compared to P(3HB) homopolymer (Tsuge et al. 2004b). These copolymers melt during the drawing of polymer material at a low temperature while avoiding thermal degradation. Various attempts have been made to characterize the bacterially produced optically active 3HB-based copolymers for improved material properties. Some of such copolymers' material properties are summarized in Table 3.4 along with material properties of P(3HB) and PP and LDPE for a comparison. The higher molecular weight gives P(3HB) a higher mechanical strength, and unlike lower molecular-weight P(3HB) polymers, ultrahigh-molecular weight-P(3HB) [UHMW-P(3HB)] can be processed into strong films or fibers by hot or cold drawing (Kusaka et al. 1999). It is expected that the improved material properties of UHMW-P(3HB) will, therefore, expand the number of applications for this polymer. It was found that the mechanical properties of the stretched P(3HB) films were markedly improved in comparison with those of the unstretched films where the elongation at break and tensile strength of the stretched films increased to 35% and 1320 MPa (Kusaka et al. 1999, Iwata 2005), respectively. As mentioned earlier, another way to improve the physical properties of P(3HB) is by the incorporation of different HA units into the P(3HB) sequence to form PHA copolymers. P(3HB-*co*-3HV) is one of the very first commonly biosynthesized copolymer, and it even was produced at a bulk amount under the trademark of Biopol™. However, isodimorphous is one of the biggest impediments in P(3HB-*co*-3HV) where 3HB and 3HV units are isodimorphous and get incorporated into the P (3HB) crystal lattice. Incorporation of 8 mol% of 3HV units (Taguchi et al. 2012) has reduced the T_m to 165 °C and T_g to 1 °C. However, the elongation at break increased to 35%. On the other hand, incorporation of 20 mol% 3HV units, T_m and T_g of the copolymer decrease to 145 °C and –1 °C, respectively, and its elongation to

Table 3.4 Comparison of material properties of few 3HB-based polymers with commercially used polypropylene and low-density polyethylene

Polymer	Monomer fraction (mol %)	T_m[a] [°C]	T_g[b] [°C]	Tensile strength [MPa]	Elongation at break [%]
PP (Taguchi et al. 2012)	–	176	−10	38	400
Low-density polyethylene (LDPE) (Taguchi et al. 2012)	–	130	−36	10	620
P(3HB) (Taguchi et al. 2012)	3HB:100	177	4	43	5
Stretched UHMW P(3HB) (Kusaka et al. 1999, Iwata 2005)	3HB:100	ca. 180	ca. 4	1320	35
P(3HB-*co*-3HV) (Taguchi et al. 2012)	3HB: 92; 3HV: 8	165	1	19	35
	3HB: 80; 3HV: 20	145	−1	50	50
P(3HB-*co*-4HB) (Taguchi et al. 2012)	3HB: 84; 4HB: 16	150	−7	20	444
	3HB: 36; 4HB: 64	50	−35	65	590
	3HB: 10; 4HB: 90	50	−42	32	1080
P(3HB-*co*-3HHx) (Taguchi et al. 2012)	3HB: 95; 3HHx: 5	160	−2	32	260
	3HB: 90; 3HHx: 10	127	−1	21	400
P(3HB-*co*-3H4MV) (Tanadchangsaeng et al. 2009)	3HB: 84; 3H4MV: 16	143	−2	17	330
P(3HB-*co*-2H4MV) (Mizuno et al. 2018)	3HB: 82; 2H4MV: 18	171	5	10	18
P(3HB-*co*-3H3PhP) (Mizuno et al. 2017)	3HB: 79; 3H3PhP: 21	n.d.	21	n.m.	n.m.
P(3HB-*co*-3H4PhB) (Mizuno et al. 2014)	3HB: 85; 3H4PhB: 15	120	10	n.m.	n.m.
P(3HB-*co*-mcl-3HA) (Matsusaki et al. 2000)	3HB: 94; MCL-3HA: 6	133	−8	17	680
P(3HD) (Hiroe et al. 2016)	3HD:100	70	−46	8	226
P(3HDD) (Hiroe et al. 2016)	3HDD:100	82	n.d.	11	270

n.m. not measured, *n.d.* not detected
[a]Melting temperature
[b]Glass transition temperature

break increases to 50% (Taguchi et al. 2012) Both the situations show that P (3HB-*co*-3HV) copolymer is more flexible than the P(3HB) homopolymer. However, better improvement in material property is, therefore, possible to be witnessed with copolymerization of 3HB with longer-chain hydroxyalkanoates (HAs). By doing so, the monomers will not fit into the crystalline lattice of each other.

Examples of these HAs include 4HB (Taguchi et al. 2012), 3HHx (Taguchi et al. 2012), 3H4MV(Tanadchangsaeng et al. 2009, 2010; Saika et al. 2015), and others (Matsusaki et al. 2000; Hiroe et al. 2016; Mizuno et al. 2014) listed in Table 3.4. Incorporation of a small amount of the second monomer unit from longer-chain HAs is enough to increase their flexibility. Interestingly, P(3HB-*co*-6 mol% 3HA) (Matsusaki et al. 2000) copolymer shows almost mechanical properties to low-density polyethylene; thus, it holds a wider opportunity to be exploited for various applications.

3.4 Remarks and Outlook

Significant human activities have created overdependence on PBP, owing to its versatility. Nevertheless, the existing situation is not sustainable and imposes a serious threat to the environment. The similitude possesses by PHA to serve as an alternative to PBP will be beneficial in circumventing this issue. Additionally, unlike PBP, PHA is able to degrade in the environment such as under composting and marine which can prevent microplastic pollution. On top of it, the use of PHA-based biomaterials can help to reduce the amount of waste produced from agricultural, industrial, and domestic. A myriad of microorganisms convert these waste materials into PHA. However, PHA has several limitations in getting it commercialized; thus, more improvement are required to diversify applications related to PHA. Comprehensive studies are needed to improve the material properties and the productivity of PHA to make it competitive to PBP. Considering the rapid increase in the development of PHA-related researches, it can be expected that PHA can get commercialized in a few decades ahead. Alongside, positive perceptions toward the use of eco-friendly material can help to dot PHA in the commercial market as expected.

References

Akiyama M, Taima Y, Doi Y (1992) Production of poly(3-hydroxyalkanoates) by a bacterium of the genus *Alcaligenes* utilizing long-chain fatty acids. Appl Microbiol Biotechnol 37 (6):698–701

Akiyama M, Tsuge T, Doi Y (2003) Environmental life cycle comparison of polyhydroxyalkanoates produced from renewable carbon resources by bacterial fermentation. Polym Degrad Stab 80(1):183–194

Albuquerque MGE, Torres CAV, Reis MAM (2010) Polyhydroxyalkanoate (PHA) production by a mixed microbial culture using sugar molasses: effect of the influent substrate concentration on culture selection. Water Res 44(11):3419–3433

Anderson AJ, Dawes EA (1990) Occurrence, metabolism, metabolic role, and industrial uses of bacterial polyhydroxyalkanoates. Microbiol Mol Biol Rev *54*(4):450–472

Antonio RV, Steinbüchel A, Rehm BH (2000) Analysis of *in vivo* substrate specificity of the PHA synthase from *Ralstonia eutropha*: formation of novel copolyesters in recombinant *Escherichia coli*. FEMS Microbiol Lett 182(1):111–117

Bhubalan K, Lee WH, Loo CY, Yamamoto T, Tsuge T, Doi Y, Sudesh K (2008) Controlled biosynthesis and characterization of poly(3-hydroxybutyrate-*co*-3-hydroxyvalerate-*co*-3-hydroxyhexanoate) from mixtures of palm kernel oil and 3HV-precursors. Polym Degrad Stab 93(1):17–23

Chek MF, Kim SY, Mori T, Arsad H, Samian MR, Sudesh K, Hakoshima T (2017) Structure of polyhydroxyalkanoate (PHA) synthase PhaC from *Chromobacterium* sp. USM2, producing biodegradable plastics. Sci Rep 7(1):5312

Chia KH, Ooi TF, Saika A, Tsuge T, Sudesh K (2010) Biosynthesis and characterization of novel polyhydroxyalkanoate polymers with high elastic property by *Cupriavidus necator* PHB^{-}4 transformant. Polym Degrad Stab 95(12):2226–2232

Choi JI, Lee SY, Han K (1998) Cloning of the *Alcaligenes latus* polyhydroxyalkanoate biosynthesis genes and use of these genes for enhanced production of poly(3-hydroxybutyrate) in *Escherichia coli*. Appl Environ Microbiol 64(12):4897–4903

Choi SY, Rhie MN, Kim HT, Joo JC, Cho IJ, Son J, Lee Y (2020) Metabolic engineering for the synthesis of polyesters: a 100-year journey from polyhydroxyalkanoates to non-natural microbial polyesters. Metab Eng 58:47–81

Dennis D, McCoy M, Stangl A, Valentin HE, Wu Z (1998) Formation of poly(3-hydroxybutyrate-*co*-3-hydroxyhexanoate) by PHA synthase from *Ralstonia eutropha*. J Biotechnol 64 (2–3):177–186

Doudoroff M, Stanier RY (1959) Role of poly-β-hydroxybutyric acid in the assimilation of organic carbon by bacteria. Nature 183(4673):1440

Fadzil M Ikhma F, MIzuno S, Hiroe A, Nomura C, Tsuge T (2018) Low carbon concentration feeding improves medium-chain-length polyhydroxyalkanoate production in *Escherichia coli* strains with defective β-oxidation. Front Bioeng Biotechnol 6:178

Fidler S, Dennis D (1992) Polyhydroxyalkanoate production in recombinant *Escherichia coli*. FEMS Microbiol Rev 9(2–4):231–235

Forsyth WGC, Hayward AC, Roberts JB (1958) Occurrence of poly-β-hydroxybutyric acid in aerobic gram-negative bacteria. Nature 182(4638):800

Fukui T, Doi Y (1998) Efficient production of polyhydroxyalkanoates from plant oils by *Alcaligenes eutrophus* and its recombinant strain. Appl Microbiol Biotechnol 49(3):333–336

Heng KS, Hatti-Kaul R, Adam F, Fukui T, Sudesh K (2017) Conversion of rice husks to polyhydroxyalkanoates (PHA) via a three-step process: Optimized alkaline pretreatment, enzymatic hydrolysis, and biosynthesis by *Burkholderia cepacia* USM (JCM 15050). J Chem Technol Biotechnol 92(1):100–108

Hiroe A, Ishii N, Ishii D, Kabe T, Abe H, Iwata T, Tsuge T (2016) Uniformity of monomer composition and material properties of medium-chain-length polyhydroxyalkanoates biosynthesized from pure and crude fatty acids. ACS Sustain Chem Eng 4(12):6905–6911

Hiroe A, Watanabe S, Kobayashi M, Nomura CT, Tsuge T (2018) Increased synthesis of poly (3-hydroxydodecanoate) by random mutagenesis of polyhydroxyalkanoate synthase. Appl Microbiol Biotechnol 102(18):7927–7934

Hokamura A, Wakida I, Miyahara Y, Tsuge T, Shiratsuchi H, Tanaka K, Matsusaki H (2015) Biosynthesis of poly(3-hydroxybutyrate-*co*-3-hydroxyalkanoates) by recombinant *Escherichia coli* from glucose. J Biosci Bioeng 120(3):305–310

Holmes PA, Collins SH, Wright LF (1984) *U.S. Patent No. 4,477,654*. U.S. Patent and Trademark Office, Washington, DC

Huang P, Okoshi T, Mizuno S, Hiroe A, Tsuge T (2018) Gas chromatography-mass spectrometry based monomer composition analysis of medium-chain-length polyhydroxyalkanoates biosynthesized by *Pseudomonas* spp. Biosci Biotechnol Biochem 82(9):1615–1623

Hyakutake M, Saito Y, Tomizawa S, Mizuno K, Tsuge T (2011) Polyhydroxyalkanoate (PHA) synthesis by class IV PHA synthases employing *Ralstonia eutropha* PHB^{-}4 as host strain. Biosci Biotechnol Biochem 75:1615–1617

Iwata T (2005) Strong fibers and films of microbial polyesters. Macromol Biosci 5(8):689–701

Kahar P, Tsuge T, Taguchi K, Doi Y (2004) High yield production of polyhydroxyalkanoates from soybean oil by *Ralstonia eutropha* and its recombinant strain. Polym Degrad Stab *83*(1):79–86

Kim J, Kim YJ, Choi SY, Lee SY, Kim KJ (2017) Crystal structure of *Ralstonia eutropha* polyhydroxyalkanoate synthase C-terminal domain and reaction mechanisms. Biotechnol J 12 (1):1600648

Kusaka S, Iwata T, Doi Y (1999) Properties and biodegradability of ultra-high-molecular-weight poly[(*R*)-3-hydroxybutyrate] produced by a recombinant *Escherichia coli*. Int J Biol Macromol 25(1–3):87–94

Lee SY (1996) Bacterial polyhydroxyalkanoates. Biotechnol Bioeng 49(1):1–14

Lee WH, Loo CY, Nomura CT, Sudesh K (2008) Biosynthesis of polyhydroxyalkanoate copolymers from mixtures of plant oils and 3-hydroxyvalerate precursors. Bioresour Technol 99(15):6844–6851

Lemoigne M (1926) Produit de déshydratation et de polymérisation de l'acide β-oxybutyrique. Bull Soc Chim Biol 8:770–782

Loo CY, Lee WH, Tsuge T, Doi Y, Sudesh K (2005) Biosynthesis and characterization of poly (3-hydroxybutyrate-*co*-3-hydroxyhexanoate) from palm oil products in a *Wautersia eutropha* mutant. Biotechnol Lett 27(18):1405–1410

Masood F, Chen P, Yasin T, Hasan F, Ahmad B, Hameed A (2013) Synthesis of poly (3-hydroxybutyrate-*co*-12 mol% 3-hydroxyvalerate) by *Bacillus cereus* FB11: its characterization and application as a drug carrier. J Mater Sci Mater Med 24(8):1927–1937

Matsusaki H, Manji S, Taguchi K, Kato M, Fukui T, Doi Y (1998) Cloning and molecular analysis of the poly(3-hydroxybutyrate) and poly(3-hydroxybutyrate-*co*-3-hydroxyalkanoate) biosynthesis genes in *Pseudomonas* sp. strain 61-3. J Bacteriol 180(24):6459–6467

Matsusaki H, Abe H, Doi Y (2000) Biosynthesis and properties of poly(3-hydroxybutyrate-*co*-3-hydroxyalkanoates) by recombinant strains of *Pseudomonas* sp. 61–3. Biomacromolecules 1 (1):17–22

Mizuno K, Ohta A, Hyakutake M, Ichinomiya Y, Tsuge T (2010) Isolation of polyhydroxyalkanoate-producing bacteria from a polluted soil and characterization of the isolated strain *Bacillus cereus* YB-4. Polym Degrad Stab 95(8):1335–1339

Mizuno S, Katsumata S, Hiroe A, Tsuge T (2014) Biosynthesis and thermal characterization of polyhydroxyalkanoates bearing phenyl and phenylalkyl side groups. Polym Degrad Stab 109:379–384

Mizuno S, Hiroe A, Fukui T, Abe H, Tsuge T (2017) Fractionation and thermal characteristics of biosynthesized polyhydoxyalkanoates bearing aromatic groups as side chains. Polym J 49(7): 557–565

Mizuno S, Enda Y, Saika A, Hiroe A, Tsuge T (2018) Biosynthesis of polyhydroxyalkanoates containing 2-hydroxy-4-methylvalerate and 2-hydroxy-3-phenylpropionate units from a related or unrelated carbon source. J Biosci Bioeng 125(3):295–300

Mukai K, Yamada K, Doi Y (1993) Enzymatic degradation of poly(hydroxyalkanoates) by a marine bacterium. Polym Degrad Stab *41*(1):85–91

Murugan P, Chhajer P, Kosugi A, Arai T, Brigham CJ, Sudesh K (2016) Production of P(3HB-*co*-3HHx) with controlled compositions by recombinant *Cupriavidus necator* Re2058/pCB113 from renewable resources. Clean Soil Air Water 44(9):1234–1241

Ng KS, Wong YM, Tsuge T, Sudesh K (2011) Biosynthesis and characterization of poly (3-hydroxybutyrate-*co*-3-hydroxyvalerate) and poly(3-hydroxybutyrate-*co*-3-hydroxyhexanoate) copolymers using jatropha oil as the main carbon source. Process Biochem 46(8):1572–1578

Pohlmann A, Fricke WF, Reinecke F, Kusian B, Liesegang H, Cramm R, Strittmatter A (2006) Genome sequence of the bioplastic-producing "Knallgas" bacterium *Ralstonia eutropha* H16. Nat Biotechnol 24(10):1257

Reinecke F, Steinbüchel A (2009) *Ralstonia eutropha* strain H16 as model organism for PHA metabolism and for biotechnological production of technically interesting biopolymers. J Mol Microbiol Biotechnol 16(1–2):91–108

Ruddiman WF (2010) Plows, plagues, and petroleum: how humans took control of climate. Princeton University Press, Princeton, p 46

Ryu HW, Hahn SK, Chang YK, Chang HN (1997) Production of poly(3-hydroxybutyrate) by high cell density fed-batch culture of *Alcaligenes eutrophus* with phosphate limitation. Biotechnol Bioeng 55(1):28–32

Saika A, Watanabe Y, Sudesh K, Tsuge T (2014) Biosynthesis of poly(3-hydroxybutyrate-*co*-3-hydroxy-4-methylvalerate) by recombinant *Escherichia coli* expressing leucine metabolism-related enzymes derived from *Clostridium difficile*. J Biosci Bioeng 117(6):670–675

Saika A, Ushimaru K, Mizuno S, Tsuge T (2015) Genome-based analysis and gene dosage studies provide new insight into 3-hydroxy-4-methylvalerate biosynthesis in *Ralstonia eutropha*. J Bacteriol 197(8):1350–1359

Sato S, Nomura CT, Abe H, Doi Y, Tsuge T (2007) Poly[(*R*)-3-hydroxybutyrate] formation in *Escherichia coli* from glucose through an enoyl-CoA hydratase-mediated pathway. J Biosci Bioeng 103(1):38–44

Stapp C (1924) Über die reserveinhaltstoffe und den schleim von *Azotobacter chroococcum Centr.* Bakt Parasitenk 2:61–276

Steinbüchel A (1992) Biodegradable plastics. Curr Opin Biotechnol 3(3):291–297

Sudesh K, Abe H, Doi Y (2000) Synthesis, structure and properties of polyhydroxyalkanoates: biological polyesters. Prog Polym Sci 25(10):1503–1555

Taguchi S, Nakamura H, Kichise T, Tsuge T, Yamato I, Doi Y (2003) Production of polyhydroxyalkanoate (PHA) from renewable carbon sources in recombinant *Ralstonia eutropha* using mutants of original PHA synthase. Biochem Eng J 16(2):107–113

Taguchi K, Taguchi S, Sudesh K, Maehara A, Tsuge T, Doi Y (2005) Metabolic pathways and engineering of polyhydroxyalkanoate biosynthesis. In: Biopolymers online. Wiley-VCH Verlag GmbH & Co. KGaA. https://doi.org/10.1002/3527600035.bpol3a07

Taguchi S, Iwata T, Abe H, Doi Y (2012) Poly(hydroxyalkanoate)s. In: Moeller M, Matyjaszewski K (eds) Polymer science: a comprehensive reference, vol 9, pp 157–182

Tanadchangsaeng N, Kitagawa A, Yamamoto T, Abe H, Tsuge T (2009) Identification, biosynthesis, and characterization of polyhydroxyalkanoate copolymer consisting of 3-hydroxybutyrate and 3-hydroxy-4-methylvalerate. Biomacromolecules 10(10):2866–2874

Tanadchangsaeng N, Tsuge T, Abe H (2010) Comonomer compositional distribution, physical properties, and enzymatic degradability of bacterial poly(3-hydroxybutyrate-*co*-3-hydroxy-4-methylvalerate) copolyesters. Biomacromolecules 11(6):1615–1622

Thellen C, Coyne M, Froio D, Auerbach M, Wirsen C, Ratto JA (2008) A processing, characterization and marine biodegradation study of melt-extruded polyhydroxyalkanoate (PHA) films. J Polym Environ 16(1):1–11

Tomizawa S, Hyakutake M, Saito Y, Agus J, Mizuno K, Abe H, Tsuge T (2011) Molecular weight change of polyhydroxyalkanoate (PHA) caused by the PhaC subunit of PHA synthase from *Bacillus cereus* YB-4 in recombinant *Escherichia coli*. Biomacromolecules 12(7):2660–2666

Tsuge T (2002) Metabolic improvements and use of inexpensive carbon sources in microbial production of polyhydroxyalkanoates. J Biosci Bioeng 94(6):579–584

Tsuge T, Fukui T, Matsusaki H, Taguchi S, Kobayashi G, Ishizaki A, Doi Y (2000) Molecular cloning of two (*R*)-specific enoyl-CoA hydratase genes from *Pseudomonas aeruginosa* and their use for polyhydroxyalkanoate synthesis. FEMS Microbiol Lett 184(2):193–198

Tsuge T, Taguchi K, Taguchi S, Doi Y (2003) Molecular characterization and properties of (*R*)-specific enoyl-CoA hydratases from *Pseudomonas aeruginosa*: metabolic tools for synthesis of polyhydroxyalkanoates via fatty acid ß-oxidation. Int J Biol Macromol 31(4–5):195–205

Tsuge T, Saito Y, Narike M, Muneta K, Normi YM, Kikkawa Y, Doi Y (2004a) Mutation effects of a conserved alanine (Ala510) in type I polyhydroxyalkanoate synthase from *Ralstonia eutropha* on polyester biosynthesis. Macromol Biosci 4(10):963–970

Tsuge T, Saito Y, Kikkawa Y, Hiraishi T, Doi Y (2004b) Biosynthesis and compositional regulation of poly[(3-hydroxybutyrate)-*co*-(3-hydroxyhexanoate)] in recombinant *Ralstonia eutropha* expressing mutated polyhydroxyalkanoate synthase genes. Macromol Biosci 4(3):238–242

Tsuge T, Yano K, Imazu SI, Numata K, Kikkawa Y, Abe H, Doi Y (2005) Biosynthesis of polyhydroxyalkanoate (PHA) copolymer from fructose using wild-type and laboratory evolved PHA synthases. Macromol Biosci 5(2):112–117

Tsuge T, Watanabe S, Shimada D, Abe H, Doi Y, Taguchi S (2007) Combination of N149S and D171G mutations in *Aeromonas caviae* polyhydroxyalkanoate synthase and impact on polyhydroxyalkanoate biosynthesis. FEMS Microbiol Lett 277(2):217–222

Tsuge T, Yamamoto T, Yano K, Abe H, Doi Y, Taguchi S (2009) Evaluating the ability of polyhydroxyalkanoate synthase mutants to produce P(3HB-*co*-3HA) from soybean oil. Macromol Biosci 9(1):71–78

Tsuge T, Hyakutake M, Mizuno K (2015) Class IV polyhydroxyalkanoate (PHA) synthases and PHA-producing *Bacillus*. Appl Microbiol Biotechnol 99(15):6231–6240

Tsuge T (2016) Fundamental factors determining the molecular weight of polyhydroxyalkanoate during biosynthesis. Polym J 48(11):1051–1057

Valdés J, Kutralam-Muniasamy G, Vergara-Porras B, Marsch R, Pérez-Guevara F, López-Cuellar MR (2018) Heterologous expression of *phaC2* gene and poly-3-hydroxyalkanoate production by recombinant *Cupriavidus necator* strains using canola oil as carbon source. New Biotechnol 40:200–206

Verlinden RA, Hill DJ, Kenward MA, Williams CD, Piotrowska-Seget Z, Radecka IK (2011) Production of polyhydroxyalkanoates from waste frying oil by *Cupriavidus necator*. AMB Express 1(1):11

Volova TG, Boyandin AN, Vasil'ev AD, Karpov VA, Kozhevnikov IV, Prudnikova SV, Gitel'Zon II (2011) Biodegradation of polyhydroxyalkanoates (PHAs) in the South China Sea and identification of PHA-degrading bacteria. Microbiology 80(2):252

Wallen LL, Rohwedder WK (1974) Poly-beta-hydroxyalkanoate from activated sludge. Environ Sci Technol 8(6):576–579

Wang Q, Yu H, Xia Y, Kang Z, Qi Q (2009) Complete PHB mobilization in *Escherichia coli* enhances the stress tolerance: a potential biotechnological application. Microb Cell Factories 8 (1):47

Xia Y, Larock RC (2010) Vegetable oil-based polymeric materials: synthesis, properties, and applications. Green Chem 12(11):1893–1909

Advanced Technologies for Ecological Reconstruction and Bioremediation of Degraded Land

4

Marian Butu, Ioan Sarac, Mihaela Corneanu, and Monica Butnariu

Abstract

The need of knowledge on bioremediation and ecological reconstruction technologies and their application in environmental degradation consists in achieving of the objectives regarding the remediation of soils and waters contaminated by anthropogenic activities, especially industrial ones. The general goal is sustainable use and development of natural resources and of the patrimony, represented by the soil quality, in the evolution of global environmental conditions. Heavy metals (HMs) contamination from mining activities is considered to be a serious environmental problem for many regions around the world. The property of HMs to accumulate in the vegetal and animal organisms, including the human, as well as the pathology they determine, justifies the interest that these pollutants arise. Considering the acute toxicity of these contaminants, there is an urgent need to develop sustainable and cost-effective methods for the accumulation, dissipation, immobilization, and degradation of plant pollutants. The efficiency of biological depollution activity and the success of bioremediation technologies result from the conservation of microbiota and phytocoenosis biodiversity, from the ecological reconstruction of anthropically degraded/polluted areas and from the bioremediation of the soil as an integrated ecologic biotechnology system.

Keywords

Soil · Microorganisms · Contamination · Heavy metals · Bioremediation · Ecological reconstruction · Degraded land

M. Butu
National Institute of Research and Development for Biological Sciences, Splaiul Independentei, Bucharest, Romania

I. Sarac · M. Corneanu · M. Butnariu (✉)
Banat's University of Agricultural Sciences and Veterinary Medicine "King Michael I of Romania" from Timisoara, Timis, Romania

R. Prasad (ed.), *Environmental Pollution and Remediation*, Environmental and Microbial Biotechnology, https://doi.org/10.1007/978-981-15-5499-5_4

4.1 Introduction

In the soil, microorganisms (MOs) establish the balance between organic and inorganic substances, between living and nonliving matter. The formation of CO_2 under the action of biodegradation–mineralization of MOs is the basic stage of the biogeochemical cycle of carbon. The carbon cycle is carried out simultaneously and in combination with the oxygen and hydrogen cycle: by photosynthesis of green plants from CO_2 and H_2O organic compounds containing H which are formed and O_2 is released; during the total decomposition of organic substances (OSs), C and H are released by the participation of O in the form of CO_2 and H_2O, respectively. The cycles of the other elements (N, P, S, etc.) are also related to the C cycle.

All of the microbial species are omnivorous capable of degradation of all OSs of biological origin. Several synthetic, xenobiotic organic compounds (nonexistent compounds, human creations) are not biodegradable. Such compounds are pesticides, detergents, medicines, and some plastics. They pose a great danger to the environment by long-term pollution (Behera and Prasad 2020a).

Soil MOs play a role in soil fertility, transforming OSs into inorganic substances. MOs also play a role in the breakdown of rocks and the formation of sulfates and nitrates. Some of them fix atmospheric nitrogen, and others produce the necessary CO_2 in the primary production of higher plants. MOs are widespread in nature everywhere, with greater plasticity and adaptability than any other living organism. The activity of MOs results in the degradation of the substrate through their vital activity, inducing a series of biochemical transformations of their living environment (Behera and Prasad 2020b).

The term degradation of the substrate is used since MOs use complex substances in the substrate to make them simpler. All activity of MOs is based on biochemical transformations, under favorable environmental conditions (temperature, pH, presence or absence of oxygen, and substrate), determined by the catalysts represented by enzymes secreted by MOs.

All reactions to the transformation of substances through MOs are enzymatic reactions that occur within their cell or in the environment under the action of enzymes secreted on the substrate. For human activity, this natural activity of MOs is important. It is represented by the decomposition of organic matter (OM) from the soil, the rotting of natural products, etc., as well as their activity in some polluted, anthropically degraded areas, artificial arrangements, and industrial premises.

In nature, MOs live in complex biocenoses along with other types of living creatures in complex interrelations whose knowledge is important for the proper control and use of these processes in practical applications.

4.2 Natural Bioresources Involved in Bioremediation

4.2.1 Enzymatic Soil Activity

Enzymatic soil activity is a soil quality parameter and is the result of the activity of accumulated enzymes and enzymes from proliferating MOs. The sources of the enzymes accumulated in the soil are MOs cells and organic residues (vegetal and animal). Enzymes accumulate in the soil as ***free enzymes*** (exoenzymes released from living cells and endoenzymes released from disintegrated cells) and ***enzymes linked*** to cellular constituents, i.e., enzymes present in disintegrating cells (in cellular fragments) in living but nonproliferating cells.

Enzymes of proliferating MOs are both those that are released from living cells undergoing multiplication as well as those found within cells undergoing multiplication. Free enzymes are adsorbed by organic and mineral particles, soil, or/and complexed with humic substances (HSs).

The amount of free enzymes is much lower in the soil solution than in the adsorbed and/or complexed state. Enzymatic activities (EAs) in the [nmol/min/mg] soil plays a role in decomposition and mineralization processes.

The main method by which this problem is studied is based on the comparison of the transformation of the enzyme substrate in the soil samples containing proliferating MOs with that of samples of the same soil, where the proliferation of the MOs was stopped. Soil is an enzymatic system where the accumulated enzymes and enzymes of the proliferating MOs have biological significance, participating in the biological cycles of the elements and contributing to the fertility of the soil, creating favorable conditions for the nutrition of the superior plants, and, implicitly, perpetuating life on our planet (Wang et al. 2013).

4.2.2 The Importance of Soil Microorganisms for Biological Depollution

A dynamic soil quality parameter is the activity of biodegradation and mineralization of organic soil substances. This includes carbohydrate biodegradation (decomposition of cellulose, glucose, pectic substances, starch, chitin, lignin, etc.) and biodegradation of aliphatic hydrocarbons (methane, ethane, propane, etc.), and aromatic (benzene, toluene, etc.). Processes such as lipid decomposition, pesticide biodegradation and decomposition of proteins from plant debris, animals and MOs, and organic nitrogen compounds occur in the presence of soil macroorganisms. The microbial population of the soil is made up of representatives of all MOs: bacteria, actinomycetes, fungi (micromycetes, yeasts), microscopic algae and protozoa, and soil viruses, also. Bacteria are the most important group of soil MOs. The intensity of processes by which MOs participate in the biological cycles of the elements is higher in bacteria; the decomposition of chitin (a nitrogen-containing polysaccharide composed of fungi or insect exoskeleton), where the action of actinomycetes is more intense; as well as the decomposition of lignin and cellulose in which the micromycetes predominate. Bacteria in soil are of two types: autochthonous and zymogenous (Zeng et al. 2018).

Autochthonous or indigenous bacteria are characterized by the fact that their number is not subject to large fluctuations and that they use as feed the OSs that belong to the soil (HSs). Zymogenous bacteria produce fermentations (ferment the cellulose, pectic substances, starch, etc.) and their number increases after the addition of OS to the ground and decreases as these substances deplete. A perfect delimitation between autochthonous and zymogenous bacteria is not possible. Bacteria develop on the surface of the soil particles and not in the soil solution but especially those with a diameter of at least 100–200 μm (Sage et al. 2014).

Although the representatives of most bacterial genus are present in the soil, only some specific bacteria are important to soil quality. These specific bacteria are present in the soil in a large number and contribute to the biological activity of soil to a high degree. Soil is the friendliest environment for MOs growth, being the natural reservoir of a huge number of MOs. The diversity of soil MOs is very high. In the soil, MOs have favorable conditions for development and propagation, processing OM by using different inorganic substances. Soil conditions vary depending on the type of soil, season, and geographical region.

4.2.3 Nature of Microorganisms in the Soil

Soil microbiology varies quantitatively and qualitatively depending on soil type, pH, vegetation, climatic conditions, etc. The soil microflora is made up of native MOs, permanently present and alohtones (the exotic), whose presence depends on the intake of mineral and OSs from the outside and which survive to the decomposition of these substances. The native microbiota is specific to each type of soil, being adapted to the metabolism of certain types of relatively stable substances in the soil type. It is involved in all microbiological processes that take place in the soil. The alohthonic microbiota is introduced by humans through fertilization or waste water spills. It occurs in some soil processes. MOs in manure participate in the mineralization of OM. Their catabolism substances are subsequently metabolized by different groups of native MOs. The majority, however, is microbiological soil pollution (Horemans et al. 2018).

4.2.4 Types of Microorganisms From Soil

The composition of the microbiota is influenced by the season and the degree of soil processing.

The Bacteria Predominant are bacteria in the form of microcolonies adsorbed on soil particles in all neutral and alkaline soil types. Some produce polymers of a mucilaginous nature that cause aggregation of soil particles so that the soil becomes more granular, which increases aeration and fertility. The most common genera are

Bacillus, Arthrobacter, Agrobacterium, Pseudomonas, Flavobacterium, and *Alcaligenes.*

Bacterial activity in the soil determines the fertility, soil properties, and evolution, interfering with ammo-nitrification processes, decomposing resistant nitrate compounds (e.g., chitin), forming HSs, and balancing the microbiota of this medium by secretion of antibiotics. In the soil substrate, there are groups of specialized bacteria that develop on specific substrates: cellulosic bacteria degrade cellulose from the remaining vegetal matter left in the soil, and nitrifying bacteria produce NH_3 oxidation resulting from metabolism processes, with the formation of nitrate (NO_3^-) and nitrite (NO_2^-), form by which plants take nitrogen from the soil. Sulfurous ferruginous bacteria play a role in the S and Fe circuit in nature and putrefaction bacteria-degrade organic N and cause rotting corpses. Actinomycetes (filamentous bacteria) represent approx. 30% of the total soil microflora. Together with bacteria, they determine the mineralization of OSs and take part in the formation of humus, which is the reservoir of nutrients for plants. Soil closure is due to actinomycetes. This type of soil retains better heat and has superior agrotechnical qualities. They occur in all soil types and are more numerous in soils rich in OSs. It develops mainly in alkaline soils. They are the main ammonium and chitinolysis agents, having a role in the formation of humus. Some species, producing antibiotics, also play a role in maintaining balance in the soil microflora. Excessive humidity and acidity limit their development (Okie et al. 2015).

The most common genera are *Actinomyces, Actinoplanes, Mycobacterium, Mycococus, Micromonospora, Nocardia,* and *Streptomyces.* Pathogenic sporulated bacteria can be found in the soil: *Bacillus anthracis* (produce the anthrax of the sheep), *Clostridium tetani* (produces tetanus), *Clostridium botulinum, Clostridium hystoliticum,* and *Clostridium sporogenes* (toxicogenic bacteria). They accidentally land in the soil by manure or dead bodies: *Salmonella typhi, Shigella dysenteriae, Vibrio cholerae,* and *Pseudomonas aeruginosa.* The durability of these bacteria in the soil varies according to species, soil type, and climatic conditions. Pathogenic bacteria do not find favorable multiplication conditions in the soil due to the low temperature relative to their thermal optimum (37 °C) and inappropriate nutritional environment and due to the competition of other bacteria, actinomycetes, and protozoan predator (*Acanthamoeba castellanii*). Cyanobacteria form a large group of photosynthetic MOs from the superficial soil layers. Dominant are the *Nostoc* and *Oscillatoria* genera. They are important because they are the only living organisms capable of using both carbon in CO_2 and molecular nitrogen in the synthesis of OSs. The molecular nitrogen fixative species are *Anabaena, Calothrix, Chroococcus, Nostoc, Plectonema,* and *Schizothrix* (Maier et al. 2018).

The viruses are especially present in the form of bacteriophages and actinophages, living in a dynamic balance with the bacterial and actinomycete species they parasite, having a weak lytic action on them. Pathogenic viruses cannot be propagated through the soil. ***Yeasts and microfungi*** (molds) are represented in the superficial soil layers that are better aerated, wet, and rich in OSs. The most common genera are *Candida, Cryptococcus, Debaryomyces, Hansenula, Lipomyces,* and

Pichia. ***The molds*** predominate in acidic soils (forest soils, podzolic soils), where they produce a biomass of 500–1000 Kg/ha. The main genera are *Mucor, Rhizopus, Aspergillus,* and *Penicillium*, which form 50% of the fungal biomass accompanied by *Fusarium, Alternaria, Trichoderma, Trichothecium,* and *Cladosporium*. ***The autotrophic algae***, encountered in superficial layers of soil, synthesize OM. ***The protozoa***, mainly represented by flagellates and rhizopodes, are conditioned by their soil moisture and the presence of OSs. They interfere in limiting the multiplication of the MOs they use as food. ***Number and density of soil microorganisms.*** In soil the MOs are in concentrations of $9 \cdot 10^7$–$3 \cdot 10^{11}$ cells/g dry soil. Their number is dependent on the type of soil (about 10^8 cells/gram in clay soils, a poorer microbe in acidic, alkaline, and sandy soils). These MOs form a total biomass of approx. 0.5–6 t/ha. Of the total of these MOs, approx. 30% are Gram-negative bacteria. The bacteria number oscillates between $3.6 \cdot 10^7$ and 10^{11} cells/g soil, totaling a biomass of up to 0.7 t/ha. Dominant are filamentous bacteria. Yeasts reach 10^3 cells/g soil, growing on the plant debris. Fungi form a biomass 1–1.5 t/ha. The huge number is due to the fact that there is always a sufficient amount of organic and mineral substances in the soil, satisfactory humidity, sufficient oxygen, and a suitable pH. The largest number of MOs is found in the upper layers of the soil. With the exception of superficial layers, MOs in the soil are protected from the harmful effects of ultraviolet rays (Somova et al. 2005).

4.2.5 The Distribution of Microorganisms in the Soil

The bacteria in the soil is in the form of associations of 5–25 cells, aggregated by the soil particles. Spreading in the soil is dependent on the degree of oxygen access. There are several aerobic bacteria on the surface of the soil, then some anaerobic bacteria, and anaerobic bacteria in depth. Yeasts and microfungi are only found in the superficial layers of the soil, being aerobic MOs. The vertical distribution of the microbiota in the soil is unequal. The richest is the area up to 30 cm depth. The cell concentration differs according to the nutrients the soil contains.

A higher concentration is found in the area of the rhizosphere (the plant root area). The horizontal distribution of the native microbiota is dependent on the type of soil that determines the selection of different physiological types adapted to the metabolism of certain types of substances present in the soil composition (Dutta and Podile 2010).

4.2.6 Factors that Determine the Distribution of Microorganisms in the Soil

The type of MOs in the soil, as well as their number, varies depending on the physicochemical structure of the soil but especially on the organic and inorganic substances in the soil, the soil pH, and the amount of oxygen. The type of soil is thus the main factor for their selection.

Humidity allows the development of MOs at the soil surface, and in more arid soils, the MOs are more abundant in deep layers. There is a seasonal variation of soil MOs due to moisture, organic matter, and temperature concentrations.

Maximum density occurs in spring and minimum in winter. A secondary peak occurs in autumn. Summer dryness reduces the number of MOs in the soil but contributes favorably to the transformation of OSs that will be metabolized more easily in the autumn. Due to their large variety, soil MOs develop under the conditions of strong competition for the available food supply.

Each species must bear decontamination products of other species that may be toxic to them. Due to the very complex interrelationships between the soil MOs, there is a self-control of the number and diversity.

Agrotechnical works and the administration of organic fertilizers also shape the soil microbiology. Plowing improves soil aeration favoring the development of aerobic MOs and allowing them to advance to deeper layers and more evenly distributes organic residues. Organic fertilizers are introduced into an additional nutrient substrate, favoring the numerical explosion of organotrophic MOs which, through the mineralization process, lead to soil fertilization.

These fertilizers also introduce an additional inoculum of allotone MOs that temporarily participate in the mineralization of OSs. Thus, 25% of the manure mass is represented by MOs (Lax et al. 2019).

In conclusion, MOs in soil are of particular importance reflected in the following:

The Mineralization Processes Plants, through the processes of exosmosis, eliminate amino acids and sugars, which favor the development of the MOs for which they are the source of food. When plants die, their roots are decomposed by MOs, and the base substances, such as C, N, P, and S, are fed into the circuit serving as plant nutrients.

Formation of Humus In connection with this decomposition process occurs the formation of HSs in the soil, which also functions as a reservoir for the future activity of the MOs and also as a material for structuring the soil particles. Therefore, mineral nutrition of plants and soil fertility are closely related to the activity of MOs. The soil microbiota determines the biochemical dynamics of the soil (Torsvik et al. 2002).

4.3 Monitoring Soil Pollution

Soils and sediments are solid components of aquatic or terrestrial ecosystems that can serve as sources and receptors for nutrients and toxic chemicals. The use of soil for industrial, agricultural, and urban activities always involves a drastic change in its composition and eventually can cause great problems for future use. Soil monitoring involves site selection and pursues a number of objectives, including evaluating and documenting the evolution or trend of pollution to allow for the problem to be corrected before the irreversible loss of soil quality.

Soil monitoring programs are very common, targeted to inspect soil characteristics and soil fertility. Although, in many cases, monitoring programs are planned to be carried out on an ongoing basis, with a frequency of 5 or 10 years, funds are often insufficient to achieve this goal. In general, sampling monitoring strategies were based on a national grid system and/or some stratification types based on land use and soil type, or the combination of the two.

Sample handling strategies and systems vary across networks, thus preventing comparability of results. The impact of the environment on terrestrial ecosystems includes changes in their physical and chemical properties. The vulnerability of soils to pollutants depends on soil characteristics, hydrogeological situation, land use, pollutant type, and pollutant load.

The most important soil parameters that determine the mobility of pollutants and, consequently, soil vulnerability are pH, redox potential, cation exchange capacity, and soil humus content.

Key soil properties can serve as indicators of soil quality. In order to have practical use, soil indicators must be sensitive to management and able to respond to change in a relatively short time and be accessible so that measurement methodologies or data are readily available; if they are not directly measurable, it must be possible to define them using pedotransfer functions or models (Weissmannová and Pavlovský 2017).

4.3.1 Soil Characteristics

Soils are of interest as environment for the transport of fluids because they are porous and permeable. Porosity refers to the spaces of the pores forming voids between the soil grains. Soil permeability is determined by factors and parameters that are associated with the physical and chemical properties of the soil, the permeant, and the physicochemical interactions occurring during permeation. Soil permeability provides information to determine and evaluate contaminant transport across the soil. The soil permeability to the movement of a certain fluid also depends on the fraction of the pore space filled by that fluid.

If the pore space is saturated, the permeability of the environment is at its maximum. If the fluid fills only a portion of the pore space, for example, only in the upper soil layers, soil permeability decreases and moisture becomes important. Humidity is characteristic of certain fluids to adhere to soil grains and significantly influences mobility and fluid retention in the medium. Near the surface, the pores of most soils are filled with both water and air. This is the unsaturated soil area. Due to the fact that the water tends to moisten most of the soils, however, a quantity of water is retained. The amount of water typically remaining, measured on the basis of the volume of water per total soil volume, is named soil field capacity. Often, under this area, there is an aquifer in which the pore spaces are saturated or filled with water.

4.3.2 Soil Components

In general, soils are characterized by four components: inorganic substance, OM, soil water, and soil atmosphere.

4.3.2.1 Inorganic Soil Substance

Ionic solids – ion of the metals: Ag, Al, Ba, Ca, Cd, Co, Cs, Cu, Fe, Hg, K, Li, Mg, Mn, Na, Ni, Pb, Rb, Sr, Ti, Zn, and Zr. ***Trace elements*** – is one of the most important aspects of the variability of soil mineral composition. The most important trace elements are As, B, Cd, Co, Cr, Cu, Mo, Ni, Pb, Se, Ti, V, and Zn, which are found in the primary and secondary part of the soil. ***Primary silicates*** – occur in soils by the physical degradation of parental rock material. Chemical degradation of primary silicates contributes to native fertility and soil electrolyte content. ***Clay minerals*** – are the aluminosilicates that predominate in the clay fractions at the intermediation of the advanced degradation phases. ***Oxides and hydroxides*** – because of their high abundance in the lithosphere and their low solubility at pH values in the normal pH range, Al, Fe, and Mn form the most important oxides, oxide-hydroxides, hydroxides and oxyhydroxides in soils. ***Carbonates and sulfates*** – of these minerals, gypsum and calcite can be dissolved and reprecipitated in a soil profile traversed by rainwater and irrigation water and can be found as a coating on soil minerals, including calcite. Sulfates and sodium carbonates are formed at the top of the soil profile, drying by evaporation (Konstantinova et al. 2019).

4.3.2.2 The Organic Substance From Soil: Biomolecules

Soils are biological media in which there is a variety of MOs. Ten grams of fertile soil can contain a population of bacteria, equal to the human population of the globe. One kilogram of soil can contain 500 billion bacteria, 10 billion actinomycetes, and almost one trillion of fungi. The contribution of plant roots can be added to this microbial biomass. Soil MOs play a role in catalyzing oxidation–reduction reactions. Exudates released by MOs and the roots of healthy plants are important for soil acidity and contribute to the cycle of trace elements in the soil. ***Organic substances***: The common ***aliphatic organic acids*** are CH_2O_2, $C_2H_2O_4$, $C_2H_4O_2$, $C_4H_6O_6$, and $C_6H_8O_7$. The ***amino acids*** from soil, such as Gly, Ala, Asp, and Glu, can react to form peptides, which in turn can turn into proteins (polymers). Another important class of biopolymers in soils is the carbohydrate class. Phenols, with their soils by polymerization, form lignin which, together with cellulose, is an important precursor of HSs in the soil. ***The humus*** is the total of OSs in the soil, except for materials identifiable as partially modified or unmodified biomass (parts of plants and MOs). Humus plays a role in aggregate formation, soil acidity control, nutrient cycles, and detoxification of hazardous compounds. Biochemical processes involved in the formation of humus are decomposition of biomass components (including lignin) into simple organic compounds; the microbial metabolism of simple compounds; cycle C, H, N, and O between organic soil and microbial biomass; and microbial mediated polymerization of the organic compounds in the cycle. The organic material contains the precipitate (humic acid) which forms after the soil is mixed with NaOH (500 mol/m^3), brought to pH=1 with HCl concentrated, and soluble organic material (fulvic acid).

The organic substance plays a role in the buffering of the metal protons and cations in the soil solution. The buffering phenomenon is performed by cation exchange. The cationic exchange capacity of soil humus is the maximum number of moles of dissociated protons in the humus mass unit under the given conditions of temperature, pressure, soil composition, and humus concentration (Collado et al. 2018).

4.3.2.3 Soil Water

Soil water is considered to be water that is discharged from the soil sample by maintaining it at 105 °C at least 24 h. Water that is bonded to the soil mineral structure (crystallization water) is not released under these conditions and is not considered soil water. Also, a distinction is made between soil water and deep water, the second being groundwater. Soil water is mainly found as a condensed phase in the soil, although the water vapor content in the soil can be 30 mL/L in a damp soil. Soil water is a landfill for dissolved gases and solids and is therefore the soil solution. Dissolved solids that dissociate into ions (electrolytes) in the soil solution are the most important for soil chemistry. Chemical elements whose ions are in uncontaminated soil solutions below 1 mmol/m^3 are called microelements and all others are macroelements (Wiatrowska and Komisarek 2019).

4.3.2.4 Soil Atmosphere

The air that exists in the soil is constantly changing with the atmosphere above it. The composition of the soil air is of the same type as atmospheric air (78% N_2, 21% O_2, 1% Ar, 0,03% CO_2), but due to soil biological activity coupled with low exchange rate, the percentage composition of soil air may differ considerably from that of atmospheric air. For example, with regard to atmospheric air, the oxygen content in the soil air is lower and that of CO_2 is higher. Under anaerobic conditions (low oxygen content), the activity of MOs can produce significant amounts of NO, N_2O, NH_3, CH_4, and H_2S. Dissolving gas from soil air into the soil solution is an important process contributing to the cycle of chemical elements in the soil environment (Niklaus et al. 2016).

4.4 Soil Degradation

Soil degradation has high priority recently admitted by the European Environment Agency. The physical characteristics of the soil can be affected by different soil management practices: irrigation, sewerage, and sanitation, leading to degradation. The soil degrades through erosion, desertification (desert emergence), and salinization.

4.4.1 The Erosion

When biomass is removed from many plants, soil erosion is exacerbated. As a consequence of the increased soil erosion, there is not enough organic substance to penetrate the soil to maintain a sustained crop production, and the vegetation begins to degrade. The expression "desertification" has not been used critically and by definition. The debate to determine the causes of desertification, namely, the effects of demographic growth or meteorological changes, is old. There are no data records of sufficient duration to distinguish between temporary fluctuations, climate change or not, and long-term climate change. Destruction of vegetation can affect climatic conditions.

On a global scale, as a result of the greenhouse effect, climate warming began. Salinization is a major cause of the destruction of soil fertility and is reflected in a sharp deterioration of nutrient reserves. Soil salinization has great economic and social repercussions and has a severe impact on the living standards of populations in arid areas; the causes of salinization are both natural and anthropogenic. The most important prerequisites for the appearance of salinated soils are groundwater mineralization close to the surface. By evaporating the capillary moisture of the upper layers of the soil, it becomes successively more salinized. Anthropogenic contributing factors to salinization are the different irrigation practices. Anthropogenic salinization of soils destroying their fertility is often the result of poor use of land and water resources (Egamberdieva et al. 2008).

4.4.2 Compaction of Soil

Compaction (tamping) of the soil is a process that has consequences for its use. Soil pores are compressed by street traffic by moving heavy machinery onto cultivated soil. By compaction of the soil, the supply of oxygen and water is reduced and the reduction reactions of NO_3^-, SO_4^{2-}, CO_2, H^+, Mn(IV), Mn(III), and Fe(III).

Normally, manganese exists in the soil as MnO_2. Under reducing conditions, it converts to Mn^{2+}, a process described by the equation:

$$MnO_2 + 4H^+ + e^- \rightarrow Mn^{2+} + 2H_2O.$$

Manganese is one of the essentials of life, but in high concentrations manganese is toxic. The Fe^{3+} compounds can be converted to Fe^{2+} at customary pH values, according to the reaction of the equation:

$$Fe(OH)_3 + 3H^+ + e^- \rightarrow Fe^{2+} + 3H_2O.$$

When the amount of available Fe increases considerably, under anaerobic conditions, Fe toxicity can be manifested, a phenomenon that is frequently encountered in oysters. Reduction processes demonstrate that soil fertility decreases under anaerobic conditions (Velasco Ayuso et al. 2017).

4.4.3 Acidification of Soil

Natural acidification of the soil can take place during vegetation growth when plants and trees in the metabolic process form acid compounds to compensate for nutrient consumption. During decomposition, hydrogen ions are used and pH values begin to rise again. Artificial soil acidification can occur as a consequence of modern forestry and farming practices based on harvest withdrawal and immediate replanting. In such situations, there is not enough time to restore the soil in terms of pH neutralization and a gradual acidification of the soil occurs. Every year, soils receive large amounts of waste.

Many SOx and NOx emissions come into the soil as H_2SO_4 and nitric acid, respectively. The gradual increase of nitrogen, in the form of nitric acid, can have fertilizing effects, stimulating growth of vegetation including trees and shrubs. However, nitrogen overload may lead to nitrogen saturation of the soil, when nitrates can begin to percolate in deep and surface waters.

Another effect of acid deposition and subsequent increased nutrient loading, organic material, and trace elements is accelerated, eutrophication of watercourses, and lakes may occur.

Acidification of soils leads to changes in its chemical properties. A typical feature is the loss of cationic nutrients, such as potassium, calcium, and magnesium, when a significant decrease in soil pH can be observed. During acidification, there is also an increased risk for the mobilization of humus and heavy metals (HMs). At low pH, the release rate of Cd and manganese increases considerably, causing their migration to deep water, their presence constituting a potential hazard to the aquifer, especially for the supply of drinking water.

Large amounts of toxic Al are not released equally in all soils in the acidification process because not all soils contain the same amount of Al-containing minerals. When the soils are very acidic, the Al is in very high concentrations because of the adsorption balance between Al^{3+} species and the negatively charged root surfaces. The pH values have other consequences for the soil. Low pH values inhibit the growth of MOs, fungus being among the injurious organisms of the soil; they are associated with the roots of plants and support them in the process of mineral extraction. Destruction of MOs decreases the soil's ability to breathe (Wang et al. 2015).

4.4.4 Heavy Metals in the Soil

Possible soil pollution with HMs can be caused by infiltration of highly contaminated water. Investigations made for several decades for a surface infiltration system on a field study that received rainwater from high traffic streets showed a significant increase in Cu, Zn, Cd, and Pb concentrations in soil layers.

The excessive content of toxic elements in the environment is associated with the etiology of a large number of diseases. HMs and their compounds are the main problem for the characterization of soil pollution from an environmental point of view.

The first approach to assessing pollution is to determine the total concentration of metals; the assessment of their bioavailability is necessary to have proper knowledge of risk assessment. Soil parameters that influence the mobility of HMs are the redox potential, pH, and cation exchange capacity.

These parameters determine the solubility and the adsorption capacity of soils. Soil is characterized by the concentration in HMs (which depends on the type of soil and its composition) and soil contamination with these metals provided by human activity. Pb has a pronounced tendency to accumulate in soil due to minimal mobility even at low pH values.

In the phosphate-containing soil, Pb forms hardly soluble lead phosphate deposits ($Pb_3(PO_4)_2$, $Pb_4O(PO_4)_2$, and $Pb_5(PO_4)_3OH$). In carbonate-containing soils, lead carbonate is formed ($PbCO_3$). Under lead conditions, lead sulfide (PbS) is formed. High levels of Pb pollution are reported in the vicinity of industrial areas and waste incinerators, where dust particles are removed.

Because plants are more resistant to Pb than humans, it is important to avoid contamination of food from areas that are too polluted with Pb. Concentrations of Cd depend on the geological origin of soil material, its texture, the intensity of degradation processes, OM, and other factors. The accumulation of Cd in the soil is partly due to industrial activities and can be explained by the composition and geological origin of the rock material. Cd enters the soil in smaller amounts than Pb and can reach the soil through the air. It can come from incineration gases and phosphate-based fertilizers. In acidic soils with pH <6, Cd is very mobile and does not accumulate. Cd compounds with humic acids are somewhat stable. In a reducing medium and in the presence of the sulfate ion, CdS is formed. Cu is more mobile than Cd and its solubility increases to pH <5. Although Cu counts among the trace elements necessary for life, in the case of plants, 20 mg/kg of dry material produces toxic effects. Cu ions are toxic to MOs at concentrations of about 0.1 mg/L. Zn is one of the most mobile metals in the soil. The solubility of Zn in soil increases especially at pH <6. At higher pH and in the presence of phosphates, Zn allotted to plants can be significantly reduced. The pH-dependent process of adsorption on clay and various oxides is the most significant regulating process for the availability of Zn in soils. Concerning the environmental risk and the trophic chain caused by Zn remains an open question because Zn deficiency in diet can be a problem.

Furthermore, it seems that Zn has a role in controlling Cd fixation in soils contaminated with Cd when the Zn:Cd ratio is greater than 100. It has been deduced that neural networks can be considered as a tool for predicting spatial analysis of processes controlling the transfer metals with the soil–plant system. Neural calculation can support decision-making processes at different levels, to improve harvest management based on monitoring data and to assess the risk of soil metal transfer to plants (Pratush et al. 2018).

4.4.5 The Bioremediation

The definition of bioremediation is understood by the general definition as the use of living organisms (MOs, plants, etc.) to ameliorate and restore the ecological status of a polluted or degraded substrate (area, land, aquifer, etc.) good, favorable to life, undisturbed, and unpolluted, or to return and to restore it to the former state. Bioremediation is a modern pollutant treatment technology that uses biological factors (MOs) to convert certain chemicals into less harmful/dangerous endpoints, ideally CO_2 and H_2O, which are nontoxic and released into the environment without altering substantially the balance of ecosystems.

Bioremediation is based on the ability of some chemical compounds to be biodegraded. The concept of biodegradation is accepted as a summing-up of processes of decomposition of natural or synthetic constituents by activating strains of specialized MOs resulting in end products useful or acceptable from the point of view of the environmental impact. Example of bioremediation is biological treatment of wastewater by using living organisms to restore it to the original situation. In the last decades, the term "bioremediation" is used in a more specific way, reflected by the two specific definitions: the use of living organisms to degrade environmental pollutants to prevent pollution or waste treatment and application of biological treatments for the cleaning, decontamination, and degradation of dangerous substances (Lebeau et al. 2008; Prasad and Aranda 2018).

4.4.5.1 Advantages of Bioremediation

Biotechnologically addressing pollution problems through bioremediation methods has the advantage of requiring moderate investment capital, does not generate waste, and is environmentally safe and self-supporting. Biotechnological processes for toxic effluent treatment compete with existing methods in terms of efficiency and economic efficiency. Biotechnological methods for toxic waste treatment are designed to replace current methods of storage and detoxification of new xenobiotic compounds (man-made). It is important, however, to limit the generation of hazardous and nonhazardous waste as well as to use recycling methods.

Other advantages of bioremediation toward other technologies are as follows: permanent disposal of waste (limitation of compliance problems), eliminates transportation and compliance costs, minimal site disturbance, can be combined with other technologies, and can be performed on the site, positive from the point of view of public acceptance. Bioremediation provides, in many cases, a permanent solution to the problem and is cost-effective. Bioremediation helps reduce costs for treatment by treating contamination on the spot, taking advantage of natural processes, and reducing environmental disturbance. Bioremediation is generally more cost-effective than the two widely used methods, namely, burial or incineration (Asgari Lajayer et al. 2019).

4.5 Technologies Involved in Bioremediation

4.5.1 Biorestoration

A number of specific terms are used to describe the activity of MOs and the ways in which they are used in bioremediation: Biodegradation is the breakage or fragmentation of a compound or substance produced by living organisms, bacteria, or fungi that may be indigenous to that area or can be entered. Biostimulation is the method by which the populations of MOs, whether natural or introduced, are enhanced by nutritional, engineering, or other area preparation work. This increases the speed of natural remediation processes. Bioaugmentation is the method by which specific living organisms are added to a site or material to achieve a desired bioremediation effect. Biorestoration is the restoration of the original state or of a state close to the original state by the use of living MOs.

Bioremediation technologies can be applied locally ("in situ") or "ex situ" (by transporting the polluted substrate to special treatment facilities). The technology used to treat a polluted site is site-dependent and pollutant-dependent. Bioremediation can be done through biological extraction of HMs by plant and MOs and biodegradation of pollutants from soil contaminated by MOs and plants. Knowledge of microbial communities is constantly developing in the process of developing advanced bioremediation technologies. MOs play a role in the circuit of chemical elements and nutrients as matter and energy globally through their great EAs over substances. These EAs have been recognized as a way of biodegradation and removal of environmentally harmful anthropogenic residues.

Research in this field addresses the deeper knowledge of the role and ecology of plants useful for depollution. MOs have developed a variety of biochemical pathways to degrade or detoxify soils. Hydrolase and oxygenase are the most important classes of enzymes, which are responsible for catalyzing biotransformation reactions. Changes in microbial communities during bioremediation can be determined using molecular and genetic microbiology methods in researching the various roles of MOs. Soil MOs activity is influenced by the presence of microhabitats, soil type, structure and texture, assurance of OM and nutrients, and environmental factors (Liu et al. 2018a, b).

4.5.2 Bioremediation with Microbial Communities

Transformation of metals under the action of MOs is a key link of metal cycles in the biosphere. Metal transformations are dominated by oxidoreduction reactions, the complexation of organic and inorganic compounds, and the change between water soluble and water-insoluble forms. When a MO oxidizes or reduces a metal, it precipitates or becomes soluble. For example: Cr^{6+} is reduced to Cr^{3+}, which precipitates as oxides, sulfides, or chromates of phosphate. Bacteria living in high metallic environments have specific physiological mechanisms that allow them to survive under these unsuitable conditions: extracellular precipitation, binding of metal ions and their removal to the surface of the cell, and intracellular sequestration in the form of intracytoplasmic inclusions. Binding of cations to the surface of the cell has become one of the most attractive biotransformation models. Metals possessing an electron configuration containing 10–12 layers of electrons are often toxic to organisms at relatively low concentrations. In this group are Hg^{2+}, Ag^{+}, Pb^{2+}, Cd^{2+}, and Zn^{2+}. Metal/MO interactions play a role in numerous biotechnologies, such as bioremediation, biomineralization, bioleaching, and microbial corrosion.

It is intended to use bacterial strains in pure cultures or in consortia capable of mobilizing/immobilizing metal ions. Remediation of metals often involves five approaches: isolation, immobilization, mobilization, physical separation, and extraction. Immobilization and mobilization involve bacterial bioremediation processes. Immobilization is a technique for reducing the mobility of the contaminant by altering its physicochemical properties.

MOs are used, and the process can be carried out in situ or ex situ. MOs can mobilize metals by autotrophic or heterotrophic leaching by chelation by metabolites and siderophores by methylation and redox transformations (Das et al. 2007). Heterotrophic leaching takes place when the MO acidizes the environment through proton efflux (proton driving force), resulting in the release of metal cations. Autotrophic leaching occurs when the acidophilic bacteria obtain the energy required to fix CO_2 by oxidation of reduced Fe inorganic compounds (Fe^{2+}) or reduced S compounds. Siderophores are specific Fe^{3+} ligands but can bind other metals, such as Mn, Mg, and Cr. Methylation involves the methyl group that is enzymatically transferred to a metal, forming a number of different metalloids. Redox transformations allow MOs to mobilize metals, metalloids, and organometallic compounds. There are many metal mobilization techniques, the technology being chosen depending on the physical and chemical characteristics of the metal (Wu et al. 2018).

4.5.3 The Importance of Microorganisms with Depollution Potential and High Bioremediation Potential

Bacteria belonging to the genus *Pseudomonas* have a chemoorganoheterotrophic nutrition and are polar flame lates of bacilli, with strict aerobic breathing, having molecular oxygen as the ultimate acceptor of electrons in the respiratory chain. They do not grow to pH below 4.5. They produce siderophores – chelating compounds that bind and immobilize Fe but also other metal ions: Al, Cr, Zn, Cu, Mn, Pb, Cd, etc.

The main siderophore produced by *P. putida*, as well as other species of the same genus (*P. aeruginosa*, *P. fluorescens*, *P. chlororaphis*), also very useful in the processes of decontamination of polluted soils, is *Pyoverdine*. The action of siderophores is of great ecological importance due to the insolubility of its compounds. Siderophores bind Fe^{3+} compounds to form complexes that are transported inside cells where they can be used for respiration by anaerobic MOs such as Fe-reducing bacteria (*Geobacter bremensis*, *G. pelophilus*,

G. sulfurreducens). Another way to enrich the solubility of trivalent Fe compounds in soil or other habitats is to acidify the environment, a process involving the microbiota of the habitat, organic acids resulting from autolysis, and plant excretions, respectively. Some *Poaceae* (wheat, barley) produce phytosiderophores, similar to siderophores produced by bacteria (Zogg et al. 2018).

Pseudomonas genus members are recognized as having the highest aerobic degradation capacity of a large number of compounds. Genes responsible for enzyme synthesis involved in the degradation of these compounds are placed on plasmids. The *Pseudomonas* species exhibit resistance to the action of toxic metals. At *P. aeruginosa* and *P. fluorescens*, the Cu resistance was reported. For the strain of *P. putida*, the Cu^{2+} ion may accumulate at concentrations of 6.5% of the dry weight of the bacterium in precultivated crops in which the ion SO_4^{2-} was limitative. *P. stutzeri* is resistant to silver by the formation of silver sulfide complexes. A plasmid of the species *P. stutzeri* confers resistance to mercury and organomercury compounds. *P. fluorescens* detoxify Al and Fe, the trivalent ions being immobilized in a lipid metabolic complex (phosphatidylethanolamine). Resistance to Zn, Ca, and Ga is due to the association with phosphatidylethanolamine. *P. aeruginosa* is resistant to Hg^{2+}; the resistance to B, Cr, and Te is also determined by plasmid genes. *P. aeruginosa* and *P. putida* accumulate structures containing tellurium in the periplasmic space. On plasmids, there are genes that determine the resistance of As. Arsenite [As(III)] to arsenate [As(V)] is reduced to As, which is eliminated outside the cell. Placing genes responsible for the degradation of pollutant compounds and genes that confer resistance to the toxic action of HMs on transferable plasmas even interspecifically represent a great advantage. Given the particular advances in molecular genetics, plasmids or other genetic vectors possessing interspecific mobility may be constructed, the bacteria possessing high biotechnological efficiency. Of interest for environmental biotechnology are also members of the genera *Burkholderia*. *Burkholderia cepacia* (previously called *Pseudomonas cepacia*) and *Burkholderia multivorans* are aerobic chemoorganoheterotrophic bacteria but can use nitrate as the ultimate acceptor of electrons in anaerobic respiration.

The species of the genus produce countless siderophores. On Fe deficiency environments (like *P. aeruginosa* and *P. fluorescens*) synthesizes salicylic acid, a high-capacity siderophore compound, for the binding of Fe and other metals. The degradation of pollutant compounds in the environment is controlled by plasmids (Cai et al. 2016).

4.5.4 The Role of Microorganisms in Fixing or Mobilizing Metals in the Soil

The mobility of various organic or inorganic metal compounds in soil, as well as in other habitats, is closely related to their oxidation state. The energy used by living systems to carry out vital activities is provided by oxidoreduction reactions catalyzed by oxidoreductase enzymes (dehydrogenases). The tendency of a chemical compound to accept or yield electrons in biological oxidation reactions is quantitatively expressed by the oxidoreduction potential (redox) = Eh. In their life environments, including soils, different 54 ecophysiological groups of bacteria perform oxidoreduction reactions according to potential redox.

Desulfovibrio: oxidizes H_2 in the presence of SO_4^{2-}:

$$4H_2 + H_2SO_4 \rightarrow H_2S + 4H_2O + E \ \ (38 \ \ KJ/mol);$$

Beggiatoa: oxidizes H_2S in the presence of O_2:

$$H_2S + 2O_2 \rightarrow H_2SO_4 + E \ \ (200 \ \ KJ/mol).$$

Thiobacillus genus contains Gram-negative bacteria of bacillary form, some species with flagellar mobility. They are chemolithoautotrophic bacteria, so-called unpigmented sulfurous bacteria, which obtain the energy needed to fix CO_2 from the oxidation reactions of the reduced S compounds or even elemental S. It preferably oxidizes S or thiosulfate, rather than hydrogen sulfide. The reaction of *T. thiooxidans*:

$$S + 1\frac{1}{2}O_2 + H_2O \rightarrow H_2SO_4 + E.$$

As a result of the production of H_2SO_4, acidity in their natural environment may reach 1.5. They are the most acid-known bacteria. It develops optimally at pH 2–4 and is aerobically bound. Like *Beggiatoa*, heterotroph can be fed. The common natural habitats of unpigmented sulfurous bacteria are waters and soils rich in reduced S compounds, mine waters, and sulfurous springs. Aerobic is also *T. ferrooxidans*, which oxidizes Fe compounds in which it is in a bivalent state, to compounds where Fe is trivalent, as well as other ferobacteria or ferruginous bacteria (*Gallionella, Sphaerotilus*):

$$2FeSO_4 + H_2SO_4 + \frac{1}{2}O_2 \rightarrow Fe_2(SO_4)_3 + H_2O + E$$
$$2Fe(OH)_2 + \frac{1}{2}O_2 + H_2O \rightarrow 2Fe(OH)3 + E.$$

Ferruginous bacteria are found more frequently in waters and rarely in soils. The formation of oxidized compounds, especially ferric hydroxide $Fe(OH)_3$ insoluble compound, can lead to clogging of the metallic conduits by depositing in the form of red-brown coats.

Denitrifying or reducing nitrate bacteria has anaerobic respiration, using nitrate as the ultimate acceptor of electrons in the electron transport system. In the presence of O_2, they do aerobic breathing, even if nitrates are found in the environment.

There are species that, through the reactions produced, fall into two distinct ecophysiological bacterial groups: *Thiobacillus denitrificans*, optional aerobic, in the presence of O_2, it is possible to carry out the oxidation reactions of the reduced S compounds, but in the absence of O_2, for oxidation they use the oxygen from the nitrates present in the medium:

$$S + 2HNO_3 \rightarrow H_2SO_4 + N_2 + O_2 + E$$

Thus, *T. denitrificans* is a sulfurous bacterium because it oxidizes S-reduced sulfate compounds, and, on the other hand, it is a denitrifying bacterium because, under the specified conditions, it produces nitrate N_2. It is a typical example of anaerobic respiration, in which the final acceptor of electrons (H) is not O_2 but an oxidized inorganic compound (NO_3^-).

Some denitrifying bacteria can oxidize and reduce Fe compounds; others can ferment, so denitrifying bacteria have a very wide range of options with regard to alternative mechanisms of energy metabolism. Disulfide- or sulfate-reducing bacteria are all anaerobic bacteria, where the final acceptor of electrons is sulfate (SO_4^{2-}). The final product of sulfate reduction is hydrogen sulfide (H_2S). In the assimilation reduction of sulfates, H_2S is converted to organic S in the composition of some amino acids. In non-assimilable reduction, H_2S is excreted in the environment. The typical desulfurizer is *Desulfovibrio desulfuricans* (Deltaproteobacteria). Reaction produced:

$$4H_2 + H_2SO_4 \rightarrow H_2S + 4H_2O + E.$$

So, HMs have importance, but they can readily pollute the environment.

Disproportionation or dismutation is a chemical reaction whereby an element is simultaneously reduced and oxidized by forming two different products. Some reducing sulfate bacteria have the ability to cleave a S compound in an intermediate oxidation state, in two compounds – one lower and the other more oxidized than the original substrate. For example, *D. sulfodismutans* can decompose thiosulfate (intermediate oxidation state) into sulfate (more oxidized) and H_2S (mare reduced):

$$Na_2S_2O_3 + H_2O \rightarrow Na_2SO_4 + H_2S + E.$$

The process has ecological significance because it provides a way for desulfurization bacteria to recover the energy of intermediate S compounds resulting from incomplete oxidation of H_2S by S-oxidizing bacteria (*Beggiatoa, Thiobacillus*). As can occur in inorganic or organic forms. There have been discovered bacteria that can use As oxide compounds in anaerobic respiration, during which they are reduced to trivalent As compounds ($As^{5+} \rightarrow As^{3+}$) (non-assimilation reduction): *Sulfurospirillum arsenophilum* and *Sulfurospirillum barnesii*.

Bacteria *Shewanella* chemoorganoheterotrophic, optionally anaerobic, do not reduce the compounds As^{5+}, but they can release the ion in the environment, as well as reduced Fe compounds (Fe^{2+}) (Sas-Nowosielska and Pawlas 2015).

Pseudomonas arsenitoxidans grow chemolithoautotroph, obtaining the energy required to fix CO_2 from the oxidation reactions of the reduced As compounds. The bacterium is able to grow in the presence of organic substance, so it is only an optional autotroph. Growth is more intense in the presence of arsenite (AsO_3^{3-} or AsO^{2-}). By its ability to obtain energy as a result of oxidation reactions of arsenite, the strain of *P. arsenitoxidans* is a unique organism in the world of prokaryotes.

Other bacteria that oxidize As-reduced compounds (*Bacillus arsenoxydans*, *Alcaligenes faecalis*) cannot increase chemolithoautotroph.

Therefore, the discovery of this strain of *P. arsenitoxidans* represents a breakthrough in understanding the interactions between MOs and As compounds, so well-known for their strong antibiotic action (Sun et al. 2009).

Cr is present in both living organisms and in rocks, waters, and soils. There is only nature in the nature of compounds, not in elementary form. The most common forms of Cr in nature are the bivalent (Cr^{2+}), trivalent (Cr^{3+}), and hexavalent (Cr^{6+}) compounds. For the manufacture of steel, it is used Cr^0. From the activities of the metallurgical industry results Cr^{2+} and Cr^{6+}, while Cr^{2+} occurs naturally in the environment. From the point of view of the negative impact of Cr compounds on the health of the environment, the greatest interest is represented by the compounds with Cr^{6+}, most commonly encountered in contaminated sites. Cr^{6+} can be reduced to Cr^{3+}, by OM and the ions S^{2-} and Fe^{2+}, under anaerobic conditions, frequently encountered in groundwater, respectively, in flooded soils. In the presence of chromates ($CrO_4{}^{2-}$) and dichromates ($Cr_2O_7{}^{2-}$), metallic cations, such as Pb cations, precipitate. In the presence of chromates and dichromates, Fe and Al oxides are adsorbed by the soil particles. Toxicity and mobility of Cr depend on soil characteristics and the amount of OM incorporated by it. Cr hexavalent is more toxic and more mobile than all other forms.

Trivalent Cr is also mobile, but its mobility decreases with adsorption by clay minerals and a decrease in pH below 5. The increase in pH stimulates the leaching, solubilization of hexavalent Cr compounds. Normally, when Cr is discharged into natural waters, it accumulates in sediments, which can be subjected to bioremediation procedures.

The effects of minerals present in the underground soil on the efficiency of Pb removal from groundwater using biofilms composed of reducing sulfate MOs were evaluated and examined the stability of metal deposits after biofilms were temporarily exposed to air. To quantify the effects, Pb was immobilized in *Desulfovibrio desulfuricans* biofilms, grown under anaerobic conditions in two bioreactors filled with one hematite (redox-active) and the other with quartz (redox-inert). Biofilms grown on hematite were denser, thicker, and more porous than those grown on quartz. The average H_2S concentrations were higher in quartz biofilm than in hematite. Pb was more efficiently immobilized in quartz biofilm than in hematite (Shahid et al. 2017).

During the reported experiment, H_2S was produced under the action of desulfurizing bacteria, which reacted with Pb present to form precipitating PbS. It has been shown that Pb precipitates more in the presence of biofilm located near redox-inerts (quartz).

Pb deposits were partially reoxidated, especially in biofilms grown on hematite. In both bioreactors, biofilms responded to the presence of O_2 by lowering their density and by increasing the production rate of H_2S. Although the reduction of Fe^{3+} to Fe^{2+} was not quantified, it was found that Fe was continuously released from hematite throughout the experiment.

Different acid-thermophilic MOs from soil, sludge and water from polluted environments from metallurgical industries were isolated for use in the bioremediation of toxic HMs.

The tolerance of these MOs to high concentrations of Ag, As, Bi, Cd, Cr, Co, Cu, Hg, Li, Mo, Pb, Sn, and Zn was improved. Isolation and adaptation of MOs to large concentrations of metals were done by growing them on autotrophic bacteria media under different incubation conditions: pH, 2–4.5; temperature, 40–65 °C; and metal concentrations in culture media, 10^{-3}–10^{-7}M (Jonasson and Afshari 2018).

4.5.5 In Situ Bioremediation of Soils Contaminated with Heavy Metals

For bioremediation of technogenic soils, tailings dumps resulting from mining operations of Pb, Zn and Fe, bioremediation of Pb and Zn tailings dumps, and Fe, on some tailings, dumps were installed experimental plots, subject to a different treatment and seeded with *Lolium perenne* and *Trifolium pratense*. Later on, the slopes of the heap were planted *H. rhamnoides*. Applied biotechnologies have led to the formation of favorable conditions for the development of MOs, for plant growth, and for intense and sustainable enzyme activity.

The best technology for tailing bioremediation containing Pb and Zn impurities was to cover a 10 cm layer of natural soil in the vicinity of the heap, NPK mineral fertilization, and seedling with a mixture of herbaceous plants or plants from the spontaneous flora of the region. The ascending evolution of the microbial and EAs of the soils of the experimental parcels has been remarkable from year to year. The crops and the juvenile or extended plantations, so that after many years, the tailings heap, initially with a serene appearance, with no trace of vegetation, especially on the upper terraces, covered the vegetation totally.

Technogenic soil is being transformed; it already records an enzymatic and microbial potential comparable to natural soils. On the terraces and the slopes of the Fe, mine tailings were planted trees and shrubs, most of them *H. rhamnoides*. Experimental parcels were cultivated with the following herbaceous species: *Festuca rubra, Festuca arundinacea, Dactylis glomerata, L. perenne* (Poaceae family), *Onobrychis viciifolia, Trifolium repens, T. pratense, Lotus corniculatus*, and *Medicago sativa* (Fabaceae family). Both parcels and planting seedlings have undergone differentiated treatments. The evolution of the vegetation and of the microbial and EAs of the soil of the experimental plots was followed. The assessment of microbial potential was based on the values of soil quality bacterial indicators, calculated taking into account the number of aerobic, ammonifier, denitrifying, reducing, and desulfurization in mesophilic heterotrophic bacteria.

The EAs were assessed on the basis of the values of enzymatic soil quality indicators, calculated on the basis of the following EAs: catalase, sucrase, phosphatase, and actual and potential dehydrogenase. The results obtained demonstrated the efficiency of applied technologies. After only 1 year of vegetation, a remarkable biological potential has developed in the soils of the experimental plots. The vegetation had a good evolution, and the *H. rhamnoides* seedlings had a massive drainage, already providing a good cover for the waste dumps (Liu et al. 2018a, b).

4.6 The Role of Plants in Bioremediation

4.6.1 The Importance of Phytoremediation of Soils Contaminated with Heavy Metals

Phytoremediation includes the use of plants to extract, seize, and detoxify pollutants. This is an efficient, noninvasive, economically efficient, aesthetically pleasing, and socially acceptable method for remedying polluted areas (Sarma et al. 2021).

The major benefits reported for phytoremediation compared to traditional remediation technologies include the following: a possibility to generate less secondary residues, a minimal degradation of the environment, an opportunity to leave the soil in place and under conditions of use after treatment, have low design costs for the candidate land for remediation, and a method requiring very little technique because implementation requires little more than basic agricultural techniques.

Disadvantages include the following:

A long period of time required (several growing seasons).
A limited depth that can be applied (1.2 m for soil and 3 m for groundwater) because the roots can effectively clean only a limited depth.
An opportunity for pollutants to enter the food chain through plant animal consumption.
Operating characteristics and costs for a large-scale implementation have not yet been fully evaluated.
Plant residue may require deposition as hazardous waste or requires additional treatment.
Degradation by-products may be mobilized to groundwater or bioaccumulated in animals.
If the concentration of the contaminants is too high, the plants may die.
Plant growth may be seasonal depending on location.
Climatic and hydrological conditions (e.g., floods, droughts) can restrict the growth rate of the type of plant that can be used.
An area of the site may be modified to prevent floods or erosion; soil amendments, including chelating agents, may be necessary to ease the uptake of pollutants by plants by breaking the links between contaminants and soil particles.

The phytoextraction technique has a number of advantages that can make it more attractive than other rehabilitation techniques: it reduces the volumes to be deposited in the landfill (at the end of the process, the plants are burned, and so the amount of ash is much smaller than the mass of contaminated soil), saves energy (the process of cleaning is based on solar energy), operates very simply, does not disturb visually and is acceptable to public opinion, and is relatively easy to apply on extended contaminated surfaces; it is possible to recycle the products from the phytoextraction processes; and it is possible to use for a wide variety of pollutants (metals, radionuclides, OSs) (Azubuike et al. 2018).

4.7 Biomonitoring Soil Pollution with Heavy Metals

HMs, such as Fe, Cu, Pb, Zn, mercury, Cd, or nickel, are important pollutants from the extractive industry (mining, ore mining, extractive industries), but these pollutants may also result from other industries, such as the chemical industry, energy industry, etc. They are emitted by polluting agents in air, water, or soil under different physicochemical forms. Unlike the vast majority of pollutants in other categories, HMs emitted in air, water, and soil do not undergo biodegradation processes. HMs can be transformed into organometallic compounds (less toxic), can undergo a series of oxidation or reduction reactions, or can be fixed to clay minerals present in soils and sediments. These latter processes are those that contribute to the self-purification of the environment in case of heavy metal contamination.

Virtually every metal has its own environmental behavior and produces a specific biogeochemical circuit but increasingly influenced by the anthropogenic factor. There are species able to accumulate in their bodies certain substances in concentrations tens of thousands of times their concentration in the environment. The pollution study has confirmed the existence of this process also in the case of pollutants, including HMs. Thus, two concepts are used to designate this process: bioconcentration and bioaccumulation. Bioconcentration means the direct increase in the concentration of a pollutant as it moves from the biotope to a body; in terrestrial organisms, it is the passage from air or soil into the body of plants by absorption or the passage from air into the animal body by inhalation. Bioaccumulation is specific for animal organisms and includes direct absorption of the pollutant plus food accumulation. In the case of a trophic network within a biocoenosis, the bioaccumulation phenomenon can be repeated several times, at each passage from one trophic level to another, from a prey to a predatory organism. In such cases, we are dealing with a bioassay process (Boim et al. 2016).

4.8 Bioindicators for Terrestrial Environments

Lichens are strong bioaccumulators due to their ability to take up pollutants present in atmospheric air. With a particular sensitivity to pollutants, lichens are used as bioindicators for different pollutants: SO_2, NO_X, HF, Cl_2, O_3, peroxyacetate, HMs, radioactive elements, fertilizers, pesticides, and herbicides. The species of cortical lichens *Parmelia physodes*, *Parmelia caperata*, and *Evernia prunastri* have served as bioindicators to monitor Pb air pollution. *Cladonia rangiferina* and *Cladonia nitei* can be used as bioindicators for U, Fe, Pb, and Ti.

Mercury can be accumulated by the lichens: *Alectoria capillaris*, *Alectoria tremontii*, *Hypogymnia physodes*, *Cladonia* sp., and *Collema* sp.

The most widely used bioindicator is the lichen *Hypogymnia physodes*. Bryophytes (mosses) have a particular sensitivity to air pollution. As a result, the number of species of bryophytes has been greatly diminished in urban areas in highly polluted industrial centers. Some species have disappeared, and others have been reduced as a number of individuals (and biomass) and as a spreading area. For example, in the Netherlands, 15% of groundbreaking species and 13% of epiphyte species have disappeared over the last 100 years. *Pleurozium schreberi*, *Hyloconium splendens*, and *Hypnum cupressiforme* have been used in a biomonitoring project for eight-metal background pollution (As, Cd, Cr, Cu, Fe, Pb, Ni, V, and Zn) of the Northern Europe. Radioactive elements are also accumulated in mosses more intense than in higher plants. Thereby, *Pleurozium schreberi* was used to monitor La, Zn, and other elements from nuclear tests, and species *Ceratodon purpureus*, *Tortula ruralis*, and *Bryum argenteum*, for biomonitoring falls of ^{137}Cs after the accident at Chernobyl. In the case of HMs, the following sequence of their toxicity for bryophytes is established (it is similar to flowering plants): Hg > Pb > Cu > Cd >Cr > Ni > Zn. Some species have a particularly high storage capacity of metals, sometimes up to extremely high concentrations. For example, *Hyloconium splendens* originating from a Cu mine has accumulated Pb, Cd, Cu, and Zn in concentration of 17320 ppm (compared to higher plants *Picea* 349.5 ppm, *Clintonia* 548.5 ppm in the same environment).

Higher Plants, Bioindicators for Heavy Metals For plants with flowers, grassy species can be considered as accumulators. Among the most commonly known metal accumulators are the following: *Melandrium album*, 4.286 μg/g; *L. perenne*, 1.683 μg/g; *Plantago lanceolata*, 1.547 μg/g; *Lepidium draba*, 1.437 μg/g; *Polygonum aviculare*, 1.190 μg/g; and *Thlaspi*, 1000–3500 μg/g (for Ni and Zn) (Hu et al. 2019).

Indicator Species *Lolium perenne* and *L. multiflora* are very suitable for use as exposure indicators. These plants are common in parks, along roads, and on roads. Besides HMs, they are indicators for S and F. Other grassy species used as bioindicators for HMs are the following: *Melandrium album* for Pb; *Thlaspi* for Ni and Zn; *Solidago canadensis* for Pb; *Artemisia vulgaris*, *Calamagrostis epigejos*, *Chelidonium majus*, *Plantago major*, and *Poa* for HMs; *Equisetum arvense* as Hg bioindicators in the area destroyed by the St. Helens volcano in the USA; *Achillea millefolium*, *Artemisia vulgaris*, *P. lanceolata*, and *Amaranthus retroflexus* for V; *Hypericum perforatum*, *Hedera helix*, and *Urtica dioica* for Pb, Cu, Zn, Cd, and Hg; and *Vaccinium myrtillus* and *Vaccinium vitis-idaea* for Cd, Fe, Mn, and P (Ji et al. 2018).

Deciduous Trees and Shrubs Tree leaves fix HMs from polluted areas (e.g., on the roadside near plants) from both the soil and directly from the air. Sensitive species are *Betula pendula*, *Fraxinus excelsior*, *Sorbus aucuparia*, *Tilia cordata*, and *Malus domestica*. Accumulator indicators are species considered resistant: *Elaeagnus angustifolia*, *Populus canadensis*, *S. alba*, and *Sambucus nigra*. Relatively resistant species (accumulators) are *Carpinus betulus*, *Quercus robur*, *Fagus sylvatica*, *Quercus palustris*, *Acer saccharum*, and *Platanus acerifolia*. Conifers are more sensitive than trees with falling leaves, due to the fact that the life of the leaves (leaves) is 3–4 years old, and are exposed to pollution during the winter periods as well.

Pollution with SO_2 and HF can be indicated by conifers and by determining leaf content (the conifers are also accumulators). Coniferous species susceptible to SO_2 pollution are *Abies alba, Picea abies, Pinus banksiana, Pinus nigra, Pinus sylvestris, Pinus strobus, and Larix decidua*, and susceptible to HF pollution are *A. alba, P. abies, Pinus ponderosa, P. sylvestris,* and *Pinus strobus*.

Some species may also be used to indicate pollution with photochemical oxidants: *P. abies, Pinus banksiana,* and *Pinus strobus*. Indication of heavy metal pollution is done by accumulating them, especially in those conifers (the conifers are heavy metal accumulators). Thus, *P. abies, P. sylvestris, Pinus nigra, Taxus baccata,* and *Thuja occidentalis* are species suitable for indication of pollution with Fe, Mn, Cu, Pb. Zn, Cd, Ag, and Hg (Turkyilmaz et al. 2019).

Bioindicators in Freshwater Given that biological abatement is particularly active in aquatic environments, in depollution ponds, we mention a series of bioindicators and bioaccumulators that allow the assessment of pollutants, such as HMs, etc.

Aquatic Macrophites A series of algae, bryophytes, amphibian, or hydrophilic phanerogams have strong accentuation of tissue concentration of both mineral elements (HMs) and organic xenobiotic compounds in water. Aquatic mosses with bioaccumulation potential are among others: *Fontinalis, Amblystegium, Rhynchostegium, Plathyhypnidium,* and *Cindidotus*. A number of researches have determined that *Fontinalis squamosa* and *Fontinalis antipyretica* have a high capacity of bioaccumulation of Zn and other toxic metals. The last species also accumulates PCBs or other xenobiotic organic compounds or rare metals, such as Ag, Bi, and Sn.

Algae Red alga *Lemanea* sp. is one of the few macrophytes that can live in the immediate vicinity of water from a tailings pond from a Pb exploitation, it can grow in waters with high concentrations of toxic Zn metals.

Phanerogam *Typha latifolia* has proven to be an effective Zn bioaccumulator, accumulating in the roots up to 1400 mg Zn/kg when the plants grew on sediments containing 10 mg Zn/kg of sediment (Zalewska and Danowska 2017).

4.9 Research on the Possibilities of Using Plants and Microorganisms for the Biological Extraction of Heavy Metals From Contaminated Soils

Recent researches report on the possibilities of using MOs for the biological extraction of HMs from soils contaminated by anthropogenic activities, with sources of pollution, such as Pb (Pb, As), wood (As, Cr), pesticides (Pb, As, Hg), mining sites, etc. Comparison of technologies in the study of decontamination of soils polluted with HMs (As, Cd, Cr, Hg, Pb, Ni, Zn, or Cu) included the following: insulation (covering) by covering and vertical or horizontal barriers; solidification/stabilization by cementation, microencapsulation, and vitrification; and separation–concentration by soil cleaning, soil washing, pyrometallurgical, electrochemical, or phytoremediation methods.

Phytoremediation for the biological extraction of HMs from contaminated soils can be achieved by phytoextraction, phytostabilization, or rhizofiltration. Phytoextraction (followed by harvesting and plant treatment) utilizes plants that have a hyperaccumulation capacity in tissues (Sarma et al. 2021).

Hyperaccumulate plants accumulate in the leaves over 0.1% Ni, Co, Cu, and Cr or 1% Zn and Mn (dry matter) irrespective of the concentration of metals in the soil. Ni and Zn were readily absorbed in plant tissues, as well as Cu and Cd. Phytostabilization (fixing in roots) uses plants to limit the mobility and bioaccessibility of metals. It is achieved by producing compounds by plants capable of immobilizing contaminants at the soil-root interface. By increasing the pH in the soil, additional stabilization is achieved.

The method is applicable to many metals, especially Pb, Cr, and Hg. Phytostabilizing plants can tolerate metals and immobilize them in the soil by adsorption, precipitation, complexation, and reduction. Three varieties of herbaceous plants have been marketed for phytostabilization: *Agrostis tenuis*, Parys for Cu waste; *A. tenuis*, cv Coginan for waste with Zn and Pb acid; and *F. rubra*, cv Merlin for waste with Zn and Pb limestone.

Rhizofiltration uses the roots of terrestrial plants to absorb, concentrate, and precipitate metals from wastewater, including leachates of contaminated soils, translating them into tissues, more easily than by phytoextraction. Terrestrial plants develop fibrous root systems with a larger surface of action than aquatic plants.

Another type of rhizofiltration that is fully developed uses water plants in wetland or wetland plants to treat wastewater or leachate from contaminated soils. These biotechnologies use the decontamination activity of a complex system of plants and microorganisms. It is to be noted that one of the goals of phytoremediation is to achieve major reductions in the cost of decontamination (Ignatius et al. 2014).

4.9.1 Removing Heavy Metals From Polluted and Leached Soils Contaminated with the Help of Cyanobacteria

Of the 35 metals considered dangerous for human health, 23 have been classified as HMs: Ag, As, Au, Bi, Cd, Ce, Co, Cr, Cu, Fe, Ga, Hg, Mn, Ni, Pb, Pt, Sb, Sn, Te, Tl, U, V, and Zn. The most dangerous substances for health are Pb, Cd, Hg, and As.

Exposure to high concentrations of these metals can lead to poisoning, with severe effects on the nervous system or on internal organs, such as the lungs, liver, kidneys, and others. In recent years, many techniques have been developed to remove HMs from wastewater in order to reduce the amount of metal-containing wastewater resulting from industrial activities and to improve the quality of effluents.

Numerous treatments, such as chemical precipitation, coagulation–flocculation, flotation, ion exchange, or membrane filtration, can be used to remove HMs from contaminated wastewater, each method having certain advantages and limitations. A way to remove HMs by using cyanobacterial microorganisms is particularly well-researched.

Use of Microorganisms for Removing Heavy Metals Biosorption is the property of nontoxic microbial biomass to accumulate heavy metal ions, a metabolically unsustainable process. In contrast, the term bioaccumulation describes an active process in which HMs are removed by metabolic activity of living organisms. In recent years, research has focused on biosorption mechanisms because biomass can be successfully used to remove HMs from industrial effluents as well as to recover precious metals from processing solutions (Thakare et al. 2021).

Microbial cells are excellent biosorbents due to their high surface area/volume ratio due to a large number of potentially active chemosorption sites. Researches in the field of heavy metal biosorption have led to the identification of several types of microbial biomass efficient in the concentration of these metals.

Some types of biomass are waste resulting from industrial fermentation (e.g., *Bacillus subtilis* bacteria). Other types of microbes that can bind metals can be harvested from specific environments such as lakes or seas (certain types of cyanobacteria). They can accumulate considerable amounts of HMs, such as Cd, Cr, Cu, Pb, U, Zn, or others (Pan et al. 2017).

General Characteristics of Cyanobacteria Cyanobacteria represent a large group of Gram-negative bacteria, phototrophic prokaryotes characterized by the ability to achieve photosynthesis in the presence of oxygen and autotrophy as the main source of nutrition. It has been observed that certain cyanobacteria can grow in the dark on some organic substrates or under anaerobic conditions, performing photosynthesis in the absence of oxygen, using sulfides as electron donors. This trophic independence, along with the ease of their cultivation, recommends cyanobacteria in decontamination processes.

The Use of Cyanobacteria in Removing Heavy Metals In recent years, the ability of MOs to accumulate and remove HMs in water has been tested. These processes can be classified as active, metabolic processes or passive, nonmetabolic processes. In active processes, adsorption, called bioaccumulation, is due to metal transport through the cell membrane, followed by intracellular accumulation depending on cellular metabolism.

In physicochemical interactions between the metal and the functional groups present on the cell surface (based on physical adsorption, ion exchange, and complexation), the sorption process does not depend on metabolism. The cell wall, composed of polysaccharides, proteins, and lipids, can provide sites for metal bonding. Cyanobacteria have a cell wall that allows the passive adsorption of large quantities of dissolved metals. On the surface of the cell wall, there are in fact functional groups, such as carboxylate, hydroxyl, sulfate, phosphate, or amino groups.

Metallic cations can be taken from the environment by the following:

Negatively charged groups on the cell wall, with the negative charge present on the polysaccharide layers surrounding the cell wall or the negative charges of polysaccharides released into the environment (water)

Interaction of metal cations with negative charges on the cell surface

Interaction of metal cations with negative charges on exocellular polysaccharide layers

Interaction of metal cations with negative polysaccharide releases released into the environment (water).

A large number of types of cyanobacteria have been tested in connection with their ability to remove HMs (especially Cd, Hg, Ni, Zn, or Pb) from metallic solutions (Garlapati et al. 2019).

For example, *Tolypothrix tenuis* and *Calothrix parietina* have removed high amounts of Hg; *Scytonema schmidlei*, *Anabaena cylindrica*, and *A. torulosa* have removed 96–98% from Cd found in a solution of Cd with concentration 1 mg/L.; *Gloeocapsa* sp., *Nostoc paludosum*, *N. piscinale*, *N. punctiforme*, *N. commune*, *Oscillatoria agardhii*, *Phormidium molle*, and *Tolypothrix* have removed 90–96% Pb from a Pb solution with concentration 1 mg/L. The metal bonding process takes place by complexing the metal ions with the polysaccharide, mucilaginous material that covers the cell wall or is released from the cell surface.

The Ecological and Economic Efficiency of the Method As a result of numerous studies on various cyanobacteria used to remove HMs from aqueous solutions, promising results have been obtained with regard to the use of the tested species. However, this type of biomass is far from being used on an industrial scale in the treatment of HMs contaminated waste for two reasons: the biomass production costs, which are still too high for industrial applications, and the lack of sufficient research into the use of cyanobacteria to achieve some industrial lines that lead to the removal of metals by means of these.

Interactions between cyanobacteria and metals are very complex, depending on a large number of factors related to the chemical and morphological characteristics of the microbial cells, to the physical and chemical properties of the metals to be removed.

The promising results achieved so far, the increase in the price of chemicals, the possibility of using naturally occurring cyanobacteria (without production costs), and the increase in public interest in the use of "green" biotechnologies to solve the problem of water pollution, open up new perspectives in the use of cyanobacteria removing HMs from wastewater (Nguyen-The et al. 2016).

4.10 Ecological Restoration

Application concerns for ecological reconstruction, based on deep and long-lasting fundamental research on biodiversity, environmental protection, and sustainable use of the heritage of natural or man-made resources, have been important goals of many studies.

Researchers approached integrated area and global system research to address practical environmental protection problems related to the productivity, conservation, and sustainable development of natural, agricultural, forestry, and urban ecosystems; problems of integrated environmental impact management in various economic activities; as well as the reconstruction of ecosystems following the destruction of biocenosis in areas affected by natural and man-made disasters, pollution, and technological dysfunctions, current climate change, etc.

The destructive effect of disturbing factors and ecological disasters on natural or anthropogenic biocenoses requires ecological reconstruction of the area affected – through green infrastructure (parks, green areas), landscaping, agriculture or forestry, antierosion, or protection against climate change – and technological and current perspectives.

The ecological reconstruction, applied on abiotic substrates, polluted and contaminated biotopes, tailings dumps, discarding materials, and other residues from mining and other industrial, agricultural, etc. activities, operates with technical and biological methods that can ensure ecological or ecological succession secondary, respectively: establishment of phyto- and pioneering zoocenoses, concomitant with pedogenesis, restoration of microbiota, humus and soil properties; phyto- and zoocenotic reconstruction, cultivation, colonization of species and stabilization of interspecific relations favorable to the biocenostic productivity and balance of the new ecosystem.

The mediogenic and phytocoenosis functions ensure ecological reconstruction by fixing the anthropic relief, setting pedogenesis and soil protection, zoocenosis formation and conservation of the fauna, microclimate regulation, precipitation intercept and water regulation, wind speed mitigation, surface leakage, deep erosion (landslides, ravines, dungeons), deposition of alluviums, etc.

Considering that ecological reconstruction is a particularly important area for ensuring the quality of the environment and specifying the role of phytocology in the reconstruction of ecosystems, it emphasizes that it can fulfill its "ecological valences" on the basis of the documentation provided by phytosociological research. They demonstrate that viable, protective, and productive eco-cenoses are those that are subject to long-term selection under the conditions of the particular biotope, and their model can be followed in ecologically applied rebuilding.

Colonization of vegetation on the abiotic substrate resulting from the profound deterioration of the natural environment, following various economic activities, mining, metallurgy, and construction, by the accumulation of tailings after the extraction of ore, through the development of tailings ponds, highways, dams, and so on is achieved by particular stages of ecogenesis. The substrate, initially abiotic, is populated with pioneer species, which have a high environmental tolerance amplitude to variations in environmental factors (Trivedi et al. 2016).

When installing this vegetation, interspecific competition is small or null. Associations are made up of a small number of species that fix the slope, such as *Rumex scutatus* with *Galeopsis angustifolia* and, on the northern slopes, wet, *Rumex* with *Tussilago farfara*.

On the slopes, there is a vegetation with a deeply branched surface and deep root system that resists the landslides, as well as the cleavage of the forest – *Clematis vitalba*. On the metal-containing tailings, special metallophyte species (for Se, a series of legumes; Sr, grasses; Zn, calming species; etc.) are installed. Colonization and fixation of anthropogenic relief by phytocoenoses depends on the adaptive capacities of the plant communities, the ecotypic variability (genetic biodiversity) of the constituent species, and the physicochemical properties of the basic material.

Phytosociological studies can provide a scientific basis for finding optimal, environmentally sound, and economically feasible solutions for barking and fixing or for the use of tailings dumps in agriculture. In support of the idea of ecological reconstruction, on the voluminous tailings dumps, accumulated in coal, salt, ferrous and nonferrous metals, thermal power plants, and other industrial activities, the arguments relate to tailings toxicity, unsightly appearance, danger of sliding the anthropic relief, and the need for their reintegration into the economic circuit (Colin et al. 2019).

4.10.1 Biological Recultivation of Degraded Lands

Vegetation coverage is the most economical and environmentally acceptable method for stabilizing, bioremediating, and rehabilitating degraded land, reducing pollution. The remediation and rehabilitation process must be integrated into ecological and socioeconomic contexts.

The ecological remedy can be achieved by restoring the exact restoration of the original ecosystem (the ecosystem prior to the degradation of the site), the restoration of restoring an ecosystem similar to the original, but less than a complete restoration. The replacement does not propose any restoration but a replacement of the original ecosystem with another one. It can be a simpler yet more productive ecosystem (an agricultural ecosystem, a meadow, or a pasture), or it can be simpler and less productive (a forest is replaced by a grassland with a recreational area). The term ecological reconstruction may overlap with any of the terms' restoration, rehabilitation, and replacement. Ecological rebuilding means an active and guided (engineering) replacement of the original ecosystem with another for a well-defined purpose.

Mining waste presents specific rehabilitation problems. Mining waste dumps are often naturally colonized by vegetation. This depends very much on the material from which the heap was built. Many mining waste dumps have no limitations on plant growth, except for the absence of nitrogen and possibly phosphorus. It is easy to install vegetation that includes obligatory nitrogen fixation plants. For such dumps, a wide range of uses from forests to agricultural fields can be considered. Many mining waste dumps contain pyrite, and H_2SO_4 is formed by oxidation.

These dumps have restrictions for both the natural vegetation installation and for a number of other uses. When the pyrite content is low, there is a chance of slow planting of a flora composed of plants tolerant to acidity. Installation of vegetation should be aided by lime treatments. Such land cannot be used for agriculture because rehabilitation costs would be too high and their agricultural efficiency would be reduced.

Nonferrous and pyrite-containing holes are more difficult. Here, acidity also carries heavy metal ions that are toxic to vegetation and to microorganisms.

Old holes from nonferrous ores have often high metal content due to the poor efficiency of metal extraction techniques used in those times. These waste dumps remain uncolonized by vegetation and are exposed to erosion and sliding phenomena, constituting a permanent and real danger to the environment. Any public use of these lands requires coating with an inert material to isolate the area of toxicity, which is quite costly. After such treatment, the huts can be used for recreational areas (Gay-des-Combes et al. 2017).

4.10.1.1 Treatments in Ecological Reconstruction Technologies of Degraded or Industrially Polluted Sites

The treatment of a site is supposed to be two components. Immediate treatment to improve the physical, chemical, topographic, toxicity, and site stability features provides conditions for long-term treatment. Long-term treatment consists of installing and maintaining vegetation on the site, etc. To improve the physical structure of the field, immediate treatment is done: scarification or discussion, if the structure is too compact; compaction and finishing, if the structure is loose; stabilization works by treatments with various stabilizing materials; drainage works; and humidification to improve stability and humidity.

Other treatment works can be carried out in order to ensure the conditions required for the efficient installation of the vegetation: fertilization works (fertilizer additions and amendments), pH lime correction can be carried out; works to neutralize and fix toxic substances (HMs). The purpose of rehabilitation is to create an ecosystem with robust self-maintenance. The installation of vegetation and its maintenance is a decisive step in this action, and it depends on the success of the whole rehabilitation action. Biological recultivation is important; the successful realization of the self-sustaining ecosystem necessarily requires that all problems (physical, nutritional, or toxicity-related) highlighted in the analysis of the original site be solved.

For example, the development of pyrite acidity contained in the waste material can make the freshly installed new ecosystem rapidly enter degeneration. The only solution is to add sufficient amounts of calcium carbonate (limestone, chalk) and to pursue any new generation of acidity. There will be no complete ecosystem in 1 year, even if it is a simple ecosystem like a green grassland. It is very difficult to create a completely fertile, completely biological soil in a single year starting from waste, except when there is a substantial layer of vegetal soil added over preexisting waste or organic material brought elsewhere (sludge sewage, compost). Such additions can only be made on a small scale. Soil has to be built over a period of years, with particular attention being paid to soil-forming processes. The requirement is not the formation of a soil that is fully developed in the pedological sense (which can take thousands of years) but a fully developed soil in the biological sense. This means that the soil formed has adequate content of OM and especially nitrogen, associated with active mineralization processes.

Normal ecosystems in the temperate region require about 150 kg of nitrogen per hectare per year. It can usually only be produced if the nitrogen is 10 times higher in the soil, resulting in a minimum amount of nitrogen required in the soil of 1000 kg/ha, linked to a large amount of OM.

Such a soil can only be formed if in the growing ecosystem among the plant species there are legumes or other nitrogen fixation plants. These may no longer be necessary in the final part of ecosystem development, but, in the initial part, the nitrogen sequestration species should be not only present among the sown or planted ones; they should be appropriate to the concrete conditions so that they are maintained and the contribution their nitrogen fixation to be durable.

Care should be taken to treat fertilizers and also to mowing or grazing. In the case of forest ecosystems, the same actions should also be taken into account for the sapping and sowing actions of the saplings. All this will ensure the satisfactory growth rate in the developing ecosystem, a satisfactory production of OM both on the surface and inside the forming soil (about 10,000 kg/ha/year). This OM, along with the current penetration of the roots into the soil and the activity of the soil-feeding animals organic, will be particularly important in developing a proper soil structure that is most often a problem in degraded lands. Improving the structure, along with the surface carpet, will reduce the erosion of the land. In freshly populated land, it is often found that, after a very good start, the rate of growth decreases over the next period to very low levels due to inadequate care.

Some species are more tolerant than others at lack of maintenance: in the temperate zone are the leguminous *Sarothamnus scoparius* and *Ulex europaeus* (tolerant to acidity and phosphorus deficiency). The need for further maintenance of vegetation is an integral part of the rehabilitation process and even of the decision on the way of rehabilitation (Pelfrêne et al. 2015).

4.10.2 Plants Used in Recultivation of Degraded Lands

Herbaceous plants. In most cases of redevelopment of degraded lands, the formation of a grassy vegetal carpet is a final target or intermediate step to reach the final goal that can be a forest, arable land, or a park. As a rule among the herbs that live on a meadow, there are legumes that have a great capacity to fix the atmospheric nitrogen, and among them, *T. pratense* is the most widespread. In each climatic zone, there is a larger or smaller number of herbaceous plants that can be used in the rehabilitation works of the areas under rehabilitation.

In the United Kingdom, in 1980, 140 species of herbs and 70 leguminous species were identified, among which the species suitable for recultivation can be chosen. The issue is which species are suitable for land rehabilitation, how to make choices, and what are the best methods of sowing? From the point of recultivation of a degraded area, the most important consideration should be the adaptation of plants to the soil (soil) conditions. On the other hand, some of these plants are easy to obtain through the seeds produced by specialized producers; others are found only as wild plants.

Wild plants should not be ignored because they can have a number of extremely favorable features and can provide very useful services in planting and permanent planting activity. Within each species, there are usually many varieties and varieties with different attributes of strength, wintering, requirements for soil and climate conditions, etc.

Searching for the most appropriate varieties is a useful expense because a fair choice must take into account all the differences regarding the persistence and easy maintenance of the final vegetal carpet (Küpper and Andresen 2016).

Leguminous plants are a crucial component in almost all herb mixtures because they contribute to maintaining an adequate nitrogen influx and ensure the provision of an adequate amount of organic nitrogen in the new soil in formation. They eliminate the need for nitrogen treatments for maintenance by increasing the amount of mineralizable nitrogen. Thus, *T. pratense* is a better fertilizer than chemical fertilizer treatments as it provides nitrogen gradually and continuously. And in the case of legumes, the choice of species depends on soil and climate conditions.

Usually the most suitable legumes are those used in agriculture because they have a high rate of nitrogen fixation. Legumes are suitable due to symbiosis with the *Rhizobium bacterium* that fixes in the roots and forms the nitrogen fixation nodules. The use of seeds already infected with *Rhizobium* is applied. It is also possible to use the soil in which the legumes have already been grown to bring an appropriate supply of suitable *Rhizobium* bacteria.

Sometimes the so-called cultures are being made nude. The species used for this purpose must be chosen according to the intended purpose. Soil preparation and sowing are usually done using the same methods that are commonly used in agriculture and with the same machinery. The sowing rate (the amount of seed per hectare) may vary depending on the land and the site, but here is also good to follow the practice of agricultural work.

Horticulture rates (500 kg / ha) are not justified, while in agriculture 40 kg/ha is applied, in the idea that much seed ensures success. This is not true; success is only ensured by ensuring the right growth conditions. It is considered normal for the temperate zone a sowing rate of 50–100 kg/ha and a little less for the tropical climate. If vegetative material specifically adapted to particular conditions such as those in metallic hatches could be identified, for example, then cloning and vegetative propagation is an effective stabilization method, ensuring the preservation of the unmodified genetic features of the adapted material.

The vegetative material planted in the field will require the same attention with regard to the breeding and maintenance work as the sowing material. Even if the starting material is bigger (plant) and has a larger reserve of nutrients in the short term, it will in the long run require the same nitrogen and other nutrients as seeding (Carrasco et al. 2018).

Trees and Shrubs Used in the Recultivation of Degraded Lands As with herbaceous plants, the choice of species is immense. For planting on degraded land, it is necessary to choose species known to be well adapted to the local environment, with the agreement of the forest inspectorate. Generally, the most valuable species are the natural pioneers. It is possible to identify species that are particularly adapted to the difficult conditions of nutrient deficiency, the deficiency that is present in the early natural vegetation sequence on degraded lands.

Trees As with herbs, species should also be chosen according to the particular features of the site such as pH and climate. There are a number of trees in the temperate area, such as *Robinia pseudoacacia* (acacia), *Cercis siliquastrum*, *Gleditsia triacanthos*, *Gymnocladus dioica*, and *Sophora japonica*, and in the warm climate, the Acacia species, as well as other tree species, are leguminous and have the ability to fix the nitrogen ex. species of *Alnus*, *H. rhamnoides*. These species grow relatively quickly without the addition of fertilizers (nitrogen) and represent tools in the rehabilitation of degraded or technogenic land, although sometimes they require the addition of other nutrients.

Shrubs The choice of shrubs is also immense. There is no question of commercial use, specifically chosen will be those that have the most appropriate attributes to ground conditions and which perform well the fastening role. Among the very valuable shrubs with the ability to fix the nitrogen are the legumes of Ulex, Sarothamnus, and Elaeagnaceae *Elaeagnus* (shrubs) and *H. rhamnoides*.

Other shrubs often used for fixing degraded lands and landfills are dogwood (*Cornus sanguinea*), cornelian cherry (*Cornus mas*), privet (*Ligustrum vulgare*), broom (*Cytisus nigricans*) wayfarer and guelder rose (*Viburnum lanthanum, Viburnum opulus*), etc. The assortment of forest species that provides good results on the waste dumps is rich, the species used depends on the existing stationary conditions or created by the improvement works (van Kuijk et al. 2014).

Soil-improving species (nitrogen fixators), such as the locust tree (*R. pseudoacacia*), the alder (Alnus), the elaeagnus (Elaeagnus), and the

H. rhamnoides, have shown good results in all cases. They also contributed to the improvement of soil material in the waste dumps in relatively short time and stimulated the growth of other wood species when mixed with them. Among the other forest species used on the heaps were good or satisfactory results: pines, especially scots *Pinus sylvestris*, on mining and industrial dumpsters with coarse and sandy deposits, with a pH below 7, and *P. nigra* on the hills limestone or clayey with pH above 7; poplars, especially *Populus alba* and *Populus tremula*; and the willows (*S. alba, Salix caprea, Salix incana*, etc.).

On dumps consisting of sandy deposits, wetter or irrigation areas; can be planted the *Fraxinus excelsior, Acer, Cerasus avium, Ulmus* and *Fraxinus pennsylvanica, Acer tataricum* and others, on the dumps with some more favorable stationary conditions or on which a layer of fertile soil has been deposited; can be planted oak and red oak (*Q. robur* and *Q. borealis*) on dumps with the best soil conditions; can be planted *Betula verrucosa, Ailanthus altissima, Acer negundo, Fraxinus ornus*, etc.

In dumps with more difficult stationary conditions (birch in wetlands on acidic dumps, and other species on dumps in warmer and drier areas – especially mojdrean), *Ligustrum vulgare, Cornus sanguinea*, and other shrubs, on different types of dumps, mixed with the species mentioned above (Solis-Gabriel et al. 2017).

Afforestation of Heaps The analysis of the situation of some waste dumps led to the conclusion that for the characterization of the stationary hulls and the classification of the waste dumps, the following factors have to be taken into account:

The general physico-climatic conditions, with the distinction of the following series of dumpster types: heaps from the plains and hillsides, steppe, and silverside; heaps in hilly regions of the oak and oak subareas; and heaps of mountainous areas in the subareas of beech and spruce

The origin of the component material, distinguishing: mining waste dumps, industrial waste dumps, household waste dumps, etc.

Characteristics of landfills in heaps, distinguishing: heaps from coarse rock fragments (gravel, stones, boulders with or without fine material) and fine rock dumps (sand, dust); heaps of very fine clay (clay) from ashes, slags, slag waste dumps

Other factors, such as rock petrography and mineralogical composition, pH, the presence of soluble salts and/or toxic compounds, and the microrelief configuration of the dumps (waves or mounds, platforms, ponds), are added to these. All these will be the basis for designing and applying differentiated technical solutions for the stabilization and afforestation of the waste dumps. Using some of these features, we have made a stationary classification of the heaps, from which we extract some types of resorts (Bourret et al. 2009).

4.10.2.1 Works to Improve the Conditions for the Installation and Development of Forest Vegetation

Generally, the dumps provide difficult stationary conditions, so in most cases, work is needed to improve these conditions to allow forest vegetation to be installed and properly developed. Among the most commonly applied categories of works can be mentioned: works for consolidation of slopes against slopes by sloping slope and redevelopment of slopes, supporting walls, drainage and works for erosion control by linear or rhombic fences, dry masonry benches, etc., land preparation work. In the case of very heavy waste dumps (very often it is the case of mining and industrial mines), land clearance is indispensable for the installation of forest vegetation. On the surface of the terrace at the flotation tailings dumps, loosening is performed by 30–40 cm depth and/or discussion and in the case of the heaps of mine materials by scarification with tooth scarifiers up to 30–40 cm. For the installation of forest vegetation, this preparation can also be done in strips of 1.0–1.5 m width alternating with nonworked strips of 0.75–1.0 m width, or even on the scales with the sizes determined on a case-by-case basis. On the slopes of the heaps, the preparation of the land will be done in 26 m wide terraces located at 3–6 m distances.

The terraces can be supported with girders (for unstable materials) or unsupported for stable materials; works to improve soil conditions. In almost all cases, industrial waste dumps have deficiencies in nutrients and especially nitrogen. Fertilization works and other works to improve the physicochemical properties of soldering materials, particularly with regard to water and nutrient retention capacity, are therefore required. Coating with a layer of fertile soil of 20–25 cm thick is a generally valid method, also useful in the case of afforestation, but it is very rarely usable due to cost considerations and possibilities of obtaining the required soil quantities.

Covering with other materials, soil substitutes, such as dredging sludge from irrigation, from drainage channels, from port channels, etc. and sludge from municipal waste water treatment plants, are often a feasible alternative. Being rich in OM, they improve the properties of superficial material in the dump and bring in nutrients. The mulching with straw, branches, bark, leaves, sawdust, or peat is another way of improving the materials, having a positive effect in preventing erosion. Other soil improvement activities include acidification by lime or limestone treatment and the fixing or neutralization of noxious substances either by washing soluble salts or by gypsum, phosphogypsum, or other treatments. When organizing forest planting activity, it is recommended that from the typical species for each dumpsite resort to select 2–3 species among which are at least one of the soil ameliorators (nitrogen fixators), namely, acacia or alder.

The indicated plant mix is in the bouquet of 50–100 m^2 or in pure, alternating strips, 5–10 m wide. White sea buckthorn and other shrubs can also be used in crops in intimate blends with large-scale species or in purely alternating lines with them. Intimate mixing should be avoided, in alternate rows or in small pine bouquets with acacia, anchovies, or shrubs, because the aforementioned hardwoods, with higher increases in the early years, overwhelm and remove the pine from the crop. Some species such as acacia can also be used in pure cultures. For the slopes, after the specific consolidation works have been done, samplings will be planted. From the assortment of species recommended, one species will be chosen, obviously most suitable for stationary conditions. In particular, nitrogen-fixing species such as acacia and alder are used in these cases (Mishra et al. 2017).

Very rarely, to achieve landscape effects, two-species combinations are used. The density of the crops will be 0.3–1 m × 2–3 m; the lowest density will be for acacia plantations (0.8–1 m × 2–2.5 m), and the highest for the plantations in the white buckthorn (0.3–0.5 m × 2–2.5 m). Preplanting or concomitant planting is highly recommended for the prevention of erosion, especially by using perennial lump for greening the slopes.

Fertilization and finishing are just as important for trees as for grasses and legumes. If deficiencies and pH are not corrected at the beginning and maintained at normal levels, tree growth will be poor. After many years, the trees will root deep enough to provide nutrients in the necessary quantities, but poor initial growth can lead to tree death through vandalism, due to animals or competition with the surrounding vegetation. If the stock of fertile soil from planting is well fertilized and yields nutrients for the first year, the main care will be given for the second and third years. The limiting nutrients will be, as in the case of herbs, nitrogen and phosphorus.

The Accompanying Vegetation Since trees are planted at low density, much of the land remains unprotected and then a vegetal carpet can be important. This is not necessary if the terrain is flat, but in slopes exposed to erosion, the vegetal mat becomes essential. It can be made from a mixture of grasses with legumes or only legumes; the latter are also valuable by fixing the nitrogen. But the vegetable mat will compete for nutrients and water with young trees and even for light if the trees are very small. Thus, the grass species must be carefully chosen, and the seed rate should be very small, much lower than in the meadows.

Even so, it may be necessary for vegetation to be combated around the trees by cutting or herbicide in the second and third years. Such works are common for forest plantations on eroded land, and other specific maintenance works will be added depending on the characteristics of the landfill (Rana and Maiti 2018).

4.11 Biological Reclamation of the Mining Waste Deposits From the Exploitation of Nonferrous and Precious Metal Ores

Mining, nonferrous and precious metal mining, underground or underground, produces two main types of waste: tailings mines and tailings ponds, to which the (sterile) materials can be added for quarries. From a physical point of view, mine tailings and flotation tailings are different. Mining waste is a complex mixture of coarse parts, mostly over 5 cm in diameter. They are stable, but, unless they have a high proportion of fine material, they will be extremely permeable and dry, with a very low water retention capacity.

The flotation waste is a finely milled material, usually less than 2 mm. As a result, the physical characteristics of the material are appropriate for plant growth: they will usually have good water retention but are sensitive to erosion by water and wind. Both types of material are equally deficient in essential nutrients N and P. They can also be deficient in other nutrients (K, Ca, etc.). The critical issue is the content of metals. In general terms, it is usually about 1000 ppm (0.1%) of any metal in the waste to be toxic to plants.

The availability of metal depends on atmospheric factors, pH, and the presence of different cations and anions. Almost all old and new mining waste have levels of toxic metal content. If there is pyrite, it will be subjected to atmospheric factors and will be oxidized to ferric sulfate and H_2SO_4. This oxidation can be a rapid process (at neutral pH initially decreases in a few months to 4.0 or below) or may be a slow process. If the material also contains carbonates, the acid will be neutralized, but a large amount of soluble salts will be produced.

In an arid climate, evaporation will concentrate these salts on the surface in the form of a salt crust. As a result, many metalliferous dumps are impossible for plant growth and have remained bare, exposed to erosion for many years. However, some old mining waste dumps can find metal tolerant plants (metalloids) that can grow on metallic soils. These adaptations may be specific (for a specific metal) or less specific (broad tolerance to metals).

However, possessing the quality of being tolerant to a plant is not sufficient to ensure its growth in the mining tailings pond; the increase will depend on the existence of sufficient nutrients (N and P) and humidity (Sun et al. 2018).

4.11.1 Chemical and Physical Treatments

Physical methods of coating materials such as crushed stone or granulated blast furnace slag are effective but expensive and are impossible to apply for soft, dry residues. Chemical methods can be an alternative by using materials that can achieve temporary stabilization, but they do not provide a permanent solution. They are temporary solutions that do not provide for biological or landscape rehabilitation of the land.

Direct Sowing Some inferior tailings dumps (0.1%) with low toxicity, where there are no acidic or saline problems, use a vegetal carpet installation by direct sowing and addition of fertilizers. Also mix with lime if the pH is below 5.5. Sowing species will be grass and legume. On sites with higher metal concentrations, it will be necessary to use varieties of metal-tolerant plants. As a rule, metal-tolerant species roots deep into the tailings, escapes the drought, and protects the surface. Since the tolerance is rather specific, tolerant plants should be found where the tolerance matches the site toxicity. Tolerable varieties for Pb and Zn produced the fescue (*F. rubra*) and the grass of the field (*Agrostis tenuis*). Legumes are needed to provide nitrogen by fixing them to the ground. However, they are more sensitive to the toxicity of HMs than grasses.

However, experience shows that different leguminous species could be installed and stabilized in the vegetal carpet on the mining tailings dumps and made the nitrogen fixation. Sowing techniques can be ordinary farming or water sowing. Growth is helped if larger amounts of phosphate are added to the usual fertilization than usual (it fixes some metals that become so unavailable). If grazing is intended, it is necessary to track the metal content of the plants. Grass may be unusable (Różyło et al. 2016).

Surface Coatings In toxic dumps, one of the alternatives is to improve the surface with sufficient material such as sewage sludge or household waste or plant soil so as to ensure plant growth. OM has the ability to complex soluble metals and make them unavailable for plants and at the same time is a long-term source of nutrients for plants.

At Lower Swansea Valley, vegetation has been successfully stabilized on various metallic wastes using a layer of at least 10 cm of coating material. It was also used in Anaconda (Montana–USA) and in England. At some sites (e.g., Broken Hill), household wastewater was pumped to the tailings pond dike, resulting in a good increase in common species. The problem is whether this vegetation lasts after stopping the water supply.

Sometimes the toxicity of metallic waste can be so great that it is preferable to achieve complete isolation, especially where the land is to be farmed and the vegetation must not contain high levels of metals. It would seem that the most useful solution would be the use of vegetal soil cover, but this is also not ideal. Plant soil is difficult to obtain in large quantities, and a substantial thickness layer is required. The soil's ability to retain soluble ions is a disadvantage.

During the dry period, the evaporation will carry upward the toxic ions that will be retained in the soil. If a porous material (zeolites) with a low ion exchange capacity is used, toxic ions will be recirculated again after the drought, and toxicity will not occur at the surface. There are good arguments for the use of coarse inorganic materials.

They will be relatively easy to obtain, often being produced even by mining operations (sterile rock from works outside the deposit); termino heat ash can be used. The surface covered with inert material will be improved by fertilization and using nitrogen fixation legumes. In order to install the trees on such dumps, it will require a layer of inert material of a consistent depth (2 m) for the trees to roots well and grow satisfactorily for many years. Maintenance of vegetation is particularly important, no matter what the landfills were.

Except when covered with a thick layer of inert material, there will be no question of agricultural use that would require a high nutrient intake. To maintain growth after the initial installation period, it is sufficient to administer 25–30 kg/year N, P, and K. This maintenance should last for 6 years. After that, the recycling of nutrients will begin to run alone, and there may be no need for other nutrient additions (Dold et al. 2011).

Ecological rehabilitation through reforestation of tailings dumps results from the ore flotation processes, so-called "tailings ponds." Tailings, although more restricted in size, are a real danger to the surrounding population and the natural environment through strong air pollution, groundwater, or flowing water as a result of the toxic substance content.

The major danger of pollution, primarily air, is due to the presence in the tailings of substances, such as cyanide, Pb, Zn, Cu, S, etc. Following the very fine particle size of the tailings deposit, these substances are trained even by the weakest air currents. Arguments in support of this statement are the data of the measurements taken within 6 months, a 27.4-cm-thick tailings layer, or the sterile accretions in live fruit plantations; the crops around the heaps measure up to 30 cm thickness annually. This explains, in the latter case, the formation over time of sterile dunes with heights of up to 5 m. At the same time, the mentioned toxic substances participate in the chemical solution in the "soil," together with other results of the processes of flotation, limiting factors of the installation of forest vegetation (and not only of it) on these ponds.

Under such conditions, it is inconceivable to develop even herbaceous vegetation without extremely costly measures to combat deflation and to improve soil conditions. Thus, in the case of Pb and Zn, although the forest cultures accounted for only 3.7% of the recultivated areas, the remainder is being covered with perennial grasses, glue plaques, borrowed land coverings, garbage disposal, manure, sludge from sewage treatment plants, and an amendment to limestone rock. In the case of tailings ponds, it should be mentioned that the value of the investment represented 900,000 euro/ha (value in 2017), one of which was only 29% of the plantations (including the price of dung) and maintenance of the forestry crops, the rest representing stabilization and soil improvement works. It should also be noted that the use of gullies or land lending leads to the degradation of the land where they were procured.

With all these laborious and costly works, the improvement of the vegetation conditions for the forest cultures is achieved only to a small extent. Thus, the 14-year-old cultures achieved average annual growths from 23.7 cm at *P. sylvestris* to 45.5 cm at the tremulous.

Euroamerican poplars at the bottom of the slopes record average annual elevations of only 45 cm, which can be an indication of the pollutant effect of pond infiltrated waters on surrounding soils. It is worth noting the very poor development in these conditions of the black alder, which only achieves 13 cm growth. Acacia and birch are intermediate. Of the shrub species, the best growth is recorded by the shallow willow (43.3 cm), the mean height, being of 20 cm.

Some fruit trees, such as plum and sour cherry, but with all the special planting and care measures, have been tried in these conditions; most of the specimens have disappeared, and the remaining ones have a lingering growth with an annual average of only 12 and 5 cm, respectively. Already after the 4th year since planting, the state of the seedlings is laced, even in some species, such as the shrub, which had a good development in the early years. Under these conditions, with the exception of the *Alnus glutinosa*, which achieves an average annual growth of 42.1 cm, the other tree species have a poor development: *R. pseudoacacia* 17.4 cm, *Ulmus pumila* 8 cm, black pine 6.0 cm, and *Fraxinus ornus* only 2 cm. Among the species of shrubs, the best growths are recorded by the willow shrub and the white *H. rhamnoides*, 18.6 and 18.3 cm, respectively, followed by the red *H. rhamnoides* and bird cherry with 13.2 and 11.1 cm, respectively. The poorest development has, among shrubs, the dogwood with an average annual growth of 7.5 and 8.7 cm, in the special improvement conditions adopted, mentioned above, the best development in the first 3 years had *R. pseudoacacia* and trench, with increases in height of 1.18 and 1.01 m/year respectively, followed by white *H. rhamnoides* by 0.44 m. The weakest increases were registered by the *Prunus padus* (0.28 m) (Cetinkaya and Sozen 2011).

4.11.2 Models of Ecological Rehabilitation and Recultivation of Polluted and Degraded Soils From Technical Waste Dumps from Iron Mines

The enzymatic research on the catalytic and invertase activities of the reburied tailings dumps (the Staroosko Fe deposit area, the Kursk Magnetic Anomaly Region) took place in 1979. The dumps from here are of three types: sandy, luteal, and cretaceous–maroon. They were revegetated with spontaneous and introduced herbs or with forest plants (sea buckthorn, acacia, sail). After 8–10 years, reedbeds with herbs were covered with a 1.5–2 cm lawn, which protects them against erosion and deflation. In the forest plantations of 8–10 years, the litter mass reached 0.5–0.9 t/ha.

Catalase activity was measurable in all reefs. This activity, like breathing (CO_2 production) and the accumulation of humus, was more intense under grassy vegetation than in the forest. Invertebrate activity has been influenced not only by the nature of the vegetation but also by the nature of the tailings dumps. This activity was maximal in the rhizosphere of the *H. rhamnoides* growing on the cretaceous–marsh hill (Juwarkar et al. 2009).

Enzymologically, the tailings dumps of existing Fe mining from Stoilensk (located in the region of the Magnetic Anomaly in Kursk) were studied. These dumps are of two types: sandy or cretaceous. The age of spontaneous vegetation on the heaps ranges from 3 to 20 years. The invertase, urease, and catalase activities increased in the 0–5 cm layer of the soil in parallel to the vegetation age. The increase was more pronounced in the soil of sandy dumps than in the cretaceous heaps. During the vegetation period, peak activity values were obtained in the middle of the summer. In both the new and the old ponds, subjected to phytoamelioration with a mixture of six leguminous plants and grasses – common sainfoin, alfalfa, red clover, *Bromus inermis*, fescue, and couch grass or twitch grass (*Elymus repens*) – the EAs increased by 1.5–3 times.

Other studies consisted of splitting the plots of the dump materials covered with surface soil (chernozem), redeposited either in dry state (parcels I) or wet (pumping of the soil and water mixture) (parcels II). Alfalfa was used for revegetation. In the 3rd year of revegetation, samples were collected from the depths of 0–10, 10–20,

20–30, and 30–40 cm of the technogenic soil of parcels I and II and of the zonal soil (chernozem typically) unaffected by mining. Soil samples were chemically analyzed (humus, total N, etc.), enzymatic (urease activity) and microbiological (respiration – CO_2 production and nitrification capacity). For plots II, the soil samples were taken from the central area or examined separately from those in the peripheral area of these parcels. Alfalfa production in plots I and II has also been recorded. All of the parameters mentioned above showed a downward trend depending on the depth from which soil samples were collected.

The urease activity, respiration, and nitrification capacity in the 0–40 cm layer of the technogenic soil had much higher average values in plots I than in parcels II. At the same time, the average values obtained in the soil of the parcels I were similar to those recorded in the zonal soil. From here, it can be deduced that redeposition of dry soil surface is a more biologically favorable method than its redeposition in a wet state.

Within plots II, large differences between the central and the peripheral areas were not highlighted. It has also been established that urease activity and nitrification capacity in the soil of parcels I and II correlated significantly with the total N content; the urease activity in the technogenic soil also correlated significantly with the production of *M. sativa*. The dehydrogenase, catalase, and invertase activities were compared to nipple tailings or calcareous clay (pH in H_2O = 7.7–8.3), leveling for agricultural recultivation in the northern area of Fe mining and adjacent soils.

In soils, activities decreased or decreased with depth (0–70 or 0–80 cm), while in the tailings dumps, they were approximately the same in the 0–20 and 50–80 cm layers. In the 0–20 cm layer, each activity was several times lower in the heaps than in soils. In the 50–80 cm layer, the differences between heaps and soils were high in their dehydrogenase activity, but not so pronounced in their catalase and invertase activities (Wang et al. 2014).

In another study, similar results (i.e., very low and high enzyme activities) were obtained in different profiles of dumps and soils, except for a landfill profile that exhibited relatively high dehydrogenase and catalase activity. In the sterile waste dumps in the southern area of the Fe exploitation, the recultivation plots were installed. Some plots have been recultivated with *Onobrychis viciifolia* and others with a *D. glomerata*. After 3 years of recultivation, the soil material in these parcels was analyzed enzymatically.

For comparison, the 0–15 cm layer of an adjacent native soil (rendzine) and the same layer of the soil stock in a parcel that was not recultivated were also analyzed. The results show that recultivation has led to increased enzyme activities in tailings dumps during their transformation into technogenic soils. The potential dehydrogenase activity increased to a lesser extent than phosphatase activity. This activity has reached values similar to those found in native soil. Both activities were higher in the sparse plot than in the recultivated field. In each case, the 0–20 cm layer was more active than the 20–40 cm (Joniec et al. 2019).

In Fe mining, very low values of dehydrogenase, catalase, and invertase activities were found in the 0–20 and 50–80 cm layers of three leveling heap profiles for recultivation. In November 1985, this year being the 9th year of an experiment of fertilization and crop rotation in the southern area of Fe exploitation, 5–20-cm-depth samples were collected for the enzymatic analysis of parcels representing five fertilization variants: V_1, non-fertilized (control); V_2, fertilized with manure (40 t/ha); V_3, fertilized complex, organic and mineral stable (40 t/ha $+N_{100}P_{60}K_{40}$); V_4, fertilized with NPK in the dose $N_{100}P_{60}K_{40}$; and V5, fertilized with NPK in the dose $N_{300}P_{180}K_{120}$. Mineral fertilizers were administered as NH_4NO_3, superphosphate, and potassium salt. Each variant comprised plots that were cultivated in 1995 with corn, oats, or sprats.

Nonenzymatic, enzymatic, and catalytic potential (current and potential dehydrogenase activities, invertase, phosphatase, urease, and nonenzymatic cleavage capacity of H_2O), as well as the harvest of the three plants, was the highest in the soil of the complex fertilized plots and the lowest in the soil of the unfertilized plots.

Significant correlations were found between invertase activity and corn harvest, and between phosphatase activity and sparse production. But under the influence of long-term fertilization, the productive capacity of the studied technogenic soil has increased to a greater extent than its biological potential reflected by its EAs. Thus, long-term fertilization is able to increase the productive capacity of the technical soil, but its stimulating effect on the EAs of the soil is the result of much slower processes. An experiment was described in which the fermented and dried sludge was used for fertilization, coming from the middle of the 2.5 m distance between the pine trees and the algin soil under the alder. The invertase, urease, and proteinase activities were also highest in the technical soil under the alder. In a manganese quarry, several EAs were determined in different layers of a 20-year-old spontaneously re-smeared plot of land, which is currently covered by a stable phytocoenosis, *Poa angustifolia* and *Artemisia austriaca*, and found that the activities were much higher (catalase activity of 1.5, phosphatase 13, urease 36, invertase 46, and dehydrogenase 72 times) in the upper layer of 1 cm, rich in roots than at the depth of 6 cm or in the deeper layers of the dump. The number of MOs, as well as the humus content, N total, P mobile, and K exchangeable, was also higher in the upper layer (Jain et al., 2016).

In the Nikopol manganese basin (Dnepropetrovsk region), invertase, urease, and phosphatase activities were determined in fresh discovery materials (loessoid clay, red clay, and gray clay) in the quarry walls as well as in the 0–20 cm layer of parcels of discovery materials (soil, loessoid clay, red–brown clay, and gray–green clay) spontaneously revegetated for 17 years. As a control it was used in this experiment served the plot of an adjacent native soil (chernozem) uncultivated. In the discovery materials in the quarry walls, the three EAs studied were very poor or absent. Each activity was evidenced by the loessoid clay, but only the urease activity was measurable in the clays.

In the plots of revegetated discovery materials, the EAs increased significantly, providing the maximum values in the plot where the discovery material was soil. But the discovery material in any parcel did not reach the level of the EAs in the native soil: the invertase, urease, and phosphatase activities of the plots, according to the nature of the discovery materials, represented 39.4, 34.5, and 28, 4% (soil); 23.2, 14.8, and 9.30% (loessoid clay); 34.0, 22.6, and 7.6% (red–brown clay); and 37.7, 27.3, and 5.4% (gray–green clay) from native soil activities.

These data show that, compared to the level of EAs in the native soil, the accumulation of enzymes in the discovery materials, whatever their nature, occurred in the order of the following: invertase > urease > phosphatase. It has been found that in the parcels installed 18 years ago and spontaneously revegetated, not only the activity of hydrolytic enzymes (invertase, urease, and phosphatase) but also of oxidoreducting increased. Growth was more pronounced in catalase than in dehydrocerase activity. Enzyme activities were determined in other plots. These were also installed in the Nikopol area, 18 years ago, on the same discovery materials, but they were revegetated not spontaneously, but by sowing *M. sativa* or *O. viciifolia*, also known as *O. sativa* or common sainfoin. Under the action of these plants, the process of hydrolytic enzyme accumulation in all plots has been intensified. The activity of oxidoreducting enzymes increased, but in contrast to spontaneous revegetation, the recultivation with *M. sativa* and *O. viciifolia* increased the dehydrogenase activity rather than the catalase activity: the enzymatic properties of the tailings dumps recultivated to an up-to-date exploitation of the Chiatura mantle or the basin of the Kvirila River, West Georgia). Native soil had much higher EAs than uncultivated tailings, consisting of limestone sands and limestone, pH in H2O – 8.4. Activities increased in recultivated plots but did not reach the values recorded in native soil. The blend of *Lolium multiflorum–M. sativa* was more effective than vines. Covering tailings dumps with a layer of surface soil had a favorable effect on the accumulation of enzymes (invertase, phosphatase) under the conditions of cultivation with vines. The 0–20 cm layer was more active and richer in humus and MOs than the 20–40 cm. There was no link between catalase activity in tailings dumps and applied recultivation measures (de Varennes et al. 2010).

4.12 Technogenic Soils from Residues from Lead and Zinc Mines

Studying the decomposition of vegetal remains by increasing the residue of Pb and Zn mines, inter alia, carried out enzymatic analyzes. The studied residue is located around the abandoned mine in Y Fan (Powys, Wales) and contains high concentrations of Pb and Zn. Following the abandonment of the mine (1928), the residue was partially naturally colonized with *A. tenuis* tolerant to metals. For studies, a land colonized equally has been selected. The control was considered a similar but uncontaminated land, located on a pasture at about 500 m from the mine.

The vegetation of the control field is mainly composed of *A. tenuis* and *Festuca ovina*. The two lands were compared in terms of urease activity in soil, microbial populations in litter and soil, and litter microfauna. The accumulation of the lithium was more pronounced on the residue, which contained significantly less humic and fulvic acids in the soil located immediately beneath the lithium layer. Urease activity was significantly lower in mine soil than in adjacent pastureland. The number of MOs in the litter of the two fields was not remarkably different, although the number of fungi was lower in the mine litter, while the number of bacteria and actinomycetes was higher. In contrast, the number of all microorganism groups in the mine soil was considerably lower than that in grassland.

Similarly, fewer animals were found in the mine litter. Decreasing the biological activity of litter and soil with residue, caused by high concentrations of Pb and Zn, explains the delayed decomposition of vegetation remains growing on this site. Pedo-enzymological methods have been applied to determine the causes of the differences in the flora of two adjacent areas, one poor and another rich in species, on a limestone terrace in the Pb mining complex on Grassington Moor in Yorkshire Pennines (England). The northern half of the terrace receives drainage water and textured fine residue containing Pb and Zn from the abandoned mine on the slope. In this area, the vegetation is rare, has poor flora, and is composed of typical species for the heavy metal mines in the British Isles, namely: *Minuartia verna, A. tenuis*, and *F. ovina*. The southern terrain is not affected by the mine residue.

There, vegetation was rich and continuous flora, except for places with limestone outcrops. The average number of species per 0.25 m^2 was 2.4 in the species-poor and 10.1 in the species-rich area. Poor soils in the species have lower pHs and contain less humus, N–NO_3^-, N–NH_4^+, P assimilable, and K exchangeable, in comparison to those of the species-rich area. Total Pb content in poor soils, respectively rich in species was on average 78,000 and 8000 μg/g soil, respectively, well above the limit value of 350 μg/g, above which the Pb content is considered abnormally high. Consequently, the soils of both areas were expected to be toxic to all plants, except for tolerant forms. The average Pb extraction rate with ammonium acetate was 21,800 μg/g soil in the poorer species and only 311 μg/g soil in the other area. The amount of Zn was much lower than Pb in both areas, and the difference in areas in terms of total Zn and Zn extractable ammonium acetate was less marked than Pb.

Acidic, dehydrogenase, and urease phosphatase activities and breathing – CO_2 production – were measured in soil samples collected from the root area at 2–9 cm below the surface. Expressed on the basis of the dry soil weight of the soil, they are higher in the soil of the area rich in species. But, expressed on the basis of the dry weight of OM, the differences have diminished or eliminated. In each half of the terrace, there were significant correlations between species density, nutrient quantity, and EAs, all of which were in reverse relation to the amount of extractable Pb.

It has been concluded that nutrient enrichment is involved in the formation of the species-rich area on Grassington Moor; higher EAs in the soil of the area rich in species show that metal detoxification took place here and that the more abundant OM in this area is related to EAs (Hesami et al. 2018).

4.13 Conclusions and Recommendations

In nature, extreme, dangerous, natural phenomena of different origins (geological, geomorphological, atmospheric, climatic, etc.), which are part of the natural evolution of the terrestrial geosystem and which, due to their consequences, result in major environmental degradation due to disasters which cause risk-dependent damages. Soil is the shallow part of the earth's bark that allows the growth of plants and animals. It has been formed over time by the long and interdependent action of climatic and biotic factors on parental rocks. Unlike other natural resources, the soil is limited in scope and has a fixed character; once destroyed, it can no longer be restored as it was, because it cannot reproduce the conditions of its formation. Soil degradation, through loss of fertility, occurs through the following: export of nutrients from the soil with harvest, redevelopment of marshes, and erosion caused by massive deforestation or excessive grazing, acidification or salinization, or pollution. Soil pollution consists of changing the qualitative and quantitative composition, a change that affects the normal evolution of the biocenosis associated with it. The main purpose of this paper was to identify, group, and research soil conditions with regard to the forms in which it is exposed and the risk of the occurrence and extension of these processes by describing advanced technologies of ecological reconstruction and bioremediation of degraded lands. A clear differentiation was done between reconstruction, rehabilitation, degradation, and deterioration as follows:

- ***Reconstruction vs. rehabilitation***. Reconstruction involves the restoration of all processes, subsystems, and functional structures of an ecosystem affected by a major disruption, replacing or constructing if (new) destroyed structures/subsystems are needed to enable the ecosystem to perform its previous functions. Rehabilitation refers to actions aimed at restoring the main functions and processes of a system. Support is provided for the affected ecosystem to resume most of the functions it was capable of before confronting a major disruption.
- ***Degradation vs damage.*** Degradation is the process by which ecosystems are progressively contaminated, polluted, overexploited, fragmented, and sometimes even destroyed (following anthropogenic intervention). Deterioration of the main components of the ecosystem involves both the degradation of the biotope/habitats and the biotic–biocenosis component (populations, species).

The causes of various negative soil degradation processes are many and have led to a worsening of the physical, chemical, and biological properties of soils and, ultimately, their productivity status. Applying poor agricultural management without taking into account local specific conditions has increased the risk of agrophysical soil degradation and has intensified these negative processes that already affected important land areas. The only reliable procedure to address such events is integrated risk management. The components of this managerial strategy are limiting the size of the risk by applying mandatory authorities' activities, assessing the situation, reducing vulnerability through prevention, and preparing and informing the population and local authorities.

References

Asgari Lajayer B, Khadem Moghadam N, Maghsoodi MR, Ghorbanpour M, Kariman K (2019) Phytoextraction of heavy metals from contaminated soil, water and atmosphere using ornamental plants: mechanisms and efficiency improvement strategies. Environ Sci Pollut Res Int 26(9):8468–8484

Azubuike CC, Chikere CB, Okpokwasili GC (2018) Bioremediation techniques-classification based on site of application: principles, advantages, limitations and prospects. World J Microbiol Biotechnol 32(11):180

Behera BK, Prasad R (2020a) Environmental technology and sustainability. Elsevier. ISBN 9780128191033. https://www.elsevier.com/books/environmental-technology-and-sustainability/behera/978-0-12-819103-3

Behera BK, Prasad R (2020b) Strategies for soil management. In: Behera BK, Prasad R (eds) Environmental technology and sustainability. Elsevier, Amsterdam, pp 143–167

Boim AG, Melo LC, Moreno FN, Alleoni LR (2016) Bioconcentration factors and the risk concentrations of potentially toxic elements in garden soils. J Environ Manag 170:21–27

Bourret MM, Brummer JE, Leininger WC (2009) Establishment and growth of two willow species in a riparian zone impacted by mine tailings. J Environ Qual 38(2):693–701

Cai X, Zhang Z, Yin N, Du H, Li Z, Cui Y (2016) Comparison of arsenate reduction and release by three As(V)-reducing bacteria isolated from arsenic-contaminated soil of Inner Mongolia, China. Chemosphere 161:200–207

Carrasco L, Azcón R, Kohler J, Roldán A, Caravaca F (2018) Comparative effects of native filamentous and arbuscular mycorrhizal fungi in the establishment of an autochthonous, leguminous shrub growing in a metal-contaminated soil. Sci Total Environ 409(6):1205–1209

Cetinkaya G, Sozen N (2011) Plant species potentially useful in the phytostabilization process for the abandoned CMC mining site in northern Cyprus. Int J Phytoremediation 13(7):681–691

Colin Y, Goberna M, Verdú M, Navarro-Cano JA (2019) Successional trajectories of soil bacterial communities in mine tailings: the role of plant functional traits. J Environ Manag 241:284–292

Collado S, Oulego P, Suárez-Iglesias O, Díaz M (2018) Biodegradation of dissolved humic substances by fungi. Appl Microbiol Biotechnol:3497–3511

de Varennes A, Abreu MM, Qu G, Cunha-Queda C (2010) Enzymatic activity of a mine soil varies according to vegetation cover and level of compost applied. Int J Phytoremediation 12 (4):371–383

Das A, Prasad R, Srivastava A, Giang PH, Bhatnagar K, Varma A (2007) Fungal siderophores: structure, functions and regulations. In: Varma A, Chincholkar SB (eds) Microbial siderophores, vol 12. Springer-Verlag, Berlin, pp 1–42

Dold B, Diaby N, Spangenberg JE (2011) Remediation of a marine shore tailings deposit and the importance of water-rock interaction on element cycling in the coastal aquifer. Environ Sci Technol 45(11):4876–4883

Dutta S, Podile AR (2010) Plant growth promoting rhizobacteria (PGPR): the bugs to debug the root zone. Crit Rev Microbiol 36(3):232–244

Egamberdieva D, Kamilova F, Validov S, Gafurova L, Kucharova Z, Lugtenberg B (2008) High incidence of plant growth-stimulating bacteria associated with the rhizosphere of wheat grown on salinated soil in Uzbekistan. Environ Microbiol 10:1):1–1):9

Garlapati D, Chandrasekaran M, Devanesan A, Mathimani T, Pugazhendhi A (2019) Role of cyanobacteria in agricultural and industrial sectors: an outlook on economically important byproducts. Appl Microbiol Biotechnol 103(12):4709–4721

Gay-des-Combes JM, Sanz Carrillo C, Robroek BJM, Jassey VEJ, Mills RTE, Arif MS, Falquet L, Frossard E, Buttler A (2017) Tropical soils degraded by slash-and-burn cultivation can be recultivated when amended with ashes and compost. Ecol Evol 7(14):5378–5388

Hesami R, Salimi A, Ghaderian SM (2018) Lead, zinc, and cadmium uptake, accumulation, and phytoremediation by plants growing around Tang-e Douzan lead-zinc mine, Iran. Environ Sci Pollut Res Int 25(9):8701–8714

Horemans B, Breugelmans P, Saeys W, Springael D (2018) Soil-bacterium compatibility model as a decision-making tool for soil bioremediation. Environ Sci Technol 51(3):1605–1615

Hu C, Yang X, Gao L, Zhang P, Li W, Dong J, Li C, Zhang X (2019) Comparative analysis of heavy metal accumulation and bioindication in three seagrasses: which species is more suitable as a bioindicator? Sci Total Environ 669:41–48

Ignatius A, Arunbabu V, Neethu J, Ramasamy EV (2014) Rhizofiltration of lead using an aromatic medicinal plant *Plectranthus amboinicus* cultured in a hydroponic nutrient film technique (NFT) system. Environ Sci Pollut Res Int 21(22):13007–13016

Jain S, Mishra D, Khare P, Yadav V, Deshmukh Y, Meena A (2016) Impact of biochar amendment on enzymatic resilience properties of mine spoils. Sci Total Environ 544:410–421

Ji Y, Wu P, Zhang J, Zhang J, Zhou Y, Peng Y, Zhang S, Cai G, Gao G (2018) Heavy metal accumulation, risk assessment and integrated biomarker responses of local vegetables: a case study along the Le'an river. Chemosphere 199:361–371

Jonasson ME, Afshari R (2018) Historical documentation of lead toxicity prior to the 20th century in English literature. Hum Exp Toxicol 37(8):775–788

Joniec J, Gąsior J, Voloshanska S, Nazarkiewicz M, Hoivanovych N (2019) Evaluation of the effectiveness of land reclamation based on microbiological and biochemical parameters assessed in an ozokerite mining and processing landfill sown with *Trifolium hybridum* and *Dactylis glomerata*. J Environ Manag 242:343–350

Juwarkar AA, Yadav SK, Thawale PR, Kumar P, Singh SK, Chakrabarti T (2009) Developmental strategies for sustainable ecosystem on mine spoil dumps: a case of study. Environ Monit Assess 157(1–4):471–481

Konstantinova E, Minkina T, Sushkova S, Konstantinov A, Rajput VD, Sherstnev A (2019) Urban soil geochemistry of an intensively developing Siberian city: a case study of Tyumen, Russia. J Environ Manag 239:366–375

Küpper H, Andresen E (2016) Mechanisms of metal toxicity in plants. Metallomics 8(3):269–285

Lax S, Cardona C, Zhao D, Winton VJ, Goodney G, Gao P, Gottel N, Hartmann EM, Henry C, Thomas PM, Kelley ST, Stephens B, Gilbert JA (2019) Microbial and metabolic succession on common building materials under high humidity conditions. Nat Commun 10(1):1767

Lebeau T, Braud A, Jézéquel K (2008) Performance of bioaugmentation-assisted phytoextraction applied to metal contaminated soils: a review. Environ Pollut 153(3):497–522

Liu L, Li W, Song W, Guo M (2018a) Remediation techniques for heavy metal-contaminated soils: principles and applicability. Sci Total Environ 633:206–219

Liu C, Lin H, Dong Y, Li B, Liu Y (2018b) Investigation on microbial community in remediation of lead-contaminated soil by *Trifolium repens* L. Ecotoxicol Environ Saf 165:52–60

Maier S, Tamm A, Wu D, Caesar J, Grube M, Weber B (2018) Photoautotrophic organisms control microbial abundance, diversity, and physiology in different types of biological soil crusts. ISME J 12(4):1032–1046

Mishra T, Singh NB, Singh N (2017) Restoration of red mud deposits by naturally growing vegetation. Int J Phytoremediation 19(5):439–445

Nguyen-The C, Bardin M, Berard A, Berge O, Brillard J, Broussolle V, Carlin F, Renault P, Tchamitchian M, Morris CE (2016) Agrifood systems and the microbial safety of fresh produce: Trade-offs in the wake of increased sustainability. Sci Total Environ 562:751–759

Niklaus PA, Le Roux X, Poly F, Buchmann N, Scherer-Lorenzen M, Weigelt A, Barnard RL (2016) Plant species diversity affects soil-atmosphere fluxes of methane and nitrous oxide. Oecologia 181(3):919–930

Okie JG, Van Horn DJ, Storch D, Barrett JE, Gooseff MN, Kopsova L, Takacs-Vesbach CD (2015) Niche and metabolic principles explain patterns of diversity and distribution: theory and a case study with soil bacterial communities. Proc Biol Sci 282(1809):20142630

Pan X, Chen Z, Li L, Rao W, Xu Z, Guan X (2017) Microbial strategy for potential lead remediation: a review study. World J Microbiol Biotechnol 33(2):35

Pelfrêne A, Kleckerová A, Pourrut B, Nsanganwimana F, Douay F, Waterlot C (2015) Effect of Miscanthus cultivation on metal fractionation and human bioaccessibility in metal-contaminated soils: comparison between greenhouse and field experiments. Environ Sci Pollut Res Int 22 (4):3043–3054

Prasad R, Aranda E (2018) Approaches in bioremediation: the new era of environmental microbiology and nanobiotechnology. Springer International Publishing (978–3–030-02369-0). https://www.springer.com/gp/book/9783030023683
Pratush A, Kumar A, Hu Z (2018) Adverse effect of heavy metals (As, Pb, Hg, and Cr) on health and their bioremediation strategies: a review. Int Microbiol 21(3):97–106
Rana V, Maiti SK (2018) Differential distribution of metals in tree tissues growing on reclaimed coal mine overburden dumps, Jharia coal field (India). Environ Sci Pollut Res Int 25 (10):9745–9758
Różyło K, Gawlik-Dziki U, Świeca M, Różyło R, Pałys E (2016) Winter wheat fertilized with biogas residue and mining waste: yielding and the quality of grain. J Sci Food Agric 96 (10):3454–3461
Sage L, Périgon S, Faure M, Gaignaire C, Abdelghafour M, Mehu J, Geremia RA, Mouhamadou B (2014) Autochthonous ascomycetes in depollution of polychlorinated biphenyls contaminated soil and sediment. Chemosphere 110:62–69
Sarma H, Forid N, Prasad R, Prasad MNV, Ma LQ, Rinklebe J (2021) Enhancing phytoremediation of hazardous metal(loid)s using genome engineering CRISPR–Cas9 technology. J Hazard Mater. https://doi.org/10.1016/j.jhazmat.2021.125493
Sas-Nowosielska H, Pawlas N (2015) Heavy metals in the cell nucleus-role in pathogenesis. Acta Biochim Pol 62(1):7–13
Shahid M, Shamshad S, Rafiq M, Khalid S, Bibi I, Niazi NK, Dumat C, Rashid MI (2017) Chromium speciation, bioavailability, uptake, toxicity and detoxification in soil-plant system: a review. Chemosphere 178:513–533
Solis-Gabriel L, Mendoza-Arroyo W, Boege K, Del-Val E (2017) Restoring lepidopteran diversity in a tropical dry forest: relative importance of restoration treatment, tree identity and predator pressure. PeerJ 5:e3344
Somova LA, Pechurkin NS, Pisman TI (2005) Principles of biological adaptation of organisms in artificial ecosystems to changes of environmental factors. Adv Space Res 35(9):1512–1515
Sun W, Sierra-Alvarez R, Milner L, Oremland R, Field JA (2009) Arsenite and ferrous iron oxidation linked to chemolithotrophic denitrification for the immobilization of arsenic in anoxic environments. Environ Sci Technol 43(17):6585–6591
Sun W, Ji B, Khoso SA, Tang H, Liu R, Wang L, Hu Y (2018) An extensive review on restoration technologies for mining tailings. Environ Sci Pollut Res Int 25(34):33911–33925
Thakare M, Sarma H, Datar S, Roy A, Pawar P, Gupta K, Pandit S, Prasad R (2021) Understanding the holistic approach to plant-microbe remediation technologies for removing heavy metals and radionuclides from soil. Curr Res Biotechnol. https://doi.org/10.1016/j.crbiot.2021.02.004
Torsvik V, Øvreås L, Thingstad TF (2002) Prokaryotic diversity-magnitude, dynamics, and controlling factors. Science 296(5570):1064–1066
Trivedi P, Delgado-Baquerizo M, Anderson IC, Singh BK (2016) Response of soil properties and microbial communities to agriculture: implications for primary productivity and soil health indicators. Front Plant Sci 7:990
Turkyilmaz A, Sevik H, Isinkaralar K, Cetin M (2019) Use of tree rings as a bioindicator to observe atmospheric heavy metal deposition. Environ Sci Pollut Res Int 26(5):5122–5130
van Kuijk M, Anten NP, Oomen RJ, Schieving F (2014) Stimulating seedling growth in early stages of secondary forest succession: a modeling approach to guide tree liberation. Front Plant Sci 5:345
Velasco Ayuso S, Giraldo Silva A, Nelson C, Barger NN, Garcia-Pichel F (2017) Microbial nursery production of high-quality biological soil crust biomass for restoration of degraded dryland soils. Appl Environ Microbiol 83(3): pii: e02179-16
Wang G, Post WM, Mayes MA (2013) Development of microbial-enzyme-mediated decomposition model parameters through steady-state and dynamic analyses. Ecol Appl 23(1):255–272
Wang F, Yao J, Chen H, Yi Z, Choi MM (2014) Influence of short-time imidacloprid and acetamiprid application on soil microbial metabolic activity and enzymatic activity. Environ Sci Pollut Res Int 21(17):10129–10138
Wang C, Li W, Yang Z, Chen Y, Shao W, Ji J (2015) An invisible soil acidification: critical role of soil carbonate and its impact on heavy metal bioavailability. Sci Rep 5:12735

Weissmannová HD, Pavlovský J (2017) Indices of soil contamination by heavy metals-methodology of calculation for pollution assessment (minireview). Environ Monit Assess 189 (12):616
Wiatrowska K, Komisarek J (2019) Role of the light fraction of soil organic matter in trace elements binding. PLoS One 14(5):e0217077
Wu Y, Jing X, Gao C, Huang Q, Cai P (2018) Recent advances in microbial electrochemical system for soil bioremediation. Chemosphere 211:156–163
Zalewska T, Danowska B (2017) Marine environment status assessment based on macrophytobenthic plants as bio-indicators of heavy metals pollution. Mar Pollut Bull 118 (1–2):281–288
Zeng Z, Guo X, Xu P, Xiao R, Huang D, Gong X, Cheng M, Yi H, Li T, Zeng G (2018) Responses of microbial carbon metabolism and function diversity induced by complex fungal enzymes in lignocellulosic waste composting. Sci Total Environ 643:539–547
Zogg GP, Travis SE, Brazeau DA (2018) Strong associations between plant genotypes and bacterial communities in a natural salt marsh. Ecol Evol 8(9):4721–4730

The Recent Strategies Employed in Chemical Analysis of Contaminated Waters, Sediments and Soils as a Part of the Remediation Process: Extraction

5

Lucia Nemček and Ingrid Hagarová

Abstract

For the past few decades, Earth's terrestrial and aquatic ecosystems have been altered dramatically by human actions. Many industrial activities have significant adverse effects on the environment; a lot of mass-produced chemicals that play a critical role in our everyday lives present risks to living things during their manufacture, use and disposal. The pollutants often persist in the environment for a long time; therefore, there is a call for research and development of remediation procedures that lead to successful removal of emerging contaminants from water, air, sediments and soils. One of the leading sample preparation techniques in trace element analysis is the extraction. The technique is available in different modifications and it is applicable for both organic and inorganic pollutants. This chapter presents an overview of different extraction techniques which represent an important step in preparing samples for analysis of metals, organic compounds and other substances that can act as environmental pollutants. The reliability of solid-liquid extraction for separation of Cu^{2+}, Ni^{2+}, Cd^{2+} and Pb^{2+} ions from contaminated soils and river sediments prior to their determination by flame atomic absorption spectrometry is illustrated by the original experiment. The efficacy of different types of extraction agents was evaluated: Triton X-100 and Triton X-114 as synthetic non-ionic surfactants' representatives, secondary metabolites of *Aspergillus niger* and quillaja saponin as a natural surfactant that meets the green chemistry concept. Environmental samples known to contain considerable amounts of heavy metals were collected from four areas of historical mining activity in Slovakia – Šobov, Pezinok, Richňava and Jaklovce.

L. Nemček (✉) · I. Hagarová
Institute of Laboratory Research on Geomaterials, Faculty of Natural Sciences, Comenius University in Bratislava, Bratislava, Slovakia
e-mail: lucia.nemcek@uniba.sk

R. Prasad (ed.), *Environmental Pollution and Remediation*, Environmental and Microbial Biotechnology, https://doi.org/10.1007/978-981-15-5499-5_5

Keywords

Analytical · Heavy metals · Liquid-liquid extraction · Quillaja saponin · Sediment · Soil · Solid-liquid extraction · Solvent · Surfactant · Water analysis

5.1 Introduction

The use of materials, many of which are now recognized as hazardous, has been increasing continuously in different fields of applications throughout the last decades. Although production and an intentional use of certain substances has already been restricted, banned or is about to be banned and thus their release into the environment has been considerably reduced or eliminated, the issue sustains as these substances tend to persist in the environment and accumulate in animal and human tissues. Since the harmful effects and potential toxicity on living organisms of many of these compounds have long been confirmed, the concerns about their presence in the environment are rising.

The selection of suitable remediation procedure that leads to successful removal of emerging contaminants from the media stands on their precise identification and quantification. The proper chemical characterization of the industrial and municipal waters, sediments and soils is closely related to the selection of appropriate analytical techniques. Liquid- and solid-phase (micro)extraction in combination with gas chromatography, liquid chromatography and spectrometric methods, such as flame atomic absorption spectrometry, electrothermal atomic absorption spectrometry, inductively coupled plasma optical emission spectrometry and inductively coupled plasma mass spectrometry, are techniques considered as most responsible for the progress in the detection of wide range of contaminants.

Based on literature research, this chapter gives an overview of different methods currently used for sample preparation in the analysis of substances that have been accidentally or deliberately introduced into our environment and have a harming potential. For the scope of this work, the discussion will be limited to the brief description of the techniques with an emphasis on their suitability for preconcentration and determination of particular categories of environmental contaminants. In this chapter, the effectiveness of solid-liquid extraction as one of the commonly used techniques for the removal of heavy metal cations from contaminated soils and sediments is illustrated by the original experiment; for separation of Cu^{2+}, Ni^{2+}, Cd^{2+} and Pb^{2+} prior to their determination by flame atomic absorption spectrometry (FAAS), following types of extraction agents were compared: (1) two types of octyl-phenoxy-poly-ethoxy-ethanol (Triton X-100 and Triton X-114) representing the group of popular synthetic non-ionic surfactants, and (2) secondary metabolites of microscopic filamentous fungus *Aspergillus niger* and (3) saponin isolated from the bark of *Quillaja saponaria* tree as the examples of agents derived from biological material, meeting the requirements for green surfactants.

5.2 The Role of Extraction in Environmental Analysis

The effects of human activity on the environment have raised an increasing concern since the 1970s, and international policy frameworks have been developed and implemented over the past decades to control and mitigate this impact (ISDR 2007). Nowadays, our environment is constantly being sampled and analysed for a variety of purposes, including for the assessment of the quality of drinking water, groundwater, river and seawater, wastewaters, effluents and sewage sludge, sediments, soil and biota. In order to be able to detect even very low concentrations of pollutants in the ecosystems, it is necessary to follow a series of steps.

The overall process of data gathering, analysis and interpretation starts with sampling. This step involves collecting samples that faithfully represent a waste or element of the environment from which they are taken. Samples must not be contaminated during collection and handling, and analyte concentrations must not change between the time of collection and analysis (EPA Victoria 2009). If an unrepresentative sample is obtained due to choosing an inappropriate technique or cross-contamination, even the careful laboratory analysis that follows cannot ensure delivery of accurate and reliable data.

Prior to analysis performed by instrumental methods that dominate modern analytical chemistry, most samples require preparation. This may involve mechanical operations such as crushing, sieving and blending, air−/freeze−/oven−/microwave-drying and different operations that fall under three main categories of chemical and physico-chemical sample pretreatment: (1) isolation and clean-up, (2) sample preconcentration and (3) sample derivatization. The isolation of the chemical specie(s) of interest (i.e. analyte) from sample matrix (air, water, sediment, soil, biota, etc.) is achieved by various separation methods, such as (co)precipitation, filtration, adsorption on a sorbent, decantation, centrifugation, magnetic separation, evaporation, distillation, dissolution, digestion and, by one of the leading sample preparation techniques in trace element analysis, extraction. Some analytes also need to be cleaned up from co-extracted, non-target substances and/or increased in concentration prior to analysis or detection. Besides being thought of as a separation technique, extraction is also being frequently used for both of these purposes, i.e. sample purification and preconcentration.

Occasionally, it is also necessary to transform a chemical compound (educt) into another similar compound (derivative) by altering one or more of its functional groups (definition given by the Royal Society of Chemistry). This can be done for various reasons: (1) to enhance extraction efficiency, (2) to boost detection sensitivity or (3) to make compounds more amenable to a particular analysis mode (Kudlejova et al. 2012). For instance, if the analyte is not volatile enough to be conveniently determined via headspace extraction (Jeannot 2007), it can be derivatized to improve its volatility. The derivatizing agent can be added to the sample before, during or after sample extraction. The first two options may lead to enhanced sensitivity and selectivity of both extraction and detection and, in case of solid-phase microextraction, to determination of analytes that are normally not amenable by this method. Regarding post-extraction derivatization, it can only improve the chromatographic behaviour and detection properties, not the extraction efficiency (Pawliszyn 2002; Kudlejova et al. 2012). The fact that extraction and derivatization can be performed simultaneously is illustrated for example in the study on in-matrix derivatization of amphetamine-like molecules, where Mohamed (2017) presents a simple method in which both procedures can be carried out in one step.

Sample preparation is a critical step and the main source of uncertainties in the analysis of environmental samples (Ribeiro et al. 2014); it adds complexity to the analytical process, but on the other hand it is a potential source of bias, variance, contamination and mechanical loss. Therefore, sample preparation should be planned carefully and adequately documented to provide a complete record of the sample history (Keith et al. 1983). Especially derivatization should be performed only when necessary, since the derivatizing agents can interfere and introduce errors in the system (Kudlejova et al. 2012).

After pretreatment, samples are finally prepared to be measured by highly selective and sensitive instrumental analytical methods, such as molecular spectroscopic methods (UV-vis, fluorescence, infrared (IR) and Raman spectroscopy), atomic spectroscopic methods (atomic absorption flame and non-flame spectrometry, atomic fluorescence flame and non-flame spectrometry and atomic emission spectrometry (including inductively coupled plasma (ICP) and microwave plasma (MP)), chromatographic methods (gas chromatography (GC), liquid chromatography (LC), high-performance liquid chromatography (HPLC) and thin-layer chromatography (TLC)), mass spectrometric methods, electrochemical methods, thermal methods and radiological methods.

5.3 The Types of Extraction Techniques

Extraction is considered a crucial step in most chemical analyses; it entails removing the target analyte from the sample matrix and passing it into the phase required for identification and quantification performed mostly by spectroscopic and chromatographic techniques. The analyte might be one specific compound or a group of compounds of similar nature. Solubilization and separation of the analyte of interest from other material is done with an appropriate solvent, chosen with regard to the solubility of the target analyte and on the balance of cost, safety and environmental concerns. In order to perform an extraction, the analyte must have a higher solubility in the latter phase than in the starting phase (JoVE 2019). The starting phase might be the original sample matrix or a solution in which the sample has been dissolved or digested (Tissue 2013). Two most common types of laboratory extraction are liquid-liquid and solid-liquid extraction; their principles, the types and amounts of solvents used and some current trends that favour effective analyte extractions from environmental matrices are discussed in the next section. The most widely used liquid-liquid and solid-liquid extraction procedures with their advantages, drawbacks and suitability for the extraction of different compounds are presented in Table 5.1.

Table 5.1 The overview of the most commonly used liquid-liquid and solid-liquid extraction techniques

Extraction technique	Sample matrix	Pollutant state	Priority environmental pollutants	Solvent volume (ml)	Time (min)	Pros	Cons
Liquid -liquid extraction techniques							
Liquid-liquid extraction (LLE)	L	V, semi-V, non-V	PAHs, PCBs, PAEs, chlorinated volatile hydrocarbons, semi-V organic compounds, phenols, benzidines, nitrogen-containing pesticides, nitrosamines, nitroaromatics	500 (high)	60	Simple and inexpensive equipment, versatility, wide choice of solvents with diverse properties, large linear sample capacity, high processing rate while energy consumption is kept low, suitable for separation of temperature-sensitive substances, the organic extract can be directly subjected to quantitative analysis (e.g. HC, LC), carryover and other 'memory' effects are negligible	Toxic and flammable solvents, evaporation of large volumes of solvent required, laborious and time-consuming, less efficient for highly polar compounds, tendency to form emulsions which are difficult to break, difficult to automate
Single-drop liquid-phase microextraction (SD-LPME or SDME)	L, S	V, semi-V	Pesticides, PAEs, PAHs	1–3 μl	10–40	Rapid, simple, cost-effective, environmentally friendly, extensive choice of solvents, applicability to different complex matrices (headspace technique), suitable for both inorganic and organic compounds, high extraction efficacy, no	Slow kinetics, solvent drop dislodgment problems

(continued)

Table 5.1 (continued)

Extraction technique	Sample matrix	Pollutant state	Priority environmental pollutants	Solvent volume (ml)	Time (min)	Pros	Cons
						carryover problems, chromatographic and electrophoretic systems compatible, can be automated	
Hollow fibre liquid-phase microextraction (HF-LPME)	L	V, semi-V	Pesticides, PAHs, PCBs, organochlorines, aromatic amines	Several μl	15–45	Simple and inexpensive equipment, high selectivity, high sensitivity, no loss of the extraction phase when stirred, no sample carryover problems, environmentally friendly, can be automated	Low-speed passive diffusion-based extraction mechanism, possible solvent loss during extraction, air bubbles formation, lack of commercially available equipment
Dispersive liquid-liquid microextraction (DLLME)	L	V, semi-V	PAEs, PAHs, PCBs, pesticides, halogenated organic compounds	Several μl	1–15	Simple, short extraction times (fast reaching of the equilibrium), cost-effective, efficient, environmentally friendly, no sample carryover problems, high enrichment factors, high recoveries, ionic liquids as solvents, can be coupled to SPE, SFE, SBSE, nano techniques	Low selectivity, high sample preparation time, limited choice of solvents, the use of three solvents required, decreasing performance of ionic liquids with increasing concentrations of salts, centrifugation required, automation limitations

Solid-liquid extraction techniques – performed mostly on solid samples							
Shaking	S	Semi-V, non-V	Pesticides (atrazine), petroleum hydrocarbons, heavy metals, antibiotics	25–100 (moderate)	15–30	Very simple, low cost, good sensitivity	Limited selectivity, additional filtration of the extract required, solvent evaporation often required, matrix type dependent
Soxhlet extraction	S	V, semi-V, non-V	Pesticides, phenolic compounds, PAHs, PCBs, PAEs, semi-V organic compounds, nitrosamines, nitroaromatics, isophorone, chlorinated hydrocarbons	100–500 (high)	30–2880	Matrix type independent, low cost of the basic equipment, no further filtration of the extract required, many Soxhlet extractors can be set up to perform unattended	Long extraction times, large sample amounts (10–30 g), toxic solvents, solvent evaporation after extraction required, potential loss of volatile compounds
Supercritical fluid extraction (SFE)	S, L	Semi-V, non-V	Oils and fats, pesticides (organochlorine, DDT, TCDD, toxaphene, triazine), volatile toxins, polymers, PAHs, PCBs, PAEs, CBs, chlorophenols, heavy metals, organometallics, explosives, fossil fuels, pharmaceuticals and pharmaceutical metabolites (antibiotics, steroids, barbiturates, etc.)	5–20 (low)	30	High efficacy, high selectivity, short extraction times, environmentally friendly, can be directly coupled to GC or SFC, low cost and easy removal of supercritical fluid after extraction, a wide range of analytes can be extracted when using modifiers	High initial cost of the equipment, modifiers are often needed to separate polar analytes, limitations in selecting the optimum SFE parameters to obtain a lipid-free sample
Accelerated solvent extraction (ASE)	S, semi-S	Semi-V, non-V	Organochlorine and organophosphorous pesticides, PAHs, PCBs,	15–40 (low)	15–30	Rapid, fully automated, wide range of applications, easy control of extraction parameters	Initial high cost, matrix type dependent, not suitable for heat-sensitive compounds

(continued)

Table 5.1 (continued)

Extraction technique	Sample matrix	Pollutant state	Priority environmental pollutants	Solvent volume (ml)	Time (min)	Pros	Cons
			BNAs, dioxins, furans, TPH			(temperature, pressure), handles acidic and alkaline matrices, no filtration of the extract required	
Ultrasound-assisted extraction (UAE)	S, L	Semi-V, non-V	PAHs, PCBs, PAEs, phenolic compounds (chlorophenols), organochlorine pesticides, pharmaceuticals, carbon-based and organic nanoparticles	5–30 (low)	15–40	Simple, rapid, inexpensive equipment, allows simultaneous extractions, sensitive, adjustable bath temperature	Additional filtration of the extract required, matrix type dependent
Microwave-assisted extraction (MAE)	S	Semi-V, non-V	Thermostable compounds, PAHs, PCBs, PAEs, organochlorine pesticides, polybrominated diphenyl ethers, TPH, phenols, methylmercury, dioxins	10–50	5–30	Rapid, highly efficient, allows simultaneous extractions, allows full control of extraction parameters (temperature, pressure, power), easily programmable, allows high-temperature extraction, allows stirring of the sample, no drying agents needed	Additional filtration of the extract required, polar solvents are required, extract clean-up is required, moderate cost of the equipment, not suitable for heat-sensitive compounds
Solid-liquid extraction techniques – performed mostly on liquid samples							
Solid-phase extraction (SPE)	L	V, semi-V, non-V	Pesticides (carbamate), PAHs, PCBs, PAEs, lipids, semi-V organic	50–100	30	Simple, versatile, efficient, high selectivity, extraction of compounds	Complex technique, thus difficult to master, some organic solvents are toxic,

			compounds, chlorophenols, explosives (TNT, RDX), pharmaceuticals and personal care products			with different polarities, high analyte recoveries in general, small sample amounts can be processed, broad variety of commercially available sorbents, good reproducibility, can be automated, online/offline mode performance	possibility of low analyte recoveries due to interactions between sorbent and matrix, clogging/plugging of cartridges
Solid-phase microextraction (SPME)	L, S, G	V, semi-V, non-V	PAEs, PAHs, chlorophenols, nonylphenols, bisphenol A, volatile halocarbons, pesticides (organochlorine, triazines), organometallics, amines	Solvent-free	30	Simple, high sensitivity (ppt levels), convenient coupling with field analytical instruments, can be coupled to LC, GC, HPLC, HP LC-MS and CE, can be automated, broad variety of commercially available fibres and coatings	Some drawbacks of coating fibres (high cost, limited lifetime, possible carryover problems, limited commercial availability); some organic solvents used are toxic
Stir bar sorptive extraction	L, S, G	V, semi-V	Parabens, pesticides, PAHs, PCBs, VOCs, halogenated phenols preservatives, organctin compounds	Solvent-free	> 40	Predictable extraction efficiencies, high sensitivity (<1000 x more sensitive than SPME), high recoveries, reusable stir bars, large number of	'Matrix effect', expensive desorption device, difficult automation

(continued)

Table 5.1 (continued)

Extraction technique	Sample matrix	Pollutant state	Priority environmental pollutants	Solvent volume (ml)	Time (min)	Pros	Cons
						samples can be extracted simultaneously, excellent repeatability and reproducibility	

G gas, *L* liquid, *S* solid, *semi-S* semi-solid, *semi-V* semi-volatile, *non-V* non-volatile, *BNAs* base neutral acids, *CBs* chlorinated biphenyls, *DDT* dichlorodiphenyltrichloroethane, *PAEs* phthalate esters, *PAHs* polycyclic aromatic hydrocarbons, *PCBs* polychlorinated biphenyls, *RDX* cyclotrimethylenetrinitramine, *TCDD* 2,3,7,8-tetrachlorodibenzo-p-dioxin, *TNT* trinitrotoluene, *TPH* total petroleum hydrocarbons, *VOCs* volatile organic compounds, *CE* capillary electrophoresis, *GC* gas chromatography, *HPLC* high-performance liquid chromatography, *HPLC-MS* high-performance liquid chromatography-mass spectrometry, *LC* liquid chromatography, *SFC* supercritical fluid chromatography

5.3.1 Liquid-Liquid Extraction

Liquid-liquid extraction (LLE) also known as 'solvent extraction' or 'partitioning' is a method to separate compounds or metal complexes based on their relative solubilities in two different liquids that differ in their density, polar (usually water) and non-polar (organic solvent) (Berk 2013). Most organic compounds will partition into the organic phase and salts and ionic forms of solutes will remain in water (Tissue 2013). In the simplest case, the system consists of three components: solute (s), carrier liquid and solvent. The initial mixture, containing one or more solutes dissolved in the carrier liquid, is mixed with the solvent. Upon mixing, the solute is transferred from the carrier liquid to the solvent. The denser solution settles to the bottom. The location of the solute will depend on the properties of both liquids and the solute (JoVE 2019). Hexane and cyclohexane are solvents from the alkane family widely used for extracting non-polar compounds (e.g. organochlorine and some organophosphorus pesticides), whereas dichloromethane and chloroform are typical solvents used for the extraction of non-polar to medium polarity organic pollutants (Kassim et al. 2005; Chirila and Drăghici 2013).

Although LLE is probably the oldest separation technique in analytical chemistry, it still remains one of the most powerful and versatile sample clean-up and preconcentration methods (Jeannot 2007; Saha et al. 2017). It is frequently used in the chemical and petrochemical (for the separation of aromatic components and the extraction of metals from aqueous solution), mining (as a second step out of three in oxide ore processing) and pharmaceutical industry to separate products from each other (Kiezyk and Mackay 1971; Berk 2013; Stanley et al. 2015). In wastewater treatment, it is primarily used for the removal of phenols, creosols and other phenolic acids. These wastewater streams arise principally in petroleum refineries, coke-oven plants in steel industry and plastics industry (Kiezyk and Mackay 1971).

Even if LLE offers many advantages, it is gradually being set aside due to an extensive use of SPE (Saha et al. 2017). Many recent developments in LLE have focused on the reduction of the amounts of solvent used, on environmental factors as well as on the automation of the process in conjunction with chromatographic analysis. To minimize the disadvantages of LLE while preserving the advantages it offers, various modifications of the technique have been introduced recently, such as direct extraction from aqueous solution into a small volume of organic solvent, headspace techniques and microextraction followed by back-extraction into a small volume of aqueous solution. These methods are referred to as liquid-phase microextraction (LPME) or solvent microextraction techniques (Jeannot 2007; Saha et al. 2017).

There are three main variants of LPME, single-drop liquid-phase microextraction (SD-LPME or SDME), hollow fibre liquid-phase microextraction (HF-LPME) and dispersive liquid-liquid microextraction (DLLME); the latter represents the latest development in LPME. All of them are simple, rapid, cost-effective and environmentally friendly since the volume of organic solvent used is typically in the microliter range. These methods also do not have sample carryover problems associated with SPME (Quigley et al. 2016).

The DLLME technique uses somewhat bigger volumes of organic solvents than SDME, but it still only involves microliters. DLLME, much like SDME, is also often being combined with GC, HPLC, ET-AAS and FAAS methods. DLLME can also be coupled to other extraction techniques, such as SPE, SFE, SBSE and nano techniques. The use of ultrasonic waves with DLLME has gained increasing popularity in recent years because UA-DLLME provides an increased rate of mass transfer of the analyte from aqueous phase into the fine extracting droplets (Albero et al. 2015). For instance, UA-DLLME was successfully applied for the simultaneous determination of alkylphenols, parabens and bisphenols (Jain et al. 2014), polybrominated diphenyl ethers (Zhang et al. 2013) or dechlorane along with other chlorinated flame retardants (Hsieh et al. 2013) in water samples. A sample preparation technique related to DLLME (Przyjazny 2019) which uses the property of non-ionic surfactants to form micelles in aqueous solutions and to become turbid when heated to particular temperature (so-called cloud point temperature) is termed cloud point extraction (CPE). Above this temperature, the micellar solution separates into two phases: a surfactant-rich phase of a small volume and a diluted aqueous phase (Hagarová and Urík 2016). CPE is a simple, rapid, versatile, inexpensive and environmentally friendly method utilized for separation and preconcentration of trace metals from aqueous solutions which has recently become an attractive area of research and an alternative to LLE (Saha et al. 2017). Regarding HF-LPME, the extraction yields obtained by this method are higher than those gained by direct SDME, since hydrophobic hollow fibres permit the use of vigorous stirring rates to increase the speed of the extraction kinetics (Saha et al. 2017).

Supported liquid membrane extraction (SLME) and microporous membrane liquid-liquid extraction (MMLLE) are techniques that use hydrophobic porous membranes impregnated by organic solvent, which separates two immiscible phases. Liquid membranes have been successfully used in an extraction of metals and organic pollutants, mainly pesticides, such as phenoxy acid, dinitrophenol, sulfonylurea and triazine herbicides from waters and wastewaters (Bartolome et al. 2007; Msagati et al. 2008; Chirila and Drăghici 2013). A competitive advantage of membrane-based techniques is that they can be coupled to a variety of other analytical methods such as ion chromatography (IC), LC, GC, capillary electrophoresis and atomic absorption spectrometry.

5.3.2 Solid-Liquid Extraction

When the sample matrix is solid and the phase required to be analysed is of liquid nature, the process is called solid-liquid extraction. It begins with wetting of the matrix surface with a solvent or immersing the whole matrix in a solvent and letting the solvent penetrate into the pores of the sample particulates. This is followed by the dissolution of extractable material, transport of the solute(s) from inside to the surface of the solid particles and the dispersion of the solute within the bulk of the solvent surrounding solid particles by diffusion and agitation (Berk 2013). The solid phase is then removed by filtration. The key factor in a procedure is the proper selection of the solvent; however, extraction efficacy is significantly influenced also by other parameters, such as pressure, temperature, particle size, agitation, etc. Working under conditions of elevated pressure facilitates the solvent to penetrate sample pores. An increase in temperature usually increases solubility of the extractables in the solvent, diffusion coefficients and solvent's capacity to disrupt matrix-solute interactions (Barahona et al. 2019). However, in certain cases the solvent selectivity towards desired extractable is higher at lower temperatures. The extraction rate is enhanced with solid particle size reduction. In this context, samples are crushed, ground, milled, flaked or sliced into thin strips prior to extraction (Berk 2013). This gives an increased surface area per unit volume of solids to be leached and reduced distance to be traversed within the solid by the solvent and the extract (Tiwari 1995). The higher extraction yields can be often obtained by agitation which increases the contact surface between the sorbent and the analyte (Tashakkori et al. 2017). It does not affect extraction rate in case that any variations in extraction yields can be attributed to the other factors (e.g. heat treatment history) rather than to kinetic ones (Cogan et al. 1967).

The general solid-liquid extraction procedure is applicable in a range of fields including environmental monitoring. It can be used to extract a broad range of semi-volatile environmental pollutants that partition primarily on soils and sludges, such as polycyclic aromatic hydrocarbons (PAHs), and polychlorinated biphenyls (PCBs) and pesticides; it is probably the most widely used procedure in the analysis of pesticides in solid samples (Barahona et al. 2019). Because of the potential health effects, identification and quantification of these species is of academic interest and also widely practiced in the environmental consulting industry and in government agencies (JoVE 2019).

Solid-liquid extraction includes various techniques based on the contact of a certain amount of sample with an appropriate solvent (Barahona et al. 2019). Shaking is a very simple way to extract pollutants bound weakly to the sample; for strongly bound analytes it might not be effective enough. The method is applicable to a wide range of soil types and petroleum contaminants (Schwab et al. 1999). The principle of this method is based on the manual or automatic shaking of the sample in the presence of solvent for a certain period of time. The preferable organic water-miscible solvents are acetone, acetonitrile and methanol. Water-immiscible solvents (e.g. dichloromethane, hexane) can also be used, depending on the properties of the analyte. In order to make this process more efficient, shaking can be coupled to UAE. The acoustic cavitation and some mechanical effects induced by ultrasound enhance the agitation of the solvent and solute, thus increasing the contact area between solid and liquid phase, resulting in a better penetration of solvent into sample matrix. Owing to this, extraction times and the amount of solvents used in the shaking process are considerably reduced (Gutte et al. 2015; Barahona et al. 2019).

Although first described in 1879, Soxhlet extraction is still one of the most relevant and frequently used solid-liquid extraction techniques in environmental analysis. It has been used for an isolation of non-polar and semi-polar trace organic contaminants from a wide variety of solid materials, such as sediments, soils, sludges, wastes and biological structures (Wells 1978; Domini et al. 2005; Kassim et al. 2005). It is very useful when it comes to separation of poorly soluble compounds. Soxhlet extraction normally uses polar organic solvents or mixtures; non-polar solvents such as *n*-hexane are used to extract non-polar contaminants, such as OCPs and PCBs (Wells 1978). The long Soxhlet extraction times have prompted the development of modified extractors, such as revolutionary RaFaTec solvent extraction unit (Foss Tecator AB, Höganäs, Sweden). The apparatus offered considerable advantages in terms of time and solvent use, it also allowed the evaporation and collection of solvent further improving efficiency and the results were comparable to the traditional method (Bicking 2000). First introduced in 1975, it had quickly gained widespread acceptance and become an inspiration for an improved system Soxtec™ launched by FOSS 6 years later.

The more recent improvement of the method allows the application of microwaves to Soxhlet apparatus. This significantly accelerates the process of extraction of PCBs from contaminated soils (from 24 h to 70 min) and reduces the organic solvent disposal since 75–80% of the extractant can be recycled. The extractor also enables a fully automated extraction of two samples simultaneously (Luque-Garcia and Luque de Castro 2003). Other innovative methods include ultrasound-assisted Soxhlet (UASE) developed in 2004 (Luque-Garcia and Luque de Castro 2004) with an ultrasonic probe immersed in a water bath that circulates around the Soxhlet chamber or Sono-Soxhlet system with an ultrasonic probe inserted directly into the extraction chamber so an ultrasound can be applied within an extraction reactor (Djenni et al. 2013). Nevertheless, Soxhlet extraction is frequently referred to as the benchmark technique so the results obtained with newer extraction techniques are compared to those obtained by Soxhlet extraction (Dean 2000).

Traditional extraction techniques, such as liquid-liquid extraction and Soxhlet extraction, require large volumes of toxic organic solvents, thus creating new environmental hazards, increasing the risk of cancer and contributing to the ozone layer depletion. The newer extraction techniques such as SFE, MAE, ASE, SPE, SPME and hot water extraction are very attractive because they use much smaller or no amounts of solvents (O'Reilly et al. 2005; Risticevic et al. 2010), they are rapid and environmentally friendly.

Unlike Soxhlet, supercritical fluid extraction (SFE) represents a green technology that offers several advantages over conventional extraction methods. It is a process of separating the extractant from the matrix using supercritical fluids as an extracting solvent. A solute is usually separated from a solid matrix, but liquid matrices can be used as well. Supercritical fluid is any substance at a temperature and pressure above its critical point, where distinct liquid and gas phases do not exist. It can diffuse through solids like a gas, which facilitates the extraction of analytes located in not easily accessible pores, and dissolve materials into their component parts like a liquid (Sapkale et al. 2010; Chirila and Drăghici 2013). The power of supercritical fluid as a solvent depends largely on its density. Near the critical point of a fluid, even modest change in pressure or temperature may result in significant changes in the physico-chemical properties of the solvent, such as its density, diffusivity, or solubility characteristics (Padrela et al. 2009; Naresh et al. 2013).

Nowadays, more than 90% of all analytic SFE are carried out with supercritical carbon dioxide (Saha et al. 2017; Pannu et al. 2018) which has been available since the 1980s (Buckle 2015). It is an attractive alternative to organic solvents such as halogenated hydrocarbons (e.g. chloroform, dichloromethane) since it has a relatively low critical temperature (31 °C) and pressure (74 bar), it is non-flammable, accessible in high purity and very low cost (it can be captured for use with no need for synthesis and processing), it can be readily isolated from the extract and it leaves no toxic residues behind, thus posing no threat to the environment or human health (Rozzi et al. 2002; Naresh et al. 2013). In the supercritical state, CO_2 has a polarity comparable to liquid pentane what makes it suitable for the solubilization of lipophilic compounds (Saha et al. 2017). The major drawback of CO_2 as a non-polar solvent medium and a linear molecule with no net dipole moment is a significant difficulty in dissolving polar and ionic species (Peach and Eastoe 2014). The addition of a small amount of polar co-solvent (modifier) enhances the solubilizing capacity of the supercritical CO_2, making it possible to extract more polar molecules (Wrona et al. 2017). Most popular co-solvents are ethanol, water and methanol. Other fluids used in SFE include methane, ethane, ethylene, propane, propylene, ammonia, pentane, *n*-butane, *n*-hexane, acetone, nitrous oxide (N_2O), dichlorodifluoromethane, trifluoromethane, argon, xenon, etc. Other supercritical fluids that have been getting more attention in environmental SFE are freons and sulfur hexafluoride (SF_6). The latter is a non-polar but easily polarizable molecule and as a supercritical fluid it has been reported to selectively extract aliphatic hydrocarbons with a chain length of up to C-24 from mixtures containing both aliphatic and aromatic hydrocarbons (Saha et al. 2017). Due to the legal limitations of solvents and solvent residues (Vági et al. 2005) the interest in supercritical fluid technology has grown in the past two decades, being increasingly proposed as an alternative to many conventional solvent-based processes (Padrela et al. 2009). It is widely used in petroleum industry for the determination of aromatics as well as for other hydrocarbon separations.

Ultrasound-assisted extraction (UAE), also known as ultrasonic solvent extraction, is also considered a good option for the extraction of organic compounds from different matrices. The use of ultrasound enhances contact between solvent and solid due to an increase of pressure which aids penetration of the solvent into material and mass transfer, and increase of temperature which improves solubility and diffusivity (Saha et al. 2017). Even if originally developed for solid samples, the ultrasound radiation has recently been coupled to microextraction techniques to enable the analysis of liquid samples (Albero et al. 2015). Sonication can be used with other extraction techniques to enhance their efficiency. For instance, it can be combined with a stir bar sorptive extraction technique (SBSE) for an extraction and preconcentration of halogenated phenols from sediments (Tölgyessy et al. 2009).

Besides ultrasonic waves, microwaves are another form of radiation that is recognized as a simple and valuable tool in applied chemistry. Microwave-assisted extraction (MAE) is a procedure that uses microwaves that easily penetrate into the sample pores making the solvent trapped in pores to heat evenly and rapidly. In contrast to conventional heating where a certain period of time is needed to heat the vessel before heat is transferred to the solution, the extraction time is significantly reduced with MAE because with microwaves the heat is transferred directly to the solvent (Lopez-Avila 2000; Llompart et al. 2019). In their work, Hailemariam et al. (2016) compared the efficiency of MAE, Soxhlet extraction and sonication for the recovery of several classes of pollutants in soils and sediments. The average recovery value for phenolic compounds obtained by MAE was 71% vs. 52% and 57% for Soxhlet and sonication, respectively. This technique also generated higher recoveries for most of the organochlorine pesticides investigated, except for 4,4′-DDT and dieldrin which were more efficiently recovered by Soxhlet extraction. Similarly, using MAE resulted in higher recoveries of phthalate esters in comparison with traditional techniques.

The change in temperature and pressure is another way of increasing speed and efficiency of conventional extraction methods. Accelerated solvent extraction (ASE), also referred in the literature as pressurized solvent extraction (PSE), pressurized liquid extraction (PLE), pressurized fluid extraction (PFE), high-pressure solvent extraction (HPSE), high-pressure high-temperature solvent extraction (HPHTSE), pressurized hot solvent extraction (PHWE), subcritical solvent extraction (SSE) and enhanced solvent extraction (Duarte et al. 2014; Sánchez-Camargo et al. 2017), has been developed as another alternative to Soxhlet extraction, maceration, percolation or reflux, offering advantages with respect to solvent consumption, extraction yields, extraction time and automation. First ASE has been applied for the extraction of environmental hazards from solid matrices. Within a very short time it was approved by the US Environmental Protection Agency (EPA) for the extraction of semi-volatiles or base neutral acids (BNAs), PAHs, polychlorinated biphenyls (PC13s), organochlorine and organophosphorous pesticides, total petroleum hydrocarbons (TPH) and dioxins from solid samples. Especially when extracting dioxins, the time with ASE is reduced to 20 min in comparison to 18 h using Soxhlet (Höfler 2002; Stojić et al. 2018).

ASE combines elevated pressure and temperature with liquid solvents to accelerate the process of extraction which is done in a closed vessel at 50–200 °C and pressure of 1500–2000 psi. Higher temperature increases the extraction kinetics, while elevated pressure keeps the solvent in liquid phase above its boiling point (Björklund et al. 2000). The series of different sample size can be processed and the instrumentation allows extraction in unattended operation (Lopez-Avila 2000). The extraction can be performed in static and dynamic mode or as a combination of these two modes, which is often viewed as the best option. The reason is the dynamic ASE requires a larger volume of solvent than the static one which makes it less suitable for trace analysis. Static ASE, on the other hand, may lead to incomplete extraction because of the limited volume of solvent used (Mandal et al. 2015).

When it comes to solid-liquid extraction techniques performed mostly on fluids and aqueous samples, the most widely accepted one is probably a solid-phase extraction (SPE) in which a target compound is retained by a chromatographic stationary phase material packed in disposable cartridges or enmeshed in inert matrix

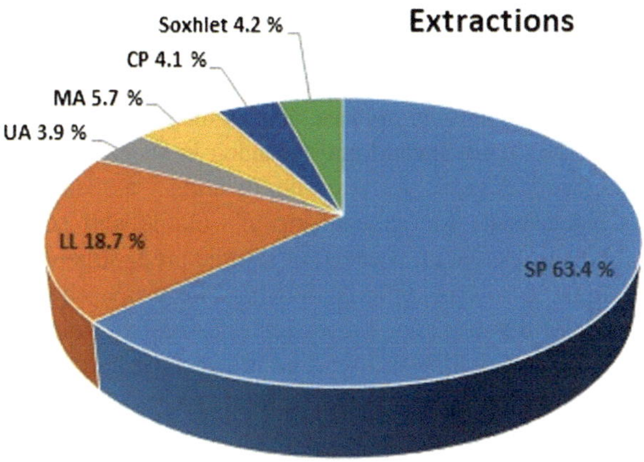

Fig. 5.1 Publications on different types of extraction procedures identified using a systematic search of scientific papers published between 1998 and 2018. (Source: ISI Web of Knowledge (Web of Science); publications tagged with the keywords "solid-phase" (SP), "liquid-liquid" (LL), "ultrasound-assisted" (UA), "microwave-assisted" (MA), "cloud point" (CP), "Soxhlet")

of an extraction disk and subsequently eluted in a selective solvent. It can be used either as a sample clean-up method or as a preconcentration method. Nowadays, the large variety of commercially available sorbents makes this technique suitable for the determination of analytes of divergent chemical structures and polarities (Escarpa et al. 2012; Chirila and Drăghici 2013). The novel adsorbent materials being used for solid-phase extraction of pollutants in environmental samples include carbon nanotubes, TiO_2 nanotubes, graphene, ion imprinting polymers, magnetic materials (or combination of ion imprinting and magnetic materials), core-shell materials, mesoporous materials, metal organic frameworks and sorbents of biological origin derived from bacteria, yeast, algae and fungi (Xiao et al. 2016).

Along with LLE, SPE is a main method for the extraction of organic pollutants and heavy metals (Fig. 5.1). However, SPE is becoming more popular for analyte preconcentration and matrix removal than LLE, due to its simplicity and economic advantage in terms of time and solvent used (Picó et al. 2007). In recent decades, the use of SPE has increased due to development of new materials that may find application as solid sorbents (Płotka-Wasylka et al. 2015)

The high capital cost of SFE, MAE and ASE and in some cases the level of expertise required to operate the instruments effectively still represent a barrier to their wide acceptance. In this context, the use of SPME, introduced in 1989, has been proposed (Belardi and Pawliszyn 1989; Dean 2000). It has been used routinely in combination with GC and GC/MS and successfully applied to a wide variety of compounds, especially for the extraction of volatile and semi-volatile organic contaminants from water samples. It was also introduced for direct coupling with HPLC and LC/MS in order to analyse weakly volatile or thermally labile compounds not amenable to GC or GC/MS (Kataoka et al. 2000). In order for this simple, efficient and solvent-free method to be used also on solid samples, the analyte must be released from solid matrix and enter either liquid or a gaseous phase. There are several approaches that can be adopted for the extraction of analytes from solid matrices. These include direct extraction of the analytes from a soil-water suspension or slurry, extraction of the analyte from the sample matrix using hot water or headspace extraction. The first two approaches assume that the analyte is highly soluble in water and that water is a suitable solvent to liberate the analyte from the matrix. The latter scenario assumes that the analytes of interest are volatile or semi-volatile so that they are available in the headspace above the sample (Dean 2000).

In 2015, Ghiasvand and Pirdadeh-Beiranvand introduced an effective and inexpensive device for cooling/heating assisted solid-phase microextraction (CHA-SPME) for an analysis of volatile and semi-volatile species in complex solid samples, with no need of sample pretreatment and with minimal manipulation compared to conventional SPME. The proposed device coupled to GC-FID was successfully used for the extraction and determination of PAHs in contaminated soils. Besides SPME, it can be utilized for simultaneous cooling of the extraction phase and heating of the sample matrix in LPME, NTD (needle trap devices) and INCAT (inside needle capillary adsorption trap) methods, all used for environmental sample preparation.

Another promising trend in the search for faster, simpler, cheaper and more environmentally friendly sample preparation and separation techniques has been the utilization of electrical driving force. In recent years, a noticeable progress has been made in this field and some new electrically assisted microextraction techniques have been introduced, such as electrically assisted SPME coupled to liquid chromatography-mass spectrometry that was applied for determination of trace levels of organophosphate insecticide in water (Yang and Lee 2010) or electrically assisted LPME used for determination of β2-receptor agonist drugs in wastewater (Rezazadeh et al. 2012). One of the very attractive areas of use, rarely pointed out in literature, is synthesizing and preparing the membranes and solid-phase sorbents for SPE or SPME (Yamini et al. 2014). Membrane-protected SPME can be used for the analysis of a wide range of environmental materials, including air, water, sediment and soil.

Stir bar sorptive extraction (SBSE) operates on a similar principle as SPME which involves partitioning of the target analyte between (usually an aqueous-based liquid) sample and a stationary phase-coated stir bar (Chirila and Drăghici 2013). In SBSE, stir bars are coated with polydimethylsiloxane (PDMS). When compared to SPME, it is characterized by a higher sorption capacity due to the presence of large volumes of coating material (Nießner and Schäffer 2017). SBSE is rapid and simple, which makes it ideal for automation and in situ measurements (Nilsson 2000) and it can be effectively applied to different kinds of environmental matrices, such as water, wastewater, soils, biological fluids and gases (Saha et al. 2017).

5.4 Surfactants of Biological Origin as Eco-friendly Extractants for Environmental Pollutants Removal

The extraction using surfactants is one of the several applicable methods for the remediation of petroleum hydrocarbons (Goswami et al. 2018) and heavy metals in soils. After their utilization, water-soluble surfactants are very difficult to recover or reuse. Generally, if not biodegradable, they are discharged as part of wastewater to the environment. In order to conserve energy resources and to reduce their environmental impact, surfactants should be recycled, particularly after industrial use. Therefore, the development of surfactants with enhanced biodegradability and recyclability using renewable resources under environmentally benign conditions has been desired (Banno et al. 2013). With respect to green chemistry principles, there is a tendency to substitute synthetic surfactants with naturally derived alternatives. Microbial surfactants offer numerous benefits over synthetic ones. Due to their versatility, biodegradability, low toxicity, high activity even under extreme conditions (such as extreme pH and salinity), high effectivity in biodegradation and solubilization of low solubility compounds and mild production conditions (Mulligan 2005; Morita et al. 2016; Costa et al. 2018; Inamuddin et al. 2021), they have been identified as promising agents in remediation technologies. Unfortunately, a substantial portion of biodegradable materials are at a production cost disadvantage against synthetic substances which limits their manufacture and large-scale application.

Biosurfactants are surface-active compounds that are produced extracellularly or as part of the cell membrane by bacteria, yeasts and fungi (Mulligan 2005) and due to their remarkable properties they are considered multifunctional biomolecules of the twenty-first century (Santos et al. 2016). All biosurfactants are amphiphilic molecules with a polar (hydrophilic) moiety consisting of mono-, oligo- or polysaccharides, peptides or proteins, and a non-polar (hydrophobic) group usually containing saturated, unsaturated and hydroxylated fatty acids or fatty alcohols (Lang 2002) that partition at liquid-liquid, liquid-gas or liquid-solid interfaces. Such structure predestines them for the use in a range of industrial applications where emulsification/de-emulsification, foaming, wetting, coating, detergency and dispersion take place (Farn 2006; Santos et al. 2016). In recent years, emulsifying, solubilizing and mobilizing (Usman et al. 2016) potential of biosurfactants has been intensely explored for the remediation of different organic and inorganic environmental contaminants.

Biosurfactant ability of changing the surface and interfacial tensions and stabilizing emulsions is a measure of its effectiveness, while critical micelle concentration (CMC) is commonly used to measure surfactant efficacy (Pacwa-Płociniczak et al. 2011). CMC is a physical characteristic that indicates the concentration of surfactant above which micelles start to form (IUPAC 1997). The lower the CMC, the less surfactant is required to decrease the surface tension (Desai and Banat 1997), i.e. to effectively emulsify, solubilize and disperse waste materials at the surface (Abhijit 2017). The CMC of biosurfactants ranges from 1 to 2000 mg L^{-1} (Santos et al. 2016) and among surfactant types classified according to the composition of their head (as non-ionic, anionic, cationic and amphoteric), the lowest CMCs are generally found in the non-ionic category (Abhijit 2017). For instance, CMCs of Triton X-100 and Triton X-114 – two high purity, inexpensive, commercially available non-ionic detergents suitable for most biological applications used in our experiment aimed at heavy metal removal from contaminated soils, are 0.24 and 0.2 mM, respectively.

To determine its appropriate application, the surfactant is being ranked by its hydrophilic-lipophilic balance (HLB) value which is a common term used in industry and reflects the degree to which a surfactant is hydrophilic or lipophilic. Surfactants with low HLB (< 10) are more lipophilic, thus more oil (fat) soluble, and create water-in-oil emulsions, whereas HLB values over 10 indicate predominantly hydrophilic molecules with better water solubility that stabilize oil-in-water emulsions. A value of 10 corresponds to a compound with equal affinity for water and oil; complete water solubility of a surfactant occurs at HLB of ~7.3 (Burch et al. 2010; Rami et al. 2017). In cleaning applications, high HLB scale valued surfactants are often used to emulsify oily soils (Király 2013).

Based on their molecular weight, biosurfactants can be classified as (1) low molecular weight (low-mass) molecules including glycolipids, phospholipids and lipopeptides efficient in lowering surface and interfacial tension and (2) high molecular weight (high-mass) polymers such as amphipathic polysaccharides, proteins, lipopolysaccharides, lipoproteins or their complex mixtures that are more effective at stabilizing oil-in-water emulsions (Rosenberg and Ron 1999; Pacwa-Płociniczak et al. 2011).

5.4.1 Rhamnolipids

Rhamnose lipids, or rhamnolipids, are glycolipids often described as 'the best-studied' representatives of bacterial surfactants (e.g. Müller et al. 2012; Thiengmag et al. 2016; Liu et al. 2018). The evolution of sensitive analytical techniques has enabled great progress to be made in the identification of a wide diversity of rhamnolipid congeners and homologues (about 60) that are produced at different concentrations by bacterial species mainly from strains of *Pseudomonas* and *Burkholderia* (Abdel-Mawgoud et al. 2010). Because of their anionic structure, rhamnolipids have been proposed as a suitable material for the removal of some heavy metals present in their cationic form, such as Cd^{2+}, Pb^{2+}, Zn^{2+}, Ni^{2+}, Ba^{2+} and Sr^{2+} from soil and water (Miller 1995; Elouzi et al. 2012). Due to their excellent emulsification properties, rhamnolipids also boost remediation of crude oil-contaminated sites. Their addition in the soil enhances bacterial colonization, growth, survival and metabolic activity (Rahman et al. 2003; Costa et al. 2010; Tahseen et al. 2016) and helps microorganisms with uptake and assimilation of many aliphatic hydrocarbons such as linear alkanes, which are poorly soluble in water but provide a good nutrient source for *P. aeruginosa* (Hommel 1994). Promising results have also been reported for the removal of PAHs, which are recognized as being carcinogenic and hard to degrade environmental pollutants (Makkar and Rockne 2003; Yu et al. 2011). PAHs make up about 85% of the composition of creosote, which is besides copper chromium arsenate and pentachlorophenol a major chemical used in the wood-preserving industry worldwide. Bruguera et al. (2017) focused their research on areas around wood preservation facilities that are often highly contaminated due to practices such as the use of unlined wood treatment cells and the disposal of wastes in unlined lagoons, and long-term leaching of preservatives from treated wood onto site soils. Their study concludes the high potential of rhamnolipids (JBR 425) to extract PAHs from such contaminated soils. However, the ionic strength and the cations present in the soil solution of creosote-polluted soils should be considered when rhamnolipids are used as extractants for remediation purposes (Madrid et al. 2019).

5.4.2 Sophorolipids

Of all currently known biological surface-active agents, rhamnolipids are believed to have the highest potential for becoming the next generation of biosurfactants. However, the first microbiological surfactants introduced on the market were another member of glycolipid group – sophorolipids (Müller et al. 2012). Sophorolipids are synthesized through the fermentation by a number of non-pathogenic yeasts such as *Candida bombicola* (Le et al. 2016) which produces high yields of sophorolipids from vegetable oils and sugars (Mulligan 2005), *C. apicola*, *Rhodotorula bogoriensis*, etc. Like rhamnolipids, they solubilize hydrophobic compounds and hence increase their bioavailability for subsequent remediation. Sophorolipids lower the surface tension of water from 73 to about 30–40 mN m^{-1} and have a CMC of 40–100 mg L^{-1} (Hubert et al. 2012).

There is a growing number of studies that found these biosurfactants to be very efficient for the remediation of petroleum hydrocarbon-polluted soils, even outperforming the conventional detergents of chemical origin (such as Triton X-100), which are in fact considered a new class of chemical pollutants in aquatic environments. Goswami et al. (2018) compared the performance of environmentally friendly sophorolipids derived from two yeast, *C. bombicola* and *Yarrowia lipolytica*, with synthetic Triton X-100 on such contaminated soils. The best results were obtained with the biosurfactants produced by *C. bombicola*, where 40–68% reduction of contaminant concentration was achieved using the biosurfactant compared to 37–38% reduction when using Triton X-100.

5.4.3 Lipopeptides

Among other categories of biosurfactants, lipopeptides are particularly interesting because of their potent surface activity (Meena and Kanwar 2015). They can act as antibiotics, antiviral and antitumor agents, immunomodulators and specific toxins and enzyme inhibitors (Rodrigues et al. 2006). Besides their application in medicine, lipopeptides can be counted among promising dispersants in oil spill remediation. The one with the greatest potential in the group is probably surfactin, a cyclic lipopeptide produced by various strains of *Bacillus subtilis*. It was identified in 1968 and named after its exceptional surfactant activity (Arima et al. 1968). It has been reported as the strongest lipopeptide surfactant so far (Liu et al. 2015) with the ability to decrease the surface tension of water from 72 to 27 mN m^{-1} at a concentration as low as 20 μM and the interfacial tension of the water/*n*-hexadecane system from 43 to values <1 mN m^{-1} (Lang 2002; Yeh et al. 2005). There has been a broad spectrum of applications proposed for this compound including its utilization in petroleum recovery and spill mitigation (Liu et al. 2015), since surfactin enhances bioavailability of water-insoluble substrates and regulates the attachment/detachment of microorganisms to and from surfaces (Rosenberg and Ron 1999). Possessing exceedingly favourable features, including low toxicity, high biodegradability and no or minimal loss of activity under harsh conditions such as extreme temperatures, pH and salinity, as well as a high surface activity and low CMC values, surfactin is decidedly a candidate for either ex situ biosurfactant injection or in situ biosurfactant production that lead to an enhancement in oil recovery from subsurface reservoirs (Banat et al. 2010; Liu et al. 2015).

Schaller et al. (2004) yielded favourable results when examining the effect of altering salt concentration (0–10% NaCl), pH (3–10) and temperature (21–70 °C) on the performance of surfactin produced by *B. subtilis* ATCC 21332 cultures; high salt concentration, high temperature and high pH describe the conditions of many oil reservoirs and can affect compound's activity. Al-Wahaibi et al. (2014) reported that biosurfactants produced by *B. subtilis* B30 identified as a mixture of lipopeptides similar to surfactin gave stable emulsions with a wide range of hydrocarbons including light and heavy crude oil, and remained stable over a broad range of pH, salinity and temperature. The crude biosurfactant preparation enhanced light oil recovery by 17–26% and heavy oil recovery by 31%.

Similarly, lipopeptides secreted by *B. subtilis* HSO121 showed good dispersion effectiveness on crude oil at low surfactant-to-oil ratios and remained active over changes in temperature (15–25 °C), pH values (5–11) and salinity (0–4% NaCl). Lipopeptides showed excellent activity in accelerating degradation of long-chain hydrocarbons after 2 days. The degradation rates of alkanes treated by lipopeptides after 1 day and 2 days were 38.78% and 71.45%, respectively, which were much higher than those of commercial dispersant-treated group (9.16% and 34.16% after 1 day and 2 days treatment) (Feng et al. 2019).

5.4.4 Saponins

Many surface-active compounds can be derived from renewable plant resources (Xu et al. 2011). Although microbial surfactants are generally considered to have more advantages than plant-based surfactants in terms of scale-up capacity, versatile properties and rapid production (Randhawa and Rahman 2014), there are several plant-based biosurfactants exhibiting excellent functional properties, among which the best known are those belonging to the structurally diverse class of compounds that derived their name from their ability to form soap-like foams in aqueous solutions – saponins (Hill 2003). Saponins can be isolated from more than a hundred plant families (Samal et al. 2017) as well as from several marine invertebrates including Antarctic starfish of the family Asteriidae and sea cucumbers; they are also present in some marine sponges, soft coral and small fish (De Marino et al. 1998; Xiao et al. 2019). Among the most studied plant materials that were found to

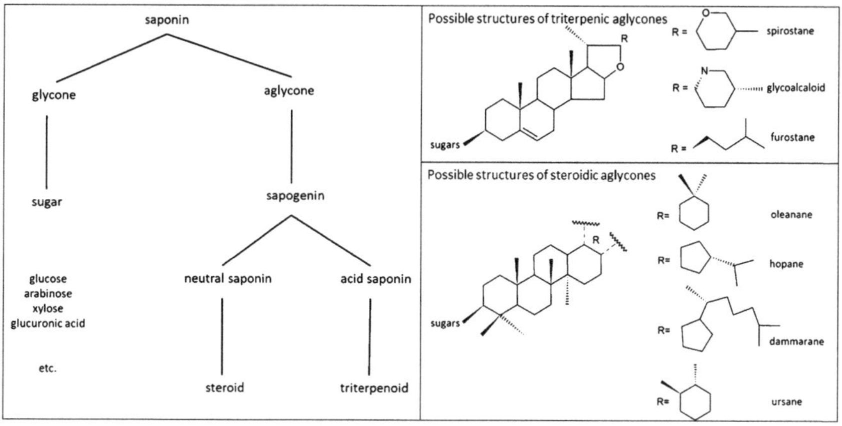

Fig. 5.2 Schematic illustration of saponin composition. These heterosides are composed of a water-soluble glucidic chain and a triterpenic or steroidic liposoluble structure; however, variation in saponin composition is high. Possible structures of triterpenic and steroidic aglycones (Chaieb 2010) are shown on the right side of the figure

be rich in saponins are ginseng, licorice root, asparagus, soybeans, sugar beet, chickpea, green pea, horse chestnut, quinoa, etc. (Price et al. 1987; Fenwick et al. 1991; Shidhaye et al. 2008; Cheok et al. 2014; Shin et al. 2015). Saponins are amphipathic glycosides containing one or more hydrophilic glycoside (sugar) chain (s) on a triterpene or steroid aglycone backbone also known as sapogenin (Fig. 5.2). Due to the presence of a lipid-soluble aglycone and water-soluble sugar chain(s) in their structure, saponins are regarded as surface active compounds with detergent, wetting, emulsifying and foaming properties (Üstündağ and Mazza 2007).

The recent advances in extraction techniques for bioactive compounds present in plant material have been thoroughly reviewed (e.g. by Sasidharan et al. 2010; Azmir et al. 2013); there are several methods to obtain saponin from a plant. Traditional extraction techniques include maceration, which is a very slow process of solid-liquid extraction, and Soxhlet and reflux extraction where a distillation process is involved (Cheok et al. 2014). In conventional extraction, the extractability of the desired compound from plant material depends mainly on the solubility of this compound (the solute) in the solvent, mass transfer kinetics of the product and matrix interactions (Spigno and De Faveri 2009). It is often necessary to use high volumes of the solvent to extract the target solute, even if the process is being aided by heating, and mechanical stirring or shaking to elevate the temperature (Cheok et al. 2014). Furthermore, the ability of the solvent to extract various substances from natural material leads to the use of excess solvent volumes in order to ensure efficient extraction of the desired bioactive compound. The green technologies employed in saponin extraction are UAE, MAE and ASE (Heng et al. 2013). Compared to conventional extraction, these techniques meet the green chemistry concept, i.e. they are environmentally friendly, offer safer chemical synthesis, allow the use of alternative and less hazardous substances, use only small amounts of catalyst, reduce energy consumption, maximize atom economy, prevent waste, avoid chemical derivatization and minimize potential for accidents (Anastas and Warner 1998; Azmir et al. 2013).

5.4.4.1 Quillaja Saponins

The soap bark tree or soapbark (*Quillaja saponaria*) is one of the major commercial sources of saponins (Cheeke 2000); saponin extracted from the bark of this tree can be used as a natural low molecular weight biosurfactant (Bai and McClements 2016), its chemical structure is outlined in Fig. 5.3. The micelle-forming properties of quillaja saponins may be affected by temperature, salt concentration and pH of the aqueous phase (CMC increases with temperature and pH and decreases with raising salt concentration; the size of micelles has been found to increase strongly with temperature but to have little dependence on salt concentration or pH) (Mitra and Dungan 1997). Nonetheless, a growing number of studies have reported that saponins could effectively remove heavy metals and complex organic chemicals such as PAHs and PCBs from various polluted environmental substrates. Gao et al. (2012) compared the efficacy of quillaja saponin and sophorolipid produced from the yeast strain *Wickerhamiella domercqiae* for the Pb^{2+}, Ni^{2+} and Cr^{3+} removal from the sludge collected from an industrial water treatment system. From the two non-ionic biosurfactants used in batch and column experiments, saponin was found to be more efficient at the removal of selected heavy metals from polluted sludge. Saponin molecule contains the carboxyl group (–COOH) on its hydrophilic head (Fig. 5.3) by which saponin forms complexes with metal ions in aqueous solutions

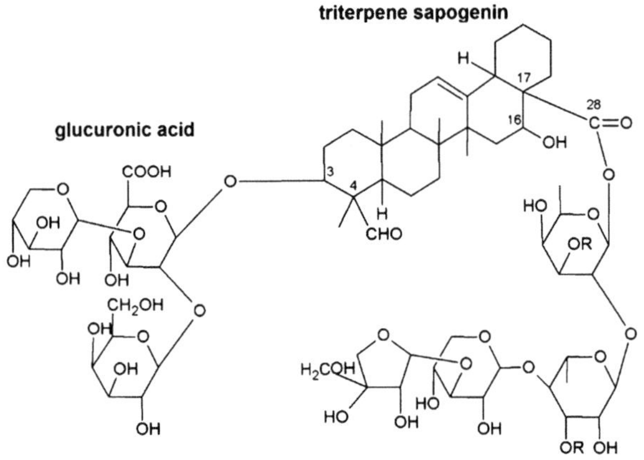

Fig. 5.3 The structure of quillaja saponin (Chen et al. 2008) according to Higuchi formula (1986). Unlike the hydrophobic tail of conventional surfactants that typically has only one long straight hydrocarbon chain, saponin molecule consists of a hydrophobic fused ring of triterpenes with two hydrophilic sugar chains attached to positions C-3 and C-28. These oligosaccharide chains carry various moieties, such as β-D-glucuronic acid, anionic carboxyl group and non-ionic glycoside groups. The ability of carboxyl groups to form complexes with metal ions in aqueous solutions appears to play an important role in their removal from contaminated materials

(Hong et al. 2002). Saponin showed selectivity for target metals and mainly reacted with heavy metals in fraction bound to carbonates and in fraction bound to Fe-Mn oxides. The subsequent metal recovery was achieved by precipitation in alkaline solution and reached 89.7%, 91.1% and 99.1% for Pb^{2+}, Ni^{2+} and Cr^{3+}, respectively.

Batch extraction is a simple separation technique used in environmental analysis. In its simplest form, a batch extraction system consists of an agitated mixing vessel where the solids are being mixed with the solvent and a solid-liquid separation device. Decanter centrifuges are advantageously used as separators (Berk 2013). Song et al. (2008) conducted the batch experiments with soils spiked with phenanthrene and cadmium to evaluate pollutant removal efficacy of saponin. Phenanthrene partitioned into a surfactant micelle, thus becoming directly bioavailable to organisms capable of degrading PAHs, whereas Cd^{2+} was effectively removed from soil via its complexation with the external carboxyl groups of saponin micelle. At saponin concentration of 3.75 g L^{-1}, a removal yield of 87.7% for Cd^{2+} and 76.2% for phenanthrene was obtained. The removal efficacy of saponin was greater than that of Triton X-100 and citric acid. EDTA removed Cd^{2+} effectively at low concentration; however, this substance is quite persistent in the environment due to its low biodegradability and may have a toxicity effect on soil microorganisms and plants (Hauser et al. 2005). The removal of Cd^{2+} was not constraining the removal of phenanthrene which demonstrates a substantial potential of saponins for the removal of heavy metals and PAHs from soils.

Cao et al. (2013) investigated the simultaneous desorption of metal elements (Pb^{2+}, Cu^{2+}) and polychlorinated biphenyl (PCB) from the combined contaminated soil with a novel combination of saponin and S,S-ethylenediamine disuccinic acid (EDDS) – an aminopolycarboxylic acid that can be used as a chelating agent, thus offering a biodegradable alternative to synthetic EDTA (Kaushik 2015). The maximal desorption of contaminants was achieved by addition of 10 mmol L^{-1} EDDS and 3 g L^{-1} saponin, and was 99.8%, 85.7% and 45.7% for Pb^{2+}, Cu^{2+} and PCB, respectively. In the presence of the mixed solution of 5 mmol L^{-1} EDTA and 20 mmol L^{-1} Triton X-100, the maximal desorption was only 88%, 74.4% and 53.3% for Pb^{2+}, Cu^{2+} and PCB, respectively. These results suggest the combination of saponin and EDDS may have practical application in remediation of soils co-contaminated by hydrophobic organic compounds and metals.

5.5 Experimental: A Solid-Liquid Extraction with Synthetic and Biological Surfactants for the Removal of Cu, Pb, Ni and Zn from Contaminated Soils and Sediments

As an example of solid-liquid extraction with the use of synthetic and biological surface-active agents for the removal of heavy metals from contaminated soils and river sediments, the extraction of Cu, Pb, Ni and Zn with two synthetic non-ionic surfactants Triton X-100 and Triton X-114 (octyl-phenoxy-poly-ethoxy-ethanol with polyoxyethylene chain length of 9.5 and 7.5, respectively), biosurfactant saponin (isolated from the bark of *Quillaja saponaria* tree) and secondary

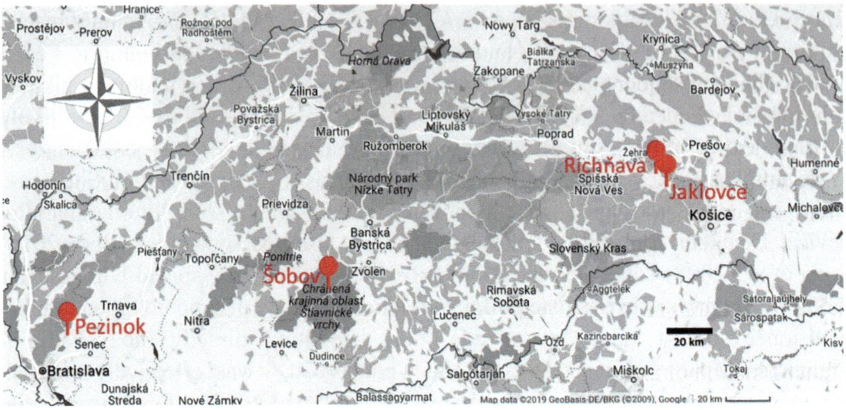

Fig. 5.4 Topographical position of the sampling sites in Slovakia. For details on soils and sediments sampled, see Table 5.2

Table 5.2 Selected characteristics of sampled soils and river sediments used in the solid-liquid extraction study (pH, TOC and EC represent mean values from three replicates)

Sample	Location	Sampling depth	pH/ H_2O	TOC %	EC µS. cm^{-1}
Soil Šobov (acid sulphate soil)	Near open quartzite mine area, Šobov, Banská Štiavnica	0–20 cm	2.99	4.66	902
Soil Pezinok (Eutric Fluvic Gleysol)	Ore deposit, Pezinok, Little Carpathian Mts.	0–20 cm	3.51	5.37	200
Sediment Richňava	Hornád alluvium, left riverside	Sediment thickness: 20 cm, Water column above the sediment surface: 40 cm	6.68	4.56	452
Sediment Jaklovce	Hnilec alluvium, right riverside	Sediment thickness: 20 cm, Water column above the sediment surface: 60 cm	7.44	3.28	385

TOC total organic carbon, *EC* electrical conductivity

metabolites of microscopic filamentous fungus *Aspergillus niger* was evaluated in laboratory-scale batch experiments.

Soil and sediment samples with elevated concentrations of heavy metals were collected from four different industrially stressed areas of Slovakia (Fig. 5.4); for characteristics, see Table 5.2.

The town of Banská Štiavnica is the oldest mining town in Slovakia, the history of mining in this district dates back to the late Bronze Age. The prosperity of the town reached its peak in the eighteenth and nineteenth centuries, by the end of the nineteenth century mining fell into decline and the last mine was closed in 2001.

However, the effects of former mining activities are numerous in their impact on the environment of the region. At this place it should also be noted that ‘closed’ or ‘abandoned’ mine status does not necessarily imply that the problems are less than those experienced in active mining areas, since the environmental standards were considerably lower in the past and the competence of the authorities could have been rather limited (Puura and D’Alessandro 2005). One of the most critical areas is Šobov, situated 3 km northeast of Banská Štiavnica city. It is an open mine where hydrothermal quartzite was extracted and then transported, crushed and grinded at a nearby plant in Banská Belá for refractory bricks. The mined ore often contained pyrite and clay minerals, mainly illite and pyrophyllite (Uhlík and Šucha 1997) which were left on site as waste and had been accumulated along over 30 years (Šottník et al. 2002). In the course of time, pyrite deposited on the pile in large amounts has been under the slow attack of weathering processes and oxidized, forming sulfuric acid. The acid attacks other minerals and accelerates migration of different toxic elements (Kubová et al. 2004, 2005; Bujdoš et al. 2005). The water that flows out from the mine site represents typical acid mine drainage with high metal concentrations and pH as low as 1.2–2.5. The major dissolved components are Al, Fe and SO_4^{2-}, followed by Zn, Mn, Cu, Pb and As (Matúš 2007; Matúš and Kubová 2005, 2006, 2008; Matúš et al. 2003, 2004, 2005, 2006). The dissolved salt concentrations remain between 15 and 50 g L^{-1} (Šucha et al. 1997; Šottník et al. 2002). The acid mine drainage is associated with enormous damage to the entire surrounding habitat, it causes significant changes in soil properties and plant species richness. In the last decades, different remediation strategies have been implemented to reduce negative influence of these waters, but with little success. In 1999, a passive treatment system was designed to treat acid mine drainage, with aerobic terrace with wetland vegetation arranged in the last stage (Šottník and Šucha 2001). In order to at least partially prevent the negative effect of acid mine drainage, a simple construction – the surface channel system – has been built to collect acid water that flows out of the dumps into an open pool from which it is subsequently delivered by an underground tube into the deep abandoned mine (Lintnerová 2002). The polluted area covers an estimated 145,000 m^2, badlands with no vegetation or humus cover spread for 35,000 m^2 (Dlapa et al. 2002; Šottník et al. 2002; Medveď et al. 2004, 2006, 2008).

The second sampling site was situated in Pezinok region (Little Carpathian Mts.), in so-called Pezinok-Pernek crystalline complex, known for its ore deposits. The sample was taken from the soil at a Sb ore dump site situated on the south-eastern side of the mountain above the town. This area has several thousand years of history in ore mining. With a series of little breaks, it was mined for antimony until the early 1990s, when an interest in antimony had dropped off due to the end of the Cold War; besides that, only residual deposits of Sb remained on the site. The ore mining and dressing have significantly contributed to Sb and As contamination of soils, sediments and waters (Čerňanský et al. 2007, 2009; Urík et al. 2007; Kubová et al. 2008a, b; Kališ and Matúš 2011; Littera 2011). Nowadays, a surface water leaking out of the mining dump has a pH of 4–6.5, and although it is partially purified by a natural process of mechanical sedimentation and sorption in the sludge pond, the water mineralization is about 1.25 g L^{-1} and water is high in content of sulphates (680 mg L^{-1}), Fe (31.2 mg L^{-1}) and heavy metals (Andráš et al. 2004a; Šlesárová 2006). Due to the polymetallic nature of the deposit, it can be assumed that the oxidation of the sulphidic material continues (pyrite is commonly found in polymetallic ore) and therefore the deposit can be classified as a potential source of acid mine drainage. The presence of carbonates in the surrounding rocks contributes to the neutralisation of the acid with calcium carbonate, thereby preventing more intense acid mine drainage formation. However, the conditions at the sludge lagoons should be monitored, as the neutralizing potential of the carbonates can be exhausted, and the character of the acid mine drainage pH may change towards markedly acid values (Luptáková and Andráš 2018). Such a change could trigger calamitous degradation process and cause the substantial contamination of the surrounding landscape (Andráš et al. 2004b).

The area in Spiš-Gemer Ore Mt. (Spišsko-Gemerské Rudohorie) chosen for stream sediment sampling was industrially stressed during the long period of time, pollution problems caused by mining operations date back to the thirteenth century (Špaldoň et al. 2006). Large deposits of precious and industrial metals and their processing characterized this region for centuries, many of these deposits are now considered as sources of acid mine drainage. Ore deposits in Jaklovce and in the surrounding area were mainly mined for iron, mercury (cinnabar), copper and silver; processing of iron ore was conducted in a nearby town of Richňava as well. A dramatic change in ore processing came with the introduction of the waterwheel, by which the industry solved the problem of power. This meant that ironworks now had to be sited next to streams and rivers where there was an adequate supply of water (Hayman 2016). The ore had to be freighted to the river. Until the second half of the fourteenth century, furnaces were built on the site of local ore mines. Relocating works contributed to the further contamination of the river systems in the area. The valley is polluted with Hg, Cu, As, S and nitrates. Hornád river suffers from a poor water quality in a long-term sense, heavy copper contamination was detected near the confluence of the Hornád and Hnilec rivers, and both rivers show high levels of As, Hg and Zn (Klinda et al. 1994). Apart from geological and industrial points of view, another reason of reduced environmental quality of the Hornád river valley is poor environmental awareness of the local community. Inadequate and uncontrolled disposal of communal and other waste (such as scrap metals, plastics, building material, chemicals, etc.), particularly discarding of waste on illegal dumps with a close distance to the river, continues to pose a significant problem (Jablonská et al. 2009).

The collection of sediment samples was carried out in accordance with the Methodical Instruction of the Slovak Ministry of Environment (1998). Chemical analyses were performed on fine-grained fluvial sediment, since a number of pollutants carried by a river tend to bind to a very fine fraction which is being delivered to the riverbanks and adjacent river channels where it settles as the water flow velocity decreases, thus allowing pollution to accumulate. The processes by which benthic organisms stir sediment within the riverbed (i.e. bioturbation) destroying physical sedimentary structures and creating biological structures occur within a 5–20 cm thick region known as the surface mixing layer (Nittrouer et al. 2007); therefore, the upper 20 cm of sediment was taken with a grab corer. Before chemical analysis, samples from stream sediments were dried at 40 °C and passed through a 0.125 mm mesh. A fraction with the particle size <0.125 mm was then ground in an agate mortar to fine powder <0.09 mm. Soil samples were collected at a depth of 0–20 cm. Under laboratory conditions, soil was air-dried, homogenized by quartering, sieved through a 2-mm mesh and subsequently processed as given above.

The batch extraction experiments were conducted at a sample to extracting solution ratio of 1:5 (5 g of soil, 25 ml of extracting solution). Five g of the sample was weighed and placed in a 100 ml high-density polyethylene (HDPE) bottle, 25 ml of the extracting solution were then added and the bottle was sealed with a HDPE top. Each bottle was hand-shaken for about 1 min to ensure the sample was fully saturated with solution, and the bottles were subsequently shaken in a mechanical shaker (Unimax 2010, Heidolph, Germany) at 250 rpm for 24 h. After shaking, the soil-solution mixture was centrifuged (MPW-360 centrifuge, Mechanika Precyzyjna, Warsaw, Poland) at 4000 rpm for 30 min. The supernatant was then poured through a glass funnel holding a Whatman 42 filter to remove any floating particles or debris, and the concentration of heavy metals (Cu, Pb, Ni, and Zn) was determined by flame atomic absorption spectrometry (FAAS) carried out on a Perkin-Elmer 1100B spectrometer (USA) with air-acetylene flame (air 8 L min^{-1}, acetylene 3.5 L min^{-1}). The batch extractions were performed in triplicates to ensure reproducibility of the results.

Total concentrations of Cu, Pb, Ni and Zn in samples were determined after decomposition as follows. 0.5 g of sample in a polytetrafluoroethylene (PTFE) vessel with a mixture of mineral acids (15 ml of 48% HF, 1 ml of 65% HNO_3 and 1 ml of 70% $HClO_4$) was heated in a water bath at 80 °C for 2 h. The mixture was then allowed to settle down overnight. Next day, the samples were evaporated to wet salts, 5 ml of HF were added and heated until the fumes of $HClO_4$ were formed. Then, 1 ml of HNO_3, 1 ml of $HClO_4$ and 10 ml of H_3BO_3 (saturated solution) were added and samples were heated to dryness. The dried mass was heated again with an addition of 10 ml of deionized water in a water bath for 1–2 h. To purify the solution, 3 ml of HNO_3 and 3–5 drops of 30% H_2O_2 were added and the solution was allowed to digest in a water bath until dissolving the residues. Then, the sample solution was transferred into a 25 ml volumetric flask and filled up to the mark with deionized water (Medveď et al. 1998, 2003). Total contents of heavy metals in contaminated soils and sediments were determined by FAAS and are presented in Table 5.3.

In soil samples, the total contents of Cu and Zn exceed limit values. Furthermore, the Šobov soil is contaminated also by Pb and soil around the Pezinok deposit shows an inadmissible degree of Zn pollution. The analysis of river sediments for Cu, Pb, Ni and Zn indicates high levels of contamination, except for Pb in Jaklovce sediment which does not exceed permitted limit value. It should be pointed out that Slovak threshold values for contaminants in sediments are unbelievably lenient, therefore, the limit concentrations established in European Union directive (EU WFD 2000) that are far more strict and closer to common standards than limits set by Slovak legislation are listed as well (see Table 5.3). Based on these limits and on the data

Table 5.3 Cu, Pb, Ni and Zn concentration in studied samples (mg kg^{-1} of dry matter) collected from contaminated soils and sediments at the locations of Šobov, Pezinok, Richňava and Jaklovce; values were calculated from three replicates (mean ± standard deviation)

Sample/code	**Cu** (mg kg^{-1})	**Pb** (mg kg^{-1})	**Ni** (mg kg^{-1})	**Zn** (mg kg^{-1})
Soil Šobov/SoS	37.0 ± 0.8	99.4 ± 2.5	5.3 ± 0.3	92.5 ± 2.2
Soil Pezinok/SoP	72.3 ± 2.0	48.7 ± 1.8	72.0 ± 2.2	210 ± 4.0
Sediment Richňava/SeR	376 ± 2.8	116 ± 2.9	56.2 ± 1.5	425 ± 2.8
Sediment Jaklovce/SeJ	304 ± 2.6	70.0 ± 3.4	40.1 ± 1.8	430 ± 3.2
[a]Slovak limits for soil with pH 5–6	20	70	15	60
[a]Slovak limits for soil with pH > 6	50	70	50	150
[b]Slovak limits for sediments	1000	750	300	2500
[c]EU WFD limits for sediments	40	85	35	200

Comparison with limit values for risk elements in [a]agricultural and forest soils of Slovakia as listed in Annex no. 4 to Act No. 188/2003 Coll, [b]river floor sediments of Slovakia as listed in Annex no. 3 to Act No. 188/2003 Coll 188/2003 and [c]sediments according to European Water Framework Directive 2000/60/EC

reported by other authors who conducted their research in the area, the results indicate a long-term Cu, Ni, Zn and to some extent also Pb contamination in Hornád and Hnilec river sediments which should be of concern from the standpoint of health of aquatic ecosystem. Regarding future directions for research, particular attention should be given to the issue of resuspension and remobilisation of hazardous elements from sediments during dry seasons when low flows are experienced because the adverse effects may become significant (Ilie et al. 2014).

The removal of heavy metals cations, such as Cu^{2+}, Ni^{2+}, Cd^{2+} and Pb^{2+}, has been receiving extensive attention; in some cases, removal yields close to 100% have been reported. During the last decade, the non-ionic surfactants and *Aspergillus niger* strain's metabolites have been successfully applied in the extraction of Zn, As, Cd, Pb, Cu, Al, Fe, Sn, Co, Au, Mn, Se, Sb and Si from various solid substrates in a series of experiments undertaken by our research team (Kolenčík et al. 2011, 2013a, b; Urík et al. 2014, 2015, 2018, 2019; Boriová et al. 2016; Milová-Žiaková et al. 2016; Polák et al. 2018, 2019). In this study, the agents mentioned above showed only a weak capability to remove accumulated metals from collected samples (removal efficacy <6%; see Table 5.4). In contrast, a natural plant-based biosurfactant saponin with the removal percentage range of 14–46% proved itself as a promising biodegradable agent for Cu, Ni and Zn remediation from contaminated soils and sediments (Fig. 5.5). Maximal removal yields were achieved at saponin concentration of 2% (m/v) and removal incubation time of 24 h. Higher concentrations and longer incubation time have brought no improvement in removal efficacy. A higher removal efficacy can though be achieved by multiple washing. This parameter is always affected by environmental variables such as soil and

Table 5.4 Removal of Cu, Pb, Ni and Zn from contaminated soils and sediments by different agents

Sample	2% Triton X-100				2% Triton X-114				Metabolites of *A. niger*			
	Cu	Pb	Ni	Zn	Cu	Pb	Ni	Zn	Cu	Pb	Ni	Zn
	%	%	%	%	%	%	%	%	%	%	%	%
Soil Šobov	5.8	0.4	5.6	1.8	5.6	0.5	1.3	3.3	4.9	0.4	4.8	1.2
Soil Pezinok	0.9	0.1	4.8	2.7	0.8	0.2	2.5	3.8	1.3	0.5	2.5	2.5
Sediment Richňava	1.8	0.2	4.4	4.1	1.9	0.3	1.8	4.6	4.0	0.2	4.7	2.8
Sediment Jaklovce	1.1	0.1	5.2	3.8	1.0	0.2	2.8	5.4	4.7	0.3	3.3	4.4

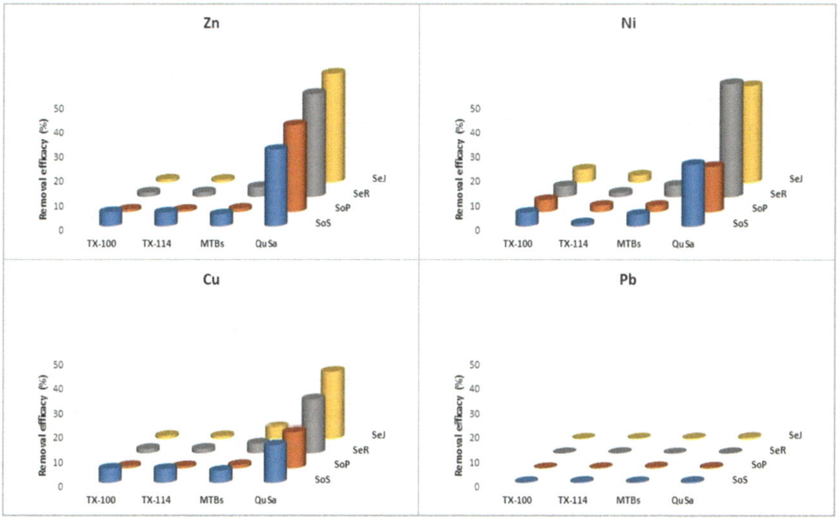

Fig. 5.5 The removal of Zn, Ni, Cu and Pb from contaminated soils and sediments by non-ionic surfactants Triton X-100 (TX-100), Triton X-114 (TX-114), metabolites of *Aspergillus niger* (MTBs) and quillaja saponin (QuSa); none of the agents were effective enough to release Pb from the structures to which it was adhered. *SoS* soil Šobov, *SoP* soil Pezinok, *SeR* sediment Richňava, *SeJ* sediment Jaklovce

sediment composition, soil pH, cation exchange capacity, particle size, permeability and the presence of other contaminants.

A modified sequential extraction procedure (SEP), originally proposed by the Commission of the European Communities Bureau of Reference (BCR) (Rauret et al. 1999), applied to the same sediment samples (Vojteková et al. 2008; Blašková et al. 2013) allowed to distinguish different fractions (water-soluble and extractable fraction, metals bound to carbonates easily extractable by a weak acid and metals bound to Mn and Fe oxides) of the heavy metals studied. If we compare recovery values for Cu, Ni and Zn given by the two extraction techniques, a percentage computed as a sum of the amounts determined in steps one and two of SEP was similar to the value obtained by the extraction with saponin. It indicates that the most mobile fractions (water soluble and acid extractable) were easily released in the presence of selected biosurfactant. Regarding Pb, no presence of this element was detected in the first two steps of SEP. The maximum recovery yields were attained in step three of SEP (0.1 mol L^{-1} $NH_2OH.HCl$), reaching about 50–80% of total Pb content in sediment samples; such a finding is consistent with the ability of Pb to co-precipitate with Mn and Fe oxides (reducible sediment fraction) (Vojteková et al. 2008) and thereby remain stable for a relatively long period of time. Saponin was not able to release Pb from such structures.

The hazard level associated with a heavy metal-contaminated site depends significantly on the chemical form and speciation of the metal. It can be concluded that if the most mobile fractions pose an increased risk of contamination and re-entering biogeochemical cycles, the current risk of Pb contamination in Richňava and Jaklovce is low. It can be also mentioned that the sites in which metal ions exist mainly in residual forms, such as Pb in phosphates (e.g. $Pb_3(PO_4)_2$, $PbHPO_4$, $Pb_5(PO_4)_3Cl$), PbS and $PbSO_4$, present little or no hazard to the environment and living organisms (Ogundiran and Osibanjo 2009).

5.6 Conclusions

The residues of organic and inorganic substances found in soils, sediments, waters and aquatic biota have been of environmental concern since the 1960–1970s. Although production of many hazardous chemicals has been gradually banned across the world since the 1970s, the amounts released in the environment can be counted in millions of tons. Moreover, new substances that are harmful to living organisms are being produced by polluting industries every day; they are commonly used in agriculture; worldwide, companies routinely pollute our natural resources with their waste. Many of these compounds are resistant to biological and chemical degradation, thus persisting in the environment for a long period of time. For their determination in different environmental matrices, sample preparation can be time-consuming and challenging but still an essential part of the whole contaminant analysis process. Due to a low concentration of pollutants and often a complex nature of collected samples, separation and pre-concentration prior to analysis become even more important. Among sample preparation methods, liquid-liquid extraction and solid-phase extraction are reported to be the most frequently used for isolation and clean-up, sample preconcentration and derivatization. Both techniques are available in different modifications and both are applicable for an extraction of organic and inorganic pollutants. It is evident that much effort has been devoted to improving extraction procedures in order to make the process more effective, faster, simpler, less expensive and eco-friendly. In this context, conventional surfactants are gradually being replaced with substances of biological origin that are equally or even more efficient and are not harmful to the environment and human health. In our study on extraction of metal ions from solid materials using liquid solvent, we tested plant-derived surfactant saponin for the recovery of selected heavy metals from soils and river sediments collected in environmentally stressed areas. The experiment conducted indicates that this approach can be successfully applied to recover Cu, Ni and Zn from contaminated soils and sediments. Good removal rates were achieved after single washing; to increase the yields, multiple washing might be needed. Although scientific studies that have been published on this topic point to a high efficiency of biosurfactants in removing metals from contaminated sites, more information is required with regard to the interactions between these compounds and contaminants in order to utilize their full potential in remediation technologies.

Acknowledgements This work was supported by the Scientific Grant Agency of the Slovak Republic Ministry of Education, Science, Research and Sport and the Slovak Academy of Sciences under contract nos. VEGA 1/0153/17, 1/0146/18.

References

Abdel-Mawgoud AM, Lépine F, Déziel E (2010) Rhamnolipids: diversity of structures microbial origins and roles. Appl Microbiol Biotechnol 86:1323–1336. https://doi.org/10.1007/s00253-010-2498-2

Abhijit SP (2017) Efficiency of surfactants in pigment dispersion. Paintindia 67:55–70

Albero B, Sánchez-Brunete C, García-Valcárcel AI et al (2015) Ultrasound-assisted extraction of emerging contaminants from environmental samples. TrAC Trends Anal Chem 71:110–118. https://doi.org/10.1016/j.trac.2015.03.015

Al-Wahaibi Y, Joshi S, Al-Bahry S et al (2014) Biosurfactant production by *Bacillus subtilis* B30 and its application in enhancing oil recovery. Colloid Surf B 114:324–333. https://doi.org/10.1016/j.colsurfb.2013.09.022

Anastas PT, Warner JC (1998) Green chemistry: theory and practice. Oxford University Press, New York, 135 p. ISBN: 9780198502340

Andráš P, Kušnierová M, Adam M et al (2004a) Biooxidácia rudných minerálov na ložisku Pezinok. Acta Montan Slovaca 9:442–453. *(in Slovak)*

Andráš P, Kušnierová M, Adam M et al (2004b) Bacterial leaching of ore minerals from waste at the Pezinok deposit (Western Slovakia). Slovak Geol Mag 12:79–90

Arima K, Kakinuma A, Tamura G (1968) Surfactin, a crystalline peptidelipid surfactant produced by *Bacillus subtilis*: isolation, characterization and its inhibition of fibrin clot formation. Biochem Biophys Res Commun 31:488–494. https://doi.org/10.1016/0006-291X(68)90503-2

Azmir J, Zaidul ISM, Rahman MM et al (2013) Techniques for extraction of bioactive compounds from plant materials: a review. J Food Eng 117:426–436. https://doi.org/10.1016/j.jfoodeng.2013.01.014

Bai L, McClements DJ (2016) Formation and stabilization of nanoemulsions using biosurfactants: Rhamnolipids. J Colloid Interface Sci 479:71–79. https://doi.org/10.1016/j.jcis.2016.06.047

Banat IM, Franzetti A, Gandolfi I et al (2010) Microbial biosurfactants production, applications and future potential. Appl Microbiol Biotechnol 87:427–444. https://doi.org/10.1007/s00253-010-2589-0

Banno T, Toyota T, Matsumura S (2013) Chapter 5 – Creation of novel green surfactants containing carbonate linkages. In: Chamy R (ed) Biodegradation – life of science. IntechOpen, pp 115–140. https://doi.org/10.5772/56163

Barahona F, Turiel E, Martín-Esteban A (2019) Human resource function and interaction with the safety and health professional. In: Tadeo JL (ed) Analysis of pesticides in food and environmental samples, 2nd edn. CRC Press/Taylor & Francis Group, Boca Raton. 423 p. ISBN: 9781138486034

Bartolome L, Lezamiz J, Etxebarria N et al (2007) Determination of trace levels of dinitrophenolic compounds by microporous membrane liquid-liquid extraction in environmental water samples. J Sep Sci 30:2144–2152. https://doi.org/10.1002/jssc.200600509

Belardi RP, Pawliszyn J (1989) Application of chemically modified fused silica fibers in the extraction of organics from water matrix samples and their rapid transfer to capillary columns. Wat Pollut Res J Can 24:179–191. https://doi.org/10.2166/wqrj.1989.010

Berk Z (2013) Chapter 11 – Extraction. In: Food process engineering and technology, Food science and technology, 2nd edn. Academic Press, Amsterdam, pp 287–309. eBook ISBN: 9780124159860

Bicking MKL (2000) Extraction I Analytical extractions. In: Wilson ID (ed) Encyclopedia of separation science. Academic Press, San Diego, pp 1371–1382. https://doi.org/10.1016/B0-12-226770-2/02271-7

Björklund E, Nilsson T, Bøwadt S (2000) Pressurised liquid extraction of persistent organic pollutants in environmental analysis. TrAC Trends Anal Chem 19:434–445. https://doi.org/10.1016/S0165-9936(00)00002-9

Blašková J, Vojteková V, Nováková J et al (2013) Sono-extraction as a pretreatment approach for the screening evaluation of element mobility of sediment samples. Cent Eur J Chem 11:1201–1212. https://doi.org/10.2478/s11532-013-0249-9

Boriová K, Urík M, Bujdoš M et al (2016) Chemical mimicking of bio-assisted aluminium extraction by *Aspergillus niger's* exometabolites. Environ Pollut 218:281–288. https://doi.org/10.1016/j.envpol.2016.07.003

Bruguera LS, González ME, Díaz MF (2017) Effects of cyclodextrins and rhamnolipids on the extraction of PAHs and metals from an aged creosote polluted soil. 16th International conference on chemistry and the environment, 18–22 June 2017 Oslo, Norway, Norwegian Chemical Society

Buckle J (2015) Chapter 3 – Basic plant taxonomy, basic essential oil chemistry, extraction, biosynthesis, and analysis. In: Essential oils in healthcare, Clin Aromatherapy, 3rd edn. Churchill Livingstone, London, pp 37–72. https://doi.org/10.1016/B978-0-7020-5440-2.00003-6

Bujdoš M, Muľová A, Kubová J et al (2005) Selenium fractionation and speciation in rocks, soils, waters and plants in polluted surface mine environment. Environ Geol 47:353–360. https://doi.org/10.1007/s00254-004-1157-2

Burch AY, Shimada BK, Browne PJ et al (2010) Novel high-throughput detection method to assess bacterial surfactant production. Appl Environ Microbiol 76:5363–5372. https://doi.org/10.1128/AEM.00592-10

Cao M, Hu Y, Sun Q et al (2013) Enhanced desorption of PCB and trace metal elements (Pb and Cu) from contaminated soils by saponin and EDDS mixed solution. Environ Pollut 174:93–99. https://doi.org/10.1016/j.envpol.2012.11.015

Čerňanský S, Urík M, Ševc J et al (2007) Biosorption and biovolatilization of arsenic by heat-resistant fungi. Environ Sci Pollut Res 14:31–35. https://doi.org/10.1065/espr2006.11.361

Čerňanský S, Kolenčík M, Ševc J et al (2009) Fungal volatilization of trivalent and pentavalent arsenic under laboratory conditions. Bioresour Technol 100:1037–1040. https://doi.org/10.1016/j.biortech.2008.07.030

Chaieb I (2010) Saponins as insecticides: a review. Tunis J Plant Prot 5:39–50

Cheeke PR (2000) Actual and potential applications of *Yucca schidigera* and *Quillaja saponaria* saponins in human and animal nutrition. J Anim Sci 77:1–10. https://doi.org/10.2527/jas2000.00218812007700ES0009x

Chen WJ, Hsiao LC, Chen KKY (2008) Metal desorption from copper(II)/nickel(II)-spiked kaolin as a soil component using plant-derived saponin biosurfactant. Process Biochem 43:488–498. https://doi.org/10.1016/j.procbio.2007.11.017

Cheok CY, Salman HAK, Sulaiman R (2014) Extraction and quantification of saponins: a review. Food Res Int 9:16–40. https://doi.org/10.1016/j.foodres.2014.01.057

Chirila E, Drăghici C (2013) Analytical approaches of environmental sampling and sample preparation for pesticides analysis in environmental, food and biological samples. In: Simeonov LI, Macaev FZ, Simeonova BG (eds) Environmental security assessment and management of obsolete pesticides in Southeast Europe, NATO Science for Peace and Security, Series C: Environmental Security. Springer Science + Business Media, Dordrecht, pp 37–54. https://doi.org/10.1007/978-94-007-6461-3

Cogan U, Yaron A, Berk Z et al (1967) Isolation of soybean protein: effect of processing conditions on yields and purity. J Am Oil Chem Soc 44:321–324. https://doi.org/10.1007/BF02635625

Costa SGVAO, Nitschke M, Lépine F et al (2010) Structure, properties and applications of rhamnolipids produced by *Pseudomonas aeruginosa* L2-1 from cassava wastewater. Process Biochem 45:1511–1516. https://doi.org/10.1016/j.procbio.2010.05.033

Costa JAV, Treichel H, Santos LO et al (2018) Chapter 16 - Solid-state fermentation for the production of biosurfactants and their applications. Curr Dev Biotechnol Bioeng 2018:357–372. https://doi.org/10.1016/B978-0-444-63990-5.00016-5

De Marino S, Iorizzi M, Palagiano E et al (1998) Starfish saponins. 55. Isolation, structure elucidation, and biological activity of the steroid oligoglycosides from an Antarctic starfish of the family Asteriidae. J Nat Prod 61:1319–1327. https://doi.org/10.1021/np970500e

Dean JR (2000) Solid-phase microextraction – overview. In: Wilson ID (ed) Encyclopedia of separation science. Academic Press, San Diego, pp 4190–4199. https://doi.org/10.1016/B0-12-226770-2/02181-5

Desai JD, Banat IM (1997) Microbial production of surfactants and their commercial potential. Microbiol Mol Biol Rev 61:47–64

Djenni Z, Pingret D, Mason TJ (2013) Sono–soxhlet: in situ ultrasound-assisted extraction of food products. Food Anal Method 6:1229–1233. https://doi.org/10.1007/s12161-012-9531-2

Dlapa P, Kubová J, Matúš P et al (2002) Chemical and physical aluminium speciation in soil solutions. Fresenius Environ Bull 11:626–630

Domini CE et al (2005) Sample preparation for chromatographic analysis of environmental samples. In: Nollet LML (ed) Chromatographic analysis of the environment, Chromatographic science series, vol 93, 3rd edn. Taylor & Francis/CRC, Boca Raton, pp 31–132. ISBN: 0824726294, 9780824726294

Duarte K, Justino CIL, Gomes AM et al (2014) Chapter 4 – Green analytical methodologies for preparation of extracts and analysis of bioactive compounds. Compr Anal Chem 65:59–78. https://doi.org/10.1016/B978-0-444-63359-0.00004-5

Elouzi AA, Akasha AA, Elgerbi AM et al (2012) Removal of heavy metals contamination by bio-surfactants (Rhamnolipids). J Chem Pharm Res 4:4337–4341

EPA Victoria (2009) Industrial waste resource guidelines. Sampling and analysis of waters, wastewaters, soils and wastes. Publication No. IWRG701. Environment Protection Authority Victoria, 36 p

Escarpa A, López MA, Ramos L (2012) Chapter 17 – Miniaturization. In: de la Guardia M, Salvador Garrigues S (eds) Handbook of green analytical chemistry. Wiley, Hoboken, pp 339–387. https://doi.org/10.1002/9781119940722.ch17

European Union. Directive 2000/60/EC (2000) of the European Parliament and of the Council of 23 October 2000 establishing a framework for Community action in the field of water policy (Water Framework Directive). Official Journal of the European Communities C, L 327 22/12/2000

Farn J (2006) Chemistry and technology of surfactants. Blackwell Publishing, Oxford, 336 p. https://doi.org/10.1002/9780470988596

Feng JQ, Gang HZ, Li DS (2019) Characterization of biosurfactant lipopeptide and its performance evaluation for oil-spill remediation. RSC Adv 9:9629–9632. https://doi.org/10.1039/C9RA01430F

Fenwick GR, Price KR, Tsukamoto C et al (1991) Saponins. In: D'Mello JPF, Duffus CM, Duffus JH (eds) Toxic substances in crop plants. The Royal Society of Chemistry, Cambridge, pp 285–327. https://doi.org/10.1017/S0021859600012296

Gao L, Kano N, Sato Y et al (2012) Behavior and distribution of heavy metals including rare earth elements, thorium, and uranium in sludge from industry water treatment plant and recovery method of metals by biosurfactants application. Bioinorg Chem Appl 2012(2):173819. https://doi.org/10.1155/2012/173819

Ghiasvand AL, Pirdadeh-Beiranvand M (2015) Cooling/heating-assisted headspace solid-phase microextraction of polycyclic aromatic hydrocarbons from contaminated soils. Anal Chim Acta 900:56–66. https://doi.org/10.1016/j.aca.2015.10.016

Goswami T, Sir M, Tack FMG (2018) Remediation of petroleum hydrocarbons with sophorolipids from *Candida bombicola* and Triton-X: a preliminary study. Conference: Sanační technologie XXI, Czech Republic

Gutte KB, Sahoo AK, Ranveer RC (2015) Effect of ultrasonic treatment on extraction and fatty acid profile of flaxseed oil. OCL 22:D606. https://doi.org/10.1051/ocl/2015038

Hagarová I, Urík M (2016) New approaches to the cloud point extraction: utilizable for separation and preconcentration of trace metals. Curr Anal Chem 12:87–93. https://doi.org/10.2174/1573411011666150601204931

Hailemariam T, Bekele T, Madhusudhan A (2016) Recent advances in sample preparation techniques using SFE; MAE & MIP in environmental matrice: a critical overview. Res J Chem Environ Sci 4:11–27

Hauser L, Tandy S, Schulin R et al (2005) Column extraction of heavy metals from soils using the biodegradable chelating agent EDDS. Environ Sci Technol 39:6819–6824. https://doi.org/10.1021/es050143r

Hayman R (2016) The iron industry. Bloomsbury Publishing, Oxford. 64 p. ISBN: 9780747814832

Heng MY, Tan SN, Yong JWH et al (2013) Emerging green technologies for the chemical standardization of botanicals and herbal preparations. TrAC Trends Anal Chem 50:1–10. https://doi.org/10.1016/j.trac.2013.03.012

Higuchi R, Tokimitsu Y, Fujioka T et al (1986) Structure of desacylsaponins obtained from the bark of *Quillaja Saponaria*. Phytochemistry 26:229–235. https://doi.org/10.1016/S0031-9422(00)81518-6

Hill GD (2003) Plant antinutritional factors / characteristics. In: Trugo L, Finglas PM (eds) Encyclopedia of food sciences and nutrition, 2nd edn. Academic Press, Amsterdam, pp 4578–4587. https://doi.org/10.1016/B0-12-227055-X/01318-3

Höfler F (2002) Accelerated solvent extraction (ASE) – a fast and automated technique with low solvent consumption for the extractions of solid samples (T 12). In: Book of abstracts – trends in sample preparation 2002, development and application. Institute for Analytical Chemistry, Micro- and Radiochemistry, Graz University of Technology, Austria, 30th of June – 3rd of July 2002, 70 p

Hommel R (1994) Formation and function of biosurfactants for degradation of water-insoluble substrates. In: Ratledge C (ed) Biochemistry of microbial degradation. Kluwer Academic Publishers, London, pp 63–87. https://doi.org/10.1007/978-94-011-1687-9

Hong KJ, Tokunaga S, Kajiuchi T (2002) Evaluation of remediation process with plant-derived biosurfactant for recovery of heavy metals from contaminated soils. Chemosphere 49:379–387. https://doi.org/10.1016/S0045-6535(02)00321-1

Hsieh HK, Chen CL, Ding WH (2013) Determination of Dechlorane compounds in aqueous samples using ultrasound-assisted dispersive liquid-liquid microextraction and gas chromatography-electron-capture negative ion-mass spectrometry. Anal Methods 5:7001–7007. https://doi.org/10.1039/C3AY41505H

Hubert J, Plé K, Hamzaoui M et al (2012) New perspectives for microbial glycolipid fractionation and purification processes. C R Chim 15:18–28. https://doi.org/10.1016/j.crci.2011.11.002

Ilie M, Marinescu F, Ghita G et al (2014) Assessment of heavy metal in water and sediments of the Danube River. J Environ Prot Ecol 15:825–833

Inamuddin, Ahamed MI, Prasad R (2021) Microbial biosurfactants: preparation, properties and applications. Springer Singapore. ISBN 978–981–15-6606-6. https://www.springer.com/gp/book/9789811566066

International Strategy for Disaster Reduction (ISDR) (2007) Environment and vulnerability, emerging perspectives. Environment and Disaster Working Group, Geneva. 34 p

IUPAC (1997) Compendium of chemical terminology, 2nd ed. (the "Gold Book"). In: McNaught AD, Wilkinson A (eds). Blackwell Scientific Publications, Oxford. XML on-line corrected version: http://goldbook.iupac.org (2006) created by Nic M, Jirat J, Kosata B; updates compiled by Jenkins A. https://doi.org/10.1351/goldbook.C01395

Jablonská J, Timčák G, Pixová L (2009) Geotourism and water quality of river Hornád (E. Slovakia). Acta Montan Slovaca 14:213–220

Jain R, Mudiam MKR, Singh R (2014) Application of ultrasound-assisted dispersive liquid-liquid microextraction and automated in-port silylation for the simultaneous determination of phenolic endocrine disruptor chemicals in water samples by gas chromatography-triple quadrupole mass spectrometry. Anal Methods 6:1802–1810. https://doi.org/10.1039/C3AY41658E

Jeannot MA (2007) Extraction I liquid-phase microextraction. In: Wilson ID (ed) Encyclopedia of separation science. Academic Press, San Diego, pp 1–5. https://doi.org/10.1016/B978-012226770-3/10677-6

JoVE Science Education Database (2019) Organic chemistry. Solid-liquid extraction. JoVE, Cambridge, MA. Available from: https://www.jove.com/science-education/5538/solid-liquid-extraction

Kališ M, Matúš P (2011) Štúdium chemických foriem a distribúcie tália. In: Urík M (ed) Kapitoly z environmentálnej geochémie 1. Vysoká škola báňská – Technická univerzita Ostrava, Ostrava, pp 425–494. ISBN: 9788024825281 *(in Slovak)*

Kassim TA, Simoneit BRT, Williamson KJ (2005) Forensic investigation of leachates from recycled solid wastes: an environmental analysis approach. In: Kassim TA, Williamson KJ (eds) Water pollution. Handb Environ Chem Vol. 5, Part F, 1:321–400, Springer-Verlag, Berlin, Heidelberg. https://doi.org/10.1007/b98265

Kataoka H, Lord HL, Pawliszyn J (2000) Solid-phase microextraction I biomedical applications. In: Wilson ID (ed) Encyclopedia of separation science. Academic Press, San Diego, pp 4153–4169. https://doi.org/10.1016/B0-12-226770-2/00971-6

Kaushik P (2015) Use of natural organic and synthetic chelating agents for efficient phytoremediation. Int J Enhanced Res Sci Technol Eng 4:99–101

Keith LH, Crummett W, Deegan J et al (1983) Principles of environmental analysis. Anal Chem 55:2210–2218. https://doi.org/10.1021/ac00264a003

Kiezyk PR, Mackay D (1971) Waste water treatment by solvent extraction. Can J Chem Eng 49:747–752. https://doi.org/10.1002/cjce.5450490607

Király Z (2013) The HLB system: Croda's time-saving guide to surfactant selection. Personal care edition. Croda, 20 p

Klinda et al (1994) Životné prostredie Slovenskej republiky v rokoch 1992–1993. Správa o stave životného prostredia. Ministerstvo životného prostredia, Bratislava, 520 p. *(in Slovak)*

Kolenčík M, Urík M, Gardošová K et al (2011) Biological and chemical leaching of arsenic and zinc from adamite. Chem Listy 105:961–965. *(in Slovak)*

Kolenčík M, Urík M, Bujdoš M et al (2013a) Leaching of Al, Fe, Sn, Co and Au from electronics wastes using organic acid and microscopic fibrous fungus *Aspergillus niger*. Chem Listy 107:182–185. *(in Slovak)*

Kolenčík M, Urík M, Čerňanský S et al (2013b) Leaching of zinc, cadmium, lead and copper from electronic scrap using organic acids and the *Aspergillus niger* strain. Fresenius Environ Bull 22:3673–3679

Kubová J, Streško V, Bujdoš M et al (2004) Fractionation of various elements in CRMs and in polluted soils. Anal Bioanal Chem 379:108–114. https://doi.org/10.1007/s00216-004-2505-5

Kubová J, Matúš P, Bujdoš M et al (2005) Influence of acid mining activity on release of aluminium to the environment. Anal Chim Acta 547:119–125. https://doi.org/10.1016/j.aca.2004.12.014

Kubová J, Matúš P, Bujdoš M et al (2008a) Optimized BCR three-step sequential and dilute HCl single extraction protocols – the suitable tools for soil-plant metal transfer predictions in polluted areas? Trans Univ Kosice 3:50–58

Kubová J, Matúš P, Bujdoš M et al (2008b) Utilization of optimized BCR three-step sequential and dilute HCl single extraction procedures for soil-plant metal transfer predictions in contaminated lands. Talanta 75:1110–1122. https://doi.org/10.1016/j.talanta.2008.01.002

Kudlejova L, Risticevic S, Vuckovic D (2012) Solid-phase microextraction method development. In: Pawliszyn J (ed) Handbook of solid phase microextraction. Elsevier, London, pp 201–249. https://doi.org/10.1016/B978-0-12-416017-0.00007-3

Lang S (2002) Biological amphiphiles (microbial biosurfactants). Curr Opin Colloid Interface Sci 7:12–20. https://doi.org/10.1016/S1359-0294(02)00007-9

Le LQ, Ngo HD, Hoang KQ et al (2016) Production and characterization of sophorolipids produced by *Candida bombicola* from coconut oil. Sci Technol Dev J 19:15–25. https://doi.org/10.32508/stdj.v19i4.627

Lintnerová O (2002) Vplyv ťažby nerastných surovín na životné prostredie. Univerzita Komenského, Bratislava. 160 p. ISBN: 802231630X *(in Slovak)*

Littera P (2011) Kontaminácia pitnej vody arzénom: prehľad najviac zaťažených oblastí sveta a niekoľko príkladov zo Slovenska. In: Urík M (ed) Arzén – výskyt, metabolizmus, toxicita, analytické stanovenie a využitie v didaktike. Vysoká škola báňská – Technická univerzita Ostrava, Ostrava, pp 6–22. ISBN: 9788024825267 *(in Slovak)*

Liu JF, Mbadinga SM, Yang SZ et al (2015) Chemical structure, property and potential applications of biosurfactants produced by *Bacillus subtilis* in petroleum recovery and spill mitigation. Int J Mol Sci 16:4814–4837. https://doi.org/10.3390/ijms16034814

Liu G, Zhong H, Yang X et al (2018) Advances in applications of rhamnolipids biosurfactant in environmental remediation: a review. Biotechnol Bioeng 115:796–814. https://doi.org/10.1002/bit.26517

Llompart M, Garcia-Jares C, Celeiro M et al (2019) Extraction I Microwave-assisted extraction. In: Worsfold P, Poole C, Townshend A et al (eds) Encyclopedia of analytical science, vol 3, 3rd edn. Elsevier, Amsterdam, pp 67–77. https://doi.org/10.1016/B978-0-12-409547-2.14442-7

Lopez-Avila V (2000) Extraction I microwave-assisted extraction. In: Wilson ID (ed) Encyclopedia of separation science. Academic Press, San Diego, pp 1389–1398. https://doi.org/10.1016/B0-12-226770-2/02221-3

Luptáková A, Andráš P (2018) Formation of acid mine drainage in sulphide ore deposits. In: Negm A, Zeleňáková M (eds) Water resources in Slovakia: part I. The handbook of environmental chemistry, vol 69. Springer, Cham, pp 259–276. https://doi.org/10.1007/698_2018_313

Luque-Garcia JL, Luque de Castro MD (2003) Extraction of polychlorinated biphenyls from soils by automated focused microwave-assisted Soxhlet extraction. J Chromatogr A 998:21–29. https://doi.org/10.1016/S0021-9673(03)00652-6

Luque-Garcia JL, Luque de Castro MD (2004) Ultrasound-assisted Soxhlet extraction: an expeditive approach for solid sample treatment: application to the extraction of total fat from oleaginous seeds. J Chromatogr A 1034:237–242. https://doi.org/10.1016/j.chroma.2004.02.020

Madrid F, Ballesteros R, Lacorte S et al (2019) Extraction of PAHs from an aged creosote-polluted soil by cyclodextrins and rhamnolipids. Side effects on removal and availability of potentially toxic elements. Sci Total Environ 653:384–392. https://doi.org/10.1016/j.scitotenv.2018.10.316

Makkar RS, Rockne KJ (2003) Comparison of synthetic surfactants and biosurfactants in enhancing biodegradation of polycyclic aromatic hydrocarbons. Environ Toxicol Chem 22:2280–2292. https://doi.org/10.1897/02-472

Mandal SC, Mandal V, Das AK (2015) Chapter 6 - Classification of extraction methods. In: Mandal SC, Mandal V, Das AK (eds) Essentials of botanical extraction. Academic Press, Amsterdam, pp 83–136. https://doi.org/10.1016/B978-0-12-802325-9.00006-9

Matúš P (2007) Evaluation of separation and determination of phytoavailable and phytotoxic aluminium species fractions in soil, sediment and water samples by five different methods. J Inorg Biochem 101:1214–1223. https://doi.org/10.1016/j.jinorgbio.2007.06.014

Matúš P, Kubová J (2005) Complexation of labile aluminium species by chelating resins Iontosorb – a new method for Al environmental risk assessment. J Inorg Biochem 99:1769–1778. https://doi.org/10.1016/j.jinorgbio.2005.06.009

Matúš P, Kubová J (2006) Complexation efficiency of differently fixed 8-hydroxyquinoline and salicylic acid ligand groups for labile aluminium species determination in soils. Anal Chim Acta 573–574:474–481. https://doi.org/10.1016/j.aca.2006.03.063

Matúš P, Kubová J (2008) Recent developments in the determination, fractionation and speciation analysis of aluminium by spectrochemical analytical methods and computer modelling. In: Dubois AN (ed) Soil contamination: new research. Nova Science, New York, pp 43–72. ISBN: 9781604561449

Matúš P, Kubová J, Streško V (2003) Utilization of chelating ion exchange for aluminium speciation. Chem Pap 57:176–178

Matúš P, Kubová J, Bujdoš M et al (2004) Chemical partitioning of aluminium in rocks, soils, and sediments acidified by mining activity. Anal Bioanal Chem 379:96–103. https://doi.org/10.1007/s00216-004-2562-9

Matúš P, Kubová J, Bujdoš M et al (2005) Determination of operationally defined fractions of aluminium in reference materials and acid attacked environmental samples. Anal Chim Acta 540:33–43. https://doi.org/10.1016/j.aca.2004.09.017

Matúš P, Kubová J, Bujdoš M et al (2006) Free aluminium extraction from various reference materials and acid soils with relation to plant availability. Talanta 70:996–1005. https://doi.org/10.1016/j.talanta.2006.05.045

Medveď J, Streško V, Kubová J et al (1998) Efficiency of decomposition procedures for the determination of some elements in soils by atomic spectroscopic methods. Fresenius J Anal Chem 360:219–224. https://doi.org/10.1007/s002160050678

Medveď J, Streško V, Kubová J et al (2003) Evaluation of atomic spectrometry methods for determination of some heavy metals in soils, soil extracts plants and biota. Chem Pap 57:169–171

Medveď J, Bujdoš M, Matúš P et al (2004) Determination of trace amounts of gold in acid-attacked environmental samples by atomic absorption spectrometry with electrothermal atomization after preconcentration. Anal Bioanal Chem 379:60–65. https://doi.org/10.1007/s00216-004-2538-9

Medveď J, Matúš P, Bujdoš M et al (2006) Gold and silver determination in waters by Spheron Thiol 1000 preconcentration and ETAAS. Chem Pap 60:27–31. https://doi.org/10.2478/s11696-006-0005-0

Medveď J, Kališ M, Hagarová I et al (2008) Thallium fractionation in polluted environmental samples using a modified BCR three-step sequential extraction procedure and its determination by electrothermal atomic absorption spectrometry. Chem Pap 62:168–175. https://doi.org/10.2478/s11696-008-0007-1

Meena KR, Kanwar SS (2015) Lipopeptides as the antifungal and antibacterial agents: applications in food safety and therapeutics. Biomed Res Int 2015:473050. https://doi.org/10.1155/2015/473050

Miller RM (1995) Biosurfactant-facilitated remediation of metal-contaminated soils. Environ Health Perspect 103:59–62. https://doi.org/10.2307/3432014

Milová-Žiaková B, Urík M, Boriová K et al (2016) Fungal solubilization of manganese oxide and its significance for antimony mobility. Int Biodeterior Biodegrad 114:157–163. https://doi.org/10.1016/j.ibiod.2016.06.011

Ministry of Environment of the Slovak Republic (1998) Methodical Instruction No. 549/1998–2 for Risk assessment posed by contaminated sediments in streams and water reservoirs. Ministry of Environment of the Slovak Republic, Bratislava, Slovak Republic. *(in Slovak)*

Mitra S, Dungan SR (1997) Micellar properties of Quillaja saponin. 1. Effects of temperature, salt, and pH on solution properties. J Agric Food Chem 45:1587–1595. https://doi.org/10.1021/jf960349z

Mohamed K (2017) One-step derivatization-extraction method for rapid analysis of eleven amphetamines and cathinones in oral fluid by GC-MS. J Anal Toxicol 41:639–645. https://doi.org/10.1093/jat/bkx046

Morita T, Fukuoka T, Imura D et al (2016) Glycolipid biosurfactants. Reference module in chemistry. Molecular sciences and chemical engineering https://doi.org/10.1016/B978-0-12-409547-2.11565-3

Msagati T, Chimuka L, Cukrowska E (2008) Sample preparation using liquid membrane extraction techniques. Water SA 34:421–428

Müller MM, Kügler JH, Henkel M et al (2012) Rhamnolipids–next generation surfactants? J Biotechnol 162:366–380. https://doi.org/10.1016/j.jbiotec.2012.05.022

Mulligan CN (2005) Environmental applications for biosurfactants. Environ Pollut 133:183–198. https://doi.org/10.1016/j.envpol.2004.06.009

Naresh M, Bhargavi S, Krishna Sowmya D et al (2013) Supercritical fluid technology: green chemistry for the 21st century. J Pharm Res 2:7–11

Nießner R, Schäffer A (2017) Organic trace analysis. Walter de Gruyter GmbH & Co KG, 369 p. ISBN: 978-3-11-044115-4

Nilsson T (2000) Environmental applications I solid-phase microextraction. In: Wilson ID (ed) Encyclopedia of separation science. Academic Press, San Diego, pp 2695–2701. https://doi.org/10.1016/B0-12-226770-2/02191-8

Nittrouer CA, Austin JA, Field ME et al (2007) Writing a Rosetta stone: insights into continental-margin sedimentary processes and strata. In: Continental margin sedimentation: from sediment transport to sequence stratigraphy. Special publication number 37 of the International Association of Sedimentologists. Blackwell Publishing, Malden, pp 1–48. https://doi.org/10.1002/9781444304398.ch1

O'Reilly J, Wang Q, Setkova L et al (2005) Automation of solid-phase microextraction. J Sep Sci 28:2010–2022. https://doi.org/10.1002/jssc.200500244

Ogundiran MB, Osibanjo O (2009) Mobility and speciation of heavy metals in soils impacted by hazardous waste. Chem Spec Bioavailab 21:59–69. https://doi.org/10.3184/095422909X449481

Pacwa-Płociniczak M, Płaza GA, Piotrowska-Seget Z (2011) Environmental applications of biosurfactants: recent advances. Int J Mol Sci 12:633–654. https://doi.org/10.3390/ijms12010633

Padrela L, Rodrigues MA, Velaga SP et al (2009) Formation of indomethacin–saccharin cocrystals using supercritical fluid technology. Eur J Pharm Sci 38:9–17. https://doi.org/10.1016/j.ejps.2009.05.010

Pannu SKV, Kaur H, Kaur K et al (2018) A comparative study of various oil extraction techniques from plants. Agriways 6:59–65. ISSN: 2454–2318 (online)

Pawliszyn J (ed) (2002) Sampling and sample preparation in field and laboratory: fundamentals and new directions in sample preparation. Elsevier Science, Boston. 1166 p. ISBN: 9780444505101

Peach J, Eastoe J (2014) Supercritical carbon dioxide: a solvent like no other. Beilstein J Org Chem 10:1878–1895. https://doi.org/10.3762/bjoc.10.196

Picó Y, Fernández M, Ruiz MJ et al (2007) Current trends in solid-phase-based extraction techniques for the determination of pesticides in food and environment. J Biochem Biophys Methods 70:117–131. https://doi.org/10.1016/j.jbbm.2006.10.010

Płotka-Wasylka J, Szczepańska N, de la Guardia M et al (2015) Miniaturized solid-phase extraction techniques. TrAC Trends Anal Chem 73:19–38. https://doi.org/10.1016/j.trac.2015.04.026

Polák F, Urík M, Bujdoš M et al (2018) Evaluation of aluminium mobilization from its soil mineral pools by simultaneous effect of *Aspergillus strains'* acidic and chelating exometabolites. J Inorg Biochem 181:162–168. https://doi.org/10.1016/j.jinorgbio.2017.09.006

Polák F, Urík M, Bujdoš M et al (2019) Fungal bioextraction of iron from kaolin. Chem Pap 73:3025–3029. https://doi.org/10.1007/s11696-019-00896-6

Price KR, Johnson IT, Fenwick GR (1987) The chemistry and biological significance of saponins in foods and feedingstuffs. CRC Crit Rev Food Sci 26:27–135. https://doi.org/10.1080/10408398709527461

Przyjazny A (2019) Extraction I Liquid-phase microextraction. In: Worsfold P, Poole C, Townshend A et al (eds) Encyclopedia of analytical science, vol 3, 3rd edn. Elsevier, Amsterdam, pp 52–62. https://doi.org/10.1016/B978-0-12-409547-2.12678-2

Puura E, D'Alessandro M (2005) A classification system for pressures related to hazardous water emissions from mine sites – comparison of pressures in ten EU Accession Countries. Mine Water Environ 1:43–52. https://doi.org/10.1007/s10230-005-0070-6

Quigley A, Cummins W, Connolly D (2016) Dispersive liquid-liquid microextraction in the analysis of milk and dairy products: a review. J Chem Article ID 4040165, 12 p. https://doi.org/10.1155/2016/4040165

Rahman KS, Rahman TJ, Kourkoutas Y et al (2003) Enhanced bioremediation of n-alkane in petroleum sludge using bacterial consortium amended with rhamnolipid and micronutrients. Bioresour Technol 90:159–168. https://doi.org/10.1016/S0960-8524(03)00114-7

Rami N, Rai S, Basu D (2017) Scale ranging 1–20 of surfactant decides the solubility of water into oil or oil into water to produce monophasic formulation. Int Educ Sci Res J 3:32–35

Randhawa KKS, Rahman PKSM (2014) Rhamnolipid biosurfactants – past, present, and future scenario of global market. Front Microbiol 5:454. https://doi.org/10.3389/fmicb.2014.00454

Rauret G, Lopez-Sanchez JF, Sahuquillo A et al (1999) Improvement of the BCR three step sequential extraction procedure prior to the certification of new sediment and soil reference materials. J Environ Monit 1:57–61. https://doi.org/10.1039/a807854h

Rezazadeh MK, Yamini Y, Seidi S (2012) Electrically assisted liquid-phase microextraction for determination of β2-receptor agonist drugs in wastewater. J Sep Sci 35:571–579. https://doi.org/10.1002/jssc.201100869

Ribeiro C, Ribeiro A, Maia AS et al (2014) New trends in sample preparation techniques for environmental analysis. Crit Rev Anal Chem 44:142–185. https://doi.org/10.1080/10408347.2013.833850

Risticevic S, Lord H, Górecki T et al (2010) Protocol for solid-phase microextraction method development. Nat Protoc 5:122–139. https://doi.org/10.1038/nprot.2009.179

Rodrigues L, Banat IM, Teixeira J (2006) Biosurfactants: potential applications in medicine. J Antimicrob Chemother 57:609–618. https://doi.org/10.1093/jac/dkl024

Rosenberg E, Ron EZ (1999) High- and low-molecular-mass microbial surfactants. Appl Microbiol Biotechnol 52:154–162. https://doi.org/10.1007/s002530051502

Royal Society of Chemistry, ID: CMO:0001485 https://www.rsc.org/publishing/journals/prospect/ontology.asp?id=CMO:0001485

Rozzi NL, Phippen W, Simon JE et al (2002) Supercritical fluid extraction of essential oil components from lemon-scented botanicals. Lebensm Wiss Technol 35:319–324. https://doi.org/10.1006/fstl.2001.0873

Saha S, Singh AK, Keshari AK et al (2017) Modern extraction techniques for drugs and medicinal agents. In: Grumezescu AM, Holban AM (eds) Ingredients extraction by physico-chemical methods, Handbook of food bioengineering, vol 4. Academic Press/Elsevier, London, pp 65–106. https://doi.org/10.1016/B978-0-12-811521-3.00002-8

Samal K, Das C, Mohanty K (2017) Eco-friendly biosurfactant saponin for the solubilization of cationic and anionic dyes in aqueous system. Dyes Pigments 140:100–108. https://doi.org/10.1016/j.dyepig.2017.01.031

Sánchez-Camargo ADP, Ibáñez E, Cifuentes A et al (2017) Chapter 2 - Bioactives obtained from plants, seaweeds, microalgae and food by-products using pressurized liquid extraction and supercritical fluid extraction. In: Ibáñez E, Cifuentes A (eds) Comprehensive analytical chemistry, vol 76. Elsevier, Amsterdam, pp 27–51. https://doi.org/10.1016/bs.coac.2017.01.001

Santos DKF, Rufino RD, Luna JM et al (2016) Biosurfactants: multifunctional biomolecules of the 21st century. Int J Mol Sci 17:401. https://doi.org/10.3390/ijms17030401

Sapkale GN, Patil SM, Surwase US et al (2010) Supercritical fluid extraction – a review. Int J Chem Sci 8:729–743

Sasidharan S, Chen Y, Saravanan D et al (2010) Extraction, isolation and characterization of bioactive compounds from plants' extracts. Afr J Tradit Complement Altern Med 8:1–10. https://doi.org/10.4314/ajtcam.v8i1.60483

Schaller KD, Fox S, Bruhn D et al (2004) Characterization of surfactin from *Bacillus subtilis* for application as an agent for enhanced oil recovery. Appl Biochem Biotechnol 113:827–836. https://doi.org/10.1385/ABAB:115:1-3:0827

Schwab AP, Su J, Wetzel S (1999) Extraction of petroleum hydrocarbons from soil by mechanical shaking. Environ Sci Technol 33:1940–1945. https://doi.org/10.1021/es9809758

Shidhaye S, Malke S, Mandal S et al (2008) Soy – a hidden treasure for therapeutic, cosmetic and pharmaceutical use. Int J Alternative Med 7:1–6

Shin BK, Kwon SW, Park JH (2015) Chemical diversity of ginseng saponins from *Panax ginseng*. J Ginseng Res 39:287–298. https://doi.org/10.1016/j.jgr.2014.12.005

Šlesárová A (2006) Problems of the mine water quality at selected Slovak localities. Acta Montan Slovaca 11:371–374

Slovak Act No. 188/2003 Coll. on the application of treated sludge and bottom sediments onto soil, Annex No. 3 and No. 4. In: Collection of Laws of the Slovak Republic *(in Slovak)*

Song S, Zhu L, Zhou W (2008) Simultaneous removal of phenanthrene and cadmium from contaminated soils by saponin, a plant-derived biosurfactant. Environ Pollut 156:1368–1370. https://doi.org/10.1016/j.envpol.2008.06.018

Šottník P, Šucha V (2001) Možnosti úpravy kyslého banského výtoku ložiska Banská Štiavnica – Šobov. Miner Slovaca 33:53–60. *(in Slovak)*

Šottník P, Dubíková M, Lintnerová O et al (2002) The links between the physico-chemical character of different mining waste in Slovakia and their environmental impacts. In: Proceedings of XVII. Congress of Carpathian-Balkan Geological Association, Bratislava, September 1–4, 2002, Geol Carpath, vol 53, 6 p

Špaldoň T, Brehuv J, Bobro M et al (2006) Mining development the Spis-Gemer ore-location. Acta Montan Slovaca 11:375–379

Spigno G, De Faveri DM (2009) Microwave-assisted extraction of tea phenols: a phenomenological study. J Food Eng 93:210–217. https://doi.org/10.1016/j.jfoodeng.2009.01.006

Stanley J, Wilkinson ST, Moreno Ramírez D et al (2015) Tribal mining educational modules: copper mining and processing [PDF document]. Retrieved from http://www.superfund.pharmacy.arizona.edu/learning-modules/tribal-modules

Stojić N, Štrbac S, Prokić D (2018) Soil pollution and remediation. In: Hussain CM (ed) Handbook of environmental materials management. Springer, Cham, pp 1–34. https://doi.org/10.1007/978-3-319-58538-3_81-1

Šucha V, Kraus I, Zlocha M et al (1997) Acidification in the Šobov region (Štiavnické vrchy Mts.). Miner Slovaca 29:407–416. *(in Slovak)*

Tahseen R, Afzal M, Iqbal S et al (2016) Rhamnolipids and nutrients boost remediation of crude oil-contaminated soil by enhancing bacterial colonization and metabolic activities. Int Biodeterior Biodegrad 115:192–198. https://doi.org/10.1016/j.ibiod.2016.08.010

Tashakkori P, Erdem P, Bozkurt SS (2017) Molecularly imprinted polymer based on magnetic ionic liquid for solid-phase extraction of phenolic acids. J Liq Chromatogr Relat Technol 40:657–666. https://doi.org/10.1080/10826076.2017.1343732

Thiengmag S, Chuencharoen S, Thasana N et al (2016) Bacterial consortium expressing surface displayed, intra- and extracellular lipases and pseudopyronine B for the degradation of oil. Int J Environ Sci Technol 13:2067–2078. https://doi.org/10.1007/s13762-016-1034-z

Tissue BM (2013) Basics of analytical chemistry and chemical equilibria. Wiley, Hoboken, 426 p. ISBN: 978-0-470-59208-3

Tiwari KK (1995) Extraction technologies related to food processing. In: Gaonkar AG (ed) Food processing: recent developments. Elsevier Science, New York, pp 269–302. https://doi.org/10.1016/B978-044481500-2/50016-5

Tölgyessy P, Vrana B, Bartal M et al (2009) Determination of chlorophenols in sediments using ultrasonic solvent extraction followed by stir bar sorptive extraction coupled to TD-GC-MS. Chromatographia 69:389–392. https://doi.org/10.1365/s10337-008-0877-y

Uhlík P, Šucha V (1997) Distribution of pyrophyllite in the Šobov deposit and comparison with pyrophyllite from Víglašská Huta deposit. Miner Slovaca 29:73–79. *(in Slovak)*

Urík M, Čerňanský S, Ševc J et al (2007) Biovolatilization of arsenic by different fungal strains. Water Air Soil Pollut 186:337–342. https://doi.org/10.1007/s11270-007-9489-7
Urík M, Bujdoš M, Milová B (2014) Biologically induced mobilization of arsenic adsorbed onto amorphous ferric oxyhydroxides in aqueous solution during fungal cultivation. Water Air Soil Pollut 225:2172. https://doi.org/10.1007/s11270-014-2172-x
Urík M, Bujdoš M, Milová-Žiaková B et al (2015) Aluminium leaching from red mud by filamentous fungi. J Inorg Biochem 152:154–159. https://doi.org/10.1016/j.jinorgbio.2015.08.022
Urík M, Bujdoš M, Gardošová K et al (2018) Selenite distribution in multicomponent system consisting of filamentous fungus, humic acids, bentonite, and ferric oxyhydroxides. Water Air Soil Pollut 229:52. https://doi.org/10.1007/s11270-018-3719-z
Urík M, Polák F, Bujdoš M et al (2019) Antimony leaching from antimony-bearing ferric oxyhydroxides by filamentous fungi and biotransformation of ferric substrate. Sci Total Environ 664:683–689. https://doi.org/10.1016/j.scitotenv.2019.02.033
Usman MM, Dadrasnia A, Lim KT et al (2016) Application of biosurfactants in environmental biotechnology; remediation of oil and heavy metal. AIMS Bioeng 3:289–304. https://doi.org/10.3934/bioeng.2016.3.289
Üstündağ ÖG, Mazza G (2007) Saponins: properties, applications and processing. Crit Rev Food Sci Nutr 47:231–258. https://doi.org/10.1080/10408390600698197
Vági E, Simándi B, Suhajda Á et al (2005) Essential oil composition and antimicrobial activity of *Origanum majorana* L. extracts obtained with ethyl alcohol and supercritical carbon dioxide. Food Res Int 38:51–57. https://doi.org/10.1016/j.foodres.2004.07.006
Vojteková V, Nováková J, Šoltýsová H et al (2008) Screening control of transport of element contaminants in the environment. Chem Listy 102:506–514. (in Slovak)
Wells DE (1978) Extraction, clean-up and recoveries of persistent trace organic contaminants from sediment and biota samples. In: Dijkstra A, Massart DL, Kaufman L (eds) Evaluation and optimization of laboratory methods and analytical procedures, vol 1, 1st ed. Elsevier Science, pp 79–109. ISBN: 9780080875484
Wrona O, Rafińska K, Możeński C et al (2017) Supercritical fluid extraction of bioactive compounds from plant materials. J AOAC Int 100:1624–1635. https://doi.org/10.5740/jaoacint.17-0232
Xiao L, Wang J, Fan H et al (2016) Recent advances of adsorbents in solid phase extraction for environmental samples. Int J Environ Anal Chem 96:407–435. https://doi.org/10.1080/03067319.2016.1150459
Xiao G, Shao X, Zhu D et al (2019) Chemical synthesis of marine saponins. Nat Prod Rep 36:769–787. https://doi.org/10.1039/C8NP00087E
Xu Q, Nakajima M, Liu Z et al (2011) Biosurfactants for microbubble preparation and application. Int J Mol Sci 12:462–475. https://doi.org/10.3390/ijms12010462
Yamini Y, Seidi S, Rezazadeh M (2014) Electrical field-induced extraction and separation techniques: promising trends in analytical chemistry – a review. Anal Chim Acta 814:1–22. https://doi.org/10.1016/j.aca.2013.12.019
Yang TJ, Lee MR (2010) Electrically assisted solid-phase microextraction combined with liquid chromatography-mass spectrometry for determination of parathion in water. Talanta 82:766–770. https://doi.org/10.1016/j.talanta.2010.05.053
Yeh MS, Wei YH, Chang JS (2005) Enhanced production of surfactin from *Bacillus subtilis* by addition of solid carriers. Biotechnol Prog 21:1329–1334. https://doi.org/10.1021/bp050040c
Yu H, Huang G, Wei J, An C (2011) Solubilization of mixed polycyclic aromatic hydrocarbons through a rhamnolipid biosurfactant. J Environ Qual 40:477–483. https://doi.org/10.2134/jeq2010.0286
Zhang Q, Liang T, Guan L (2013) Ultrasound-assisted dispersive liquid-liquid microextraction combined with gas chromatography-mass spectrometry in negative chemical ionization mode for the determination of polybrominated diphenyl ethers in water. J Sep Sci 36:1263–1269. https://doi.org/10.1002/jssc.201201049

Fluoride Remediation Using Membrane Processes

6

Vijaylaxmi and Suphiya Khan

Abstract

Water pollution due to fluoride (F) is a serious problem worldwide; effective remediation techniques are required. The World Health Organization (WHO) has set a permissible limit of 1.5 mg/L^{-1} for drinking water. F-bearing rocks in India have also been reported to be plentiful, resulting in fluoride leaching out and degradation of nearby water and soil resources. F contamination affects many life forms (both plants and humans) when they enter the food system. The researchers were fascinated by the evolution of new membrane filtration, their high selectivity, high efficiency, and elevated membrane durability. Membrane process proved great technology for F remediation from wastewater, groundwater and potable water. Membrane process is climate safe, low cost and user-friendly. This chapter reviews different membrane processes for the fluoride remediation.

Keywords

Adsorption · Defluoridation · Fluoride · Membrane processes · Remediation

6.1 Introduction

Water is an important source of human life and development. Currently 71% of water occupies the earth's atmosphere, including natural resources that can be accessed exclusively by people, such as river water, inland reservoirs and shallow groundwater, comprising just 0.03% of the overall water. Scarcity of safe drinking water and contamination in water bodies is a major issue throughout the world. More than 663 million peoples are suffering from safe drinking water problems. F is the

Vijaylaxmi · S. Khan (✉)
Department of Bioscience and Biotechnology, Banasthali Vidyapith, Banasthali Tonk, Rajasthan, India

R. Prasad (ed.), *Environmental Pollution and Remediation*, Environmental and Microbial Biotechnology, https://doi.org/10.1007/978-981-15-5499-5_6

ninth highly electronegative and reactive element in the periodic table. It mostly occurs in water in combination with sodium fluoride and hydrogen fluoride (Premathilaka and Liyanagedera 2019). Eighty-five million tons of accumulated F in the earth's crust is non-biodegradable which is toxic to humans, animals and plants (Singh et al. 2014). F accumulated in plant causes leaf necrosis and inhibits photosynthesis and critical to plant growth enzyme activity. More than 260 million people suffer from fluorosis in 25 countries worldwide and 66 million in India (Chakraborti et al. 2016).

Increasing F contamination in groundwater and soil through anthropogenic and geogenic activities, groundwater contamination is becoming a global problem as it is common and uncontrollable. F geographical belts occur and India is one of them, extending from Turkey through Iran, Iraq, Afghanistan to Japan and China (Saravanan et al. 2008). F-endemic states are Andhra Pradesh, Gujarat Tamil Nadu, Haryana, Punjab and all 33 districts of Rajasthan which are completely and partially affected by F contamination due to the presence of F-rich rocks which are basalts, mica, granites, etc. (Singh et al. 2018). Black and McKay studied first time dental fluorosis in human, the brown patch called mottled enamel. The WHO guideline for F in drinking water is 1.5 mg/L; above that guideline is unsafe for human health (WHO 2004).

Drinking water is a major way of fluoride intake in the human body. The health issue of eliminating high concentration F from water is very important for the environment and for the public (Mohapatra et al. 2004). There are several methods for removal of fluoride from water, such as chemical precipitation, ion exchange and electrochemistry (Mohapatra et al. 2009). None of the method is easy and costly to use for water purification to solve the groundwater pollution problem. Membrane separation is based on two principles which are adsorption and filtration. Membrane filters water at a lower cost, less energy consumption and reduces environmental impact. Membrane has high efficiency in separation with easy maintenance, compact modular structure and low effluent from chemical sludge. Membrane distinguishes the metal ion on the basis of size and shape of the ion. Membrane performance depends upon several factors like material used, pore size, structure, shape (design) and fabrication methods. Membrane manufacturing using a variety of processes, such as phase inversion, interfacial, polymerization, stretching, electrospinning, pyrolysis, layer by layer deposition, slip casting and sol-gel. Different types of membrane (organic, inorganic, hybrid) used in membrane technology as needed. Different membrane technology are used in defluoridation of water on the basis of pore size, types is reverse osmosis (RO), nanofiltration (NF), microfiltration (MF), ultrafiltration (UF) and dialysis and electro dialysis processes. RO removes the solutes from the stream through semipermeable membrane. Semipermeable membranes serve as salt ion barriers. In Reverse osmosis process pressure exerted to transfer molecules from one side to other side of semipermeable membrane. Membrane commonly used in the RO system is made of polyamide and cellulose acetate. But these membranes are not having a good outcome. Recently these material composites are used with advanced modification like polyamide ultra-thin membrane. RO membrane pores (0.0001–0.001 μm) are small as opposed to MF and

NF. RO has two modules which are often used such as hollow fibre and spiral wound. RO membrane can remove 95% of fluoride from water. RO is an efficient fluoride removal membrane process. RO membrane also separates salt ions, metal ions, minerals, organic, inorganic and biomolecules. MF and UF separate particles through membrane pores. MF membrane pore size (0.1–5μm) and UF (0.1–10μm) diameter usually separates organic and biological cellular particle.

Ultrafiltration and microfiltration are similar to RO processes which are also powered by pressure. Membrane material types, such as polysulfone, polyvinylidene fluoride, polypropylene, polyacrylonitrile, cellulose acetate polymers and other polymers, are commonly used. Many membrane materials have different properties with regard to degree of hydrophobicity, antifouling, chemical and thermal resistance, pH and oxidant tolerance and surface charging. Hollow fibre membrane module is used for water purification in both NF and MF. Both techniques are used in water remediation and wastewater treatment (Pellegrin et al. 2013). Nanofiltration membranes are porous and consist of polyamide materials that filter both organic and inorganic (divalent ions) and species (virus, bacteria), with a pore size of 0.001–0.01 μm. NF is widely used for softening hard water (Hilal et al. 2004). Electrodialysis also uses the same procedure as RO, using electric current instead of pressure to remove solute contaminants from water. Dialysis separates the solid particles by passing the particles through a membrane instead of using a membrane to retain the particles as water passes through it as in reverse osmosis and nanofiltration. Ions are charged particles that move through selective membranes (cation and anion) of the ions. Solute can be powered by either the Donnan effect or an electric field related to it. Concentration differential factor is relevant acting as a driving force for ion separation (Loeb 1921).

6.2 Source of Fluoride in Atmosphere

F contamination occurs in the environment due to natural activity and anthropogenic activity shown in Fig. 6.1. F deposit in large amount in earth crust and F-rich rocks and minerals are the main natural source of F contamination in water and soil. F-rich minerals are fluorspar-CaF_2 (lime stones and sandstones), cryolite-Na_3AlF_6 (granites) and fluorapatite-$Ca_3(PO_4)$ $Ca(FCl_2)$, micas, sellaite MgF_2, biotite $K(Mg, Fe)_3$ $AlSi_3O_{10}(OH, F)_2$, muscovite, topaz $Al_2SiO_4(OH, F)_2$ and amphiboles such as tremolite and villiaumite NaF (Karunanithi et al. 2019). F after separation from rock becomes mobile and enters into soil, vegetables, crops, and fruits. Other anthropogenic factors such as chemical fertilizer plants, smelting industries (metals such as zinc, steel, etc.), glass, ceramic industries and power plants are responsible for fluoride contamination in water, soil and air (Bhattacharya and Samal 2018). High carbonate and bicarbonate containing water passed through these F-rich rocks; F ions leached out and increase the F content in the groundwater (Saxena and Ahmed 2001).

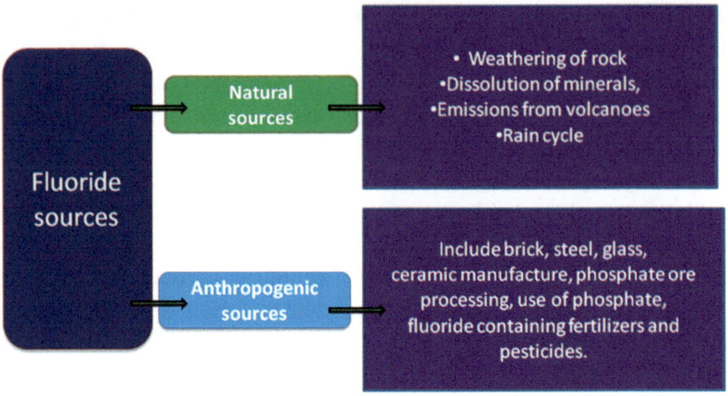

Fig. 6.1 Different sources of fluoride in atmosphere

Table 6.1 F concentrations and their biological effects on human beings

Fluoride concentration in drinking water (mgL^{-1})	Problems
1 mg L^{-1}	Less tooth decay
2 mg L^{-1} or more	Mottled enamel
3.1–6.0 mg L^{-1}	Osteoporosis
8 mg L^{-1}	10% osteoporosis
20–80 mg L^{-1}	Crippling skeletal fluorosis
50 mg L^{-1}	Thyroid disorder
100 mg L^{-1}	Reduce of growth
More than 125 mg L^{-1}	Kidney disorders
2.5–5.0 gm in actual dose	Death

6.3 Effects of Fluoride on Life Forms

Several studies found negative effects of F on human beings and plants also. F is considered beneficial to health up to a certain limit (below 1.5 mg L^{-1}). It prevents the tooth decay and considered necessary for the formation of dental enamels and growth of strong bones especially among in children (Suthar et al. 2008). After a certain level (>1.5 mg L^{-1}), it becomes a dangerous to human health and can cause many health problems. Effects of F concentrations on human body are shown in Table 6.1.

On the other side, above these concentrations and long-time exposure of this cause chronic fluorine toxicity, known as fluorosis. Fluoride endemic areas people have dental fluorosis with brown patches on teeth. Scratch and harm occur in high toxic condition, with increased tooth porosity (Ozsvath 2009). During skeleton fluorosis, joint problems increase with the mass and bone density. Extreme pain in joint, hip and spine is an early symptom. Poker back spine is a late manifestation of skeletal fluorosis in which the whole spine becomes a single bone column. In severity condition muscles become inactive and paralysed, other problems are joint crippling deformities, increase spinal cord density crippling deformities, neurological defects (Annadurai et al. 2014). Skeletal fluorosis may occur with a high F movement from water and other dietary factors. Also, thyroid disorder, growth retardation, kidney problem and even death can occur. F move from soil to root through the diffusion process. F in high concentrations has adverse effects on plants such as necrotic lesion, chlorosis, leaf tips and burning edges. Other physiological parameters also disturbed such as root length, shoot length, chlorophyll a, and chlorophyll b content, enzyme activity, seed germination rate. F adverse effects in different plants were observed such as in *Cyamopsis tetragonoloba* (cluster bean), *Populus deltoides* (poplar), *Cicer arietinum* (Bengal gram), *Oryza sativa* (paddy) and *Vigna radiata* (mung bean) (Baunthiyal and Ranghar 2015).

6.4 Available Technologies for Removing Fluoride

There are different defluoridation technologies investigated for providing safe drinking water for the F-affected populations. It is described as "the downward adjustment of level of F in drinking water to the optimal level" (Singh et al. 2016). Various techniques were available throughout the world for F removal from the water.

6.4.1 Adsorption

Adsorption is an endothermic and mass transfer process that accumulates a material or substance from liquid phase to the solid surface. The existence of special functional groups on the adsorbent surface provides interaction with heavy metal which leads to the adsorptive removal of metals from water. Popular adsorbents are natural fibres, zeolite and activated carbon; they suffer from low adsorption capacity, low selective sorption and unsatisfactory regeneration capability. Nowadays innovative adsorbents are metal oxide nanoparticles, CNT, graphene and others which have been developed to overcome these problems (Khulbe and Matsuura 2018). Filtration process requires high pressure and more expensive rather than adsorption. Adsorption process is a prominent, economic, convenient and simple activity. In an experiment, zirconium nanoparticles embedded in a hollow PSF membrane demonstrated a high adsorption potential for F removal from aqueous (He et al. 2014). Adsorption kinetics and isotherms are useful in testing adsorbent efficiency and defluoridation ability (Smitha and Thampi 2017).

6.4.2 Membrane Process

In this method, particles are separated using exceptionally composed semipermeable membrane based on molecular shape and size. In general, the semipermeable membrane consists of inorganic, organic and hybrid (organic and inorganic) material. Main factors are very important in designing an efficient membrane choice of material and physical and chemical properties of material. Membrane efficiency is also based upon membrane pore size and membrane content (Lee et al. 2016). Based on pore size membrane is classified as reverse osmosis (RO), nanofiltration (NF), ultrafiltration (UF) and microfiltration (MF) shown in Fig. 6.2.

In designing a functional membrane various factors should be considered like quality of material, greater water flux rate and separation rate, module design, chemical and physical strength, mechanical strength, membrane material choice, solute rejection increase, materials, water flux increase, module configuration, fabrication methods and less cost operating condition. Membrane is divided into two broad categories on the basis of the structure and types of materials used in the synthesis of membranes such as isotropic type of membrane and anisotropic type of membranes shown in Fig. 6.3.

Isotropic type membranes have a homogeneous chemical composition and compose of a single type of material. Isotropic membrane is again subdivided into non-porous thin film membranes, macroporous thick films and electrically charged membranes. Non-porous membranes are those membranes in which a force such as electric field gradient, concentration or pressure is applied to each side of the membrane and the separation of the solution controlled by the rate of transport (Lee et al. 2016). Macroporous membranes are also referred to as sieving membranes, separating the solutes depending on the particle size and membrane

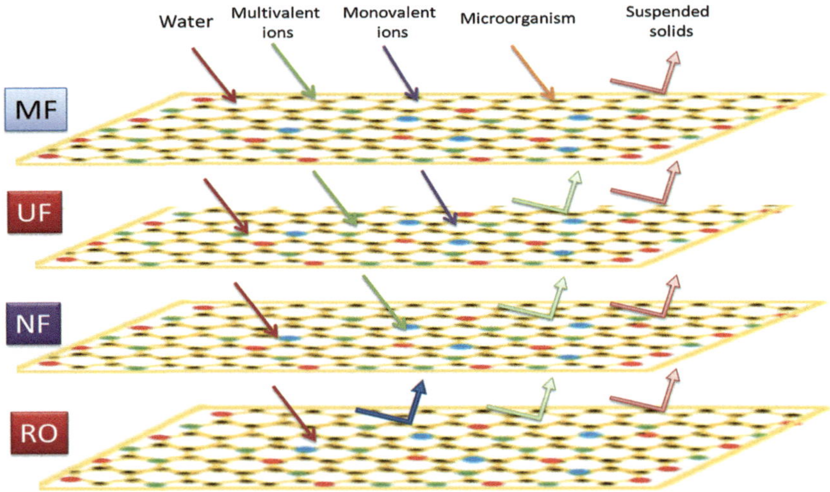

Fig. 6.2 Different types of membrane on the basis of pore size

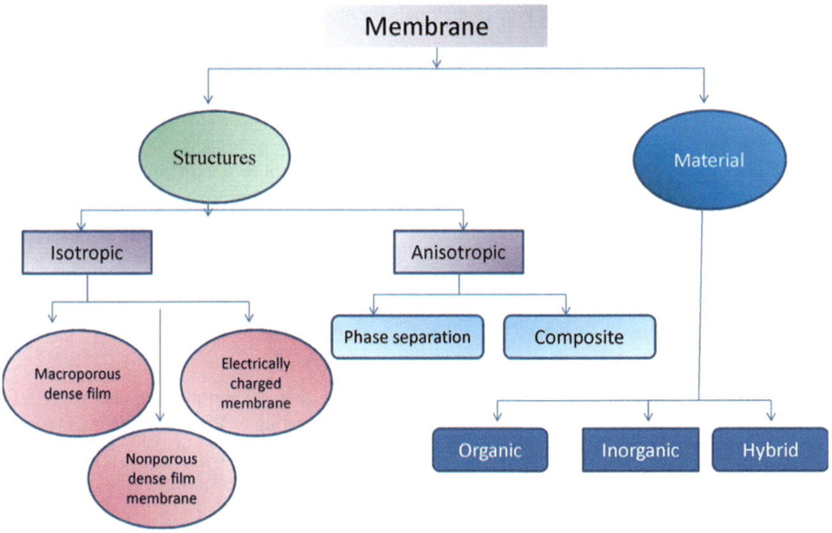

Fig. 6.3 Categories of membranes on the basis of material and structure

pore diameter. Microporous membranes are identical to traditional filters, and the size range varies only for pores. Microporous membrane pores are in size diameter range 0.1–5μm (Thakur and voicu 2016). Microporous and non-porous dense film cover by positively and negatively charged ion. The separation of the solute in this membrane is driven by the ion concentration and the amount of the charge. There are two types of anisotropic membranes, i.e. the phase separation membranes (Loeb-Sourirajan membranes) and the self-assembled hybrid membranes, which are thin films. Phase separation membranes are similar to isotropic microporous membrane that have single chemical composition membranes, but different in porosity and pore size across the membrane thickness. Structurally and chemically heterogeneous thin film is a composite membrane. Thicker porous structure provides mechanical strength to a thin surface of membrane. Traditionally membrane structure is made up of different polymeric materials. But nowadays other different materials are used in combination to enhance the efficiency of membrane. Several methods are available for membrane synthesis such as phase inversion, solution coating and plasma polymerization, etc. Membrane thin surface layer determines the solute permeation rate (Buonomenna 2016).

Membranes on the basis of materials used are mainly of three types: organic membrane, inorganic membrane and hybrid membrane.

6.4.2.1 Organic Membrane

Membrane can be divided on the basis of materials used for synthesis which are organic (polymeric), inorganic and hybrid. Polymeric membranes are those made of natural and synthetic polymers such as polysulfone, polyethersulfone, cellulose acetate, cellulose nitrate, polyamide, polyetherimide, polymethylpentene,

Table 6.2 New advanced organic type of materials for membrane synthesis, their performance and key features

Material type	Performance (L/m^2h bar)	Key features
Microfiltration		
Polyvinyl alcohol polymer with polypropylene polymer	32, 346 L/m^2 h at 0.24 bar pressure	Small range pore size membrane synthesis through electrospinning technique (Lee et al. 2016)
Polyvinylidene fluoride and hydroxyethyl methacrylate		Chitosan polymer surface charged coated with nanofibrous by electrospun, hydrophilicity increased and increased flux rate (Nasreen et al. 2014)
Ultrafiltration		
Polyimide with polyethersulfone polymer	3565–1780	Improve thermal and mechanical properties through coating of PI on PES by ALD method (Sheng et al. 2014)
Polyethylene terephthalate polymer	0.1–0.21 mL/ cm^2 s at pH 4–8	Permeation of reversible pH (Pan et al. 2014)
Polyvinyl chloride and polyvinyl formal	52–323 L/m^2 h at 0.1 MPa	Improve fouling resistance property (Fan et al. 2014)
Polysulfone (PSU) 7.5–8.9 L/m2h	7.5–8.9	Presence of biofilm over the surface of membrane increased permeability property. But it can decrease when organic composite stays for a long period (Derlon et al. 2014)
Cellulose acetate (CA) nanofiber	3540	10X increase water permeability, uniform porous structure of membrane with high porosity (Soyekwo et al. 2014)
Nano filtration		
Polysulfone (PSU)	Max 70	Through IGEPAL surfactant and amphiphilic additive increased cadmium removal and hydrophilicity property (Saljoughi and Mousavi 2012)

polycarbonate, polydimethylsiloxane, polyphenylene oxide, polyvinylidene fluoride and polyacrylonitrile and their performance shown in Table 6.2. Organic membranes are very costly, their life efficiency is lower and their range of use is lower. Polymeric membranes are commonly used by researchers owing to good chemical strength and high temperature-tolerant capability. PES polymer is hydrophobic in nature due to this lack membrane antifouling properties, chemical and mechanical ability, which limits the life of the membrane. NPs and carbon-based material (CNT) with graphene have been used to enhance fouling resistance, chemical, mechanical strength, membrane life efficiency (Guo et al. 2019). CNTs have been proved effective in membrane processing as one-dimensional additives. CNT have important properties such as wide surface area, high ratio of length to diameter and excellent mechanical resistance (Ihsanullah 2019).

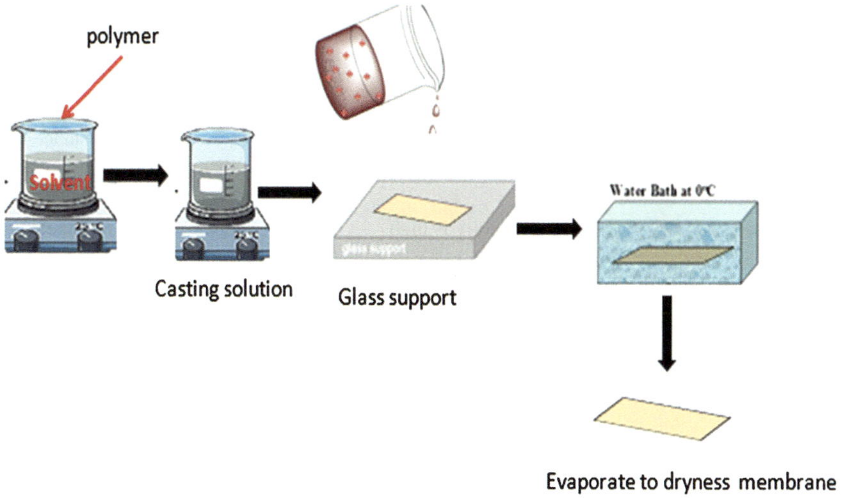

Fig. 6.4 Phase inversion method

Membranes Fabrication Methods

A method of polymeric membrane manufacture depends on polymer forms and ideal membrane structure. The most widely used methods for polymeric membrane synthesis are phase inversion, interfacial polymerization, stretching, electrospinning and track etching.

Phase Inversion

Phase inversion is a method of the regulated transformation of polymers from liquid to solid phase (shown in Fig. 6.4). Polymer powder dissolve in solution. Polymer solution added to the required support plate. Dip the casting support layer or tray into a non-solvent. Solid polymeric membrane synthesis occur. Several solvents can use as polymer casting solutions such as dimethylformamide (DMF), dimethyl sulfoxide (DMSO), N-methyl-2-pyrrolidnone (NMP). The membrane formed by DMF had a wider pore size and smaller surface area (Adamczak et al. 2019). Non-solvent can be water and other liquid solvents (Drioli and Giorno 2009). Solvent evaporate from polymeric solution during phase inversion process to synthesis dry and porus membrane. Non-solvent diffusion occurs into casting film. As a result porous membrane is synthesized from this vapour phase precipitation method (Boussu et al. 2006). Immersion precipitation method is a widely used method for polymeric membrane for water treatment process. Polymer used in this method is shown in Table 6.3. In this method polymer must be soluble into solvent. Solvent and non-solvent exchange occurs into each other and precipitation occurs. Solvent evaporates due to volatile nature. Polymeric solution precipitation and membrane synthesis occur. Polymer amount and types of solvent and non-solvent are also important factors for synthesis of membrane via immersion precipitation (Saljoughi

Table 6.3 Main polymer used in membrane formation via immersion precipitation

Polymer	Advantages	Disadvantages
Acetate cellulose (CA)	Hydrophilicity	Low chemical resistance (pH range 2–8)
	Flexibility in synthesis	
	Cost effective	Low thermal resistance (below 30 °C)
		Less resistance to chlorine
Polyamide (PA)	Thermal stability is high	Poor chlorine resistance
	pH tolerance in large range	
	Mechanical properties high	
Polyethersulfone and polysulfone	Membrane fabrication flexibility (wide pore size range)	Hydrophobicity
		Low operating pressure limits
	High thermal resistance (up to 75 °C)	Poor resistance to chlorine
	Good chlorine resistance	
	Wide pH tolerance (1–3)	
Polyvinylidene fluoride	High thermal stability (up to 75 °C)	Hydrophobicity
	High chemical resistance and mechanical strength	

and Mousavi 2012) reported polymer concentration effect on membrane water flux rate.

Interfacial Polymerization (IP)

Formation of thin composite membrane interfacial polymerization is a general method. Interfacial polymerization processed membrane used in RO and NF which are commercially available. For synthesis of inorganic membrane or hybrid membrane interfacial polymerization method can also be used. Interfacial polymerization includes soaking a microporous polysulfone polymer into polymeric amine solution and then transferring casting film into di-isocyanate with hexane solution. Prepared TFC polyurea membrane has better salt rejection and high water flux (Lalia et al. 2013). Polyamide polymer coated on a teflon support sheet, mainly polymeric membrane synthesised by this process used in the RO and NF processes. MPD (m-Phenylenediamine), trimesoyl chloride, piperazine, triethylenetetramine, N-N-diaminopiperazine, poly(ethyleneimine) and N-(2-aminoethyl)-piperazine monomer are used for TFC (thin film composite) PA (polyamide) membrane synthesis via interfacial polymerization fabrication method. Liu et al. reported a functional RO composite membrane prepared from 5-isocyanatoisophthaloyl chloride and MPD. Several membranes can be synthesized from interfacial polymerization method, i.e. TFC polyester and polyesteramide developed from this method. Oxidation resistance of membrane increases with incorporation of ester linkage; resistance against chlorine attack of membrane is increased (Tang et al. 2008).

Stretching
In streching method polymer heat above their melting point to synthesis porous and thin film shape sheet membrane (Trommer and Morgenstern 2010). Microporous membranes are generally used in MF and UF. Stretching technique is used for crystalline polymer. Stretching process is usually done in two stages, cold stretching and then hot stretching. Cold stretching is used to create micropores in film and heat stretching used to enhance or control the final pore structure of the membranes. Pores of membrane are controlled by material's physical properties (like crystallinity, tensile strength, etc.).

Track Etching This method is where thin polymer film is irradiated with a high-energy heavy ion beam to create a flat, distorted track through the irradiated polymer film. This technique is known for precise control over membrane pore size distribution. Polyamide film bombarded by heavy ions (64 Cu, 127I, 79 Br, 107Ag) then treated in the presence of oxygen with gamma-rays. Polymer most widely used in track etching is polycarbon and terephthalate polyethylene (Lalia et al. 2013).

Electrospinning This is a fibre production method which uses high voltage to draw charged threads of polymer solutions or polymer melts up to fibre diameters in the order of some hundred nanometres. The process does not require the use of coagulation chemistry or high temperatures to produce solid threads from solution. This method ensures that no solvent can be carried over into the final product. Porous membrane fabricates through this technique and is used in filtration and desalination purpose (Lalia et al. 2013). Polymer concentration affects the membrane pore size and this helps to improve salt rejection (Kaur et al. 2012).

Organic membranes have limitation of high fouling tendency due to membrane surface hydrophobicity. Fouling is generally caused by deposition of protein microbial communities and organic and inorganic components on the membrane surface. Other drawbacks are highly expensive and have irreproducible separation, less lifetime performance and smaller application range.

6.4.2.2 Inorganic Membrane

J.W. McBain (1932) first time introduces inorganic molecular sieve membrane. The inorganic membrane is classified into porous and non-porous (dense) based upon structure and morphology. This is an important factor for membrane performance. Porous inorganic membrane has porous ceramic and metal supports on the upper side with structure and morphology, with additional porous sheet. This type of membrane has various pore shapes such as conical-shaped pores, spongy pore structure, straight pores and regular shapes. Examples of inorganic membrane material are metal oxide, metal, alumina, zeolite, tin, zirconium, carbon (graphene, carbon nanotube), silica, glass, silicon carbide and nitride, etc., shown in Table 6.4. Non-porous and dense inorganic membrane contains solid electrolyte or solid layer of metals (Ag, Pd, alloys). Electrolyte layer allows moving oxide ion and hydrogen diffusion via membrane pores. In non-porous (pores block), block by immobilized

Table 6.4 New advanced inorganic type of materials for membrane synthesis, their performance and key features

Material type	Performance (L/m^2 h bar)	Key features
Microfiltration		
Nanowires of TiO_2		TiO_2 nanowire formation by hydrothermal process and photo catalytic in presence of UV
		Decomposition of pharmaceutical components (Hu et al. 2011)
ZrO_2/Al2O_3	118–1695 (UF-MF)	Membrane thickness coating controllable and pore size decrease by ALD techniques (Li et al. 2012)
Ultrafiltration		
TIO_2	Highest limits 20	Azo dye contaminant breakdown by photocatalysis under UV radiation (Athanasekou, et al. 2012)
Silver NPs	9.5 m^3/m^2 day atm	Silver NP formation by layer by layer method (Kawada et al. 2014)
Nanofiber silver/TiO_2	5–20 at 1–4 bar	80% dye breakdown in presence of solar radiation and inactivate 99.9% bacteria within 30 min. (Liu et al. 2012)
TiO_2	4.05	N, C and Ce co-doped TiO_2 (Lee et al. 2016)
Nanowire of TiO_2	12.2	Oxidation by photocatalytic and antibacterial and antifouling properties (Lee et al. 2016)
Modified TiO_2/ Al_2O_3	12	Azo dye degradation by photocatalytic and modification of TiO_2 with urea (Moustakas et al. 2014)
TiO_2 nanotube	15–33	TiO_2 inner surface of tube controlled by technique (Lee et al. 2016)
Titanium oxide-silica oxide with alumina	7–8 at 5 bar	Titanium oxide with silica oxide nanostructure formation through physical separation and photocatalytic capabilities (Tajer-Kajinebaf et al. 2014)
Nanofiltration		
Titanium oxide and silicon oxide	2.5, 5.0 and 10.0 ml/min	Nano-metallic fabric membranes are flexible, prepared by electrospinning method and can be reused via calcination and chemical and thermal stability of nanofibrous membranes (Lee et al. 2016)
TiO_2 hollow fibre	12.2	Hollow fibre membrane is synthesis by spinning sintering technique and calcination temperature (900 °C) also effect membrane features. 90% organic components released out to synthesis hollow membrane (Zhang et al. 2014)
Graphene oxide (GO)	80–276 L mh/MPa	Commercially NF cross-linked graphene oxide sheet has 4–10 times less flux rate than GO-based membrane (Lee et al. 2016)
Graphene oxide (GO)	4 mol/m^2 h	Permeation rate high for ions (Joshi et al. 2014)
Graphene	21.8	Retention rate greater for organic dyes and medium for ion salts (Lee et al. 2016)

liquid into support layer (i.e. salt in liquid state immobilized into metal porous layer or ceramic supports). Gold nickel, stable zirconia, palladium and its alloys are examples of thick membranes. Non-porous membrane separation efficiency depends on the type of material, chemical and physical strength between pollutant and membrane and nature of the pollutant to be separated. Non-porous membranes exhibited high water purification efficiency and remove most contaminants (1–10 nm) from wastewater, such as metal ion, acidic and basic salts, organic particles and microbes (bacteria, virus). The general types of inorganic membranes are liquid and fluidic membrane, glass, zeolite, iron, synthetic (inorganic) membranes, silica (SiO_2), alumina (Al_2O_3), titanium (TiO_2) and zirconia (ZrO_2). Carbon-based membranes (graphene, carbon nanotubes) are used mainly for water treatment. Various methods are available for the synthesis of inorganic membranes, e.g. chemical vapour deposition (CVD), sol-gel method, slip casting, pyrolysis technique, etc. (Fard et al. 2018). **Ceramic membrane:** The most commonly used materials for membrane synthesis are aluminium (Al_2O_3), titanium (TiO_2), zirconia (ZrO_2), glass (SiO_2), silicon carbide (SiC) or a composite of these metal oxides. Other materials include suitable non-oxide (carbide, nitrides, boride), oxides and non-oxide composites. Ceramic membranes have different porosity layers and asymmetrical structure. Multichannel monolith ceramic, flat and tubular membranes are commercially available in markets. Compared to other membrane structures tubular membranes have a high surface ratio. Tabular ceramic membrane synthesis includes three primary phases: (a) paste or suspension preparation from ceramic powder, (b) paste or suspension converted into flat or tubular geometry and (c) calcination and sintering heat treatment (Li 2007).

6.4.2.3 Hybrid Membrane

Inorganic-organic hybrid membranes are typically turned into a polymeric matrix structure by adding inorganic materials (metals, metal oxide or carbon-based materials) shown in Table 6.5. Hybrid membranes have fewer drawbacks as compared to polymeric membrane. Incorporate inorganic material into polymeric matrix to form multipurpose use and improve antifouling, water permeability, mechanical strength and rejection rate. Membrane surface structure and pore structure are modified by inorganic material (CNT). Several methods are available for hybrid membrane synthesis, such as blending (properly combining nanomaterial and polymer matrix), interfacial polymerization, surface coating with nanocomposite material, layer by layer deposition and surface grafting of nanoparticles (Wang et al. 2012).

6.5 Membrane-Based Techniques for Fluoride Removal

Most utilizing membrane-based techniques for fluoride removal are reverse osmosis (RO), nanofiltration (NF) and electrodialysis (ED).

The RO membrane cycle reverses the normal osmosis arising from the hydraulic pressure applied to the high concentration side of the solution. Through

Table 6.5 Hybrid membrane material types, organic and inorganic (composite), their performance and key feature

Material type	Performance (L/m^2 h at 1 bar)	Key features
Micro filtration		
Silver nanoparticle with polysulfone polymer	0.181–0.086 kg/m^2 h1Pa1 at 1000 kPa (UF + MF)	Increase microbial resistance property through loading silver nanoparticle by cold spray jet (Dumee et al. 2015)
Ultrafiltration		
Carbon nanotubes with polyethersulfone polymer	Max 10–90	Composite of CNTs and PES have greater flux, hydrophilicity and antifouling rate higher than unhybridized PES (Celik et al. 2011)
Silver-SiO_2/polyethersulfone polymer	Max 140	Incorporation of Ag into membrane by AgNP accumulates on upper layer of silica, fouling resistance and antimicrobial properties (Huang et al. 2014)
Al_2O_3/polycarbonate	Max 11	Better resistivity to organic liquids and acids and hydrophilicity, pore size deposition of Al_2O_3 control by ALD method (Li et al. 2011)
Titanium oxide/ polypropylene and polyvinylidene fluoride polymer	Max 190, 300–420	By decreasing pore size and increasing hydrophilicity improves water flux and retention rate. Uniform TiO_2 deposition by ALD method (Xu et al. 2013)
Anodic alumina membranes	~2–7 g/h at 1–4 bar pressure	Water increased in hydrophilic area with decreased pore size diameter (Lee et al. 2016)
Polyvinylidene fluoride polymer and graphene oxide carbon material	271–346	Better mechanical hydrophilic and permeation property (Wang et al. 2013)
Polyethersulfone (PES)/TiO_2	365–596	Incorporation of TiO_2 improves hydrophilicity, mechanical and thermal, fouling resistivity properties but no change in membrane structural property (Wu et al. 2008)
Modified TiO_2/ polyvinylidene fluoride (PVDF)	82.5	TiO_2 NP additive add in a limited quantity to achieve a small pore size and a high hydrophilicity membrane. TiO_2 and LiCl. H_2O are also additives (Yuliwati and Ismail 2011)

(continued)

Table 6.5 (continued)

Material type	Performance (L/m^2 h at 1 bar)	Key features
AgNPs/polysulfone(PSU)	6–12 ml/min	AgNP (6% and 10%) gives greater antifouling property and better antibacterial and hydrophilicity (Bernardes et al. 2014)
Silver nanoparticles with Polyethyleneimine and poly (sodium styrenesulfonate)	9 m^3/m^2 day atm	Polymer and silver NP stabilize the electrostatic interaction and layer by layer deposition (Kawada et al. 2014)
Nanofiltration		
(NH_2-MWCNTs)/ polyethersulfone (PES)	23.7	Increase concentration of NH_2 and MWCNT; improve hydrophilicity pure water flux rate. High fouling recovery ratio, high negative charge for bovine serum albumin flow. Salt retention rate for disodium sulfate (60–65%) and sodium chloride (18–20%) (Vatanpour et al. 2014)
Polyethersulfone polymer with O-carboxymethyl chitosan polymer with iron nanoparticles	20–40 kg/m^2 h at 4 bar	Very less irreversible antifouling, enhance antifouling and rejection, improve pure water flux rate and hydrophilicity (Lee et al. 2016)
ZnO/PES	Max 60	ZnO optional nanoparticle, greater permeability, high dye rejection and antifouling (Balta et al. 2012)

semipermeable membrane pass the solvent with a force, against pressure gradient into the lower concentration solution. Hydraulic pressure is applied to the high concentration side of column as is the reverse of natural osmosis. In this, using mechanical pump pressure is applied on the solution to overcome inalienable osmotic pressure. Using RO membrane process from industrial wastewater 98% F is removed (Ndiaye et al. 2005). In RO system used polymeric membrane which remove 95 to 98% fluoride from ground water of Moradgaon village chandrapur district (Gedam et al. 2012).

Advantages

- RO membrane is a powerful technique for fluoride removal.
- Above 98% F can be removed through this process.
- RO membrane removes different types of ions at the same time.
- No hindrance by different ions.

- It efforts under wide pH range.
- Chemicals not use, less laborious work, least expensive operational work.
- Water treatment and purification both allow in one time.
- RO gives high quality of contaminant-free water.

Disadvantages

- Not affordable for rural areas.
- Expensive technique.
- Remove the necessary minerals from water during the process, resulting in a decline in the consistency of the water.
- During process use chemicals as result water get acidic and need to improvement water quality.
- Noisy process due to use of high-pressure pump.

Nanofiltration

Nanofilteration (NF) is a newly proposed pressure-driven technique. NF membrane have large pore size and less friction for passing ions from this membrane, NF membrane advantages over other methods is remove all dissolve solute with less pressure and faster flow rate (Lhassani et al. 2001). NF90 membrane is used to remove fluoride from brackish water in south Morocco, Tantan city (Waghmare and Arfin 2015).

Advantages

- Better productivity
- Chemicals not used
- Effective impermeable to suspended solids, all inorganic metal, organic micropollutants, pesticides, fertilizer and microbes

Disadvantages

- Highly expensive techniques from other defluoridation techniques like adsorption.
- It removes all important ions which are essential concentration to health.
- Membrane faces fouling, short life time problems.

Electrodialysis Removal of solute ions from water in the presence of electric-driven force through ion exchange membrane. Electrodialysis is like RO process but current use as a driving force instead of pressure. Electrodialysis is not used in rural areas because of the need of electricity in operating this technique.

Advantages

- Less expensive pre- and post-treatment
- Less chemical use
- Water recovery in high amount

Disadvantages

- Necessity of electricity
- Separation only ionic components
- Necessity of concentrate treatment

6.6 Conclusion

Fluoride pollutant is considered to be highly life-threatening and the development of cost-efficient, environmentally safe and effective method of remediation methods is needed. Recently the defluoridation technology has not met the optimal conditions. Because of their special properties, the application of membrane method use for defluoridation remediation is proven advantageous. Several researchers reported high defluoridation capacities by the use of various polymers, metal and nanomaterial composite membranes. This chapter concludes use of membrane technologies in remediation procedures for the reduction of fluoride. In fact, there are several drawbacks that need change in technology.

Acknowledgement We acknowledge Bioscience and Biotechnology Department. We are deeply grateful to prof. Aditya Shastri for the research amenities and the Bioinformatics Centre, Banasthali Vidyapith, Rajasthan (India) for use of computational facilities. This study was funded by DST major project entitled "Low Cost-Renewable Energy Driven (LC-RED) Water Treatment Solution Centre" Ref No. DST/TM/WTI/WIC/2K17/124.

References

Adamczak M, Kamińska G, Bohdziewicz J (2019) Preparation of polymer membranes by in situ interfacial polymerization. Int J Polym Sci 2019:1–13

Annadurai ST, Rengasamy JK, Sundaram R, Munusamy AP (2014) Incidence and effects of fluoride in Indian natural ecosystem: a review. Adv Appl Sci Res 5(2):173–185

Athanasekou CP, Romanos GE, Katsaros FK, Kordatos K, Likodimos V, Falaras P (2012) Very efficient composite titania membranes in hybrid ultrafiltration/photocatalysis water treatment processes. J Membr Sci 392:192–203

Balta S, Sotto A, Luis P, Benea L, Van der Bruggen B, Kim J (2012) A new outlook on membrane enhancement with nanoparticles: the alternative of ZnO. J Membr Sci 389:155–161

Baunthiyal M, Ranghar S (2015) Accumulation of fluoride by plants: potential for phytoremediation. Clean Soil Air Water 43(1):127–132

Bernardes PC, de Andrade NJ, da Silva LHM, de Carvalho AF, Fernandes PÉ, Araújo EA, de Sá JPN (2014) Modification of polysulfone membrane used in the water filtration process to reduce biofouling. J Nanosci Nanotechnol 14(8):6355–6367

Bhattacharya P, Samal AC (2018) Fluoride contamination in groundwater, soil and cultivated foodstuffs of India and its associated health risks: a review. Res J Recent Sci 7(4):36–47

Boussu K, Vandecasteele C, Van der Bruggen B (2006) Study of the characteristics and the performance of self-made nanoporous polyethersulfone membranes. Polymer 47 (10):3464–3476

Buonomenna MG (2016) Smart composite membranes for advanced wastewater treatments. In: Smart composite coatings and membranes. Woodhead Publishing, Sawston, pp 371–419

Celik E, Park H, Choi H, Choi H (2011) Carbon nanotube blended polyethersulfone membranes for fouling control in water treatment. Water Res 45(1):274–282

Chakraborti D, Rahman MM, Chatterjee A, Das D, Das B, Nayak B, Sengupta MK (2016) Fate of over 480 million inhabitants living in arsenic and fluoride endemic Indian districts: magnitude, health, socio-economic effects and mitigation approaches. J Trace Elem Med Biol 38:33–45

Derlon N, Mimoso J, Klein T, Koetzsch S, Morgenroth E (2014) Presence of biofilms on ultrafiltration membrane surfaces increases the quality of permeate produced during ultra-low pressure gravity-driven membrane filtration. Water Res 60:164–173

Drioli E, Giorno L (eds) (2009) Membrane operations: innovative separations and transformations. Wiley, Weinheim

Dumée LF, He L, King PC, Le Moing M, Güller I, Duke M, Kong L (2015) Towards integrated anti-microbial capabilities: novel bio-fouling resistant membranes by high velocity embedment of silver particles. J Membr Sci 475:552–561

Fan X, Su Y, Zhao X, Li Y, Zhang R, Zhao J, Liu Y (2014) Fabrication of polyvinyl chloride ultrafiltration membranes with stable antifouling property by exploring the pore formation and surface modification capabilities of polyvinyl formal. J Membr Sci 464:100–109

Fard KA, McKay G, Buekenhoudt A, Al Sulaiti H, Motmans F, Khraisheh M, Atieh M (2018) Inorganic membranes: preparation and application for water treatment and desalination. Materials 11(1):74

Gedam VV, Patil JL, Kagne S, Sirsam RS, Labhasetwar P (2012) Performance evaluation of polyamide reverse osmosis membrane for removal of contaminants in ground water collected from Chandrapur district. J Membr Sci Technol 2(3):2–5

Guo X, Li C, Li C, Wei T, Tong L, Shao H, Liao Y (2019) G-CNTs/PVDF mixed matrix membranes with improved antifouling properties and filtration performance. Front Environ Sci Eng 13:1–11

He J, Siah TS, Chen JP (2014) Performance of an optimized Zr-based nanoparticle-embedded PSF blend hollow fiber membrane in treatment of fluoride contaminated water. Water Res 56:88–97

Hilal N, Al-Zoubi H, Darwish NA, Mohamma AW, Arabi MA (2004) A comprehensive review of nanofiltration membranes: treatment, pretreatment, modelling, and atomic force microscopy. Desalination 170(3):281–308

Hu A, Zhang X, Oakes KD, Peng P, Zhou YN, Servos MR (2011) Hydrothermal growth of free standing TiO_2 nanowire membranes for photocatalytic degradation of pharmaceuticals. J Hazard Mater 189(1–2):278–285

Huang J, Wang H, Zhang K (2014) Modification of PES membrane with ag–SiO_2: reduction of biofouling and improvement of filtration performance. Desalination 336:8–17

Ihsanullah (2019) Carbon nanotube membranes for water purification: developments, challenges, and prospects for the future. Sep Purif Technol 209:307–337

Joshi RK, Carbone P, Wang FC, Kravets VG, Su Y, Grigorieva IV, Nair RR (2014) Precise and ultrafast molecular sieving through graphene oxide membranes. Science 343(6172):752–754

Karunanithi M, Agarwal R, Qanungo K (2019) A review of fluoride removal from groundwater. Period Polytech Chem Eng 63(3):425–437

Kaur S, Sundarrajan S, Rana D, Matsuura T, Ramakrishna S (2012) Influence of electrospun fiber size on the separation efficiency of thin film nanofiltration composite membrane. J Membr Sci 392:101–111

Kawada S, Saeki D, Matsuyama H (2014) Development of ultrafiltration membrane by stacking of silver nanoparticles stabilized with oppositely charged polyelectrolytes. Colloids Surf A Physicochem Eng Asp 451:33–37

Khulbe KC, Matsuura T (2018) Removal of heavy metals and pollutants by membrane adsorption techniques. Appl Water Sci 8(1):19

Lalia BS, Kochkodan V, Hashaikeh R, Hilal N (2013) A review on membrane fabrication: structure, properties and performance relationship. Desalination 326:77–95

Lee A, Elam JW, Darling SB (2016) Membrane materials for water purification: design, development, and application. Environ Sci Water Res Technol 2(1):17–42

Lhassani A, Rumeau M, Benjelloun D, Pontie M (2001) Selective demineralization of water by nanofiltration application to the defluorination of brackish water. Water Res 35(13):3260–3264
Li K (2007) Ceramic membranes for separation and reaction. Wiley, Hoboken
Li F, Li L, Liao X, Wang Y (2011) Precise pore size tuning and surface modifications of polymeric membranes using the atomic layer deposition technique. J Membr Sci 385:1–9
Li F, Yang Y, Fan Y, Xing W, Wang Y (2012) Modification of ceramic membranes for pore structure tailoring: the atomic layer deposition route. J Membr Sci 397:17–23
Liu L, Liu Z, Bai H, Sun DD (2012) Concurrent filtration and solar photocatalytic disinfection/degradation using high-performance ag/TiO_2 nanofiber membrane. Water Res 46(4):1101–1112
Loeb J (1921) Donan equilibrium and the physical properties of proteins: I. membrane potentials. J Gen Physiol 3(5):667–690
McBain JW (1932) The sorption of gases and vapours by solids. J Phys Chem 37(1):149–150
Mohapatra D, Mishra D, Mishra SP, Chaudhury GR, Das RP (2004) Use of oxide minerals to abate fluoride from water. J Colloid Interface Sci 275(2):355–359
Mohapatra M, Anand S, Mishra BK, Giles DE, Singh P (2009) Review of fluoride removal from drinking water. J Environ Manag 91(1):67–77
Moustakas NG, Katsaros FK, Kontos AG, Romanos GE, Dionysiou DD, Falaras P (2014) Visible light active TiO_2 photocatalytic filtration membranes with improved permeability and low energy consumption. Catal Today 224:56–69
Nasreen SAAN, Sundarrajan S, Nizar SAS, Balamurugan R, Ramakrishna S (2014) In situ polymerization of PVDF-HEMA polymers: electrospun membranes with improved flux and antifouling properties for water filtration. Polym J 46(3):167
Ndiaye PI, Moulin P, Dominguez L, Millet JC, Charbit F (2005) Removal of fluoride from electronic industrial effluent by RO membrane separation. Desalination 173(1):25–32
Ozsvath DL (2009) Fluoride and environmental health: a review. Rev Environ Sci Biotechnol 8 (1):59–79
Pan K, Ren R, Liang B, Li L, Li H, Cao B (2014) Synthesis of pH-responsive polyethylene terephthalate track-etched membranes by grafting hydroxyethyl-methacrylate using atom-transfer radical polymerization method. J Appl Polym Sci 131(20)
Pellegrin ML, Aguinaldo J, Arabi S, Sadler ME, Min K, Liu M, Owerdieck C (2013) Membrane processes. Water Environ Res 85(10):1092–1175
Premathilaka RW, Liyanagedera ND (2019) Fluoride in drinking water and nanotechnological approaches for eliminating excess fluoride. J Nanotechnol 2019:1–15
Saljoughi E, Mousavi SM (2012) Preparation and characterization of novel polysulfone nanofiltration membranes for removal of cadmium from contaminated water. Sep Purif Technol 90:22–30
Saravanan S, Kalyani C, Vijayarani MP, Jayakodi P, Felix AJW, Nagarajan S, Krishnan V (2008) Prevalence of dental fluorosis among primary school children in rural areas of Chidambaram taluk, Cuddalore district, Tamil Nadu, India. Ind J Commun Med 33(3):146
Saxena V, Ahmed S (2001) Dissolution of fluoride in groundwater: a water-rock interaction study. Environ Geol 40(9):1084–1087
Sheng T, Chen H, Xiong S, Chen X, Wang Y (2014) Atomic layer deposition of polyimide on microporous polyethersulfone membranes for enhanced and tunable performances. AIChE journal, 60(10), 3614–3622. Membr Sci 454:339–345
Singh N, Verma KG, Verma P, Sidhu GK, Sachdeva S (2014) A comparative study of fluoride ingestion levels, serum thyroid hormone & TSH level derangements, dental fluorosis status among school children from endemic and non-endemic fluorosis areas. Springerplus 3(1):7
Singh J, Singh P, Singh A (2016) Fluoride ions vs removal technologies: a study. Arab J Chem 9 (6):815–824
Singh G, Kumari B, Sinam G, Kumar N, Mallick S (2018) Fluoride distribution and contamination in the water, soil and plants continuum and its remedial technologies, an Indian perspective–a review. Environ Pollut 239:95–108

Smitha K, Thampi SG (2017) Experimental investigations on fluoride removal from water using nanoalumina-carbon nanotubes blend. J Water Resour Prot 9:760

Soyekwo F, Zhang QG, Deng C, Gong Y, Zhu AM, Liu QL (2014) Highly permeable cellulose acetate nanofibrous composite membranes by freeze-extraction. J Membr Sci 454:339–345

Suthar S, Garg VK, Jangir S, Kaur S, Goswami N, Singh S (2008) Fluoride contamination in drinking water in rural habitations of northern Rajasthan, India. Environ Monit Assess 145 (1–3):1–6

Tajer-Kajinebaf V, Sarpoolaky H, Mohammadi T (2014) Sol–gel synthesis of nanostructured titania–silica mesoporous membranes with photo-degradation and physical separation capacities for water purification. Ceram Int 40(1):1747–1757

Tang B, Huo Z, Wu P (2008) Study on a novel polyester composite nanofiltration membrane by interfacial polymerization of triethanolamine (TEOA) and trimesoyl chloride (TMC): I. preparation, characterization and nanofiltration properties test of membrane. J Membr Sci 320(1–2):198–205

Thakur VK, Voicu SI (2016) Recent advances in cellulose and chitosan based membranes for water purification: a concise review. Carbohydr Polym 146:148–165

Trommer K, Morgenstern B (2010) Nonrigid microporous PVC sheets: preparation and properties. J Appl Polym Sci 115(4):2119–2126

Vatanpour V, Esmaeili M, Farahani MHDA (2014) Fouling reduction and retention increment of polyethersulfone nanofiltration membranes embedded by amine-functionalized multi-walled carbon nanotubes. J Membr Sci 466:70–81

Waghmare SS, Arfin T (2015) Fluoride removal from water by various techniques. Int J Innov Sci Eng Technol 2(9):560–571

Wang Z, Yu H, Xia J, Zhang F, Li F, Xia Y, Li Y (2012) Novel GO-blended PVDF ultrafiltration membranes. Desalination 299:50–54

Wang P, Ma J, Shi F, Ma Y, Wang Z, Zhao X (2013) Ind Eng Chem Res 52:10355–10363

WHO - World Health Organization (2004) Guidelines for drinking water quality, World Health Organization, Third Edition, Vol. 1, Geneva

Wu G, Gan S, Cui L, Xu Y (2008) Preparation and characterization of PES/TiO_2 composite membranes. Appl Surf Sci 254(21):7080–7086

Xu Q, Yang J, Dai J, Yang Y, Chen X, Wang Y (2013) Hydrophilization of porous polypropylene membranes by atomic layer deposition of TiO_2 for simultaneously improved permeability and selectivity. J Membr Sci 448:215–222

Yuliwati E, Ismail AF (2011) Effect of additives concentration on the surface properties and performance of PVDF ultrafiltration membranes for refinery produced wastewater treatment. Desalination 273(1):226–234

Zhang Y, Yu H, Sun X, Liu J, Zhang H (2014) Improving the antifouling property of polyethersulfone ultrafiltration membrane by incorporation of dextran grafted halloysite nanotubes. Chem Eng J 237:322–328

A Two-Stage Constructed Wetland Design Integrating Artificial Aeration and Sludge Mineralization for Municipal Wastewater Treatment

7

Alexandros I. Stefanakis

Abstract

The technology of constructed wetlands is well known for its multiple benefits in terms of cost-efficiency and environmental performance. The main advantage arises during the operational phase for which the related costs can be significantly lower compared to common conventional/mechanical solutions. One of the main limitations is, however, the higher area demand, which usually limits the application range of these systems to areas where adequate land is available. Over the last years, various modifications have been tested towards the optimization of the system design in terms of both pollutant removal efficiency and reducing area demands. The introduction of air using artificial means has gained significant attraction among researchers and engineers mainly due to the reduced area requirements. Furthermore, the implementation of such systems in arid and dry climates is always a technical and operational challenge. Bauer Nimr LLC designed and constructed in 2017–2018 a new constructed wetland facility in Oman using a novel combination of different wetland designs. The first stage is a vertical flow constructed wetland with integrated sludge accumulation and dewatering, and the second stage is an aerated horizontal subsurface flow constructed wetland. This hybrid system is designed to treat 350 m^3/day of domestic wastewater from a nearby residential area with approx. 1800 persons equivalent. This facility introduces not only an innovative design with minimum footprint (1.6 m^2/PE), but it is also designed to provide a treated effluent quality that meets the strict national standards for irrigation reuse. This is the first hybrid wetland system of that design implemented in the GCC region under desert climatic conditions, which also acts as a first full-scale demonstration of the

A. I. Stefanakis (✉)
School of Environmental Engineering, Technical University of Crete, Crete, Greece

Bauer Nimr LLC, Muscat, Sultanate of Oman
e-mail: astefanakis@enveng.tuc.gr

R. Prasad (ed.), *Environmental Pollution and Remediation*, Environmental and Microbial Biotechnology, https://doi.org/10.1007/978-981-15-5499-5_7

aerated wetland technology and the feasibility of these systems even in harsh environments.

Keywords

Constructed wetlands · Vertical flow · Aerated wetland · Domestic wastewater · Hybrid system · Wastewater reuse

7.1 Introduction

Natural processes have been used in waste/wastewater treatment processes for a long time; the most known technologies and methods utilize these processes, for example, sedimentation, filtration, biological activity, etc., under intensified conditions; hence they typically include complex and energy-consuming mechanical equipment. The main difference that separates the so-called natural treatment systems from the conventional ones is that only natural components and processes are used for the treatment. Naturally occurring processes are utilized under a controlled environment in natural treatment systems, without the need for external energy input for these processes to occur. Typically, energy is consumed only for few mechanical parts, e.g., pump operation. Among the various natural treatment systems, such as waste stabilization ponds, constructed wetlands (CWs) appear as one of the most promising eco-tech methods, which have been attracting increasing worldwide interest. Their excellent treatment performance, combined with their environmentally friendly (ecological) character and the typically reduced overall costs, has placed constructed wetlands in the forefront of the scientific and marketing interest (Stefanakis 2019, 2020).

CWs are already an established ecological treatment method for municipal (Stefanakis et al. 2014; Stefanakis 2015) and various industrial wastewaters (Wu et al. 2015; Stefanakis 2018; Stefanakis et al. 2019). The main driver for the tremendous expansion of applications worldwide and the intensification of international research is the multiple advantages these systems offer compared to conventional treatment methods such as activated sludge systems or membrane bioreactors.

Conventional and mechanical treatment methods are used for more than a century, based on a different approach of wastewater treatment, which consists in the installation of large, centralized facilities and respectively extended sewer networks for the collection and transport of wastewater from its source. This translates to large end-of-the-pipe centralized wastewater treatment facilities and a related long sewer network, with respectively high required investment. These facilities are also typically characterized by high operation and maintenance costs (Stefanakis 2015). It is true that such facilities provide a wastewater treatment solution for large cities, capitals and populated urban areas. For that perspective, the high investment cost is difficult to avoid. The question that arises is whether this centralized approach is a proper and feasible solution for smaller cities, communities and peri-urban and rural areas. Hence, over the last years an alternative approach of wastewater treatment has risen, the so-called decentralized approach.

CWs belong to this category and serve this philosophy as point-of-use treatment systems that deliver effluents of the desired quality. CWs can be installed in most areas and regions with reduced overall costs. It should be noted that the technology of constructed wetlands should not be viewed as a direct competitor of conventional treatment systems. A total replacement of centralized facilities with CWs and/or other natural treatment systems is practically infeasible, given that the required land area would be very high (Stefanakis 2015). Such a demand cannot be met in large densely populated urban areas. However, CW systems can be an ideal solution for small and medium communities, small cities, villages, single households or block of flats and generally rural, mountainous and remote areas, where sewer network does not exist, and the construction of a conventional treatment plant is economically unfeasible.

CWs are in general easy to build and provide treatment that is robust to flow fluctuations and pollutant concentrations. Among the advantages of wetland technology are also its simple and easy operation and minimal mechanical equipment (Stefanakis et al. 2014). CW facilities do not include large and multiple mechanical parts. The only mechanical equipment is usually pumps installed to transfer the wastewater from one stage to the other, which can even be avoided with a proper design and exploitation of the topography of the installation area. All these translate to minimum operational costs, which probably represent the main advantage of CW technology (Stefanakis 2015). In CWs, the energy input needed is very small, usually only for any pumps that may exist in the facility. Energy requirements for the treatment processes are covered by renewable energy sources used by plants (solar, wind energy). Maintenance needs are also low, since the operation of a CW facility is practically autonomous; at the stage of full operation, a typical maintenance scheme applied is one visit at the site on a monthly basis. Additionally, there is limited need for specialized personnel to run the CW facility. Global experience indicates that operational costs of CW facilities can be up to 90% lower compared to conventional treatment methods. Moreover, CWs operate without the need for the addition of any chemical substances, which is not the case in conventional treatment plants.

Based on the above, it is obvious that these various and multiple characteristics of CW provide this treatment technology an ecological character. They are often characterized as sustainable systems due to the low energy consumption and the use of natural materials (gravel, soil, sand and plants). The minimum energy consumption in CW facilities saves natural resources and minimizes pollution generation, especially when the energy source is non-renewable (e.g. fossil fuels). These environmental benefits can be translated to low levels of greenhouse gas emissions (CO_2, CH_4 and N_2O) (Langergraber 2013; Stefanakis et al. 2014; Stefanakis 2019).

The main limitation of CW technology is the higher area demand compared to mechanical solutions. However, continuous research and development can now deliver improved designs in terms of area demand and operational mode. The type of vertical flow constructed wetlands (VFCW) is widely used for raw wastewater treatment providing integrated sludge accumulation and mineralization, thus avoiding the need for additional sludge dewatering equipment (Stefanakis and Tsihrintzis 2012; Stefanakis et al. 2014; Morvannou et al. 2015; Stefanakis 2020). VFCWs also have lower area demand than horizontal subsurface flow CWs (HSFCW). Moreover, over the last years, a new design modification has been tested and applied, that of the artificially aerated constructed wetlands (ACW). In these systems, artificial means are used to maximize the air supply to the system in order to improve the oxygen availability. For this, an air pump (a blower or injector) is used to provide compressed air to the CW bed. This design mode has been tested in HSFCWs where oxygen availability is limited (Ouellet-Plamondon et al. 2006; Chazarenc et al. 2009; Stefanakis et al. 2009, 2011; Zhang et al. 2010; Uggetti et al. 2016), as well as in VFCW systems (Dong et al. 2012; Stefanakis et al. 2014, 2019; Murphy et al. 2016; Liu et al. 2019).

Although wastewater aeration is a common practice in other treatment methods (e.g., in activated sludge systems and membrane bioreactors), aeration of gravel beds is not that old. In tanks and ponds, aeration creates air bubbles, creating this way a hydrodynamic mixing effect, which results in a more uniform distribution of the dissolved oxygen at a considerable distance from the air diffuser (Nivala et al. 2012; Stefanakis et al. 2014). Therefore, aeration in conventional wastewater treatment methods is provided via high-volume devices. On the contrary, in subsurface flow CWs containing gravel media, the hydrodynamic mixing is limited and the distance an air bubble can reach does not exceed approx. 30–40 cm. For this reason, effective aeration in subsurface flow CWs can be implemented with uniform distribution of small air quantities across the bottom of the bed. The main advantage of bottom additional aeration especially in VFCWs is that the combined action of the vertical downward drainage of the wastewater with the upflow movement of air bubbles results in a very good water mixture within the bed (Stefanakis 2015). Additional aeration of the bed has been promoted as a positive alternative to enhance nitrogen removal, especially in cold climate systems where ambient temperatures are not favourable for nitrification (Wallace et al. 2000). Artificial aeration has shown a tremendous treatment potential with significantly reduced area requirements (Dong et al. 2012; Wu et al. 2014), indicating that they reach high levels of organic degradation and enhanced nitrogen removal.

The existing international published literature already focuses on the investigation of the various operational parameters and the functions of these advanced designs. The optimization of the system's efficiency is of course always the goal. Such information can be derived, however, only through the verification of experimental designs in full-scale systems operating under real conditions. Under this frame, the technological status of CWs indicates that the combination of two different wetland designs, i.e. of a VFCW and an ACW, could be a new approach to wetland technology, providing an advanced wetland design with further enhanced benefits. This concept aims at a dual target: (a) integrating sludge mineralization within the system without the need for additional sludge dewatering equipment and (b) minimizing the area demand using the aerated wetland system to provide a treated effluent of high quality. Hence, this chapter presents one of the first full-scale constructed wetland systems using a two-stage design, operating under a warm and dry climatic condition.

7.2 Description of the Constructed Wetland Facility

The full-scale hybrid CW facility uses a combination of a VFCW and an ACW stage and has been implemented in the Sultanate of Oman (Stefanakis 2020). This facility has been designed and built by Bauer Nimr LLC in 2017 under the harsh climatic conditions of the Arabic peninsula and Middle East (i.e. dry and hot climate). The site conditions can be characterized as severe; there are frequent sandstorm incidents, while the atmosphere can be laden with airborne dust particles (even <2 microns). The daily average temperature exceeds 40 °C in summer (i.e., period from May to August) and reaches up to 30 °C in winter (i.e. period from December to January). Precipitation is almost negligible in the area. The following Table 7.1 summarizes the weather conditions in the area of the wetland facility. The harsh environmental conditions already pose a challenge for any mechanical equipment that is to be used, to ensure its proper and continuous function, as well as for the establishment of wetland plants.

The design of the wetland facility was completed in 2017. The construction started in late 2017 and was completed in the first half of 2018. Since Spring 2018, the commissioning phase of the wetland system started to allow for the establishment and growth of the plants. The new wetland facility is planned to replace an existing conventional treatment plant. The CW system receives municipal wastewater from a remote settlement in the interior under desert environmental conditions. The settlement is a medium-sized community installed by an oil and gas company, hosting on the average 1800 persons on a daily basis. The wastewater

Table 7.1 Statistical data of various climatic parameters in the area of the constructed wetland facility

Parameter	Unit	Average	Maximum	Minimum
Temperature				
Summer	°C	40	60	
Winter		15		5
Design		55		
Precipitation				
Design	mm	25 (1 h)		
Annual average		Negligible		
Relative humidity	%		98	30
Wind speed	km/h	80	123 (3 s)	
Wind direction				
March–April		NW		
May–June		SW		
Rest of the year		SE		

origin is domestic sewage from the residential and accommodation areas as well as from offices and kitchen facilities. Based on the population of the settlement, the design daily flow rate is 350 m^3/day.

Raw wastewater is collected in a collection tank and lifting station within the settlement, where after simple screening (coarse screen) it is pumped and sent to the new CW facility. The facility is located next to the settlement area perimeter fence, i.e. approx. 150 m away from the first buildings. The CW facility is designed to generate a final treated effluent that complies with the water quality requirements of the MD 145/93 Oman Standard A (Table 7.2). It was also a requirement to have a final disinfection stage of the final effluent so that it does not constitute a biological aerosol hazard in order to comply with requirements of the MD 145/93 Oman Standard. The treated effluent will be reused for irrigation of green public areas within the settlement.

The design of the wetland facility is based on a raw wastewater quality of BOD 400 mg/L, COD 700 mg/L, TSS of 400 mg/L, ammonia nitrogen of 65 mg/l and oil and grease 20 mg/L. The influent pH range is from 6 to 9. It should also be mentioned that the proposed technical solution for wastewater treatment should be odourless, environmentally friendly, easy to install, with minimal chemical dosing, less space requirements, minimal operation and maintenance cost and sludge-free operation. This wetland system is designed to meet the irrigation Standard A, Oman Ministerial Decision 145/93 (Table 7.3; i.e. $BOD_5 < 10$ mg/L, NH_4-N < 5 mg/L, TN < 21 mg/L, TSS < 15 mg/L, $NO_3 < 50$ mg/L, faecal coliform < 5 CFU/100 mL, intestinal helminth/nematode eggs < 1 egg/L).

The overall design and concept of the constructed wetland facility is shown in Fig. 7.1. The CW consists of two stages; the first stage is a vertical flow constructed wetland (VFCW) and the second stage is an aerated constructed wetland (ACW) with horizontal subsurface flow. Figure 7.2 shows an aerial overview of the two-stage constructed wetland facility.

The first stage is a VFCW filter (surface of 2100 m^2) and receives raw wastewater (after coarse screening) (Fig. 7.3). The VFCW surface is separated in three parallel cells. Each day, only one cell is loaded with the raw wastewater, which is applied at regular intervals across the surface of the cell through a feeding pipe network, in order to ensure uniform distribution of the organic solids in the raw wastewater (Stefanakis et al. 2014). Each bed receives the full organic load and the solid content, resulting in a decreased total surface area needed. The feeding strategy is the intermittent loading and comprises feeding (1–2 days) and resting periods (2–4 days) (Stefanakis and Tsihrintzis 2012). The appropriate duration of the resting periods allows for aerobic conditions restoration and minimizes the risk of clogging due to solid retention and the gradual creation of a residual solid layer on top of the gravel media surface. The operational cycle is controlled by opening or closing three manual valves located at the front end of the VFCW in a valve chamber.

With this setup, a primary settling tank is avoided and there is no need for additional management tasks and equipment for sludge handling and dewatering. The fast introduction of wastewater onto each cell bed surface creates temporary water ponding of 3–5 cm depth and temporary saturated conditions in the top gravel

Table 7.2 Wastewater maximum quality limits (mg/L except where otherwise stated) in Oman

Parameter	Standards (See Table 7.3)	
	A	B
Biochemical oxygen demand (BOD) (5d@20 °C)	15	20
Chemical oxygen demand (COD)	150	200
Suspended solids (SS)	15	30
Total dissolved solids (TDS)	1500	2000
Electrical conductivity (EC) (micro S./cm)	2000	2700
Sodium absorption ratio (SAR) (*The effect of Sodium on soil absorption*)	10	10
pH (within range)	6–9	6–9
Aluminum (as Al)	5	5
Arsenic (as As)	0.100	0.100
Barium (as Ba)	1	2
Beryllium (as Be)	0.100	0.300
Boron (as B)	0.500	1
Cadmium (as Cd)	0.010	0.010
Chloride (as Cl)	650	650
Chromium (total as Cr)	0.050	0.050
Cobalt (as Co)	0.050	0.050
Copper (as Cu)	0.500	1
Cyanide (total as CN)	0.050	0.100
Fluoride (as F)	1	2
Iron (total as Fe)	1	5
Lead (as Pb)	0.100	0.200
Lithium (as Li)	0.070	0.070
Magnesium (as Mg)	150	150
Manganese (as Mn)	0.100	0.500
Mercury (as Hg)	0.001	0.001
Molybdenum (as Mo)	0.010	0.050
Nickel (as Ni)	0.100	0.100
Nitrogen: Ammoniacal (as N)	5	10
: Nitrate (as NO_3)	50	50
: Organic (Kjeldahl) (as N)	5	10
Oil and grease (total extractable)	0500	0.500
Phenols (total)	0.001	0.002
Phosphorus (total as P)	30	30
Selenium (as Se)	0.020	0.020
Silver (as Ag)	0.010	0.010
Sodium (as Na)	200	300
Sulfate (as SO_4)	400	400
Sulfide (total as S)	0.100	0.100
Vanadium (as V)	0.100	0.100
Zinc (as Zn)	5	5
Faecal coliform bacteria (per 100 mL)	200	1000
Viable nematode ova (per litre)	<1	<1

Table 7.3 Average influent and effluent concentrations and respective mean removal rates of the various physicochemical parameters

Parameter	Influent	Effluent	Removal (%)
BOD_5 (mg/L)	213	2	99.0
COD (mg/L)	534	24	95.1
TSS (mg/L)	98.2	3.0	96.8
Total nitrogen – TN (mg/L)	50.8	17.5	66.4
Ammonia nitrogen – NH_3-N (mg/L)	38.1	0.1	99.7
Total Kjeldahl nitrogen – TKN (mg/L)	39.4	0.5	98.7
Nitrate nitrogen – NO_3-N (mg/L)	0.6	16.4	–
Total phosphorus – TP (mg/L)	20.6	3.1	85.9
pH (–)	7.1	8.0	–
Electrical conductivity – EC (μS/cm)	1125	1492	–
Total dissolved solids – TDS (mg/L)	678	932	–
Total coliform – TC (CFU/100 mL)	962	Not detected	100
Faecal coliform – FC (CFU/100 mL)	720	Not detected	100
Intestinal helminth/nematode eggs – IH (eggs/L)	57	Not detected	100
Free oil (% v/v)	<0.01	<0.01	–
Dissolved and emulsified oil (mg/L)	20.5	<10	–

Data correspond to the first months of operation

layers. The trapped air is forced by the supernatant water to move downward. The water is then drained vertically by gravity through the porous media layers of the bed in an unsaturated flow mode (Stefanakis and Tsihrintzis 2012). As the water drains, air from the atmosphere enters the system and fills the void space of the media (Maier et al. 2009), replacing the water volume that drains. In this way, the bed aeration is enhanced, and the microbial activity is stimulated (Stefanakis et al. 2014). Once the water reaches the bottom of the bed, a network of collecting pipes collects it and sends it with gravity to the second stage wetland. The VFCW basin has a total depth of 1 m, of which 65 cm is the gravel layer and the rest the freeboard. The bottom of the bed is also lined with an HDPE membrane of 1.5 mm thickness. A layer of compacted clay sand was installed below the HPDE liner in order to eliminate the potential leakage of polluted water into the ground soil during the operation of the new wetland facility.

The second stage is an ACW with horizontal subsurface flow mode (surface of 800 m^2) gravitationally receiving the partially treated water of the first-stage wetland bed and providing final wastewater treatment reaching the desired quality (Fig. 7.4). Step-feeding of raw wastewater (screened inflow from the settlement lift station) is also applied in the ACW, with a small volume being routed to the second stage to provide additional carbon source for denitrification. The practice of wastewater step-feeding has been found to improve the treatment performance in HSFCWs (Stefanakis et al. 2011). Water is distributed through a PVC pipe with orifices, which extends along the upstream width side and is placed on top of the gravel media layer. The total depth of the bed is 1.3 m, of which 1 m is the depth of the gravel media layer and the rest is freeboard.

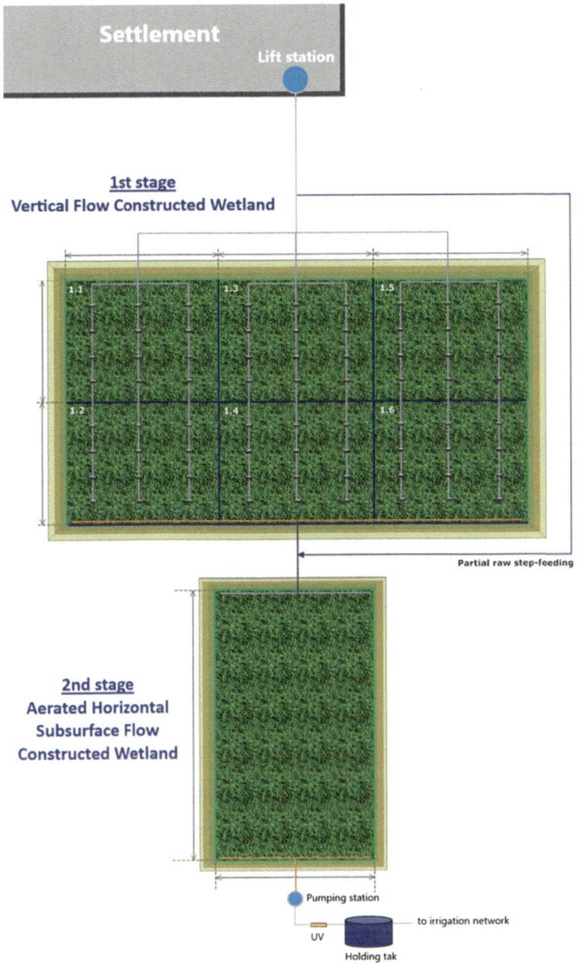

Fig. 7.1 A schematic representation of the treatment stages of the constructed wetland facility

Fig. 7.2 An aerial overview of the two-stage constructed wetland facility just before the completion of the construction phase. (Courtesy of Bauer Nimr LLC)

Fig. 7.3 The first-stage vertical flow constructed wetland for sludge mineralization during operation. (Courtesy of Bauer Nimr LLC)

As in the first stage, the second stage wetland is also lined with an HDPE membrane of 1.5 mm thickness, with a layer of compacted clay sand placed below the HPDE liner. Artificial aeration is provided in this wetland bed through a network of aeration drip lines placed at the bottom of the bed connected to a side channel blower (one working, one standby) that delivers an airflow 330 m^3/h. An alternate operation of the two blowers is conducted every 24 h to avoid any mechanical problems due to ambient air condensation when the blower is off for a long time period. The first blower supplies air continuously for 24 h to the whole network of pipe and drip lines at the bottom of the ACW, while the second blower is off. An automatic switch alternates daily the blower operation. The aeration pipes are placed at short distances at the bottom, providing small bubbles of air in the saturated gravel layer to enhance the aerobic processes, such as organic matter biodegradation and nitrification. The water level is adjustable through a water level control device and is set few centimetres below the surface of the gravel layer of the constructed wetland, so that no water surface is exposed to the atmosphere and in order to eliminate mosquitos and bad odour issues. The ACW effluent is collected through a perforated pipe placed at the bottom of the bed above the HDPE liner along the downstream width side.

Both CW stages are planted with native wetland plant species, which can be found in the country. The first stage VFCW bed is planted with common reeds

Fig. 7.4 The second-stage horizontal subsurface flow aerated constructed during the second month of operation. (Courtesy of Bauer Nimr LLC)

(*Phragmites australis*) and the second stage is a polyculture planted with *Typha domingensis*, *Schoenoplectus littoralis* and *Cyperus laevigatus*. In total, more than 17,000 plant stems were initially planted in both wetland beds.

Finally, the ACW effluent is collected with gravity in a downstream pumping station, equipped with two pumps (one working, one standby). The pumps push the treated effluent through a UV unit of 30 m^3/h capacity, placed in a 20 ft container with control temperature, for final disinfection. The final disinfected effluent is filling and stored in a treated effluent tank with a retention time of 1 day. The water stored in the effluent tank is further pumped and reused for irrigation of green spaces in the adjacent settlement.

7.3 Results and Discussion

This novel constructed wetland facility has one of the lowest area requirements (1.6 m^2/PE) for this high effluent quality and demonstrates the current advances in wetland technology. Such designs enable the implementation of wetland systems even in areas with limited available land (e.g. peri-urban areas).

Table 7.1 presents the results of the first operational months, while Figs. 7.5 and 7.6 show the variations of influent and effluent values for the various parameters

Fig. 7.5 Variations of main pollutants (COD, BOD_5, Nitrogen, TP and TSS) in the influent and effluent of the constructed wetland facility

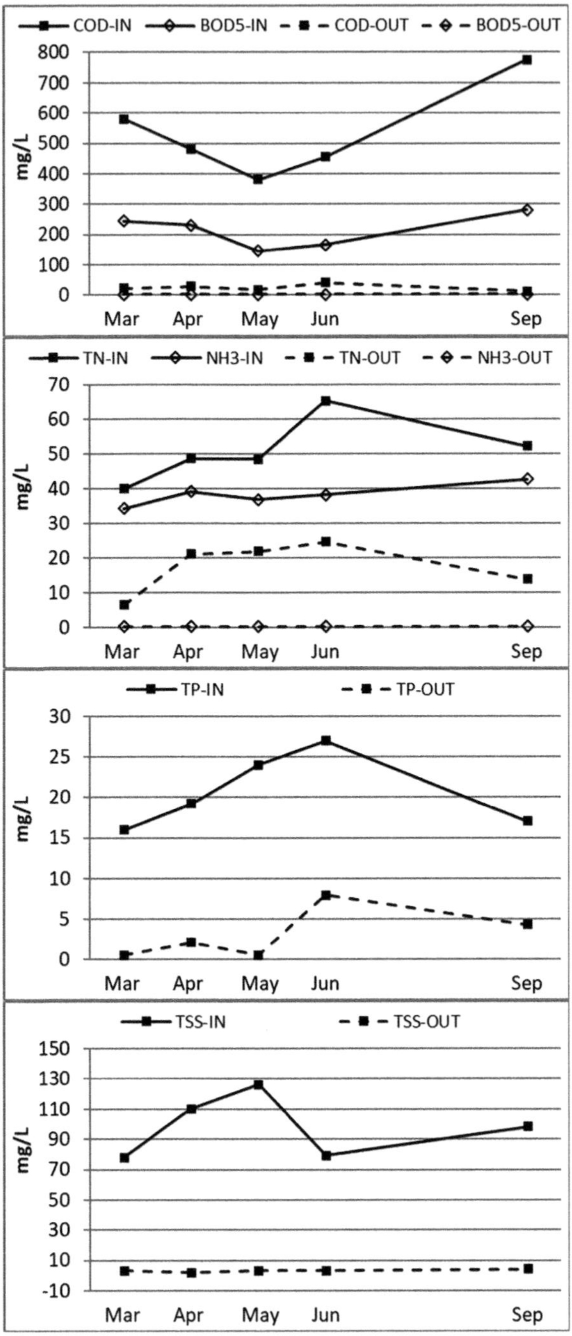

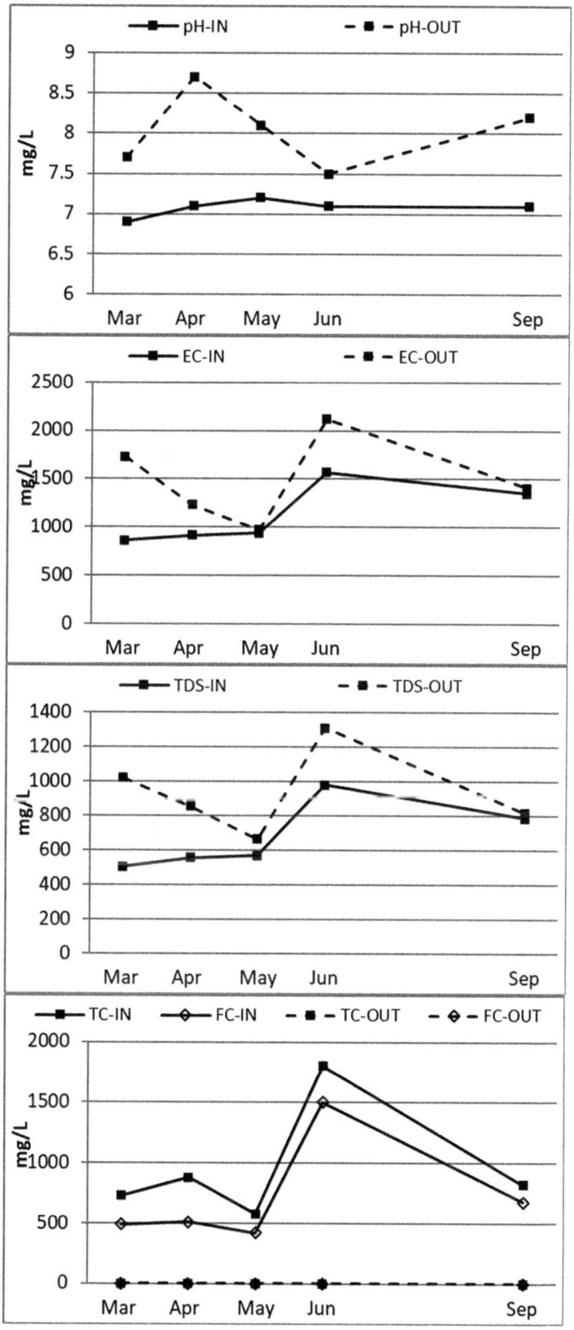

Fig. 7.6 Variations of physicochemical parameters (pH, EC and TDS) as well as microbiological indicators (TC and FC) in the influent and effluent of the constructed wetland facility

tested. The wetland system seems to possess a buffering capacity, as the pH is stabilized in the slightly basic region. The increase in the electrical conductivity and also TDS is more or less expected, due to both the interactions between the wastewater and the substrate media that possibly results in the release of salts in the water, as well as to the water loss through evapotranspiration in the two wetland beds. It is calculated that this water loss in the wetland system through both evaporation and plant transpiration accounts for approximately 15% of the inflow water as an average annual value. Higher values are expected during summer months and lower in winter.

As it is obvious, the effluent quality complies with the national irrigation standards (Table 7.2). Most of the organic solids are retained above the filter layer, as indicated by the low effluent TSS concentration. Moreover, the system achieves almost complete organic matter degradation and ammonia nitrification, mainly as a result of the high oxygen provided via the artificial aeration means. Similar high efficiency results are also reported in other studies (Butterworth et al. 2013, 2016; Fan et al. 2013). The complete and fast degradation of organic matter might be attributed to the easily biodegradable carbon in the domestic wastewater and the high temperatures of the area (Li et al. 2014). As it is known, the intensity of the metabolic activities of microorganisms that degrade organic matter is largely dependent on oxygen supply (Stefanakis et al. 2014). Artificial aeration provides sufficient oxygen amount, which favours these microbial activities, resulting in high COD and BOD removal.

The removal of ammonia nitrogen was almost complete and remained quite stable during the whole monitoring period. Nitrification is an aerobic process; thus the higher oxygen supply via the artificial aeration provides adequate dissolved oxygen for ammonia nitrogen oxidation (Li et al. 2014; Stefanakis et al. 2014). The nitrifying bacteria obtain energy from oxidizing NH4+-N and use oxygen as an electron acceptor. The ambient temperatures are also ideal for nitrification, since it is known that this process has an optimum temperature range between 25 °C and 35 °C (Kuschk et al. 2003; Stefanakis et al. 2014), as well as optimum pH values around 7.5 to slightly alkali values close to 7.8 (Cooper et al. 1997; Stefanakis et al. 2014).

In addition to that, it is also noticeable that nitrate effluent concentration remains below the legal limit. This implies that despite the aerobic conditions in the system and the almost complete nitrification, nitrate removal also takes place (denitrification). This interesting result has also been found in other studies (Tee et al. 2011; Saeed and Sun 2011; Fan et al. 2013). Denitrification is a known anaerobic process, since excess oxygen suppresses the enzyme system required for this process. More, sufficient organic carbon supply is crucial as an electron donor for nitrate reduction and provides an energy source for denitrification microorganisms. In this design, sufficient carbon supply is maintained through the step-feeding of raw wastewater to the ACW, which apparently contributes to the nitrate transformation. Additionally, the aeration regime possibly alters the microbial community composition and characteristics, enabling nitrate removal even under these conditions. Such changes in the operational mode and even wastewater composition have been found to impact the microbial community composition and patterns (Stefanakis et al. 2016). Further investigation on this issue will provide better insight on the processes taking place within the system.

The disinfection step also proves to be very effective, since practically no microbiological contamination is detected in the final treated effluent, after the UV stage. In another study on aerated constructed wetland used as tertiary stage, it is reported that the aerated system alone provided significant log reduction of pathogenic bacteria, satisfying the local standard for effluent discharge to the environment (Stefanakis et al. 2019). This is another significant potential that aerated wetland technology can possess that is worthy of further investigation.

7.4 Conclusions

The presented full-scale constructed wetland facility is one of the first in Middle East using this novel design, which combines a vertical flow constructed wetland and an aerated constructed wetland. The advantage of this design is that it integrates sludge accumulation and mineralization, i.e. eliminating the need for additional sludge management and handling equipment and related costs, while it provides an effluent quality appropriate for reuse (i.e. irrigation of green areas). Furthermore, this wetland design has one of the lowest footprints (1.6 m^2/PE), reducing thus the area demand when compared to passive wetland systems. This makes such designs more attractive for areas where space availability is an issue. The overall performance of the facility under real operating conditions proves the effectiveness of the applied design even under harsh environmental, such as dry and arid areas.

References

Butterworth E, Dotro G, Jones M, Richards A, Onunkwo P, Narroway Y, Jefferson B (2013) Effect of artificial aeration on tertiary nitrification in a full-scale subsurface horizontal flow constructed wetland. Ecol Eng 54:236–244

Butterworth E, Richards A, Jones M, Mansi G, Ranieri E, Dotro G, Jefferson B (2016) Performance of four full-scale artificially aerated horizontal flow constructed wetlands for domestic wastewater treatment. Water 8:365. https://doi.org/10.3390/w8090365

Chazarenc F, Gagnon V, Comeau Y, Brisson J (2009) Effect of plant and artificial aeration on solids accumulation and biological activities in constructed wetlands. Ecol Eng 35:1005–1010

Cooper P, Smith M, Maynard H (1997) The design and performance of a nitrifying vertical-flow reed bed treatment system. Water Sci Technol 35(5):215–221

Dong H, Qiang Z, Li T, Jin H, Chen W (2012) Effect of artificial aeration on the performance of vertical-flow constructed wetland treating heavily polluted river water. J Environ Sci 24 (4):596–601

Fan J, Liang S, Zhang B, Zhang J (2013) Enhanced organics and nitrogen removal in batch-operated vertical flow constructed wetlands by combination of intermittent aeration and step feeding strategy. Environ Sci Pollut Res 20:2448–2455

Kuschk P, Wiebner A, Kappelmeyer U, Weibbrodt E, Kaestner M, Stottmeister U (2003) Annual cycle of nitrogen removal by a pilot-scale subsurface horizontal flow in a constructed wetland under moderate climate. Water Res 37:4236–4242

Langergraber G (2013) Are constructed treatment wetlands sustainable sanitation solutions? Water Sci Technol 67(10):2133–2140

Li F, Lu L, Zheng X, Zhang X (2014) Three-stage horizontal subsurface flow constructed wetlands for organics and nitrogen removal: effect of aeration. Ecol Eng 68:90–96

Liu F-F, Fan J, Du J, Shi X, Zhang J, Shen Y (2019) Intensified nitrogen transformation in intermittently aerated constructed wetlands: removal pathways and microbial response mechanism. Sci Total Environ 650:2880–2887

Maier U, De Biase C, Baeder-Bederski O, Bayer P (2009) Calibration of hydraulic parameters for large-scale vertical flow constructed wetlands. J Hydrol 369:260–273

Morvannou A, Forquet N, Michel S, Troesch S, Molle P (2015) Treatment performances of French constructed wetlands: results from a database collected over the last 30 years. Water Sci Technol 71(9):1333–1339

Murphy C, Rajabzadeh AR, Weber KP, Nivala J, Wallace SD, Cooper DJ (2016) Nitrification cessation and recovery in an aerated saturated vertical subsurface flow treatment wetland: field studies and microscale biofilm modeling. Bioresour Technol 209:125–132

Nivala J, Wallace S, Headley T, Kassa K, Brix H (2012) Oxygen transfer and consumption in subsurface flow treatment wetlands. Ecol Eng 61(B):544–554

Ouellet-Plamondon C, Chazarenc F, Comeau Y, Brisson J (2006) Artificial aeration to increase pollutant removal efficiency of constructed wetlands in cold climate. Ecol Eng 27:258–264

Saeed T, Sun GZ (2011) Enhanced denitrification and organics removal in hybrid wetland columns: comparative experiments. Bioresour Technol 102:967–974

Stefanakis AI (2015) Constructed wetlands: description and benefits of an eco-tech water treatment system. In: McKeown AE, Bugyi G (eds) Impact of water pollution on human health and environmental sustainability. Information Science Reference (an imprint of IGI Global), Hershey, pp 281–303

Stefanakis AI (2018) Constructed wetlands for industrial wastewater treatment. Wiley-Blackwell, Chichester

Stefanakis AI (2019) The role of constructed wetlands as green infrastructure for sustainable urban water management. Sustain 11:6981. https://doi.org/10.3390/su11246981

Stefanakis AI (2020) Constructed wetlands for sustainable wastewater treatment in hot and arid climates: opportunities, challenges and case studies in the Middle East. Water 12(6):1665. https://doi.org/10.3390/w12061665

Stefanakis AI, Tsihrintzis VA (2012) Effects of loading, resting period, temperature, porous media, vegetation and aeration on performance of pilot-scale vertical flow constructed wetlands. Chem Eng 181–182:416–430

Stefanakis AI, Akratos CS, Gikas GD, Tsihrintzis VA (2009) Effluent quality improvement of two pilot-scale, horizontal subsurface flow constructed wetlands using natural zeolite (clinoptilolite). Microporous Mesoporous Mater 124(1-3):131–143

Stefanakis AI, Akratos CS, Tsihrintzis VA (2011) Effect of wastewater step-feeding on removal efficiency of pilot-scale horizontal subsurface flow constructed wetlands. Ecol Eng 37 (3):431–443

Stefanakis AI, Akratos CS, Tsihrintzis VA (2014) Vertical flow constructed wetlands: eco-engineering systems for wastewater and sludge treatment. Elsevier Inc, Amsterdam

Stefanakis AI, Seeger E, Dorer C, Sinke A, Thullner M (2016) Performance of pilot-scale horizontal subsurface flow constructed wetlands treating groundwater contaminated with phenols and petroleum derivatives. Ecol Eng 95:514–526

Stefanakis AI, Bardiau M, Silva D, Taylor H (2019) Presence of bacteria and bacteriophages in full-scale trickling filters and an aerated constructed wetland. Sci Total Environ 659:1135–1145

Tee HC, Lim PE, Seng CE, Nawi MAM (2011) Newly developed baffled subsurface-flow constructed wetland for the enhancement of nitrogen removal. Bioresour Technol 104:235–242

Uggetti E, Hughes-Riley T, Morris RH, Newton MI, Trabi CL, Hawes P, Puigagut J, García J (2016) Intermittent aeration to improve wastewater treatment efficiency in pilot-scale constructed wetland. Sci Total Environ 559:212–217

Wallace S, Parkin G, Cross C (2000) Cold climate wetlands: design and performance. In: Proceedings of the 7th international conference on wetland systems for water pollution control, 11–16 November, Lake Buena Vista, FL

Wu S, Kuschk P, Brix H, Vymazal J, Dong R (2014) Development of constructed wetlands in performance intensifications for wastewater treatment: a nitrogen and organic matter targeted review. Water Res 57(15):40–55

Wu H, Zhang J, Ngo HH, Guo W, Hu Z, Liang S, Fan J, Liu H (2015) A review on the sustainability of constructed wetlands for wastewater treatment: design and operation. Bioreour Technol 175:594–601

Zhang L-Y, Zhang L, Liu Y-D, Shen Y-W, Liu H, Xiong Y (2010) Effect of limited artificial aeration on constructed wetland treatment of domestic wastewater. Desalination 250:915–920

Persistent Organic Pollutants (POPs): Sources, Types, Impacts, and Their Remediation

8

Amtul Bari Tabinda Akhtar, Samra Naseem, Abdullah Yasar, and Zunaira Naseem

Abstract

Persistent organic pollutants are of greater concern due to their toxic, persistent, carcinogenic, lipophilic nature, ability of long-range transport, and bioaccumulation in food chain. These organic pollutants include organochlorine pesticides (OCPs), polybrominated diphenyl ethers (PBDEs), polyaromatic hydrocarbons (PAHs), polychlorinated naphthalenes (PCNs), polychlorinated biphenyls (PCBs) and dechlorane plus (DP), etc. Such pollutant's exposure results in negative ecological impacts along with many human health problems including neurological, immunological, and reproductive disorders. These pollutants are reported to significantly resist the conventional treatment methods including filtration, use of flocculants, coagulation, or chemical treatment such as chlorination. Therefore, various efficient removal methods and technologies are being analyzed and applied to remove these toxic pollutants from different environmental compartments such as water, wastewater, and soil. Methods including bioremediation, reverse osmosis, adsorption process, membrane technology, and advanced oxidation processes (AOPs) are documented to efficiently remove these recalcitrant contaminants. However, removal potential, economic, and technical feasibilities are important considerations for selection of prime treatment method. This chapter provides an outlook of various persistent organic pollutants, their sources, ecological impacts, and treatment methods used for their removal.

A. B. T. Akhtar (✉) · S. Naseem · A. Yasar
Sustainable Development Study Centre, Government College University, Lahore, Pakistan
e-mail: amtulbaritabinda@gcu.edu.pk

Z. Naseem
Department of Civil Engineering, Monash University, Clayton, VIC, Australia
e-mail: zunaira.naseem@monash.edu

R. Prasad (ed.), *Environmental Pollution and Remediation*, Environmental and Microbial Biotechnology, https://doi.org/10.1007/978-981-15-5499-5_8

Keywords

Persistent organic pollutants (POPs) · Bioremediation · Adsorption · Membrane technology · Reverse osmosis · Advanced oxidation processes (AOPs)

8.1 Introduction: Persistent Organic Pollutants (POPs)

The global industrialization has resulted in the manufacturing of various chemicals including dyes, food items, pesticides, medicines, cosmetic products, and many more (Arslan et al. 2017). These chemicals apart from their benefits are found to be harmful for environment and human health. Among these harmful chemicals, persistent organic pollutants (POPs) are of greater concern due to their toxic, persistent, carcinogenic, hydrophobic, and lipophilic nature, ability of long-range transport, long half-lives, and bioaccumulation in food chain (Baqar et al. 2017). These pollutants have strong tendency to change into gaseous form under prime environmental conditions such as temperature rendering them to transport to larger distances thus often found in the environments of the areas where these chemicals have never been used. As most of these compounds are composed of strong carbon-halogen bonds, they cannot be easily hydrolyzed or degraded and hence persist in the environment (Ministry of Environment and Water 2006). Another important fact about these persistent pollutants is their bioaccumulation potential in the living organisms due to the lipophilic and resistant nature to the metabolism resulting in transportation of these pollutants in food webs (Kelly et al. 2007).

These pollutants are now considered as contaminants of emerging concern (CEC) both in terms of their biological and economic benefits and their perilous health impacts and endocrine disrupting characteristics (Clara et al. 2005). The effects of these pollutants on the environment and living organisms were first reported by Racheal Carson in the book *Silent Spring* in 1960 in which the issue of premature deaths of several birds, due to consumption of contaminated food with dichlorodiphenyltrichloroethane (DDT), was highlighted (Carson et al. 1962). Several research studies observed that these persistent pollutants are presenting major threats to the environment, humans, and food webs at each trophic level. Due to the critical environmental and health impacts being caused by POPs, United Nations Environment Program (UNEP) in 2001 conducted a Stockholm Convention on Persistent Organic Pollutants (POPs), for reduction and ultimately elimination of production and use of these persistent compounds. Initially in 2001, 12 compounds were banned known as *dirty dozen*, then in 2009 nine other compounds were added to the list, and in 2011 and 2013, two more compounds were added to the list by Conference of the Parties (COP) to Stockholm Convention (Fiedler et al. 2019). Despite of the ban imposed on the use and production of POPs they are still in use in various parts of the world. Their historical as well as present inputs into different environmental compartments are leading to severe ecological damage arising the need of their remediation.

8.2 Sources of POPs

POPs can be produced as a result of natural processes such as volcanic eruptions, wildfires, etc., or anthropogenic activities such as industrial and agricultural activities. The common examples of persistent organic pollutants include organic dyes, pesticides, synthetic chemicals and personal care products, etc. The anthropogenic sources of persistent organic pollutants can be categorized into two major categories, i.e., intentionally produced and non-intentionally produced pollutants (El-Shahawi et al. 2010).

8.2.1 Intentionally Produced POPs

The compounds which are intentionally produced in industries to meet the essential needs are called intentionally produced POPs. These intentionally produced compounds can be further divided into other subcategories including many groups of brominated and chlorinated aromatic such as polychlorinated naphthalenes (PCNs), polybrominated diphenyl ethers (PBDEs), organochlorine pesticides (OCPs), polychlorinated biphenyls (PCB) and polychlorinated dibenzo-p-dioxins/furans (PCDD/Fs), etc.

8.2.2 Non-intentionally Produced POPs

The compounds produced unintentionally/accidentally as by-products of industrial processes or activities are called non-intentionally produced POPs. For example, polychlorinated biphenyls (PCB) as well as hexachlorobenzene (HCB) can be produced as industrial by-products and released into the environment. Other examples include the production of polycyclic aromatic hydrocarbons (PAHs), polychlorinated dibenzo-P-dioxins (PCDDs), and polychlorinated dibenzofurans (PCDFs).

8.3 Environmental Impacts of POPs

Persistent organic pollutants cause direct or indirect impacts on the environment due to the interactions among biotic and abiotic elements. POP contamination causes imbalance in ecological systems and threatens the environmental integrity and health of living organisms (Fig. 8.1). During the 1980s to 1990s, several studies indicated the severe impacts of POPs on the aquatic ecosystems due to the presence of these toxic contaminants in different species of Great Lakes, Arctic Sea, Baltic Sea, and North Sea. Thus, the POP pollution in ecosystems results in poisoning of birds, animals, and insects leading to imbalance in fertility and shifts in sex ratios and physical abnormalities (Alharbi et al. 2018).

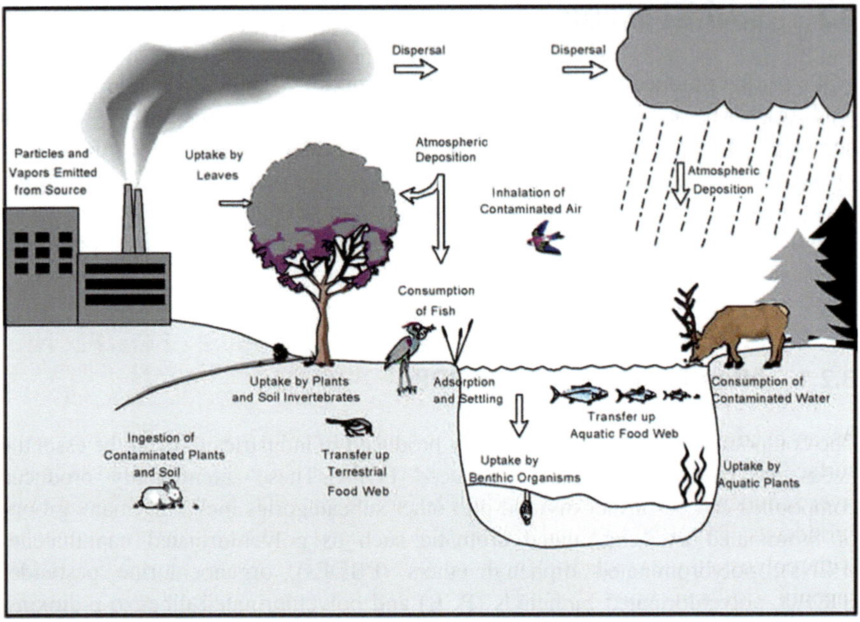

Fig. 8.1 Environmental fate processes of a released chemical and its transformation products. (Source: US EPA 1990)

Extensive research studies have documented the harmful impacts of POPs on wildlife including aberration of eggshells to extinction of certain species of birds, skeletal deformations to deaths of many beluga whales, reproductive failure, immunotoxicity, and hormone disruption among alligators of Florida's lake and other wildlife (Adeola 2004). Similarly, the harmful impacts of POPs on various species of upper trophic levels such as polar bears, seabirds, sled dogs, and polar fox are also reported, which include disruption in immune systems, reproduction failure, changes in tissues, and development process (Letcher et al. 2010).

Pesticides may cause airborne pollution as a result of volatilization after their application and pesticide drift (Rull and Ritz 2003). In indoor environments, the cooling, heating, and ventilation practices can also lead to the dispersion of pesticides being used indoors (Coxall 2014). These volatized POPs often accumulate on the aerosols and dust particles and eventually inhaled by living organisms. Inhaling and ingesting these dust particles are considered to be the one of the main sources of POPs (Walker 2008). Several studies have also reported that pollutant's derived climate change has led to the alteration in temperature, weather patterns, carbon cycle, and precipitation levels resulting in enhanced volatilization and wet deposition of POPs, thus increasing the contamination of atmosphere and water bodies (Noyes et al. 2009). Concisely, persistent organic pollutants are significantly affecting the entire ecosystem (Behera and Prasad 2020).

8.4 Health Impacts of POPs

The toxic, persistent, and bioaccumulative nature of POPs has made them highly dangerous pollutants for wildlife as well as human health (Sarwar 2015). Humans can be exposed to these pollutants through inhalation of contaminated air, ingestion of contaminated food, and direct contact with the pesticides. The concentrations of POPs are reported to be present in individuals of all age groups but children can be more susceptible to these pollutants rather than adults due to their physiology and behavior (Mascarelli 2013).

The intended or unintended long-term exposure to POPs leads to several health issues including cancer, asthma, diabetes, allergies, endocrine disruption, learning disabilities, failure of reproduction systems, neurological disorders, and cardiovascular diseases (Van Maele-Fabry et al. 2010). Several researches in the literature have also reported the presence of POPs in the wombs of mothers, childbirth complications, reduction in birth weight, and fetal deaths (Wickerham et al. 2012). Even the minute concentrations of these pollutants can be extremely hazardous for early development and health (Damalas and Eleftherohorinos 2011).

8.5 Types of Persistent Organic Pollutants

Several types of persistent organic pollutants being produced for different purposes, such as organochlorine pesticides (OCPs) used in agriculture to control pests, manufacturing of polychlorinated biphenyls (PCBs) to use in transformers and capacitors, similarly polybrominated diphenyl ethers (PBDEs) and dechlorane plus (DP) used as flame retardants in many products, and polychlorinated naphthalenes (PCNs) used as dielectrics, plasticizers, paper, and fabric preservatives, are of the major concern. Some of the highly harmful persistent organic pollutants are discussed here.

8.5.1 Polychlorinated Naphthalenes (PCNs)

Polychlorinated naphthalenes (PCNs) were first commercially produced about 100 years ago, for their applications in industrial processes as preservatives of wood, fabric, and paper; as plasticizers, to cut and grind fluids; as lubricants in dielectrics; and as electric cable insulation. These compounds are the members of a group of chlorinated aromatic hydrocarbons and consist of almost 75 congeners (Fig. 8.2). They have been recognized as persistent, toxic, and bioaccumulative in nature and hence considered as the harmful contaminants causing extremely severe effects on humans as well as wildlife (Fernandes et al. 2017). Ingestion of these contaminants is generally recognized as the main exposure route of these contaminants except there is direct or specific exposure.

Due to a vast number of similar properties as well as industrial use and applications, PCNs are generally associated with PCBs. Though, the occurrence

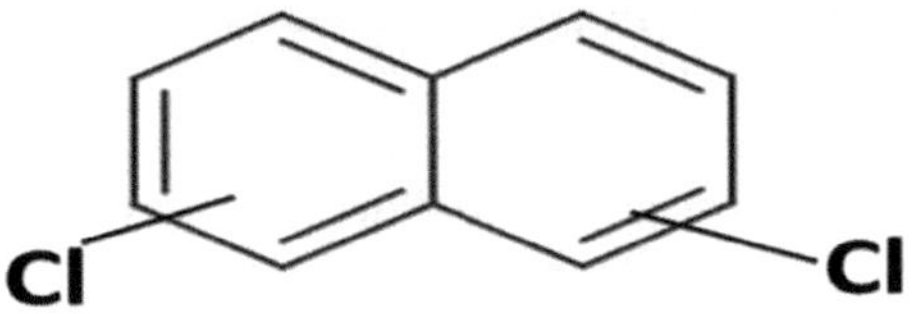

Homologue Group	Number of congeners
Monochloronaphthalene	2
Dichloronaphthalene	10
Trichloronaphthalene	14
Tetrachloronaphthalene	22
Pentachloronaphthalene	14
Hexachloronaphthalene	10
Heptachloronaphthalene	2
Octachloronaphthalene	1

Fig. 8.2 Generic structure and congener distribution of PCNs. (Source: Fernandes et al. 2017)

level of PCNs is comparatively lower than PCBs but toxicity level of PCNs is mostly similar to many congeners of PCBs (Helm et al. 2002). In addition to their intentional commercial production, PCNs are also unintentionally released as by-products from used electric equipment; as thermodynamic processes, e.g., in fly ash produced from incineration; and as unintentional presence as contaminant in the industrial chemical. PCNs are also found to be produced as a result of cooking process through heating of oils such as olive and peanut oil (Dong et al. 2013). A number of experimental studies have reported several toxic impacts of PCNs including carcinogenicity, hepatotoxicity, embryotoxicity, immunotoxicity, and mortality. Owing to their dioxin-like toxicity, PCNs are recognized by the Stockholm Convention as persistent organic pollutants, primarily to reduce their unintentional production and eventually to eliminate completely (Fernandes et al. 2017).

8.5.2 Polybrominated Diphenyl Ethers (PBDEs)

Polybrominated diphenyl ethers (PBDEs) are members of organobromine compounds, being produced to use as flame retardants in various products for reduction of fire risks. They are manufactured as a result of bromination of polymers' matrix generating the mixtures of isomers and homologues. Depending upon the required bromination degree, three most important PBDE mixtures are produced and sold in markets including penta-BDE, octa-BDE, and deca-BDE (Vonderheide et al. 2008). These mixtures are used for different purposes such as penta- and octa-BDE which are used in electronic housing and plastics; however deca-BDE is majorly used in textiles and polyurethane foams, PUFs (Lorber 2008). The ability of PBDEs to retard formation of flammable gases is the main reason of their widespread use in industrial products making them prevalent in the environment. Their existence in

Fig. 8.3 The general structural formula of PBDEs

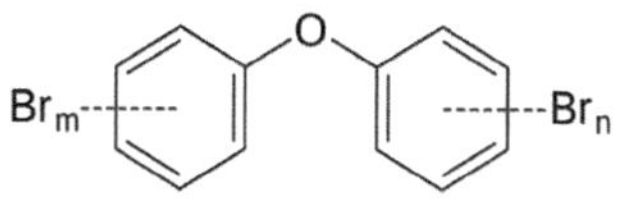

different environmental compartments has been confirmed by several studies (de Wit et al. 2010) (Fig. 8.3).

Due to the persistent, lipophilic, and global emergence of PBDEs in the environment, the levels of these compounds in marine and terrestrial ecosystems have been highly increased, making PBDEs the contaminants of emerging concern (Alaee 2006). These pollutants are probably being entered into the environment as a result of their volatilization from many products including televisions, furnishing foams, computers, and polyurethane (Jones-Otazo et al. 2005). Excessive e-waste recycling taking place in developing countries is also a major primary source of PBDE emissions into the environment (Wong et al. 2007). The occurrence of PBDEs in various compartments of the environment has been reported such as in water, sediments (Pei et al. 2018; Umulor et al. 2018; Trinh et al. 2019a, b), air (Bennett et al. 2015; Kurt-Karakus et al. 2017; Ghimire et al. 2019), and humans (Giulivo et al. 2018; Parry et al. 2018; Drobná et al. 2019). Briefly these pollutants are posing highly hazardous risks to humans and biota. European Union imposed ban on the commercial production of penta- and octa-BDE in august 2004, and eventually they were banned by Stockholm Convention (2009) because of their extensive harmful impacts on humans and other living organisms (Li et al. 2016). As for deca-BDE, it was recently listed in POP category in 2017 but still its production and usage is going on worldwide (Pei et al. 2018).

8.5.3 Polycyclic Aromatic Hydrocarbons (PAHs)

Polycyclic aromatic hydrocarbons (PAHs) are among those several chemicals that are toxic and persistent in nature having varying structures. These compounds are often colorless, pale, or white solids and mainly comprised of only hydrogen and carbon, having single or more than one benzene rings in angular, cluster, or linear orientation. The benzene rings more than one are mostly fused with each other through carbon atom pair shared in molecules of PAHs. They are categorized as "small PAHs," i.e., having aromatic rings up to six, or "large PAHs," i.e., consisting of aromatic rings more than six (International Agency for Research on Cancer 2010). The chemical structures of some mostly studied PAH compounds are shown in Fig. 8.4.

Different types of PAHs are used for different purposes such as acenaphthene which is used in manufacturing of pigments, plastics, dyes, and pharmaceuticals; anthracene is used to dilute wood preservatives; phenanthrene is for producing pesticides and resins; fluoranthene is for manufacturing agrochemicals. Similarly some of the PAHs are also present in roofing tar and asphalts for construction of

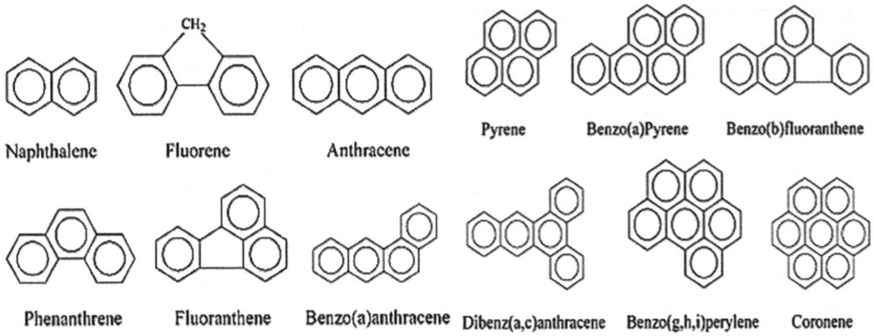

Fig. 8.4 Structural formulas of some common compounds of PAHs. (Source: Haritash and Kaushik 2009)

roads. Additionally, some products of PAHs are also used in liquid crystals and electronics (Abdel-Shafy and Mansour 2016).

PAHs generally have high boiling and melting point, low solubility in water, and low vapor pressure. The solubility of PAHs in water decreases with addition of each benzene ring (Masih et al. 2010). However, these compounds due to their lipophilic nature are highly soluble in organic solvents. Several functions are exhibited by PAHs such as physiological actions, light sensitivity, heat, and corrosion resistance (Akyüz and Çabuk 2010).

Some of the PAHs are produced in industries; however some of them are produced as a result of biological processes or as by-products of incomplete burning, either from natural sources (natural wildfires) or from anthropogenic sources of combustion such as smoke of cigarettes and emissions from automobiles. Hence, these persistent PAHs are reportedly found in soil, air, water, and lead to the exposure to humans causing severe health problems (Fig. 8.5) and living organisms (Baklanov et al. 2007).

8.5.4 Dechlorane Plus

Dechlorane plus (DP) is a member of polychlorinated compounds being used as flame retardant. It was first manufactured as alternative of dechlorane also known as mirex which was commercially used as flame retardant and pesticide but due to its toxicity potential its production was banned in 1970s (Hoh et al. 2006). Commercially produced dechlorane plus contains two isomers, i.e., syn-DP and anti-DP (Fig. 8.6), in 1:3 ratio approximately (Luo et al. 2013). DP is widely being used in domestic and industrial products as an additive flame retardant. The main usage of DP is in polymer materials with percentage composition of about 10% to 35% in industrial polymer goods (Xian et al. 2011).

DP detection in multiple environmental matrices indicates its abundance in the environment leading to its accumulation in humans. Long-range atmospheric

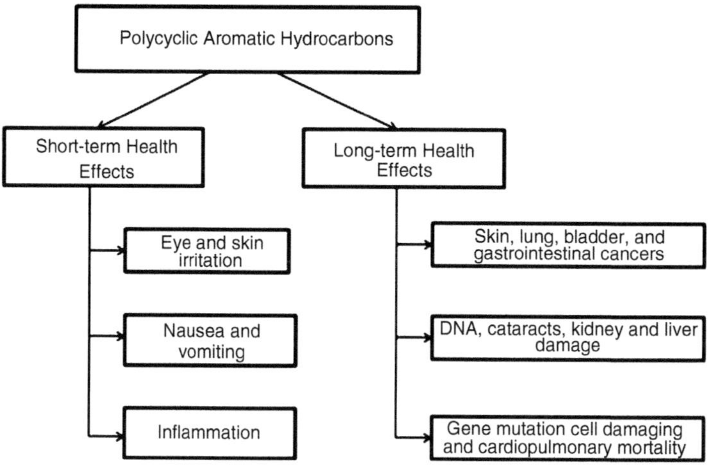

Fig. 8.5 Flowchart showing short- and long-term health effects of exposure to PAHs. (Source: Kim et al. 2013)

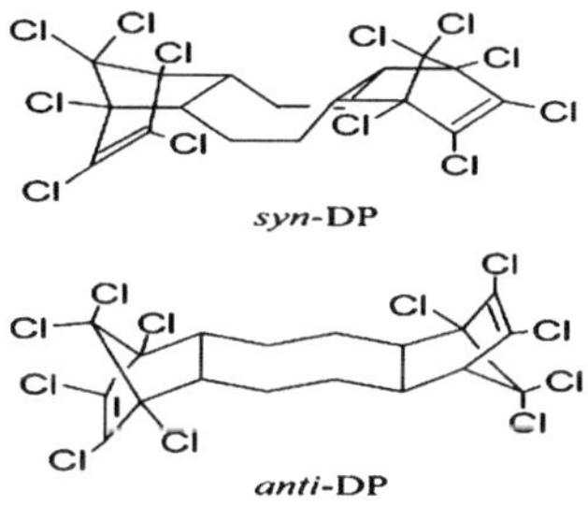

Fig. 8.6 The general structures of syn-DP and anti-DP

transport of the chemical may well explain the detection of DP in distant areas including the Arctic. Existing data on the concentrations of DP in marine ecosystem and ambient air suggests that DP could become a worldwide emerging contaminant for the environment. Although the environmental level of DP contamination is small, a clear source-related increase in DP concentrations is apparent in nearly all environmental matrices including soil (Wu et al. 2009), sediments (Shen et al. 2011), and air (Yu et al. 2015).

8.5.5 Polychlorinated Biphenyls (PCBs)

Polychlorinated biphenyls (PCBs) are industrial chemicals primarily produced during 1930 and 1977 in the United States (Hopf et al. 2009). PCBs were mainly used as lubricants and coolants in electrical equipment such as transformers and capacitors due to their general characteristics of being heat stable and chemically inert. Owing to their unique properties such as chemical stability, electrical insulating ability, high

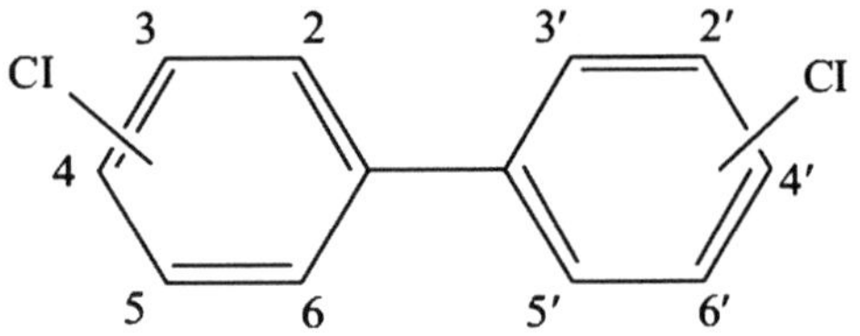

Fig. 8.7 The general structue of PCBs. (Source: Faroon and Ruiz 2016)

boiling point, and nonflammability, PCBs have been utilized in numerous industrial applications such as plasticizers in rubber, plastic, and paints, heat transfer, hydraulic and electrical equipment, carbonless papers, and several other commercial applications (U.S. EPA 2011). They have been made so useful because of their high dielectric properties. PCBs were produced as a mixture of PCB congeners and have around 209 congeners, the most common being Aroclor (Fig. 8.7).

PCBs have very high hydrophobic nature, i.e., having high octanol-water partition coefficient and low solubility in water that makes them highly affinitive toward suspended solids, which are ultimately deposited as surface sediments. These surface sediments are considered as main source of PCBs in riverine ecosystem (Desmet et al. 2012). The surface sediments indicate fresh input of PCBs whereas the core sediments suggest the historical levels of contamination. However, the intensity of PCB pollution in riverine ecosystem is measured by the surface waters (Hong et al. 2003; Montuori et al. 2016). The exposure of PCBs to humans has caused severe health impacts, endocrine disruption, carcinogenicity, immunological and pathological complications, and neurotoxic and reproductive failures (Sharma et al. 2009; Baqar et al. 2017). Polychlorinated biphenyls (PCBs) are included in those persistent organic pollutants that have been subjected to restrictions internationally for their production, usage, and emissions.

8.5.6 Organochlorine Pesticides (OCPs)

Pesticides are such chemical compounds that are used to destroy unwanted and harmful organisms, i.e., pests for the preservation of agricultural yield (EPA 2009). They can be categorized on the basis of their chemical characteristics. The several types of pesticides include organochlorines, carbamates, and organophosphates. Organochlorines are reported to be the most hazardous, harmful, and persistent organic pesticide among them all. The organochlorine pesticides (OCPs) mainly composed of chlorinated hydrocarbons have been used widely in control of mosquitos and pests in agriculture. Some important OCPs are dieldrin, DDT, chlordane, HCHs, methoxychlor, mirex, and endosulfans (Choi et al. 2016).

OCP pesticides are the most hazardous, semi-volatile organic compounds and compounds of greater concern worldwide. Due to their highly toxic, persistent, and lipophilic nature, bioaccumulation potential and the ability of transportation to the long distant areas, the production and usage of these pesticides have been banned internationally (Ali et al. 2014). These pesticides are also known as endocrine disrupters which are carcinogenic chemicals having highly detrimental impacts on the endocrine, nervous, and reproduction system. Due to these harmful health impacts, long-term persistence, and accumulation in the environment, these pesticides have been recognized as persistent organic pollutants in Stockholm Convention on persistent organic pollutants (Rani et al. 2017).

The high weight of chlorine in these pesticides makes them denser than water. These compounds are mostly prepared from chlorine or chlorine-containing chemicals such as hydrogen chloride. These pesticides make their way into organism's bodies through skin, lungs, or even absorbed through the gut wall. Endosulfans, HCHs, lindane, and cyclodienes can penetrate through skin easily; however DDT, methoxychlor, mirex, and dicofol have less absorption ability (Singh et al. 2016).

8.6 Remediation of Persistent Organic Pollutants

As the pollution caused by POPs in aquatic and terrestrial ecosystems has posed serious environmental and health impacts. It is necessary to remove these pollutants from environmental compartments such as from water, air, soil, and sediments. The river water is considered to be the direct measure of POPs due to the agricultural and industrial runoff. These highly toxic and hazardous chemicals in contaminated water cause a number of diseases in humans and other living organisms. So it is necessary to clean the water to be used by animals and humans. Such treatments should be adopted that not only transform the harmful complex chemicals into harmless products but also ensure compliance with legal threshold standards.

The conventional wastewater treatment methods are the common methods used to treat wastewater or drinking water and this treatment includes different stages such as preliminary, primary, secondary, and tertiary treatment processes to decontaminate water. These methods regulate pH, COD, BOD, removal of heavy metals, etc.; however these methods cannot remove POPs from wastewater effectively as these pollutants are highly resistant to conventional treatment methods such as filtration, coagulation, flocculation, and chlorine treatment (Benitez et al. 2001). However, several advanced treatment methods have been applied and analyzed for the removal of POPs such as bioremediation, adsorption process, membrane technology, and advanced oxidation processes (AOPs) (Ribeiro et al. 2015). Some of the methods that are proven to successfully remove POPs are discussed here.

8.6.1 Bioremediation

The biological treatment process using microorganisms, enzymes, or plants for degradation/decomposition and immobilization of organic waste compounds such as hydrocarbons, pesticides, or persistent organic pollutants (POPs) present in water or soil is referred to as bioremediation; it is an evolving technique for quality improvement of water and soil (Singh and Ward 2004). The bioremediation process depends upon various factors including the bioavailability, accessibility, and the potential of applied microorganisms for detoxifying, transforming to less harmful chemicals, or degrading completely (Prasad and Aranda 2018). The various reactions taking place during the biodegradation process include dehalogenation, oxidation, sulfur replacement by oxygen, hydrolysis, reduction or oxidation of nitrogen compounds, or OH radicle addition in benzene ring (Ortiz-Hernández et al. 2013). Bioremediation can take place either aerobically or anaerobically. The important factors that can affect the growth/metabolism of microorganisms include pH, nutrient availability, temperature, moisture content, and substrate availability. These factors in turn can affect the biodegradation process, thus limiting the efficiency of bioremediation. Some microorganisms also require other substrates for degradation of pollutants which is termed as co-metabolism (Hayatsu et al. 2000).

Bioremediation applied for the removal of persistent organic pollutants can be categorized as in situ and ex situ bioremediation. Treatment of the waste material at contaminated site is known as in situ bioremediation whereas treating the contaminated materials at some other place is known as ex situ bioremediation (Aggarwal et al. 1990). In situ bioremediation involves the degradation of complex and harmful organic pollutants into simpler compounds such as carbon dioxide and water or other less harmful substances under natural conditions. In situ bioremediation is much more sustainable, cost effective, and eco-friendly technique for the cleaning of the contaminated areas; however in case of ex situ bioremediation the costs are generally higher because of the excavation and transfer of contaminated materials. There are also differences in the biodegradation rates and uniformity of both in situ and ex situ bioremediation processes; however these methods largely depend upon metabolic potential of microorganisms (Jørgensen 2007). The in situ bioremediation processes include bio-attenuation, bio-stimulation, and bio-augmentation whereas the ex situ processes include composting, land farming, phytoremediation, bio-stimulation, and bio-restoration (Madsen 1991; Alexander 1999). However, selecting and using a bioremediation technique broadly depends upon the characteristics of pollutants, metabolic potential, and population of microorganisms and site conditions (Megharaj et al. 2011).

The bioremediation process can be classified on the basis of organisms used as bioremediator such as bacterial bioremediation (bacteria), mycoremediation (fungi) (Prasad 2017, 2018; Prasad et al. 2021), phytoremediation (plants) (Sarma et al. 2021), and phycoremediation (algae) (Velázquez-Fernández et al. 2012). Although every microorganism is capable of eliminating contaminants, some of the commonly used microorganisms (naturally occurring or bioengineered) include *Klebsiella*, *Alcaligenes*, *Shigella*, *Escherichia*, *Streptococcus*, *Bacillus*, *Enterobacter*, *Corynebacterium*, *Phanerochaete chrysosporium*, and *Staphylococcus* (Haritash and Kaushik 2009).

8.6.1.1 In Situ Bioremediation

The techniques of in situ bioremediation include bio-attenuation, bio-stimulation, and bio-augmentation. *Bio-attenuation* is the natural process of bioremediation in which the toxic contaminants are degraded into the less harmful substances by microorganisms, or by the reactions among naturally existing chemicals and contaminants, or pollutant's adsorption on geologic media. The process is highly contaminant specific such as for treatment of fuel compounds but not general for many groups of pollutants (Smets and Pritchard 2003). *Bio-stimulation* is the process of bioremediation in which the microorganisms are provided with essential nutrients (nitrogen, carbon, phosphorous, and potassium) and optimum conditions such as pH and availability of oxygen to enhance their growth and metabolism (Carberry and Wik 2001). *Bio-augmentation* is the process in which bacterial cultures are added for increasing the biodegradation rate of pollutants (Lamberts et al. 2008). This process is based on the concept of "soil activation" (by Otte et al. 1994) in which the biomass/microorganisms are cultivated from a contaminated soil to be used as inoculum for the treatment of the same contaminated soil.

8.6.1.2 Ex Situ Bioremediation

The commonly used ex situ bioremediation methods include composting, slurry phase treatment, bioreactors, etc. In *composting* the organic wastes or contaminated soil is excavated, layered in composting pad, and biologically degraded at higher temperatures ranging from 55 to 65°. To increase the biodegradation the bulking agents such as wood chips, manure, or vegetable waste are added. Heat production during the degradation process leads to the contaminant solubility and increase in the metabolism of organisms in compost. Presence of high substrate levels in compost can result in the co-metabolism of organic pollutants (Niti et al. 2013). In *bioreactors* the polluted soil/organic waste is agitated with nutrients and water in a mechanically operated bioreactor to enhance the microbial activity of microbes. The bioreactors of varying configuration can be used such as fermenters, bioslurry reactors, and prepared bed reactors. The biodegradation extent and rate are generally higher compared to in situ bioremediation, as these reactors are more manageable and controllable (Sharma 2012). In *slurry-phase treatment* the slurry of organic waste, soil sediments with water, and other bulking additives is prepared. The slurry is agitated in an engineered confined system to keep solids in suspension and to ensure the contact of microorganisms with pollutants (Sharma 2012).

8.6.1.3 Removal of Persistent Organic Pollutants

Several studies have reported the potential of removing the persistent organic pollutants from contaminated soils, sediments, and wastewater. Matsumoto et al. (2008, 2009) reviewed and reported the removal of endrin and dieldrin through bioremediation and phytoremediation from soil. Cao et al. (2013) reported the effective degradation of organochlorine (HCH) and organophosphate (parathion) pesticides from soil by using genetically modified *Sphingobium japonicum* UT 26. The detoxification and biodegradation of endosulfan an organochlorine pesticide by Alcaligenes faecalis strain JBW4 was reported by Kong et al. (2013, 2014). The Stain JBW4 effectively degraded about 87.5% and 83.9% of α-endosulfan and β-endosulfan, respectively. In another study by Odukkathil and Vasudevan (2016) soil contaminated with endosulfans was bioaugmented with a bacterial consortium (Bordetella petrii I GV 34, B. petrii II GV 36 and Achromobacter xyloxidans GV 47) in a glass reactor. Complete removal of α- and β-endosulfan was observed within

25 days. The degradation of bentazone, mecoprop, and dichlorprop in groundwater through biostimulation, i.e., by providing additional oxygen, was observed in a study by Levi et al. (2014). The 98% biodegradation and transformation of methoxychlor and DDT through reductive dechlorination by using human intestinal bacterium *Eubacterium limosum* under strict anaerobic conditions was observed within 16 days (Yim et al. 2008). The degradation of DDT, DDD, and DDE by a bacterial strain DDT-6 (D6), *Sphingobacterium* sp., was also reported by Fang et al. (2010). The 94.4 % degradation of DDT by fungal culture (*Fusarium oxysporum*) was observed in the study by Kulshrestha and Kumari (2010). Recently, Mansouri et al. (2017) have provided a detailed overview about different bioremediation approaches and their potential degrading of DDTs and its metabolites.

8.6.2 Adsorption Process

Adsorption process is considered as the most effective and highly recognized method of wastewater as compared to other wastewater methods because of the cost effectiveness and flexible and simple operations and designs. In addition, no harmful by-products are generated in the adsorption process. This is an effective method, based on equilibrium separation phenomenon (Fig. 8.8), to remove such waste pollutants that are nonbiodegradable and these pollutants include metal ions, pesticides, phenolic wastes, dyes, and herbicides discharged from domestic, agricultural, and industrial sources. Several types of adsorbents used for adsorption process include activated carbons, wood and fly ash, biomaterials, and clay minerals (Rasalingam et al. 2014).

The adsorption process largely depends upon the porosity, surface area, and number of sites available. Carbonaceous materials are considered as the major adsorbents for being able to adsorb various organic compounds. Activated carbon

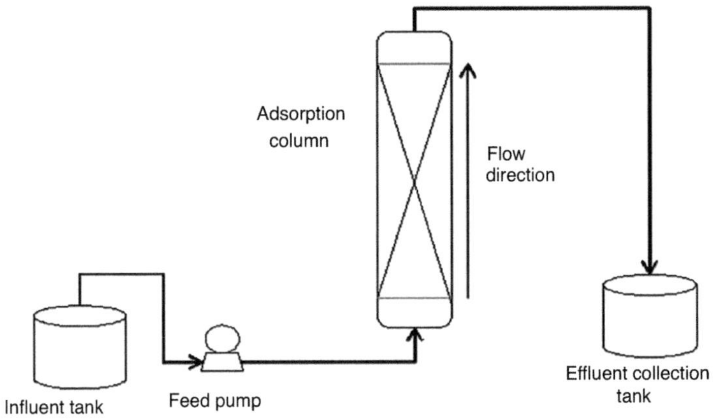

Fig. 8.8 Schematic representation of adsorption process in wastewater treatment. (Source: Jalani et al. 2016)

is highly efficient for removing pesticides and other organic compounds from wastewaters due to its large surface area and porosity. Vast varieties of activated carbons have been used including powdered activated carbon (PAC), granular activated carbon (GAC), carbon filters, carbon cloth, and black carbon. Among them PAC and GAC have been used widely because it is considered that they possess great potential to remove a various pesticides effectively. Thus, today several types of activated carbons are being produced using cheap and renewable precursors from agricultural and industrial by-products (Ahmad et al. 2010).

A number of studies have reported the ability of lignocellulosic materials to accumulate heavy metals and organic compounds, e.g., pesticides and dyes. The adsorption process through these adsorbents generally occurred as a result of interactions of pesticides with carboxyl and hydroxyl groups especially abundant in lignin and polysaccharides, both containing almost 90% lignocellulosic materials (Ofomaja 2008). Similarly, many industrial waste materials have been investigated such as fly ash, carbon slurry, and sludge because they are less costly and locally available to be used as adsorbents for removal of organic pollutants. Some of cost-effective industrial and agriculture-based absorbents reported in literature are mentioned in Table 8.1.

Natural clay minerals are also considered as the strong adsorbents as they are less costly and abundant in nature and have potential for high sorption and ion exchange. The interest in using clay minerals as adsorbents has been increased from some recent years for the removal of inorganic as well as organic pollutants and these clay materials include clinoptilolite, kerolite, cloisite, montmorillonite, and faujasite. These materials have been recognized as adsorbents due to their potential of immobilizing organic contaminants (Sanchez-Martin et al. 2006; Suciu and Capri 2009), removing pesticides from wastewater (Cruz-Guzmán et al. 2005), and photostablizing sorbed pesticides (Margulies et al. 1992). Adsorption process has easily operated design, low cost, and less land requirements and is broadly used for the removal of organic and inorganic waste materials form industrial wastewaters. Also several low-cost adsorbents have been identified such as waste materials from commercial, agricultural, and industrial activities, thus reducing the cost of adsorbent materials and waste management as well (Rashed 2013). However, the fouling of adsorbents, generation of toxic sludge, and high maintenance cost are some demerits of this process (Dağdelen et al. 2014)

8.6.3 Membrane Technology

Membrane technology is recognized as the efficient and alternative treatment technology to conventional methods in water and wastewater treatment which has progressed since last 25 years. Membrane technology includes variety of processes for separation of several chemical substances through synthetic membranes to purify water (Li et al. 2008). Membranes act as barriers and separate two phases, i.e., allowing certain components to pass; however other components are retained from given mixture of feed. Electrical or chemical potential, temperature, and gradient

Table 8.1 Different types and examples of commonly used adsorbents

Types of adsorbents	Examples of adsorbents	References
Carbonaceous adsorbents	Powdered activated carbons (PAC)	Kouras et al. (1998), Choi et al. (2008a, b), Bonvin et al. (2016) and Streicher et al. (2016)
	Granular activated carbons (GAC)	Sotelo et al. (2002), Pham et al. (2013) and Meinel et al. (2015)
	Carbon cloth	Ayranci and Hoda (2005)
	Commercial activated carbon (CAC)	Hamadi et al. (2004) and Shanthi and Mahalakshmi (2012)
	Carbon fibers	Liu et al. (2010)
Agricultural waste adsorbents	Rice straw	Wang et al. (2007)
	Pine sawdust	Sidiras et al. (2011)
	Rice husk	Ahmaruzzaman (2009) and Ahmaruzzaman and Gupta (2011)
	Wood sawdust	Dulman and Cucu-Man (2009)
Industrial waste adsorbents	Coal fly ash	Singh (2009) and Li et al. (2013)
	Blast furnace slag	Gupta et al. (2002) and Kuwahara et al. (2013)
	Bagasse fly ash	Ngo et al. (2015) and Deokar et al. (2016)
Inorganic adsorbents	Organo-zeolite	Lemić et al. (2006) and Lule and Atalay (2014)
	Calcined hydrotalcite, HT500	Pavlovic et al. (2005) and Pérez et al. (2017)
	Activated clay	Hameed (2007)
Polymeric adsorbents	Methacrylonitrile (MAN)/DVB	Trochimczuk et al. (2003)
	Biomimetic fat cell (BFC)	Liyan et al. (2007, 2009)

pressure or concentration act as the driving forces in membranes for the transport of materials (Mulder 1996). These processes are based on the physical separation and efficiently work without addition of any chemicals in the injecting feeds or changing the phases, hence known as the best alternatives of conventional methods such as flocculation, coagulation precipitation, distillation, or biological treatment (Li et al. 2008). From some previous years membranes have considerably improved with enhanced performance and they are rapidly growing worldwide.

There are basically four types of membrane processes used for separation of liquid phases which include reverse osmosis (RO), ultrafiltration (UF), nanofiltration (NF), and microfiltration (MF). These processes are based on the hydraulic pressure application which act as driving forces for transportation of mass. The permeability and retention of elements greatly depend upon the nature of membranes as the separation of these components depends upon their interaction with membranes, particle size, molar masses, and chemical affinity. Membrane technology can be used for different types of wastewater and has high processing and separation efficiency. There is no need of additives, thus limiting the generation of additional waste products. The membrane devices occupy less surface area and can be easily maintained and operated. However, membrane technology also has some limitations such as membrane fouling, high operational and maintenance cost, expansive membrane materials, and their quick exhaustion (Gao 2016). The mechanisms of these membrane processes are discussed here.

8.6.3.1 Reverse Osmosis

Reverse osmosis (RO) is one of the important and commonly used methods of membrane technology. The first RO process was applied in 1960s for desalination of water (Li et al. 2008). This is a pressure-driven method to remove smaller particles, dissolved solids, and organic pollutants from water by using such selective membranes that only allow molecules of water or solvents to pass and reject all other particles such as larger molecules or ions. Such types of membranes are called as semipermeable membranes (Warsinger et al. 2016).

Reverse osmoses is opposite of osmosis process; in this process the solvent is forced from the area of concentrated solution to the less concentrated or dilute solution by applying the mechanical force greater than the osmotic pressure (Fig. 8.9). The osmotic pressure depends upon the temperature, concentration, and species of solution irrespective of the semipermeable membranes. The RO membranes in high-pressure conditions only allow the water molecules to pass through it but it rejects other organic and inorganic particles and bacteria. RO membranes have very small pores and capable of removing particles of 0.1 nm size or even smaller (Yan et al. 2016). There are also some delimitations of this process such as it is costly as compared to other membrane technologies. It has high energy demand for application of high pressure, e.g., greater than 2000 kPa for maintenance of the permeate flux (Sch€afer et al. 2005), and fouling of membranes often occurs that affects the efficiency of the process (Liu et al. 2017).

8.6.3.2 Microfiltration (MF)

Microfiltration is a process similar to reverse osmosis as it also works with application of pressure. The microporous membranes in microfiltration can remove particles

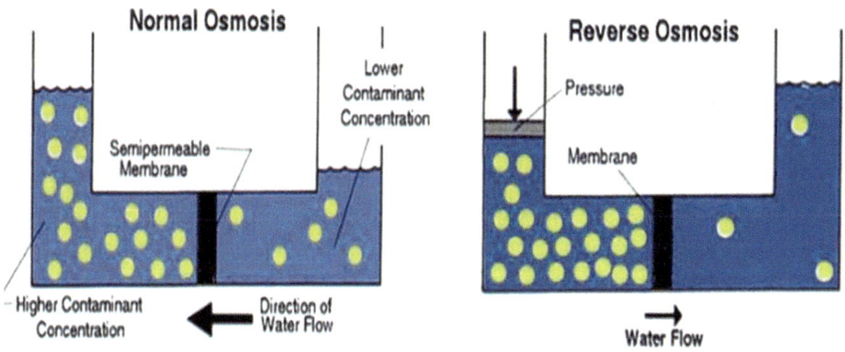

Fig. 8.9 Schematic diagram of the normal and reverse osmotic membranes. (http://www.rosmosis.com)

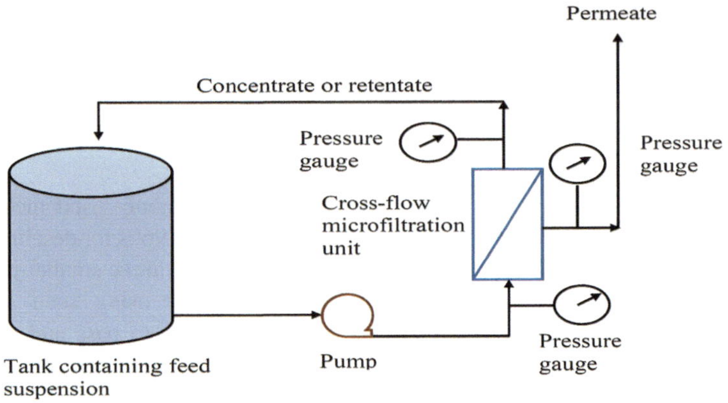

Fig. 8.10 Schematic diagram of microfiltration/ultrafiltration plant. (Source: Idan and Chatterjee 2015)

of about 0.1–0.2 μm in size which are mainly microparticles or nanoparticles (Werber et al. 2016). It is often regarded as the pretreatment process of nanofiltration and reverse osmosis processes. Along with the application of pretreatment and disinfection processes, this process can remove the organic contaminants and viruses efficiently from water (Torki et al. 2017). Schematic configuration of microfiltration or ultrafiltration plant in wastewater treatment is shown in Fig. 8.10.

The process of microfiltration is being used in the treatment of municipal wastewater (Xing et al. 2003), pharmaceutical wastewater (Hassan et al. 2016), anoxic pond effluent (Al-Malack et al. 1998), oxidation pond effluent (Katayon et al. 2007), oil and latex emulsions (Trinh et al. 2019a, b), and removal of toxic compounds from the drinking water (Han et al. 2002).

8.6.3.3 Ultrafiltration (UF)

Ultrafiltration is another type of membrane technology which is similar to RO and MF in operation (See Fig. 8.10). The membranes used in UF can retain the particles of 0.005~10 μm size and they are broadly used as water filters for the removal of suspended maters, pathogenic microorganisms, and macromolecules (Krüger et al. 2016). UF membranes are mostly defined in terms of the molecular weight of the rejected molecules by the pores of membranes whereas the membranes of MF are recognized based on the pore size (μm). In UF for the separation of particles (of molar masses around 1–300 kDa), the pressure >1 MPa is applied. The particles, solute molecules, and suspend solids with molecular weight less than 300 kDa are passed through the membranes and others are rejected (de Morais Coutinho et al. 2009).

8.6.3.4 Nanofiltration (NF)

Nanofiltration (NF) is a separation process using membranes with pore sizes between RO and UF membranes with application of pressure. This separation method has been developed recently around 1980s (Greenlee et al. 2009). The NF has ability to remove the ions from the feed solution that aid to enhance the osmotic pressure. These membranes have high potential of rejecting the organic materials and can reject the particles with 350–1000 Da molecular masses with the application of 4–20 MPa pressure (Koros et al. 1996; de Morais Coutinho et al. 2009).

It is being used for treatment of surface and groundwater (Teixeira and Rosa 2005), softening water by removing polyvalent cations (Causserand et al. 2005), removing the by-products of disinfection (Uyak et al. 2008) and inorganic compounds (Leo et al. 2011), and for removal of natural and synthetic organic matters (Choi et al. 2008a, b) such as pharmaceuticals (Nghiem et al. 2005), pesticides (Zhang et al. 2004; Plattner et al. 2018), and herbicides (Plakas et al. 2006; Benitez et al. 2009). To increase the treatment efficiency, it is suggested to apply pretreatment to reduce membrane fouling and increase pollutant removal.

8.6.4 Advanced Oxidation Processes (AOPs)

Advanced oxidation processes (AOPs) are the processes used for the treatment of water and wastewater which involve the generation of reactive hydroxyl (OH) radicals at room temperature and pressure in concentrations sufficient for decontaminating water (Oturan and Aaron 2014). These processes are based on aqueous oxidation systems in which OH radicals react and oxidize the organic pollutants to effectively degrade the complex toxic organic compounds into simple/less harmful and biodegradable compounds (Fig. 8.11). AOPs are now being considered as the most efficient, effective, and environmentally friendly methods for the removal of POPs. The factors that affect the degradation potential of OH radicals include volume of organic compounds, pH and turbidity of water, reaction period, and existence of hydroxyl ions' scavengers in the water (Bethi et al. 2016). AOPs were primarily used for the purification of drinking water in 1980s but then they were investigated and applied for their potential for treatment of wastewater from different sources. The generation of OH radicles (strong oxidants) in these processes is capable of removing inorganic pollutants and degrading the obstinate organic pollutants present in the wastewater (Deng and Zhao 2015).

Considering the wastewater characteristics and objectives of treatment, AOPs can be used as single or combined process with conventional or other oxidation processes. The commonly applied AOPs include ozone-based AOPs, Fenton and Fenton-like processes, and UV-based AOPs. Efficient degradation of persistent compounds, no generation of secondary waste materials, and limitation of hazardous chemical production in effluent are the most important advantages of these

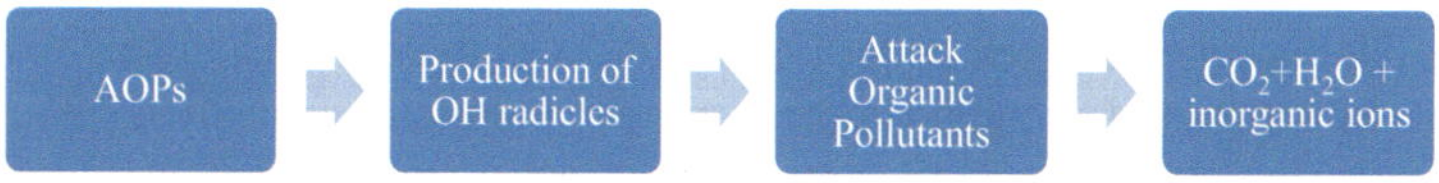

Fig. 8.11 The general working scheme of AOPs

processes. These advantages make AOPs better in comparison of other competing treatment methods such as membrane technology or chlorination of organic materials which result in the generation of secondary waste or organochlorinated compounds in effluents (Pablos et al. 2013; Deng and Zhao 2015). Some of the important AOPs categorized on the basis of methodology used for the production of oxidizing agents such as $O^{\bullet}$, $^{\bullet}OH$, and $^{\bullet}HO_2$ are briefly discussed here.

8.6.4.1 Chemical-Based AOPs

Ozone is considered as the most strong oxidizing agent (Oxidation potential = 2.07 V) and it reacts with the dissociated or ionized organics. Ozone generates the OH radicles in the effluents under certain circumstances, which then effectively degrade the organic compounds in the effluents by readily reacting with them (Gottschalk et al. 2009). Ozonation can be applied single or in combination with other methods such as peroxonation O_3/H_2O_2. In peroxonation the ozone gas is coupled with H_2O_2 leading to the production of more OH radicles (Eq. 8.1).

$$2O_3 + H_2O_2 \rightarrow 2OH^{\bullet} + O_2 \tag{8.1}$$

It is observed that this process is more efficient than simple ozonation as hydrogen peroxide has great potential of degrading ozone, thus producing OH in large quantities. The peroxonation had been applied by many researchers for the elimination of toxic compounds and micropollutants from the drinking, industrial, and groundwaters. Several studies have reported that ozone-based AOPs have ability to remove the POPs from polluted soil (Balawejder et al. 2014, 2016), model wastewater (Šimkovič et al. 2017), demineralized water (Derco et al. 2013, 2015; Giri et al. 2010), and natural water of river (Ormad et al. 2008, 2010). The important benefit of ozone-based AOPs is their potential to effectively degrade the vast variety of pollutants and easy mechanism, but high cost and being selective oxidants are their significant limitations. Similarly the presence of ionic compounds such as carbonates, nitrates, and bromides can limit the efficiency of these processes by reducing the potential of ozone decomposition and OH radicle generation.

The conventional Fenton method is a water treatment method which was first applied by Henry Fenton in 1894 using the mixture of H_2O_2 and iron salts in low quantity at room temperature and pressure without special equipment installations. The combination of hydrogen peroxide (H_2O_2) and ferric ions (Fe^{2+}) is termed as Fenton's reagent, in which iron reacts with H_2O_2 (Eq. 8.2) to break it down into OH radicles required for the degradation of organic pollutants (Oturan and Aaron 2014; Ribeiro et al. 2015).

$$Fe^{2+} + H_2O_2 \rightarrow OH^{\bullet} + OH^{-} + Fe^{3+} \tag{8.2}$$

The Fenton process, even being known for its complex mechanisms, was being used for oxidation and decomposition of organic pollutants since the 1960s. Several studies have reported the effectiveness of this process such as treatment of wastewater (Ozdemir et al. 2008), decomposition of harmful organic compounds when combined with ascorbic acid (Hou et al. 2016), and discoloration of effluents by removing dyes (Nidheesh et al. 2013).

8.6.4.2 UV-Based AOPs

In general, the UV radiations are now being applied on effluents in the presence of various oxidants such as ozone (UV/O_3), hydrogen peroxide H_2O_2, photo-Fenton (UV/Fe^{2+}/H_2O_2), or catalysts such TiO_2. The photolysis of ozone (O_3/UV) is ozone-based AOP which has been broadly applied in the treatment of contaminated drinking water or wastewater to remove noxious POPs, e.g., phenolic compounds and pesticides, etc.

$$O_3 + hv + H_2O \rightarrow 2OH^{\bullet} + O_2 \tag{8.3}$$

The ozone has ability to absorb UV radiations; thus its photolysis results in the generation of significant amount of highly reactive, efficient, and oxidizing OH radicles (Eq. 8.3) (Oturan and Aaron 2014). These O_3/UV-based AOPs have been used to decontaminate various effluents from pesticides (Lafi and Al-Qodah 2006), pharmaceutical compounds (Gebhardt and Schröder 2007), endocrine disruptors (Cesaro and Belgiorno 2016), dyes (Hsing et al. 2007), and surfactants (Ikehata and El-Din 2004).

Similarly the photolysis of hydrogen peroxide (H_2O_2/UV) is another process of generating OH radicles to treat the wastewater or drinking water. H_2O_2 can absorb the UV radiations with wavelengths of about 200–300nm, which breaks down the oxygen bond of hydrogen peroxide (H_2O_2) producing OH radicles (Eq. 8.4).

$$H_2O_2 + hv \rightarrow 2OH^{\bullet} \tag{8.4}$$

Titanium dioxide (TiO_2) is a naturally occurring substance that has a potential of being used as a photocatalyst in water photocatalysis treatment. This material is cheap, easily produced, biologically inert semiconductor and has high chemical stability. TiO_2 when bombarded with UV radiations becomes photoexcited and eventually split down the water molecules into oxygen and hydrogen, leading to formation of electron-donating and electron-accepting species which permits the redox reactions. This phenomenon of titanium oxide has resulted in the development of photocatalysis as novel AOP technology to be employed in different environmental applications (Oturan and Aaron 2014).

There is also possibility of photoexciting the Fenton reagents by irradiation of UV to enhance the catalytic reduction of Fe^{3+} into Fe^{2+} in the aqueous solution of H_2O_2 and Fe^{3+} oxalate compounds, resulting in the formation of large amount of OH radicles (Eq. 8.5); this is termed as photo-Fenton process. In this process UV radiations are also capable of decomposing hydrogen peroxide directly into OH radicles as happened in H_2O_2/UV.

$$Fe^{2+} + H_2O_2 + UV \rightarrow OH^{\bullet} + OH^{-} + Fe^{3+} \quad (8.5)$$

These photo-Fenton AOPs have been applied for degradation of various pesticides (organochlorine, phenylurea, and organophosphorous) as well as pharmaceutical compounds such as diclofenac (Ribeiro et al. 2015). This combined process is found to be more efficient in terms of generating OH radicles, thus increasing the degradation of organic materials in the effluents in comparison of using only photolysis or Fenton process (Valcárcel et al. 2012). In addition, the decrease in the requirements of catalysts can compensate the overall cost of UV installation and reduce the sludge volume making this process more effective treatment method (Umar et al. 2010).

8.6.4.3 Sonochemical AOPs

The sonochemical AOPs are another emerging types of advanced treatment process use of ultrasounds for the degradation of organic pollutants. In these processes reactive oxidizing species (ROS) can be produced either directly through physical procedure or indirectly through a chemical mechanism. In physical mechanism the oxidizing radicles are generated as a result of direct sonolysis or sonication of water molecules (Eq. 8.6).

$$2H_2O + US \rightarrow 2OH^{\bullet} + 2H \quad (8.6)$$

This is also referred to as cavitation in which water bubbles are generated and then collapsed resulting in the highly powerful breaking forces. This process creates extreme conditions such as high temperature and pressure (T = 2000–5000 K, $P = 6 \times 104$ kPa) leading to the sonolysis of water that results in the formation of highly reactive radicles (Fig. 8.12). In indirect chemical mechanism homolytic

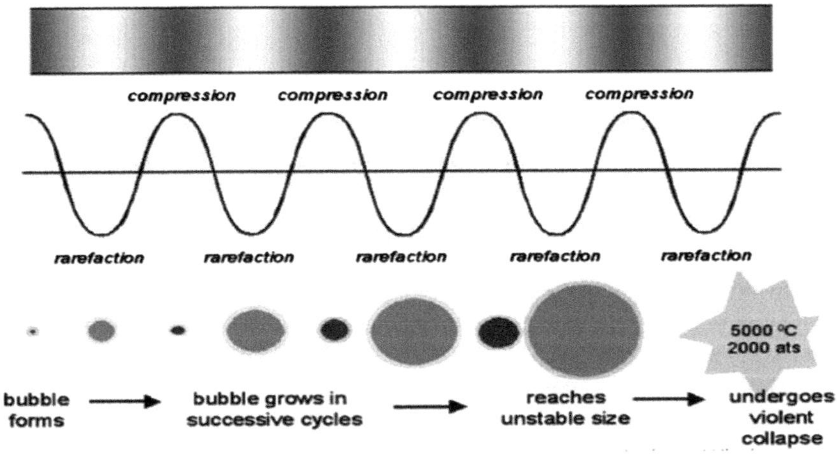

Fig. 8.12 The generation of oxidizing radicals through acoustic cavitation. (Source: Larpparisudthi et al. 2018)

cleavage of dioxygen and water molecules happens forming O, OH, and O_2H radicles (Oturan and Aaron 2014; Deng and Zhao 2015).

Recently, the sonochemical AOPs have been used broadly in wastewater, sludge, and water treatment to degrade organic pollutants as reported by Zouaghi et al. (2011); Ma (2012), and Dalhatou et al. (2015). Insufficient formation of OH radicles is the major drawback of this process; thus to overcome this problem it is being combined with other AOPs such as UV irradiation or by adding other oxidizing agents (O_2, H_2O_2) or Fenton reagent. These hybrid processes have led to the improvement of organic pollutants' degradation (Oturan and Aaron, 2014), and several studies have been conducted to investigate the efficiency of these hybrid processes such as Neppolian et al. (2012), Badmus et al. (2018), and Barik and Gogate (2018).

8.6.4.4 Electrochemical AOPs

Electrochemistry is considered as an efficient, effective, and environmental method that includes the use of electric energy and based on electron transfer. It can produce sufficient OH radicles in the effluent to degrade the organic contaminants effectively. These radicles can be generated electrochemically through electrically inducing Fenton's reagent (H_2O_2 + Fe ions) and this process is known as electro-Fenton process (EF process). In situ electrogeneration of H_2O_2 is occurred in the presence of iron ions by reduction of oxygen dissolved (O_2) in an acidic medium (Brillas et al. 2009). EF process is one of the environmentally friendly electrochemical AOPs which consists of electrically aided Fenton process. The process can be run either by adding Fenton reagent directly in the reactor along with an inert anode or by adding only hydrogen peroxide and providing Fe^{2+} ions through cast iron anode. The accumulation or transport of hydrogen peroxide can be prevented if its continuous supply to contaminated effluent is being ensured through two-electron redox reaction in acidic medium (Nidheesh and Gandhimathi 2012; Ribeiro et al. 2015).

The process can take place in either combined or separate electrochemical cells, though combined cells are more beneficial as oxidation reactions occurring on both cathode and anode can add up to the efficiency of the process to degrade/decompose organic pollutants. Additionally, the efficiency of the process can also be improved by coupling other AOPs with the EF process such as photolysis (photo-electro-Fenton), peroxi-coagulation, and solar-photo-electro-Fenton (Pipi et al. 2014; Ribeiro et al. 2015). These electrochemical Fenton processes have been used for degradation of several POPs including atrazine (Oturan et al. 2012), endosulfans (Errami et al. 2012), diruron (Pipi et al. 2014), pentachlorophenol (Govindan et al. 2014), and hexachlorocyclohexanes (HCHs) such as lindane (Wacławek et al. 2016; Dominguez et al. 2018a, b).

8.7 Conclusion

The contamination of the environment by persistent organic pollutants and their perilous impacts on human and other living organisms are evident from literature. Despite of the ban imposed on the use and production of POPs they are still in use in various parts of the world. Their historical as well as present inputs into different environmental compartments are leading to severe ecological damage arising the need of their remediation. These pollutants are reported to significantly resist the conventional treatment methods including filtration, use of flocculants, coagulation, or chemical treatment such as chlorination. Therefore various efficient removal methods and technologies have been analyzed and applied to remove these toxic pollutants from different environmental compartments such as water, wastewater, and soil. Methods including bioremediation, adsorption process, membrane technology (microfiltration, nanofiltration, and reverse osmosis), and advanced oxidation processes (AOPs) which include ozonation, peroxonation, photolysis of O_3, or H_2O_2 electrochemical and sonochemical AOPs are documented to efficiently remove these recalcitrant contaminants.

References

Abdel-Shafy HI, Mansour MS (2016) A review on polycyclic aromatic hydrocarbons: source, environmental impact, effect on human health and remediation. Egypt J Pet 25(1):107–123

Adeola FO (2004) Boon or bane? The environmental and health impacts of persistent organic pollutants (POPs). Hum Ecol Rev:27–35

Aggarwal PK, Means JL, Hinchee RE, Headington GL, Gavaskar AR (1990) Method to select chemicals for in-situ biodegradation of fuel hydrocarbons. Final report, November 1988–January 1990 (No. AD-A-227541/0/XAB). Battelle Columbus Labs., OH (USA)

Ahmad T, Rafatullah M, Ghazali A, Sulaiman O, Hashim R, Ahmad A (2010) Removal of pesticides from water and wastewater by different adsorbents: a review. J Environ Sci Health C 28(4):231–271

Ahmaruzzaman M (2009) Role of fly ash in the removal of organic pollutants from wastewater. Energy Fuel 23(3):1494–1511

Ahmaruzzaman M, Gupta VK (2011) Rice husk and its ash as low-cost adsorbents in water and wastewater treatment. Ind Eng Chem Res 50(24):13589–13613

Akyüz M, Çabuk H (2010) Gas–particle partitioning and seasonal variation of polycyclic aromatic hydrocarbons in the atmosphere of Zonguldak, Turkey. Sci Total Environ 408(22):5550–5558

Alaee M (2006) Recent progress in understanding of the levels, trends, fate and effects of BFRs in the environment

Alexander M (1999) Biodegradation and bioremediation. Gulf Professional Publishing

Alharbi OM, Khattab RA, Ali I (2018) Health and environmental effects of persistent organic pollutants. J Mol Liq 263:442–453

Ali U, Syed JH, Malik RN, Katsoyiannis A, Li J, Zhang G, Jones KC (2014) Organochlorine pesticides (OCPs) in South Asian region: a review. Sci Total Environ 476:705–717

Al-Malack MH, Anderson GK, Almasi A (1998) Treatment of anoxic pond effluent using crossflow microfiltration. Water Res 32(12):3738–3746

Arslan M, Imran A, Khan QM, Afzal M (2017) Plant–bacteria partnerships for the remediation of persistent organic pollutants. Environ Sci Pollut Res 24(5):4322–4336

Ayranci E, Hoda N (2005) Adsorption kinetics and isotherms of pesticides onto activated carbon-cloth. Chemosphere 60(11):1600–1607

Badmus KO, Tijani JO, Massima E, Petrik L (2018) Treatment of persistent organic pollutants in wastewater using hydrodynamic cavitation in synergy with advanced oxidation process. Environ Sci Pollut Res 25(8):7299–7314

Baklanov A, Hänninen O, Sløordal LH, Kukkonen J, Bjergene N, Fay B et al (2007) Integrated systems for forecasting urban meteorology, air pollution and population exposure. Atmos Chem Phys 7(3):855–874

Balawejder M, Antos P, Czyjt-Kuryło S, Józefczyk R, Pieniazek M (2014) A novel method for degradation of DDT in contaminated soil. Ozone Sci Eng 36(2):166–173. https://doi.org/10.1080/01919512.2013.861324

Balawejder M, Józefczyk R, Antos P, Pieniążek M (2016) Pilot-scale installation for remediation of DDT-contaminated soil. Ozone Sci Eng 38(4):272–278

Baqar M, Arslan M, Sadef Y, Mahmood A, Qadir A, Ahmad SR (2017) Persistent organic pollutants in Pakistan: potential threat to ecological integrities in terms of genotoxicity and oxidative stress. Hum Ecol Risk Assess Int J 23(6):1249–1271

Barik AJ, Gogate PR (2018) Hybrid treatment strategies for 2, 4, 6-trichlorophenol degradation based on combination of hydrodynamic cavitation and AOPs. Ultrason Sonochem 40:383–394

Behera BK, Prasad R (2020) Environmental technology and sustainability. Elsevier. ISBN 9780128191033. https://www.elsevier.com/books/environmental-technology-and-sustainability/behera/978-0-12-819103-3

Benitez FJ, Acero JL, Real FJ, Rubio FJ, Leal AI (2001) The role of hydroxyl radicals for the decomposition of p-hydroxy phenylacetic acid in aqueous solutions. Water Res 35 (5):1338–1343

Benitez FJ, Acero JL, Real FJ, Garcia C (2009) Removal of phenyl-urea herbicides in ultrapure water by ultrafiltration and nanofiltration processes. Water Res 43(2):267–276

Bennett DH, Moran RE, Wu X, Tulve NS, Clifton MS, Colón M et al (2015) Polybrominated diphenyl ether (PBDE) concentrations and resulting exposure in homes in C alifornia: relationships among passive air, surface wipe and dust concentrations, and temporal variability. Indoor Air 25(2):220–229

Bethi B, Sonawane SH, Rohit GS, Holkar CR, Pinjari DV, Bhanvase BA, Pandit AB (2016) Investigation of TiO2 photocatalyst performance for decolorization in the presence of hydrodynamic cavitation as hybrid AOP. Ultrason Sonochem 28:150–160

Bonvin F, Jost L, Randin L, Bonvin E, Kohn T (2016) Super-fine powdered activated carbon (SPAC) for efficient removal of micropollutants from wastewater treatment plant effluent. Water Res 90:90–99

Brillas E, Sirés I, Oturan MA (2009) Electro-Fenton process and related electrochemical technologies based on Fenton's reaction chemistry. Chem Rev 109(12):6570–6631

Cao X, Yang C, Liu R, Li Q, Zhang W, Liu J et al (2013) Simultaneous degradation of organophosphate and organochlorine pesticides by Sphingobium japonicum UT26 with surface-displayed organophosphorus hydrolase. Biodegradation 24(2):295–303

Carberry JB, Wik J (2001) Comparison of ex situ and in situ bioremediation of unsaturated soils contaminated by petroleum. J Environ Sci Health A 36(8):1491–1503

Carson R, Darling L, Darling L (1962) Silent spring. Houghton Mifflin, Boston

Causserand C, Aimar P, Cravedi JP, Singlande E (2005) Dichloroaniline retention by nanofiltration membranes. Water Res 39(8):1594–1600

Cesaro A, Belgiorno V (2016) Removal of endocrine disruptors from urban wastewater by advanced oxidation processes (AOPs): a review. Open Biotechnol J 10(1)

Choi JH, Fukushi K, Yamamoto K (2008a) A study on the removal of organic acids from wastewaters using nanofiltration membranes. Sep Purif Technol 59(1):17–25

Choi KJ, Kim SG, Kim SH (2008b) Removal of tetracycline and sulfonamide classes of antibiotic compound by powdered activated carbon. Environ Technol 29(3):333–342

Choi M, Lee I-S, Jung R-H (2016) Rapid determination of organochlorine pesticides in fish using selective pressurized liquid extraction and gas chromatography–mass spectrometry. Food Chem 205:1–8
Clara M, Kreuzinger N, Strenn B, Gans O, Kroiss H (2005) The solids retention time – a suitable design parameter to evaluate the capacity of wastewater treatment plants to remove micropollutants. Water Res 39(1):97–106
Coxall M (2014) Ethical eating: a complete guide to sustainable food. Malcolm Coxall, Cornelio Books
Cruz-Guzmán M, Celis R, Hermosin MC, Koskinen WC, Cornejo J (2005) Adsorption of pesticides from water by functionalized organobentonites. J Agric Food Chem 53(19):7502–7511
Dağdelen S, Acemioğlu B, Baran E, Koçer O (2014) Removal of remazol brilliant blue R from aqueous solution by pirina pretreated with nitric acid and commercial activated carbon. Water Air Soil Pollut 225(3):1899
Dalhatou S, Pétrier C, Laminsi S, Baup S (2015) Sonochemical removal of naphthol blue black azo dye: influence of parameters and effect of mineral ions. Int J Environ Sci Technol 12(1):35–44
Damalas CA, Eleftherohorinos IG (2011) Pesticide exposure, safety issues, and risk assessment indicators. Int J Environ Res Public Health 8(5):1402–1419
de Morais Coutinho C, Chiu MC, Basso RC, Ribeiro APB, Gonçalves LAG, Viotto LA (2009) State of art of the application of membrane technology to vegetable oils: a review. Food Res Int 42(5–6):536–550
de Wit CA, Herzke D, Vorkamp K (2010) Brominated flame retardants in the Arctic environment – trends and new candidates. Sci Total Environ 408(15):2885–2918
Deng Y, Zhao R (2015) Advanced oxidation processes (AOPs) in wastewater treatment. Curr Pollut Rep 1(3):167–176
Deokar SK, Mandavgane SA, Kulkarni BD (2016) Adsorptive removal of 2, 4-dichlorophenoxyacetic acid from aqueous solution using bagasse fly ash as adsorbent in batch and packed-bed techniques. Clean Techn Environ Policy 18(6):1971–1983
Derco J, Valičková M, Šilhárová K, Dudáš J, Luptáková A (2013) Removal of selected chlorinated micropollutants by ozonation. Chem Pap 67(12):1585–1593. https://doi.org/10.2478/s11696-013-0324-x
Derco J, Dudáš J, Valičková M, Šimovičová K, Kecskés J (2015) Removal of micropollutants by ozone-based processes. Chem Eng Proc Proc Intensif 94:78–84. https://doi.org/10.1016/j.cep.2015.03.014
Desmet M, Mourier B, Mahler BJ, Van Metre PC, Roux G, Persat H et al (2012) Spatial and temporal trends in PCBs in sediment along the lower Rhône River, France. Sci Total Environ 433:189–197
Dominguez CM, Oturan N, Romero A, Santos A, Oturan MA (2018a) Optimization of electro-Fenton process for effective degradation of organochlorine pesticide lindane. Catal Today 313:196–202
Dominguez CM, Oturan N, Romero A, Santos A, Oturan MA (2018b) Removal of lindane wastes by advanced electrochemical oxidation. Chemosphere 202:400–409
Dong S, Liu G, Zhang B, Gao L, Zheng M (2013) Formation of polychlorinated naphthalenes during the heating of cooking oil in the presence of high amounts of sucralose. Food Control 32 (1):1–5
Drobná B, Fabišiková A, Čonka K, Gago F, Oravcová P, Wimmerová S et al (2019) PBDE serum concentration and preschool maturity of children from Slovakia. Chemosphere
Dulman V, Cucu-Man SM (2009) Sorption of some textile dyes by beech wood sawdust. J Hazard Mater 162(2–3):1457–1464
El-Shahawi MS, Hamza A, Bashammakh AS, Al-Saggaf WT (2010) An overview on the accumulation, distribution, transformations, toxicity and analytical methods for the monitoring of persistent organic pollutants. Talanta 80(5):1587–1597
EPA (2009) What is a pesticide? http://www.epa.gov/opp00001/about/

Errami M, Salghi R, Zarrouk A, Chakir A, Al-Deyab SS, Hammouti B et al (2012) Electrochemical combustion of insecticides endosulfan and deltamethrin in aqueous medium using a boron-doped diamond anode. Int J Electrochem Sci 7:4272–4285
Fang H, Dong B, Yan H, Tang F, Yu Y (2010) Characterization of a bacterial strain capable of degrading DDT congeners and its use in bioremediation of contaminated soil. J Hazard Mater 184(1–3):281–289
Faroon O, Ruiz P (2016) Polychlorinated biphenyls: new evidence from the last decade. Toxicol Ind Health 32(11):1825–1847
Fernandes A, Rose M, Falandysz J (2017) Polychlorinated naphthalenes (PCNs) in food and humans. Environ Int 104:1–13
Fiedler H, Kallenborn R, de Boer J, Sydnes LK (2019) The Stockholm convention: a tool for the global regulation of persistent organic pollutants. Chem Int 41(2):4–11
Gao J (2016) Membrane separation technology for wastewater treatment and its study progress and development trend. In: 2016 4th international conference on mechanical materials and manufacturing engineering. Atlantis Press
Gebhardt W, Schröder HF (2007) Liquid chromatography–(tandem) mass spectrometry for the follow-up of the elimination of persistent pharmaceuticals during wastewater treatment applying biological wastewater treatment and advanced oxidation. J Chromatogr A 1160(1–2):34–43
Ghimire A, dela Cruz ALN, Wong R, Navasumrit P, Cormier S, Lomnicki SM (2019) Polybrominated diphenyl ethers (PBDEs) in ambient air samples at the electronic waste (e-waste) reclamation site. Waste Dispos Sustain Energy:1–11
Giri RR, Ozaki H, Ota S, Takanami R, Taniguchi S (2010) Degradation of common pharmaceuticals and personal care products in mixed solutions by advanced oxidation techniques. Int J Environ Sci Technol 7(2):251–260
Giulivo M, Suciu NA, Eljarrat E, Gatti M, Capri E, Barcelo D (2018) Ecological and human exposure assessment to PBDEs in Adige River. Environ Res 164:229–240
Gottschalk C, Libra JA, Saupe A (2009) Ozonation of water and waste water: a practical guide to understanding ozone and its applications. Wiley
Govindan K, Raja M, Noel M, James EJ (2014) Degradation of pentachlorophenol by hydroxyl radicals and sulfate radicals using electrochemical activation of peroxomonosulfate, peroxodisulfate and hydrogen peroxide. J Hazard Mater 272:42–51
Greenlee LF, Lawler DF, Freeman BD, Marrot B, Moulin P (2009) Reverse osmosis desalination: water sources, technology, and today's challenges. Water Res 43(9):2317–2348
Gupta VK, Jain CK, Ali I, Chandra S, Agarwal SJWR (2002) Removal of lindane and malathion from wastewater using bagasse fly ash – a sugar industry waste. Water Res 36(10):2483–2490
Hamadi NK, Swaminathan S, Chen XD (2004) Adsorption of paraquat dichloride from aqueous solution by activated carbon derived from used tires. J Hazard Mater 112(1–2):133–141
Hameed BH (2007) Equilibrium and kinetics studies of 2, 4, 6-trichlorophenol adsorption onto activated clay. Colloids Surf A Physicochem Eng Asp 307(1–3):45–52
Han B, Runnells T, Zimbron J, Wickramasinghe R (2002) Arsenic removal from drinking water by flocculation and microfiltration. Desalination 145(1–3):293–298
Haritash AK, Kaushik CP (2009) Biodegradation aspects of polycyclic aromatic hydrocarbons (PAHs): a review. J Hazard Mater 169(1–3):1–15
Hassan SS, Abdel-Shafy HI, Mansour MS (2016) Removal of pharmaceutical compounds from urine via chemical coagulation by green synthesized ZnO-nanoparticles followed by microfiltration for safe reuse. Arab J Chem
Hayatsu M, Hirano M, Tokuda S (2000) Involvement of two plasmids in fenitrothion degradation by Burkholderia sp. strain NF100. Appl Environ Microbiol 66(4):1737–1740
Helm PA, Bidleman TF, Stern GA, Koczanski K (2002) Polychlorinated naphthalenes and coplanar polychlorinated biphenyls in beluga whale (Delphinapterus leucas) and ringed seal (Phoca hispida) from the eastern Canadian Arctic. Environ Pollut 119(1):69–78
Hoh E, Zhu L, Hites RA (2006) Dechlorane Plus, a chlorinated flame retardant, in the Great Lakes. Environ Sci Technol 40(4):1184–1189

Hong SH, Yim UH, Shim WJ, Oh JR, Lee IS (2003) Horizontal and vertical distribution of PCBs and chlorinated pesticides in sediments from Masan Bay, Korea. Mar Pollut Bull 46(2):244–253

Hopf NB, Ruder AM, Succop P (2009) Background levels of polychlorinated biphenyls in the US population. Sci Total Environ 407(24):6109–6119

Hou X, Huang X, Ai Z, Zhao J, Zhang L (2016) Ascorbic acid/Fe@ Fe2O3: a highly efficient combined Fenton reagent to remove organic contaminants. J Hazard Mater 310:170–178

Hsing HJ, Chiang PC, Chang EE, Chen MY (2007) The decolorization and mineralization of Acid Orange 6 azo dye in aqueous solution by advanced oxidation processes: a comparative study. J Hazard Mater 141(1):8–16

Idan G, Chatterjee SG (2015) Application of a surface-renewal model to permeate-flux data for constant pressure cross-flow microfiltration with dean vortices. Braz J Chem Eng 32 (2):609–627

Ikehata K, El-Din MG (2004) Degradation of recalcitrant surfactants in wastewater by ozonation and advanced oxidation processes: a review. Ozone Sci Eng 26(4):327–343

International Agency for Research on Cancer (2010) Some non-heterocyclic polycyclic aromatic hydrocarbons and some related exposures, vol 92. IARC Press, International Agency for Research on Cancer

Jalani NF, Aziz AA, Wahab NA, Hassan WHW, Zainal NH (2016) Application of palm kernel shell activated carbon for the removal of pollutant and color in palm oil mill effluent treatment. J Earth Environ Health Sci 2(1):15

Jones-Otazo HA, Clarke JP, Diamond ML, Archbold JA, Ferguson G, Harner T et al (2005) Is house dust the missing exposure pathway for PBDEs? An analysis of the urban fate and human exposure to PBDEs. Environ Sci Technol 39(14):5121–5130

Jørgensen KS (2007) In situ bioremediation. Adv Appl Microbiol 61:285–305

Katayon S, Noor MMM, Tat WK, Halim GA, Thamer AM, Badronisa Y (2007) Effect of natural coagulant application on microfiltration performance in treatment of secondary oxidation pond effluent. Desalination 204(1–3):204–212

Kelly BC, Ikonomou MG, Blair JD, Morin AE, Gobas FA (2007) Food web–specific biomagnification of persistent organic pollutants. Science 317(5835):236–239

Kim KH, Jahan SA, Kabir E, Brown RJ (2013) A review of airborne polycyclic aromatic hydrocarbons (PAHs) and their human health effects. Environ Int 60:71–80

Kong L, Zhu S, Zhu L, Xie H, Su K, Yan T et al (2013) Biodegradation of organochlorine pesticide endosulfan by bacterial strain Alcaligenes faecalis JBW4. J Environ Sci 25(11):2257–2264

Kong L, Zhu S, Zhu L, Xie H, Wei K, Yan T et al (2014) Colonization of Alcaligenes faecalis strain JBW4 in natural soils and its detoxification of endosulfan. Appl Microbiol Biotechnol 98 (3):1407–1416

Koros WJ, Ma YH, Shimidzu T (1996) Terminology for membranes and membrane processes (IUPAC recommendations 1996). Pure Appl Chem 68(7):1479–1489

Kouras A, Zouboulis A, Samara C, Kouimtzis T (1998) Removal of pesticides from aqueous solutions by combined physicochemical processes – the behaviour of lindane. Environ Pollut 103(2–3):193–202

Krüger R, Vial D, Arifin D, Weber M, Heijnen M (2016) Novel ultrafiltration membranes from low-fouling copolymers for RO pretreatment applications. Desalin Water Treat 57 (48–49):23185–23195

Kulshrestha G, Kumari A (2010) Simultaneous degradation of mixed insecticides by mixed fungal culture isolated from sewage sludge. J Agric Food Chem 58(22):11852–11856

Kurt-Karakus PB, Alegria H, Jantunen L, Birgul A, Topcu A, Jones KC, Turgut C (2017) Polybrominated diphenyl ethers (PBDEs) and alternative flame retardants (NFRs) in indoor and outdoor air and indoor dust from Istanbul-Turkey: levels and an assessment of human exposure. Atmos Pollut Res 8(5):801–815

Kuwahara Y, Tamagawa S, Fujitani T, Yamashita H (2013) A novel conversion process for waste slag: synthesis of calcium silicate hydrate from blast furnace slag and its application as a versatile adsorbent for water purification. J Mater Chem A 1(24):7199–7210

Lafi WK, Al-Qodah Z (2006) Combined advanced oxidation and biological treatment processes for the removal of pesticides from aqueous solutions. J Hazard Mater 137(1):489–497
Lamberts RF, Johnsen AR, Andersen O, Christensen JH (2008) Univariate and multivariate characterization of heavy fuel oil weathering and biodegradation in soil. Environ Pollut 156 (2):297–305
Larpparisudthi OA, Mason TJ, Paniwnyk L (2018) Ultrasound wave-based AOPs. IWA Publishing, London
Lemić J, Kovačević D, Tomašević-Čanović M, Kovačević D, Stanić T, Pfend R (2006) Removal of atrazine, lindane and diazinone from water by organo-zeolites. Water Res 40(5):1079–1085
Leo CP, Chai WK, Mohammad AW, Qi Y, Hoedley AFA, Chai SP (2011) Phosphorus removal using nanofiltration membranes. Water Sci Technol 64(1):199–205
Letcher RJ, Bustnes JO, Dietz R, Jenssen BM, Jørgensen EH, Sonne C et al (2010) Exposure and effects assessment of persistent organohalogen contaminants in arctic wildlife and fish. Sci Total Environ 408(15):2995–3043
Levi S, Hybel AM, Bjerg PL, Albrechtsen HJ (2014) Stimulation of aerobic degradation of bentazone, mecoprop and dichlorprop by oxygen addition to aquifer sediment. Sci Total Environ 473:667–675
Li NN, Fane AG, Ho WW, Matsuura T (2008) Advanced membrane technology and applications. Wiley
Li H, Huang G, An C, Hu J, Yang S (2013) Removal of tannin from aqueous solution by adsorption onto treated coal fly ash: kinetic, equilibrium, and thermodynamic studies. Ind Eng Chem Res 52(45):15923–15931
Li WL, Ma WL, Jia HL, Hong WJ, Moon HB, Nakata H et al (2016) Polybrominated diphenyl ethers (PBDEs) in surface soils across five Asian countries: levels, spatial distribution, and source contribution. Environ Sci Technol 50(23):12779–12788
Liu QS, Zheng T, Wang P, Jiang JP, Li N (2010) Adsorption isotherm, kinetic and mechanism studies of some substituted phenols on activated carbon fibers. Chem Eng J 157(2–3):348–356
Liu G, Han K, Ye H, Zhu C, Gao Y, Liu Y, Zhou Y (2017) Graphene oxide/triethanolamine modified titanate nanowires as photocatalytic membrane for water treatment. Chem Eng J 320:74–80
Liyan S, Youcai Z, Guojian W, Bing L, Dongjie N, Xiaoli C (2007) Biomimetic fat cell (BFC) preparation and for lindane removal from aqueous solution. J Hazard Mater 146(1–2):289–294
Lorber M (2008) Exposure of Americans to polybrominated diphenyl ethers. J Expos Sci Environ Epidemiol 18(1):2
Liyan S, Youcai Z, Weimin S, Ziyang L (2009) Hydrophobic organic chemicals (HOCs) removal from biologically treated landfill leachate by powder-activated carbon (PAC), granular-activated carbon (GAC) and biomimetic fat cell (BFC). J Hazard Mater 163(2–3):1084–1089
Lule GM, Atalay MU (2014) Comparison of Fenitrothion and Trifluralin adsorption on Organo-zeolites and activated carbon. Part I: Pesticides adsorption isotherms on adsorbents. Part Sci Technol 32(4):418–425
Luo X, Wu J, Chen S, Mai B (2013) Species-specific bioaccumulation of polybrominated diphenyl ethers, hexabromocyclododecan and dechlorane plus in biota: a review. Sci Sin Chim 43 (3):291–304
Ma YS (2012) Short review: current trends and future challenges in the application of sono-Fenton oxidation for wastewater treatment. Sustain Environ Res 22(5):271–278
Madsen EL (1991) Determining in situ biodegradation. Environ Sci Technol 25(10):1662–1673
Mansouri A, Cregut M, Abbes C, Durand MJ, Landoulsi A, Thouand G (2017) The environmental issues of DDT pollution and bioremediation: a multidisciplinary review. Appl Biochem Biotechnol 181(1):309–339
Margulies L, Stern T, Rubin B, Ruzo LO (1992) Photostabilization of trifluralin adsorbed on a clay matrix. J Agric Food Chem 40(1):152–155
Mascarelli, A. (2013). Growing up with pesticides

Masih J, Masih A, Kulshrestha A, Singhvi R, Taneja A (2010) Characteristics of polycyclic aromatic hydrocarbons in indoor and outdoor atmosphere in the North central part of India. J Hazard Mater 177(1–3):190–198

Matsumoto E, Kawanaka Y, Yun SJ, Oyaizu H (2008) Isolation of dieldrin-and endrin-degrading bacteria using 1, 2-epoxycyclohexane as a structural analog of both compounds. Appl Microbiol Biotechnol 80(6):1095–1103

Matsumoto E, Kawanaka Y, Yun SJ, Oyaizu H (2009) Bioremediation of the organochlorine pesticides, dieldrin and endrin, and their occurrence in the environment. Appl Microbiol Biotechnol 84(2):205–216

Megharaj M, Ramakrishnan B, Venkateswarlu K, Sethunathan N, Naidu R (2011) Bioremediation approaches for organic pollutants: a critical perspective. Environ Int 37(8):1362–1375

Meinel F, Ruhl AS, Sperlich A, Zietzschmann F, Jekel M (2015) Pilot-scale investigation of micropollutant removal with granular and powdered activated carbon. Water Air Soil Pollut 226(1):2260

Ministry of Environment and Water (2006) Basic characteristics of persistent organic pollutants (POPS)

Montuori P, Aurino S, Garzonio F, Triassi M (2016) Polychlorinated biphenyls and organochlorine pesticides in Tiber River and Estuary: occurrence, distribution and ecological risk. Sci Total Environ 571:1001–1016

Mulder M (1996) Basic principles of membrane technology. J Membr Sci 72(3):564

Neppolian B, Ashokkumar M, Tudela I, González-García J (2012) Hybrid sonochemical treatment of contaminated wastewater: sonophotochemical and sonoelectrochemical approaches. Part I: Description of the techniques. In: Advances in water treatment and pollution prevention. Springer, Dordrecht, pp 267–302

Nghiem LD, Schäfer AI, Elimelech M (2005) Pharmaceutical retention mechanisms by nanofiltration membranes. Environ Sci Technol 39(19):7698–7705

Ngo HH, Guo W, Zhang J, Liang S, Ton-That C, Zhang X (2015) Typical low cost biosorbents for adsorptive removal of specific organic pollutants from water. Bioresour Technol 182:353–363

Nidheesh PV, Gandhimathi R (2012) Trends in electro-Fenton process for water and wastewater treatment: an overview. Desalination 299:1–15

Nidheesh PV, Gandhimathi R, Ramesh ST (2013) Degradation of dyes from aqueous solution by Fenton processes: a review. Environ Sci Pollut Res 20(4):2099–2132

Niti C, Sunita S, Kamlesh K, Rakesh K (2013) Bioremediation: an emerging technology for remediation of pesticides. Res J Chem Environ 17:4

Noyes PD, McElwee MK, Miller HD, Clark BW, Van Tiem LA, Walcott KC et al (2009) The toxicology of climate change: environmental contaminants in a warming world. Environ Int 35 (6):971–986

Odukkathil G, Vasudevan N (2016) Residues of endosulfan in surface and subsurface agricultural soil and its bioremediation. J Environ Manag 165:72–80

Ofomaja AE (2008) Kinetic study and sorption mechanism of methylene blue and methyl violet onto mansonia (Mansonia altissima) wood sawdust. Chem Eng J 143(1–3):85–95

Ormad MP, Miguel N, Claver A, Matesanz JM, Ovelleiro JL (2008) Pesticides removal in the process of drinking water production. Chemosphere 71(1):97–106. https://doi.org/10.1016/j.chemosphere.2007.10.006

Ormad MP, Miguel N, Lanao M, Mosteo R, Ovelleiro JL (2010) Effect of application of ozone and ozone combined with hydrogen peroxide and titanium dioxide in the removal of pesticides from water. Ozone Sci Eng 32(1):25–32. https://doi.org/10.1080/01919510903482764

Ortiz-Hernández ML, Sánchez-Salinas E, Dantán-González E, Castrejón-Godínez ML (2013) Pesticide biodegradation: mechanisms, genetics and strategies to enhance the process. Biodegrad Life Sci:251–287

Otte MP, Gagnon J, Comeau Y, Matte N, Greer CW, Smason R (1994) Activation of an indigenous microbial consortium for bioaugmentation of pentachlorophenol/creosate contaminated soils. Appl Microbiol Biotechnol 40(6):926–932

Oturan MA, Aaron JJ (2014) Advanced oxidation processes in water/wastewater treatment: principles and applications. A review. Crit Rev Environ Sci Technol 44(23):2577–2641

Oturan N, Brillas E, Oturan MA (2012) Unprecedented total mineralization of atrazine and cyanuric acid by anodic oxidation and electro-Fenton with a boron-doped diamond anode. Environ Chem Lett 10(2):165–170

Ozdemir C, Sahinkaya S, Onucyildiz M (2008) Treatment of pesticide wastewater by physicochemical and Fenton processes. Asian J Chem 20(5):3795

Pablos C, Marugán J, van Grieken R, Serrano E (2013) Emerging micropollutant oxidation during disinfection processes using UV-C, UV-C/H2O2, UV-A/TiO2 and UV-A/TiO2/H2O2. Water Res 47(3):1237–1245

Parry E, Zota AR, Park JS, Woodruff TJ (2018) Polybrominated diphenyl ethers (PBDEs) and hydroxylated PBDE metabolites (OH-PBDEs): a six-year temporal trend in Northern California pregnant women. Chemosphere 195:777–783

Pavlovic I, Barriga C, Hermosín MC, Cornejo J, Ulibarri MA (2005) Adsorption of acidic pesticides 2, 4-D, Clopyralid and Picloram on calcined hydrotalcite. Appl Clay Sci 30 (2):125–133

Pei J, Yao H, Wang H, Li H, Lu S, Zhang X, Xiang X (2018) Polybrominated diphenyl ethers (PBDEs) in water, surface sediment, and suspended particulate matter from the Yellow River, China: levels, spatial and seasonal distribution, and source contribution. Mar Pollut Bull 129 (1):106–113

Pérez A, Otero R, Esquinas AR, Jiménez JR, Fernández JM (2017) Potential use of modified hydrotalcites as adsorbent of bentazon and metazachlor. Appl Clay Sci 141:300–307

Pham TT, Nguyen VA, Van der Bruggen B (2013) Pilot-scale evaluation of GAC adsorption using low-cost, high-performance materials for removal of pesticides and organic matter in drinking water production. J Environ Eng 139(7):958–965

Pipi AR, Sirés I, De Andrade AR, Brillas E (2014) Application of electrochemical advanced oxidation processes to the mineralization of the herbicide diuron. Chemosphere 109:49–55

Plakas KV, Karabelas AJ, Wintgens T, Melin T (2006) A study of selected herbicides retention by nanofiltration membranes-the role of organic fouling. J Membr Sci 284(1–2):291–300

Plattner J, Kazner C, Naidu G, Wintgens T, Vigneswaran S (2018) Pesticide and microbial contaminants of groundwater and their removal methods: a mini review. J Jaffna Sci Assoc (1):1

Prasad R (2017) Mycoremediation and environmental sustainability, vol 1. Springer International Publishing. ISBN 978–3–319-68957-9. https://link.springer.com/book/10.1007/978-3-319-68957-9

Prasad R (2018) Mycoremediation and environmental sustainability, vol 2. Springer International Publishing. ISBN 978–3–319-77386-5. https://www.springer.com/us/book/9783319773858

Prasad R, Aranda E (2018) Approaches in bioremediation: the new Era of environmental microbiology and nanobiotechnology. Springer International Publishing (978–3–030-02369-0). https://www.springer.com/gp/book/9783030023683

Prasad R, Nayak SC, Kharwar RN, Dubey NK (2021) Mycoremediation and environmental sustainability, vol 3. Springer International Publishing. ISBN 978–3–030-54421-8. https://www.springer.com/gp/book/9783030544218

Rani M, Shanker U, Jassal V (2017) Recent strategies for removal and degradation of persistent & toxic organochlorine pesticides using nanoparticles: a review. J Environ Manag 190:208–222

Rashed MN (2013) Adsorption technique for the removal of organic pollutants from water and wastewater. Org Pollutants-monitoring, risk and treatment, pp 167–194

Rasalingam S, Peng R, Koodali RT (2014) Removal of hazardous pollutants from wastewaters: applications of TiO2-SiO2 mixed oxide materials. J Nanomater 2014:10

Ribeiro AR, Nunes OC, Pereira MF, Silva AM (2015) An overview on the advanced oxidation processes applied for the treatment of water pollutants defined in the recently launched Directive 2013/39/EU. Environ Int 75:33–51

Rull RP, Ritz B (2003) Historical pesticide exposure in California using pesticide use reports and land-use surveys: an assessment of misclassification error and bias. Environ Health Perspect 111 (13):1582–1589

Sanchez-Martin MJ, Rodriguez-Cruz MS, Andrades MS, Sanchez-Camazano M (2006) Efficiency of different clay minerals modified with a cationic surfactant in the adsorption of pesticides: influence of clay type and pesticide hydrophobicity. Appl Clay Sci 31(3–4):216–228

Sarma H, Forid N, Prasad R, Prasad MNV, Ma LQ, Rinklebe J (2021) Enhancing phytoremediation of hazardous metal(loid)s using genome engineering CRISPR–Cas9 technology. J Hazard Mater. https://doi.org/10.1016/j.jhazmat.2021.125493

Sarwar M (2015) The dangers of pesticides associated with public health and preventing of the risks. Int J Bioinform Biomed Eng 1(2):130–136

Shanthi S, Mahalakshmi T (2012) Studies on the removal of malachite green and methylene blue dyes from aqueous solutions of their binary mixture by adsorbtion over commercial activated carbon and tamarind kernel powder. Int J Res Pharm Chem 2(2):289–298

Sharma S (2012) Bioremediation: features, strategies and applications. Asian J Pharm Life Sci ISSN 2231:4423

Sharma CM, Rosseland BO, Almvik M, Eklo OM (2009) Bioaccumulation of organochlorine pollutants in the fish community in Lake Årungen, Norway. Environ Pollut 157 (8–9):2452–2458

Shen L, Reiner EJ, Helm PA, Marvin CH, Hill B, Zhang X et al (2011) Historic trends of Dechloranes 602, 603, 604, Dechlorane Plus and other norbornene derivatives and their bioaccumulation potential in Lake Ontario. Environ Sci Technol 45(8):3333–3340

Sidiras D, Batzias F, Schroeder E, Ranjan R, Tsapatsis M (2011) Dye adsorption on autohydrolyzed pine sawdust in batch and fixed-bed systems. Chem Eng J 171(3):883–896

Šimkovič K, Derco J, Dudáš J, Urminská B (2017) Removal of selected organochlorine compounds by ozone-based processes. Chem Biochem Eng Q 31(2):161–165. https://doi.org/10.15255/CABEQ.2016.1025

Singh N (2009) Adsorption of herbicides on coal fly ash from aqueous solutions. J Hazard Mater 168(1):233–237

Singh A, Ward OP (2004) Biotechnology and bioremediation – an overview. In: Biodegradation and bioremediation. Springer, Berlin/Heidelberg, pp 1–17

Singh Z, Kaur J, Kaur R, Hundal SS (2016) Toxic effects of organochlorine pesticides: a review. Am J Biosci 4(3–1):11–18

Smets BF, Pritchard PH (2003) Elucidating the microbial component of natural attenuation. Curr Opin Biotechnol 14(3):283–288

Sotelo JL, Ovejero G, Delgado JA, Martınez I (2002) Adsorption of lindane from water onto GAC: effect of carbon loading on kinetic behavior. Chem Eng J 87(1):111–120

Streicher J, Ruhl AS, Gnirß R, Jekel M (2016) Where to dose powdered activated carbon in a wastewater treatment plant for organic micro-pollutant removal. Chemosphere 156:88–94

Suciu NA, Capri E (2009) Adsorption of chlorpyrifos, penconazole and metalaxyl from aqueous solution by modified clays. J Environ Sci Health B 44(6):525–532

Teixeira MR, Rosa MJ (2005) Microcystins removal by nanofiltration membranes. Sep Purif Technol 46(3):192–201

Torki M, Nazari N, Mohammadi T (2017) Evaluation of biological fouling of RO/MF membrane and methods to prevent it. Eur J Adv Eng Technol 4(9):707–710

Trinh MM, Tsai CL, Chang MB (2019a) Characterization of polybrominated diphenyl ethers (PBDEs) in various aqueous samples in Taiwan. Sci Total Environ 649:388–395

Trinh TA, Han Q, Ma Y, Chew JW (2019b) Microfiltration of oil emulsions stabilized by different surfactants. J Membr Sci 579:199–209

Trochimczuk AW, Streat M, Malik DJ (2003) Sorption of pesticides and herbicides on highly polar polymeric sorbents. Sep Sci Technol 38(8):1813–1827

U.S. Environmental Protection Agency (2001) Use of bioremediation at superfund sites. EPA 542-R-01-019, Washington, DC

Umar M, Aziz HA, Yusoff MS (2010) Trends in the use of Fenton, electro-Fenton and photo-Fenton for the treatment of landfill leachate. Waste Manag 30(11):2113–2121

Umulor SO, Obanya HE, Amaeze NH, Okoroafor CU (2018) Levels, spatial and seasonal distribution of polybrominated diphenyl ethers (PBDEs) in water, sediments and sarotherodon melanotheron from Ologe Lagoon. Asian J Environ Ecol:1–13

United State Environmental Protection Agency (US. EPA) (2011) Pesticides: international activities, 26 September 2011

US EPA (1990) Multimedia fate and transport modeling – overview. https://www.epa.gov/fera/multimedia-fate-and-transport-modeling-overview

Uyak V, Koyuncu I, Oktem I, Cakmakci M, Toroz I (2008) Removal of trihalomethanes from drinking water by nanofiltration membranes. J Hazard Mater 152(2):789–794

Valcárcel Y, Martínez F, González-Alonso S, Segura Y, Catalá M, Molina R et al (2012) Drugs of abuse in surface and tap waters of the Tagus River Basin: heterogeneous photo-Fenton process is effective in their degradation. Environ Int 41:35–43

Van Maele-Fabry G, Lantin AC, Hoet P, Lison D (2010) Childhood leukaemia and parental occupational exposure to pesticides: a systematic review and meta-analysis. Cancer Causes Control 21(6):787–809

Velázquez-Fernández JB, Martínez-Rizo AB, Ramírez-Sandoval M, Domínguez-Ojeda D (2012) Biodegradation and bioremediation of organic pesticides. In: Pesticides-recent trends in pesticide residue assay. IntechOpen

Vonderheide AP, Mueller KE, Meija J, Welsh GL (2008) Polybrominated diphenyl ethers: causes for concern and knowledge gaps regarding environmental distribution, fate and toxicity. Sci Total Environ 400(1–3):425–436

Wacławek S, Antoš V, Hrabák P, Černík M, Elliott D (2016) Remediation of hexachlorocyclohexanes by electrochemically activated persulfates. Environ Sci Pollut Res 23 (1):765–773

Walker CH (2008) Organic pollutants: an ecotoxicological perspective. CRC Press

Wang SL, Tzou YM, Lu YH, Sheng G (2007) Removal of 3-chlorophenol from water using rice-straw-based carbon. J Hazard Mater 147(1–2):313–318

Warsinger DM, Tow EW, Nayar KG, Maswadeh LA (2016) Energy efficiency of batch and semi-batch (CCRO) reverse osmosis desalination. Water Res 106:272–282

Werber JR, Deshmukh A, Elimelech M (2016) The critical need for increased selectivity, not increased water permeability, for desalination membranes. Environ Sci Technol Lett 3 (4):112–120

Wickerham EL, Lozoff B, Shao J, Kaciroti N, Xia Y, Meeker JD (2012) Reduced birth weight in relation to pesticide mixtures detected in cord blood of full-term infants. Environ Int 47:80–85

Wong MH, Wu SC, Deng WJ, Yu XZ, Luo Q, Leung AOW et al (2007) Export of toxic chemicals – a review of the case of uncontrolled electronic-waste recycling. Environ Pollut 149(2):131–140

Wu JP, Zhang Y, Luo XJ, Wang J, Chen SJ, Guan YT, Mai BX (2009) Isomer-specific bioaccumulation and trophic transfer of Dechlorane Plus in the freshwater food web from a highly contaminated site, South China. Environ Sci Technol 44(2):606–611

Xian Q, Siddique S, Li T, Feng YL, Takser L, Zhu J (2011) Sources and environmental behavior of dechlorane plus – a review. Environ Int 37(7):1273–1284

Xing CH, Wen XH, Qian Y, Sun D, Klose PS, Zhang XQ (2003) Fouling and cleaning of microfiltration membrane in municipal wastewater reclamation. Water Sci Technol 47 (1):263–270

Yan HM, Cao CY, Bai G, Bai W (2016) Seawater desalination technology route and analysis of production capacity for large commercial nuclear power plant. In: International conference Pacific Basin nuclear conference. Springer, Singapore, pp 865–872

Yim YJ, Seo J, Kang SI, Ahn JH, Hur HG (2008) Reductive dechlorination of methoxychlor and DDT by human intestinal bacterium Eubacterium limosum under anaerobic conditions. Arch Environ Contam Toxicol 54(3):406–411

Yu Y, Hung H, Alexandrou N, Roach P, Nordin K (2015) Multiyear measurements of flame retardants and organochlorine pesticides in air in Canada's western sub-arctic. Environ Sci Technol 49(14):8623–8630

Zhang Y, Van der Bruggen B, Chen GX, Braeken L, Vandecasteele C (2004) Removal of pesticides by nanofiltration: effect of the water matrix. Sep Purif Technol 38(2):163–172

Zouaghi R, David B, Suptil J, Boutiti A, Guittonneau S (2011) Sonochemical and sonocatalytic degradation of monolinuron in water. Ultrason Sonochem 18(5):1107–1112

Emerging Microfiber Pollution and Its Remediation

9

Sunanda Mishra, Prasant Kumar Rout, and Alok Prasad Das

Abstract

Extensive allocation of synthetic microfiber (SMF) particles in the environment has harmful ecological impacts. These are tiny threads of synthetic fabrics having diameter less than 10 μm and are categorized under secondary microplastics, which are mainly composed of polyester, nylon, acrylic, polyethylene terephthalate, and polypropylene. These micropollutants are of increasing concern, especially due to their release into aquatic environments, including rivers and ocean. Primary sources of these tiny particles are laundering of apparel products, household sludge, direct throwing away of unused garments into the rivers and oceans, cosmetics, and cleaning agents. Approximately 13 million tons MFs all over the world along with coastal waste are entering the ocean yearly from which around 3 million tons are entering through rivers. Once these particles enter into aquatic sources, these are mistakenly ingested by various aquatic species and finally enter into the human body through consumption of MF-contaminated aquatic foods. Advanced microscopic and spectroscopic techniques are inspected for the successful detection of these synthetic micropollutants. Recent researches are focusing on the degradation of synthetic fibers using microbial enzymes that can be employed as an effective method. Strategies for plummeting microfiber pollution mainly focus on capturing these particles at source and consequently scheming of cost-effective remediation technologies. In the future, investigations are also required to develop techniques for an in situ remediation of these pollutants by enhancing natural attenuation using native microorganisms.

S. Mishra · A. P. Das (✉)
Department of Life Science, Rama Devi Women's University, Bhubaneswar, Odisha, India

P. K. Rout
Department of Material Science and Engineering, Tripura Central University, Agartala, Tripura, India

R. Prasad (ed.), *Environmental Pollution and Remediation*, Environmental and Microbial Biotechnology, https://doi.org/10.1007/978-981-15-5499-5_9

Keywords

Microfiber · Pollution · Aquatic ecosystem · Human health · Microbial remediation

9.1 Introduction

Microfibers (MFs) are small synthetic apparel products having less than 10 micrometers in diameter and length approximately 0.5–2 cm. Polyester, nylon, acrylic, polypropylene, and rayon are the universally used synthetic fibers for manufacturing garments (Gago et al. 2018). Commercial and domestic laundering process is the principal sources of this pollution in a variety of water bodies. Due to lack of filtration system, these tiny fabric particles are discharged into rivers by the sludge treatment plants (Mason et al. 2016a, b). Fabrics experience disintegration when clothing is worn or mishandled, particularly following the exposure to the environments. Waste-fractured macrofiber which is prone to physical abrasion and mechanical disintegration due to ultraviolet, thermal, and oxidative weathering results in formation of microfragments of fiber. Every year, the clothing industries manufacture million tons of garment, and the manufacture of these artificial fabrics has been escalating annually. Synthetic filaments are now included into almost every contemporary apparel product; this trend is chiefly conspicuous in the modern-style industries. The global production of textile fibers can be categorized as synthetic fibers, cotton, and blend fibers, which account for about 60%, 30%, and 10%, respectively. Polyester fibers are predominantly used for the manufacturing of apparels and other textile products. The magnitude of fibers produced per annum is increasing day by day, and prevalent integration of these modern fabrics justifies greater concentration in the world's mark. The amount of polyester manufactured increased yearly from approximately 5.5 million tons to 75 million tons in just 40 years (Brown et al. 2011). The actual and estimated amount of MF pollution is shown in Fig. 9.1. When these synthetic fibers undergo laundering, small filaments of garments are discharged into drainages. Synthetic microfleece particles are major pollutants, as they can be up taken by many small aquatic lives. These tiny constituent parts are then absorbed in the intestine of these innocent animals. Fragments of these particles in water bodies can be the basis of diverse health problems such as esthetic issues, entanglement, and other reported health diseases (Carr 2017).

Recently, employment of microorganisms for the degradation of synthetic plastic particles has been developed. These biological agents and their enzyme machineries can be utilized as an effective means for degrading polymer squalor. Mostly bacterial and fungal species have distinct degradation capability for degrading synthetic polymers. Some of the microorganisms which are associated with polymer degradation has been reported such as some species of *Pseudomonas* sp., *Streptomyces sp.*, *Rhodococcus* sp., *Clostridium* sp., and *Butyrivibrio* sp. Similarly, *Aspergillus* sp., *Aspergillus* sp., and *Fusarium* sp. are prevalent fungal species (Pathak and Navneet 2017). On the other hand, the problem of this pollution has shaped a prospect for the

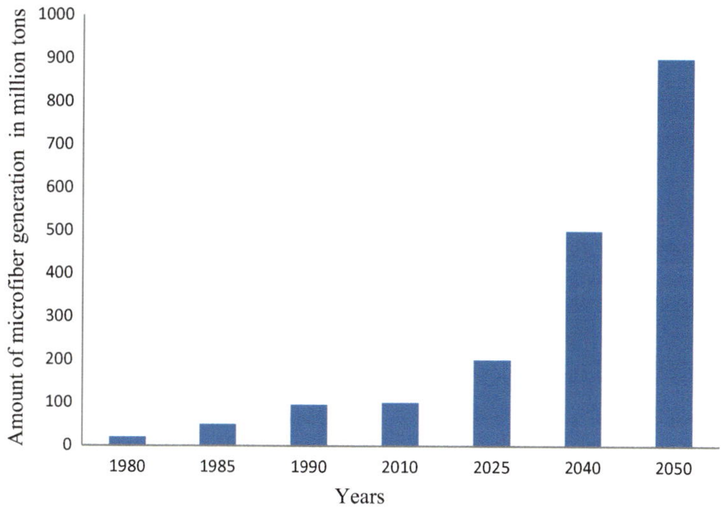

Fig. 9.1 Microfiber pollution: historical, actual, and projected rate

advancement of several unique solutions. Products like Guppyfriend Bag and Cora Balls have already been developed and urbanized (Gear of the future 2016; Rozalia Project 2017) for filtering the MF particles. The pollution requires solution at manifold levels with apparel product industries, customers themselves, and sewage water treatment plants.

9.2 Sources of Microfiber Pollution

The quality of the river and ocean water is deteriorating due to the direct release of urban, industrial, and other pollutants which can't be captured by wastewater treatment plants (WWTP). Some of the key and emerging sources of aquatic pollution are domestic laundering, textile and tire industries, personal care products, and household drainages (Almroth et al. 2017). Man-made textile fibers in contrast to natural fibers are nonbiodegradable, so these particles are persistent in nature (Mishra et al. 2019; Reed et al. 2018). Coastal tourism and commercial fishing all are sources of these plastic particles that can directly invade into various water streams (Cole et al. 2011). Various sources and the pathways of their entry into the environment are represented in Table 9.1.

Synthetic apparel products have developed an economical, long-lasting, and multipurpose nature over natural fabric (Henry et al. 2019). Additionally, the natural fabrics like cotton and wool are easily accessible, have lower prices, and contribute smaller amount of pollution to various water sources. The growing synthetic fabric waste in landfill and as garbage that are noticeable on shorelines and coastal waterways has come forward as a key global concern, predominantly in various water streams (Jambeck et al. 2015). The key sources of viscose are personal hygiene

Table 9.1 Sources and pathways of microfiber entry into the environment

Sl. no.	Types of product	Sources	Pathway of their entry into the environment	References
1	Apparel	Laundering	Domestic drainages, wastewater treatment plants	Mintenig et al. (2017) and Mishra et al. (2019)
2	Non-apparel	Fishing net, packaging, flags, sails	Directly to water sources	Carr (2017) and Yang et al. (2011)
	Non-apparel	Plastic bags, bottle cap, and bottles	From nearby landfills	Smith et al. (2018)
3	Non-apparel	Friction of tires	Storm water and wind	Carr (2017) and Galloway et al. (2017)
4	Non-apparel	Synthetic curtains and carpet	Domestic drainages, wastewater treatment plants	Talvitie et al. (2017a, b)
5	Non-apparel	Personal care products	Directly dumping off into water sources	Carr (2017)
6	Non-apparel	The old indoor wall painting, flakes, chips, and mattresses	Through air current	Carr (2017), Marklund et al. (2003) and Rauert et al. (2014)

products and modern clothing (Woodall et al. 2015; Henry et al. 2019). Separated MFs are arbitrarily dispersed, before settling in the atmosphere (Dris et al. 2016). Till date, these microparticles have been reported in water and mud samples from seashore areas throughout the world (Browne et al. 2011).

Leading synthetic fibers present abundantly in the environment were polyester, acrylic, polypropylene, polyethylene, and polyamide. Only a small portion, i.e., approximately 6%, of visible pieces of fabric corresponds to the total mass of fiber particles inflowing the oceans. Around 34.8% of secondary microplastics are releasing into the ocean through garment washing, and roughly 28.3% MF particles are releasing into the environment from the abrasion of tires (Boucher and Friot 2017). In 2016, approximately 65 million tons of plastics were produced for manufacturing synthetic garments (The Fiber Year, 2017). According to IUCN report 2017, approximately 700,000 microparticles are emitted per piece of clothing through domestic laundering, and each time synthetic jackets are washed, about 2 grams of microfleeces are released. Around 0.2 million tons of MFs are released per annum during manufacruring processes and household laundering. Subsequently these pollutants find their way to the ocean which generates high alert for the aquatic organisms. In a scientific investigation, it was concluded that nature and amount of MF discharge are largely dependent on the characteristics of the garments and approximately 300 mg of particles will be discharged per one kg of washing garment, that is, 1,500,000 numbers of particles. In the United States, approximately 8 billion microbead pieces are released from the municipal wastewater treatment plants per day (Mason et al. 2016a, b). It has been projected that approximately 5.25 trillion of micro-sized plastic particles having biomass of 270,000 tons are suspended in the world's oceans (Eriksen et al. 2013) from which 1.4 trillion are MF particles, and these values seem to accelerate in future (Horton et al. 2017).

The MF pollution, a man-made disaster, is increasing gradually and expected to envelop the total ocean surface. In actuality, micropieces of fabrics are the main repeatedly identified type of garbage in different water streams. Oceanic microplastic fibers occur in a variety of forms, size, and shape ranging from rough fragments to rods. Their incidence and occurrence have been already recognized in the oceans and coastal areas (Zalasiewicz et al. 2016; Thushari et al. 2017; Taylor et al. 2016). Although a few papers deal with MFs in ocean waters, the presence of these particles has been identified in many seawater samples (Waller et al. 2017). The presence of microplastic particles in Ross Sea of Antarctica was recently noticed from water samples (Cincinelli et al. 2017). In another investigation by Absher et al. (2018), they identified the presence of these synthetic MF particles in zooplankton samples collected from the Admiralty Bay of Antarctica (Isobe et al. 2017; Reed et al. 2018). Identifying the sources and pathways of the MF particles into the environment is necessary to manage the pollution and develop an effective remediation measure. Although the exact source of MF release and its pathway is still unknown to all of us, it was generally accepted that household gray waters from washing machine is the most important source.

9.3 Possible Pathways of Microfiber Entry into the Environment

Primary source includes fiber particle less than 5 mm in dimension emitted during making and extensive use of apparel products, and secondary sources include disintegration and crumbling of larger particles such as used garments that have been disposed. These synthetic fibers and particles having diameter $< 10\mu m$ that can easily escape through the filtration system of washing machine are entering oceans via rivers (Carr 2017). Once released into the surroundings, these synthetic plastic particles can be dispersed through air current and rain waters, especially with rainstorm runoff water. These small particles of fabric can be noticed on garments after laundry, uninterrupted floors, and display monitors of television, laptop, and mobile phones. The fiber particles are discarded while laundering and are emitted into various water bodies along with washing effluents. Due to lack of proper filtration system in presently used indoor washing machines, these tiny fragments can't be captured (Brannan 2014). MF litters having terrestrial source supply about 80–85% of the pollutants found in oceanic debris (Andrady 2011) including primary and secondary MFs along with cosmetics, fragments improperly disposed, and used plastics. These secondary microplastic particles used both in cosmetics can enter rivers via household sewage (Fendall and Sewell 2009). Small fabric particles can pierce into rivers directly or indirectly through wastewater effluent treatment plant (Browne et al. 2010).

As a consequence of the widespread utilization of artificial textiles and nonexistence of suitable management skills, invisible micro-sized pollutants are releasing into water streams, and food chains are becoming contaminated. When these particles with gray waters are emitted into household drainage systems, they are ultimately released into ocean water passing through rivers. Then these minute-sized particles with low density are extensively transported and distributed in marine water by the flow current. They are generally detected on the far-off places, such as sandy beaches, away from their sources. MF particles have already been reported in the body of a broad range of oceanic organisms and mammal (Choudhari 2018; Ghosh and Das 2018). The presence of these anthropogenic pollutants has been observed in seafood, commercial food-grade salts, and also in human stool (Thevenon et al. 2011; Wagner et al. 2014; Rochman et al. 2015; Kim et al. 2018).

Fibers are very prone to entrap in the gut components of aquatic living creatures that consume them (Lusher et al. 2017). The propensity for usual aquatic pollutants to attract to the surface of small microfiber particles has been previously reported (Bakir et al. 2014). These microfleeces can cause many injurious physical effects on humans and living organisms like entanglement and ingestion. After their emission into the environment, most of these MF particles were intertwined in a variety of small fish species, which are the chief food source for large marine animals. This is the key concern specified because of their crucial ecological role in marine food chain and webs. These particles can operate as transporter of contaminant such as fabric additives during manufacturing methods. When these marine animals are up taken by humans, the toxins accumulated in their flesh cause great health problems (Hankett et al. 2016). Aquatic environment has received more systematic attention than terrestrial, but the MF contamination on soil might be much more than marine pollution (Horton et al. 2017; Cobbing and Vicaire 2016). In reality, farming soils might accumulate more microparticles than deep-sea environment (Nizzetto et al. 2016). Understanding these pathways is significant to develop a sustainability evaluation that would be useful in scrutinizing and organizing the threats of this pollution from garments and other textile products. A detailed pathway of MF particle's entry into the environment is represented in Fig. 9.2.

9.4 Microfiber Pollution and Its Hazardous Effects

The composition of synthetic MFs and their fundamental relationship with anthropogenic activities result in significant impacts on the functioning of ecosystems. Within a WWTP, the outer surface of MF may be supplemented with pathogenic and other harmful microorganisms. These particles that can't be captured by the sludge management plants might flow into freshwater ecosystems (Talvitie et al. 2017a, b), and consequently, aquatic germs and microorganisms become inhabitants on MF surfaces. These microparticles might play as a vector for many diseases in aquatic species as well as in human species. Some of the obscure problems in the marine environment caused by these synthetic polymer wastes are represented in Fig. 9.3.

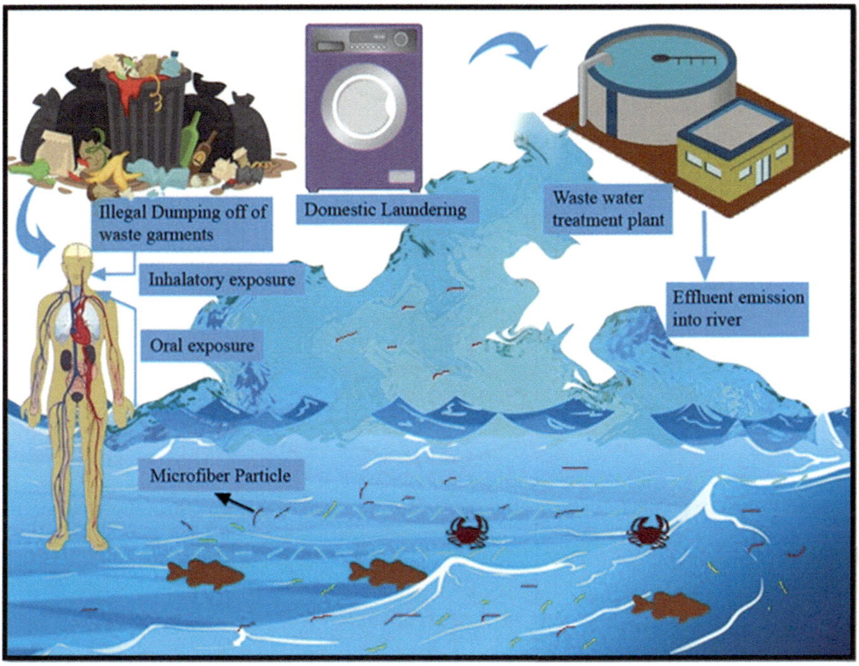

Fig. 9.2 A detailed pathway of MF particle's entry into the environment

The ultimate fate of washing machine effluent after domestic laundering is river water via sewage treatment plants. Studies concluded that the emission from sewage effluent indicates that approximately 70% of these particles can be retained in the sediment and sand filters or membrane bioreactors of sewage treatment plants (Talvitie et al. 2017a, b; Almroth et al. 2017; Bal et al. 2018). However, 1770 particles per hour or approximately 0.009 particles/L microplastic fibers are still detected in the outgoing effluent which will reach aquatic systems directly via the spreading of sludge (Magnusson and Norén 2014; Dris et al. 2015; Murphy et al. 2016). These tiny fragments have spongelike nature that they soak up other poisonous contaminants, and those particles are erroneously up taken by small aquatic animals. Marine organisms frequently consume these pollutants unintentionally while targeting plankton. In fish, these tiny particles can cause declined feeding rate, reproduction abnormalities, and poorer growth associated with liver toxicity (Sussarellu et al. 2016; Tosetto et al. 2016). The harmful effects of MF pollution on fish are depicted in Fig. 9.4.

Many researchers have been concluded that marine species consume microplastic fibers directly from ocean water (Cauwenberghe and Janssen 2014). According to some reports, microfabric particles were considerably smaller than the usual food of seabird. They can up take MF particles either accidentally or via food chain.

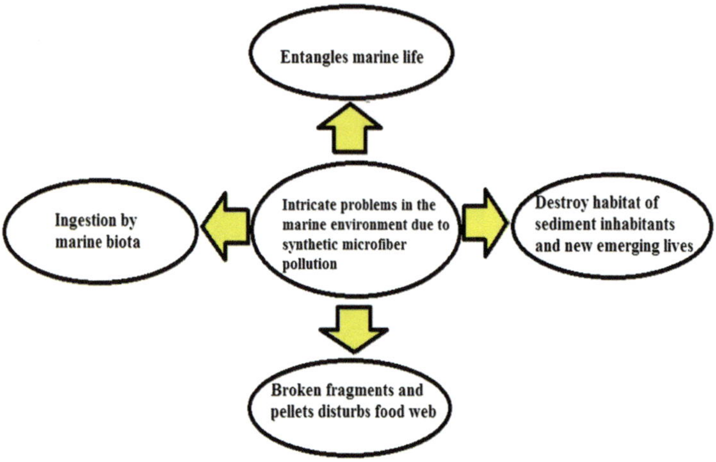

Fig. 9.3 Intricate problems in the marine environment caused by synthetic microfiber pollution

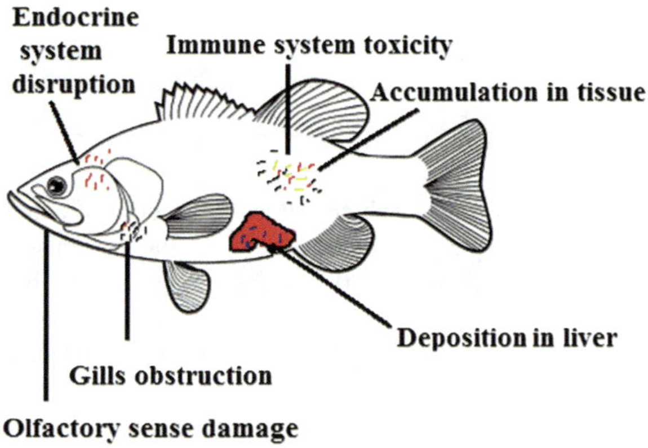

Fig. 9.4 Harmful effects of MF pollution on fish

Moreover, a primary numerical evaluation of trophic-level transfer of these particles found growing concentration in soils, earthworm casts, and chicken feces. The translocation of particles generally occurs by means of the intestinal lymphatic system, widely reported entrance for these particles into the animal body.

Transfer of ingested MF particles transversely trophic levels affects aquatic species such as fish used as human food (Nelms et al. 2018). Accidental MF particle ingestion might indeed be of concern to human health. However, the effects of elevated levels of pollution in human tissues require further research. In general, microfabric ingestion in place of feed will lead to gastrointestinal blockages. Chronic exposure of humans to microfiber particles from textiles may result in a level of bioaccumulation; as a result, these harmful toxic chemicals which are associated

Table 9.2 Harmful effect of microfiber exposure

Sl. no.	Animals affected by microfibers	Sources	Hazardous effects	References
1	Fish	River, marine water	Starvation, reproductive issues, blockage in gills and the digestive tract	Cauwenberghe and Janssen (2014) and Smith et al. (2018)
2	Bivalves	Marine environment	Gut blockage, decreased nutrition	Smith et al. (2018)
3	Crustaceans	Microfibers present in marine environment	Egg size will be reduced	Smith et al. (2018)
4	Shellfish	Microfibers present in marine environment	Reduced growth rate	Smith et al. (2018) and Messinger (2016)
5	Crabs	Microfibers present in marine environment	Obstructions in their gills	Cauwenberghe et al. (2015)

with toxicity, carcinogenicity, and mutagenicity may expose humans (Revel et al. 2018). Airborne textile microfibers are generally very large to be inhaled, but smaller particles can be inhaled easily (Dris et al. 2017). Airborne tiny textile fibers released along with dust are likely to be inhaled, and recently, plastic micro-sized fibers have been reported in human pulmonary tissues, signifying that some tiny fibers may enter through the respiratory tract (Gasperi et al. 2018). Harmful effects of this MF pollution on some aquatic and terrestrial organisms are represented in Table 9.2.

In this perspective, pollutants with broad-spectrum contamination may adversely influence many numbers of species and environmental purposes. Exposure of these pollutants is not restricted to oceanic environments. However, the effects of microbiome-colonized microplastic fiber particles in the aquatic environment remain mostly unexplored. There is presence of a number of evidences which conclude small microplastic fibers can be up taken by a variety of aquatic inhabitants of diverse ecosystems. Because there is a large number of organisms having a variety of habits and habitat can be affected by these pollutants, a number of techniques are employed for sampling procedure. As many researchers did not explain the efficient method for sampling, various methods are adopted. Some of the techniques for sampling large marine animals are represented in Table 9.3.

Table 9.3 Some of the techniques for sampling large marine animals

Sl. no.	Marine organisms	Sampling techniques	References
1	Fish	Local fisherman or fishing net	Karlsson et al. (2017)
2	Crustaceans	Bongo net or trawl net	Desforges et al. (2015)
3	Shrimp	Shrimp trawl or net	Murray and Cowie (2011)
4	Bivalves	Fauna	Oerlikon (2009)
		Trawl	
5	Lugworms	Bait pump	Gusmao et al. (2016)

9.5 Recent Remediation Strategies

All the ecological units are presently polluted with these synthetic polymer particles along with other harmful pollutants (Das et al. 2015; Das and Singh 2011; Ghosh et al. 2017). However, due to the scarcity of uniform, successful, and rapid sampling methods, very less research can be carried out to know about the geographical distribution in various habitats required to evaluate organism exposures. For illustration, there is no standard filter, aperture dimension leading to different sizes of microfabric particles for investigation. A variety of detection methodologies are the vital requirement for the ever-increasing pollution of synthetic microfleeces. Sophisticated optical, microscopic, and spectroscopic techniques are being successfully developed for quick detection of MF particles. The techniques that are frequently employed for the detection and characterization of micro-sized fabric are optical and stereomicroscope, FTIR, and Raman spectroscopy (Bradley et al. 2017). Almost up to 75% of the small MF particles, which are detected after visual characterization as particles, are usually confirmed as synthetic polymer by spectroscopic techniques. Spectroscopic technique like FTIR and Raman are strongly suggested against visual inspection for the detection of the chemical composition of the polymer sample (Mishra et al. 2019; Yang et al. 2011; Mohanty et al. 2018). Some currently developed methods for identification and characterization of microplastic particles are represented in Table 9.4.

To remove the little amount of these tiny microplastic particles from the sediment and ocean water samples, the most frequently described procedure is density separation and filtration. This process of separation of particles according to their densities was employed in most of studies scrutinized. This removal method normally included the following steps such as employment of aqueous solvent having different densities and agitating for a long period of time. The density partition was employed to separate the micropolymer particles of different densities. Filtration technique also can be used for separation process. The most common filtration techniques that are used for filtration are sieving and vacuum filtration. Visual categorization followed by sieving can be used prior to the density separation technique to lessen and divide them according to their magnitudes.

Around 80% of MPs were distinguished in effluent water emitted from wastewater treatment plant as compared to unprocessed water samples (Mintenig et al. 2019;

Table 9.4 Currently developed methods for identification and characterization of microplastic particles

Sl. no.	Methods	Techniques	Advantages	Disadvantages	References
1	Spectroscopic	FTIR	Characterizing irregular microplastic particles and applicable for small-sized microplastic	Time-consuming and skilled operator	Prata et al. (2019)
2	Spectroscopic	Raman spectroscopy	Detection of containing organic or inorganic fillers and pigments and those cannot be detected by FTIR spectroscopy	Long processing time, polymer heating and degradation, and fluorescence inference	Strungaru et al. (2019)
3	Thermal degradation	Pyrolysis-gas chromatography coupled with mass spectrometry (Py-GC/MS)	Determining of polymer compositions and organic plastic additives	Only applicable for certain selected types of synthetic polymers	Li et al. (2018)
4	Staining	Nile red staining	Short incubation time and high recovery efficiency	The microfibers are reported to be very difficult to stain by Nile red dye	Catarino et al. (2018)
5	Spectroscopic	Focal plane array (FPA)-assisted reflectance micro-Fourier-transform imaging	Pre-treatment using 30% H_2O_2; FPA is a detector that is useful for identifying the small-sized microplastic particles	Effective for the characterization of some polymers	Mintenig et al. (2017)

Woodall et al. 2015). Aquatic contamination due to SMFs and other man-made sources is a serious topic of concern, and important management equipment for the abolition of these impurities are of enormous significance (Das and Mishra 2008, 2010; Ghosh et al. 2016; Das et al. 2014). Prohibition of these particles from entering into freshwater streams by means of present methodology is exclusively high costs and operating expenditure. Membrane bioreactor, cross-flow filtration, density separation, and centrifugation are the developed methods that can exterminate these pollutants, but these methods consume a high level of energy. These tiny particles from large amount of samples can also be confiscated, but these methods cannot be employed to sieve water on a daily basis. Recently, some products like LUV-R, Guppyfriend Bag, and Cora Ball has been developed which can remove microplastic particle efficiently. If we use a Cora Ball along with other clothes in each washing cycle, it will strain approximately 26% minute fragments of MFs (Ahimsa Eco Solutions, 2018). An additional product was also developed as a solution of this pollution, that is, Guppyfriend Bag, designed specifically as a filter bag which can restrain tiny MF particles.

Other pollutants which are present in water sources are very prone to attach on the surface of MF particles (Chua et al. 2014; Lamichhane 2018). Additionally, a variety of fabric polymer chemical supplements also enter into freshwater ecosystems. These fabric particles from various sources enter into domestic drainage systems via wastewater management system which acts not only as an obstacle but also as entry path for MFs (Murphy et al. 2016) into various water sources. Although traditional methods that are used by WWTPs as waste management methods can eliminate these tiny pollutants from the sludge waters, still most of the particles are also retained in it (Carr et al. 2016). In spite of elevated diminution capability, these WWTPs can act as potential sources of these micropollutants, as a large quantity is discharged into river or directly into ocean water. But since the last few years, some advance treatment technologies were developed to improve the quality of discharged water from WWTPs and capture more amounts of marine debris particularly small microplastic particles. Some recent physical remediation technologies that were developed and employed in final stage of waste treatment processes are represented in Table 9.5.

Nowadays, emerging investigation is now beginning to enhance capability for degrading plastic polymers. However, a commercial-scale solution is required because of the amount of plastics accumulating in the environment, where they persist for many years (Hemalakshita et al. 2021). Observations of fabric marine debris in the United States (Ocean Conservancy 2013) specify that whereas natural fiber like cotton MF particles will vanish in few months and a wool fiber in few years, synthetic fiber particles take ten to thousands of years. When ingested, there is data available that, unlike man-made, natural fibers from textiles can be disintegrated within the body of the organism (Zhao et al. 2016). Further strong information on biodegradation of synthetic fabric fibers in diverse-type ecological and biological conditions is a significant requirement. Some commonly microplastic-polymer-particle-degrading microorganisms are represented in Table 9.6.

Currently, there is lack of neither government boundary for these small pieces of fibers present in drinking water nor any treatment technology for removing these particles. However, drinking water treatment plants are planned to remove these particles by developing potential hurdle for MF transfer of freshwater to drinking water for direct human consumption. But additionally, it is the required that wastewater treatment plants should have the capability to capture MPs releasing from household laundering (Napper and Thompson 2016), cosmetics, and other personal care products. Wastewater treatment plants purify the domestic drainage wastewaters in three steps including coagulation or flocculation, flotation, and

Table 9.5 Physical removal techniques in WWTPs

Sl. no.	Sources	Methods	Removal efficiency (in %)	References
1	Household drainages	Disc filter with pore size 10 mm	40	Talvitie et al. (2017a, b)
2	Domestic wastewater treatment plant	Tertiary gravity filters	97	Carr et al. (2016)
3	WWTPs	Rapid sand filters	97	Coppock et al. (2017)
4	WWTPs	Dissolved air flotation	95	Talvitie et al. (2017a, b)
5	Municipality sludge treatment plant (MSTP)	Membrane bioreactor	99	Mintenig et al. (2017)
6	WWTPs	Disc filter with pore size 10 mm	98	Talvitie et al. (2017a, b)
7	DWTP	Pile cloth media filtration	95	Mintenig et al. (2017)
8	Floatation extraction technique	Decantation	35	Coppock et al. (2017)
9	Floatation extraction technique	Elutriation, aeration, and centrifuge	98	Wessel et al. (2016)
10	Extraction technique	Separation funnel	80	Fries et al. (2013)

sand filtration, but unfortunately, these steps can't filter these MP fiber particles completely.

Global cooperation is desired to manage and clean synthetic fabric wastes from the ocean and to remove the primary source of marine microplastic pollution. Recent investigations offer reason for future hope to assess whether microbial involvement in microfabric biodegradation has begun to spread in the environment. Strategies should be developed for enhanced and effective bioremediation of synthetic MFs by application of microorganisms in natural environment.

9.6 Conclusion

The occurrence of these MF particles in all the spheres of earth is not easy to control, and the cross infectivity has already affected all the habitats. Methods providing a minimum infectivity are dangerous for a suitable assessment of their quantity in the environment. As these tiny fragments after their release into various water sources interact with planktons and sediments, both suspension and deposit feeders may accidentally consume these tiny particles. Research on microplastic and microfibers, specifically, is still in its infancy stage as there are many more unanswered problems. Solution to these questions is a prerequisite to extend a brighter depiction of their effect on the oceanic biota and developing an innovative technology.

Table 9.6 Microplastic-particle-degrading microorganisms

Sl. no.	Microorganisms	Types of polymer	References
1	*Pseudomonas aeruginosa*	Polyethylene	Kale et al. (2015)
2	*Pseudomonas fluorescens*	Polyvinyl chloride	Shah et al. (2008)
3	*Paucimonas lemoignei*	Poly(3-hydroxybutyrate)	Kale et al. (2015)
4	*Fusarium solani, Aureobasidium pullulans* sp.	Polyurethane	Shimao (2001)
5	*Acinetobacter gerneri, Aspergillus versicolor*	Polyurethane	Howard et al. (2012)
6	*Aspergillus niger, Penicillium* sp.	Plasticized PVC	Sachin and Mishra (2013)
7	*Aspergillus* sp.	Low-density polyethylene (LDPE)	Sachin and Mishra (2013)
8	*Streptomyces* sp.	LDPE	Howard et al. (2012)
9	*Pseudomonas fluorescence, P. aeruginosa, Penicillium simplicissimum*	Poly(ethylene adipate) (PEA) and polycaprolactone (PCL)	Ahmed (2018)
10	*Agromyces* sp.	Nylon 6 (oligomers)	Negoro (2012)

Microorganisms which are proficient in degrading inorganic materials should be incorporated to degrade plastic polymers and synthetic fabric particles. *Pseudomonas aeruginosa* is one of the extensively stated microorganisms for polymer degradation through biofilm formation. Mishandled plastic fiber waste is known to exert a variety of stresses on the environment, and as awareness of these stresses has grown, labors have been made by municipality, industry, government, and the general public to reduce fiber wastes. The discharged waters of the fabric industries have also long been documented as key sources of chemical pollutants, but microplastic textile fibers have received little environmental attention. Synthetic textile MFs are separated in each stair of a fabric life cycle, specifically when during laundering process (Boucher and Friot 2017).

Potential sources of MF's entry into the environment were determined, but quantitative estimation of the quantity of these marine litters being emitted from them was found to be a difficult assignment involving a large degree of ambiguities. Determining the accurate quantities that actually enter into the sea is more hard and complicated than estimating the discharges from various sources. Laundering of synthetic clothes is as the potential source from where the largest amounts of particles are emitted. For a few MF sources, the estimations of both emissions of particles and the transfer to the sea are quite straightforward. For example, these inconspicuous fragments are shed from particles released from laundering and personal care products and finally almost washed out into the rivers through wastewater treatment plant. It is, however, important to note that the smallest particles, which are microplastic fibers having less than 10 μm, probably surpass through the filtration system of WWTP to an advanced level and can be projected to have a negative ecological impact.

Control at source has been advocated by every researcher and in documents on this marine debris for decades. However, mechanical confirmation and in-print judgments have failed to initiate control of this MF pollution at all. Precise proceedings should have the intention to completely remove the primary and secondary contributors of large items and these tiny particles before entering into the ocean water. Unfortunately, based on current fashion, aquatic to terrestrial animals will continue to be in danger, and disasters will happen before these objectives are reached. Collection of all these micro-sized particles will not be possible, and even so, it would not be successful. However, scientists should propose an initiative for developing a control measure that can be followed by institutions, civilizations, and industries. Antarctic and Arctic regions are continuously trying to improvise investigational techniques to detect, enumerate, and characterize MF particles and develop biological degradation methods. These ideas will necessitate execution of educational curriculum, the cooperation of municipal as well as rural facilities, and influence through realistic examples that effortlessly and straightforwardly demonstrate proper management of waste.

An instant and required subsequent footstep is the additional development of the investigation for accessible and potable technology for controlling this pollution. This will necessitate supplementary information gathering and additional deep understanding of the sources and impacts. It is also required to improve our understanding of the growing effects of their release from time to time. In addition, demonstration of our representation with experimental data practically will be more advantageous to authenticate the method. However, this is not practicable till now, particularly their position in the literature. Sufficient investigational system should be urbanized in order to carry out this assessment.

Despite the fact that supplementary research is in progress, this field of research can participate in a very vital and significant responsibility in open debate concerning how to concentrate on these small microplastic particles away from the conventional waste management methods. The increasing occurrence of these tiny fragments in the aquatic sources needs a novel opinion and idea about how to diminish both the primary and secondary sources equally.

References

Absher TM, Ferreira SL, Kern Y, Ferreira JAL, Christo SW, Ando RA (2018) Incidence and identification of microfibers in ocean waters in Admiralty Bay, Antarctica. Environ Sci Pollut Res:1–7. https://doi.org/10.1007/s11356-018-3509-6

Ahmed T (2018) Biodegradation of plastics: current scenario and future prospects for environmental safety. Environ Sci Pollut Res Int 25:7287–7298

Almroth BMC, Aström L, Roslund S, Petersson H, Johansson M, Persson N (2017) Quantifying shedding of synthetic fibers from textiles; a source of microplastics released into the environment. Environ Sci Pollut Res 25(2):1191–1199

Andrady AL (2011) Microplastics in the marine environment. Mar Pollut Bull 62:1596–1605

Bakir A, Rowlan S, Thompson R (2014) Enhanced desorption of persistent organic pollutants from microplastics under simulated physiological conditions. Environ Pollut 185:16–23

Bal B, Ghosh S, Das AP (2018) Microbial recovery and recycling of manganese waste and their future application: a review. Geomicrobiol J. https://doi.org/10.1080/01490451.2018.1497731.

Boucher J, Friot D (2017) Primary microplastics in the oceans: a global evaluation of 413 sources. (IUCN, 2017). https://doi.org/10.2305/IUCN.CH.2017.01.en

Bradley M, Sukumaran S, Lowry S, Woods S (2017) Tracking microplastics in the environment via FT-IR microscopy 32(8):17–23. http://www.spectroscopyonline.com/

Brannan A (2014) Microplastic traces leaching into food chain. https://www.earthsfriends.com/microplastic/

Browne MA, Galloway TS, Thompson RC (2010) Spatial patterns of plastic debris along estuarine shorelines. Environ Sci Technol 44:3404–3409

Browne MA, Crump P, Niven SJ, Teuten E, Tonkin A, Galloway T, Thompson R (2011) Accumulation of microplastic on shorelines worldwide: sources and sinks. Environ Sci Technol 45(21):9175–9179

Carr SA, Liu J, Tesoro AG (2016) Transport and fate of microplastic particles in wastewater treatment plants. Water Res 91:174–182. https://doi.org/10.1016/j.watres.2016.01.002

Carr S (2017) Sources and dispersive modes of micro-fibers in the environment. Integr Environ Assess Manag 13(3):466–469

Catarino AI, Macchia V, Sanderson WG et al (2018) Low levels of microplastics (MP) in wild mussels indicate that MP ingestion by humans is minimal compared to exposure via household fibres fallout during a meal. Environ Pollut 237:675–684. https://doi.org/10.1016/j.envpol.2018.02.069

Cauwenberghe VL, Janssen CR (2014) Microplastics in bivalves cultured for human consumption. Environ Pollut 193:65–70

Cauwenberghe LV, devriese L, Galgani F, Robbens J, Janssen CR (2015) Microplastics in sediments: a review of techniques, occurrence and effects. Mar Environ Res 111:5–17

Choudhari M (2018) Microfiber pollution in environment and its impact. Chemarc.com. https://www.chemarc.com/. . ./microfiberpollution. . .impact/5aec55072d35eb266a4cc3. . .

Chua E, Shimeta J, Nugegoda D, Morrison PD, Clarke B (2014) Assimilation of polybrominated diphenyl ethers from microplastics by the marine amphipod, Allorchestes compressa. Environ Sci Technol 48:8127–8134

Cincinelli A, Scopetani C, Chelazzi D, Lombardini E, Martellini T, Katsoyiannis A, Fossi MC, Corsolini S (2017) Microplastic in the surface waters of the Ross Sea (Antarctica): occurrence, distribution and characterization by FTIR. Chemosphere 175:391–400

Cobbing M, Vicaire Y (2016) Timeout for fast fashion. Greenpeace e.v, Hamburg. www.greenpeace.org/international/en/campaigns/detox/fashion

Cole M, Lindeque P, Halsband CS, Galloway T (2011) Microplastics as contaminants in the marine environment: a review. Mar Pollut Bull 62(12)

Coppock RL, Cole M, Lindeque PK, Queiros AM, Galloway TS (2017) A small-scale, portable method for extracting microplastics from marine sediments. Environ Pollut 230:829–837

Das AP, Mishra S (2008) Hexavalent chromium (VI): health hazards & environmental pollutant. J Environ Res Dev 2:386–392
Das AP, Mishra S (2010) Biodegradation of the metallic carcinogen hexavalent chromium Cr (VI) by an indigenously isolated bacterial strain. J Carcinog 9:6
Das AP, Singh S (2011) Occupational health assessment of chromite toxicity among Indian miners. Indian J Occup Environ Med 15(1):6–17
Das AP, Kumar PS, Swain S (2014) Recent advances in biosensor based endotoxin detection. Biosens Bioelectron 51:62–75
Das AP, Ghosh S, Mohanty S, Sukla LB (2015) Consequences of manganese compounds: a review. Toxicol Environ Chem 96:981–997
Desforges JP, Galbraith M, Ross PS (2015) Ingestion of microplastics by zooplankton in the Northeast Pacific Ocean. Arch Environ Contam Toxicol 69:320–330
Dris R, Gasperi J, Rocher V, Saad M, Renault N, Tassin B (2015) Microplastic contamination in an urban area: a case study in greater Paris. Environ Chem 12:592–599
Dris R, Gasperi J, Saad M, Mirande C, Assin B (2016) Synthetic fibers in atmospheric fallout: a source of microplastics in the environment? Mar Pollut Bull 104(1–2):290–293
Dris R, Gasperi J, Mirande C, Mandin C, Guerrouache M, Langlois V, Tassin B (2017) A first overview of textile fibers, including microplastics, in indoor and outdoor environments. Environ Pollut 221:453–458
Eriksen M, Mason S, Wilson S, Box C, Zellers A, Edwards W, Farley H, Amato S (2013) Microplastic pollution in the surface waters of the Laurentian Great Lakes. Mar Pollut Bull 77 (1–2):177–182
Fendall LS, Sewell MA (2009) Contributing to marine pollution by washing your face: microplastics in facial cleansers. Mar Pollut Bull 58:1225–1228
Fries E, Dekiff JH, Willmeyer J, Nuelle MT, Ebert M, Remy D (2013) Identification of polymer types and additives in marine microplastic particles using pyrolysis-GC/MS and scanning electron microscopy. Environ Sci Process 15(10):1949. https://doi.org/10.1039/c3em00214d
Gago J, Carretero O, Filgueiras AV, Viñas L (2018) Synthetic microfibers in the marine environment: a review on their occurrence in seawater and sediments. Mar Pollut Bull 127:365–376
Galloway TS, Cole M, Lewis C (2017) Interactions of microplastic debris throughout the marine ecosystem. Nat Ecol Evol 1.0116. https://doi.org/10.1038/s41559-017-0116
Gasperi J, Wright SL, Dris R, Collard F, Mandin C et al (2018) Microplastics in air: are we breathing it in? Curr Opin Environ Sci Health 1:1–5. https://doi.org/10.1016/j.coesh.2017.10.002
Gear of the Future (2016) The great outdoors. http://www.tgomagazine.co.uk/gear-news/gear-of-the-future
Ghosh S, Das AP (2018) Metagenomic insights into the microbial diversity in manganese-contaminated mine tailings and their role in biogeochemical cycling of manganese. Sci Rep 8:8257
Ghosh S, Mohanty A, Sukla LB, Das AP (2016) A greener approach for resource recycling: manganese bioleaching. Chemosphere 154:628–639
Ghosh S, Bal B, Das AP (2017) Enhancing manganese recovery from low grade ores by using mixed culture of indigenously isolated bacterial strains. Geomicrobiol J 35:242–246
Gusmao F, Domenico MD, Amaral AC, Martinez A, Gonzalez BC, Worsaae K, Ivar do Sul JA, Cunha Lana P (2016) In situ ingestion of microfibres by meiofauna from sandy beaches. Environ Pollut 216:584–590
Hankett JM, Collin WR, Yang P, Chen Z, Duhaime M (2016) Low-volatility model demonstrates humidity affects environmental toxin deposition on plastics at a molecular level. Environ Sci Technol 50:1304–1312
Hemalakshita T, Dubey AR, Pande A, Wagh NS, Prasad R, Gupta AS, Pandit S (2021) Bioplastics: fundamentals to application. In: Prasad R, Singh J, Kumar V, Upadhyay CP (eds) Recent developments in microbial technologies. Springer Nature Singapore, Singapore, pp 301–321

Henry B, Laitala K, Klepp IG (2019) Microfibres from apparel and home textiles: prospects for including microplastics in environmental sustainability assessment. Sci Total Environ 652:483–494

Horton AA, Walton A, Spurgeon DJ, Lahive E, Svendsen C (2017) Microplastics in freshwater and terrestrial environments: evaluating the current understanding to identify the knowledge gaps and future research priorities. Sci Total Environ 586:127–141

Howard GT, Norton WN, Burks T (2012) Growth of Acinetobacter gerneri P7 on polyurethane and the purification and characterization of a polyurethane enzyme. Biodegradation 23:561–573

Isobe A, Uchiyama-Matsumoto K, Uchida K, Tokai T (2017) Microplastics in Southern Ocean. Mar Pollut Bull 114(1):623–626

Jambeck JR, Geyer R, Wilcox C, Siegler TR, Perryman M, Andrady A, Narayan R, Law KL (2015) Plastic waste inputs from land into the ocean. Science 347(6223):768–771. https://doi.org/10.1017/CBO9781107415386.010

Kale SK, Deshmukh AG, Dudhare MS, Patil VB (2015) Microbial degradation of plastic: a review. J Biochem Technol 6(2):952–961

Karlsson TM, Vethaak AD, Almroth BC, Ariese F, Velzen MV, Hassellov M, Leslie HA (2017) Screening for microplastics in sediment, water, marine invertebrates and fish: method development and microplastic accumulation. Mar Pollut Bull 122:403–408

Kim JS, Lee HJ, Kim SK, Kim HJ (2018) Global pattern of microplastics (MPs) in commercial food-grade salts: sea salt as an Indicator of seawater MP pollution. Environ Sci Technol 52:12819–12828. https://doi.org/10.1021/acs.est.8b04180

Lamichhane G (2018) Analysis of microfibers in waste water from washing machines, Metropolia University of Applied Sciences. https://www.theseus.fi/bitstream/handle/10024/141278/

Li J, Liu H, Paul Chen J (2018) Microplastics in freshwater systems: a review on occurrence, environmental effects, and methods for microplastics detection. Water Res 137:362–374. https://doi.org/10.1016/j.watres.2017.12.056

Lusher A, Hollman P, Mendoza-Hill J (2017) Microplastics in fisheries and aquaculture: status of knowledge on their occurrence and implications for aquatic organisms and food safety, FAO fisheries and aquaculture technical paper: 615. FAO, Rome

Magnusson K, Norén F (2014) Screening of microplastic particles in and down-stream a wastewater treatment plant. Swedish Environmental Protection Agency

Marklund A, Andersson B, Haglund P (2003) Screening of organophosphorus compounds and their distribution in various indoor environments. Chemosphere 53:1137–1146

Mason SA, Garneau D, Sutton R, Chu Y, Ehmann K, Barnes J, Fink P, Papazissimos D, Rogers DL (2016a) Microplastic pollution is widely detected in US municipal wastewater treatment plant effluent. Environ Pollut 218:1045–1054

Mason SA, Garneau D, Sutton R, Chu Y, Ehmann K, Barnes J, Rogers DL (2016b) Microplastic pollution is widely detected in US municipal wastewater treatment plant effluent. Environ Pollut 218:1045–1054

Messinger L (2016) How your clothes are poisoning our oceans and food supply. The Guardian (US Ed.). Sect: Environment. https://www.theguardian.com/environment/2016/jun/20/microfibers-plastic-pollution-oceans-patagonia-synthetic-clothesmicro beads

Mintenig SM, Int-Veena I, Lodera MGJ, Primpkea S, Gerdtsa G (2017) Identification of microplastic in effluents of waste water treatment plants, using focal plane array-based micro-Fourier-transform infrared imaging. Water Res 108:365–372

Mintenig SM et al (2019) Low numbers of microplastics detected in drinking water from ground water sources. Sci Total Environ 648:631–5. https://doi.org/10.1016/j.scitotenv.2018.08.178

Mishra S, Rath CC, Das AP (2019) Marine microfiber pollution: a review on present status and future challenges. Mar Pollut Bull 140:188–197

Mohanty S, Ghosh S, Bal B, Das AP (2018) A review of biotechnology processes applied for manganese recovery from wastes. Rev Environ Sci Biotechnol 17:791–811

Murphy F, Ewins C, Carbonnier F, Quinn B (2016) Wastewater treatment works (WwTW) as a source of microplastics in the aquatic environment. Environ Sci Technol 50:5800–5808

Murray F, Cowie PR (2011) Plastic contamination in the decapod crustacean Nephrops norvegicus (Linnaeus, 1758). Mar Pollut Bull 62:1207–1217

Napper IE, Thompson RC (2016) Release of synthetic microplastic plastic fibres from domestic washing machines: effects of fabric type and washing conditions. Mar Pollut Bull 112:39–45

Negoro S (2012) Three-dimensional structure of nylon hydrolase and mechanism of nylon-6 hydrolysis. J Biol Chem 287:5079–5090

Nelms SE, Galloway TS, Godley BJ, Jarvis DS, Lindeque PK (2018) Investigating microplastic trophic transfer in marine top predators. Environ Pollut 238:999–1007

Nizzetto L, Futter M, Langaas S (2016) Are agricultural soils dumps for microplastics of urban origin? Environ Sci Technol 50:10777–10779. https://doi.org/10.1021/acs.est.6b04140

Ocean Conservancy (2013) N.O.A.A. Ocean Conservancy. Talking trash & taking action. Publication of the Talking Trash & Taking Action Educational Program of the Ocean Conservancy and National Oceanic and Atmospheric Administration Marine Debris Program. Accessed January 2018

Oerlikon (2009) The fiber year 2008/2009: a world survey on textile and nonwovens industry. Oerlikon, Freienbach. https://www.oerlikon.com/ecomaXL/get_blob.php?name=Layout_Fibre_Year_08_09_290509__s_screen.pdf. Accessed 25 Sept 2018

Pathak VM, Navneet (2017) Bioresour Bioprocess 4:15. https://doi.org/10.1186/s40643-017-0145-9

Prata JC, da Costa JP, Duarte AC, Rocha-Santos T (2019) Methods for sampling and detection of microplastics in water and sediment: a critical review. Trends Anal Chem 110:150–159. https://doi.org/10.1016/j.trac.2018.10.029

Rauert C, Lazarov B, Harrad S, Covaci A, Stranger M (2014) A review of chamber experiments for determining specific emission rates and investigating migration pathways of flame retardants. Atmos Environ 82:44–55

Reed S, Clerk M, Thompson R, Hughes KA (2018) Microplastics in marine sediments near Rothera Research Station, Antarctica. Mar Pollut Res 133:460–463

Revel M, Châtel A, Mouneyrac C (2018) Micro(nano)plastics: a threat to human health? Curr Opin Environ Sci Health 1:17–23

Rochman CM, Tahir A, Williams SL, Baxa DV, Lam R, Miller JT, Teh SJ (2015) Anthropogenic debris in seafood: plastic debris and fibers from textiles in fish and bivalves sold for human consumption. Sci Rep 5

Rozalia Project (2017) A human-scale solution to the biggest pollution problem facing our ocean: Microfibers. http://rozaliaproject.org/stop-microfiber-pollution/

Sachin SS, Mishra RL (2013) Screening and identification of soil fungi with potential of plastic degrading ability. Indian J Appl Res 3(11):34–36

Shah AA, Hasan F, Akhter JI, Hameed A, Ahmed S (2008) Degradation of polyurethane by novel bacterial consortium isolated from soil. Anal Microbiol 58(3):381–386

Shimao M (2001) Biodegradation of plastics. Curr Opin Biotechnol 12:242–247

Smith M, Love DC, Rochman CM, Neff RA (2018) Microplastics in seafood and the implications for human health. Curr Environ Health Rep 5(3):375–386

Strungaru S, Jijie R, Nicoara M et al (2019) Micro- (nano) plastics in freshwater ecosystems: abundance, toxicological impact and quantification methodology. Trends Anal Chem 110:116–128. https://doi.org/10.1016/j.trac.2018.10.025

Sussarellu R, Suquet M, Thomas Y, Lambert C, Fabioux C, Pernet MEJ, Corporeau C (2016) Oyster reproduction is affected by exposure to polystyrene microplastics. Proc Natl Acad Sci 113(9):2430–2435

Talvitie J, Mikola A, Koistinen A, Setälä O (2017a) Solutions to microplastic pollution—removal of microplastics from wastewater effluent with advanced wastewater treatment technologies. Water Res 123:401–407

Talvitie J, Mikola A, Setala O, Heinonen M, Koistinen A (2017b) How well is microlitter purified from wastewater? A detailed study on the stepwise removal of microlitter in a tertiary level wastewater treatment plant. Water Res 109:164–172

Taylor ML, Claire G, Robinson LF, Woodall LC (2016) Plastic microfiber ingestion by deep-sea organisms. Sci Rep 6:33997. https://doi.org/10.1038/srep33997

Thevenon F, Carroll C, Sousa J (2011) Plastic debris in the ocean: the characterization of marine plastics and their environmental impacts, situation analysis report, UNEP year book 2011: emerging issues in our global environment. IUCN, Gland. https://doi.org/10.1073/pnas.1314705111

Thushari GGN, Senevirathna JDM, Yakupitiyage A, Chavanichc S (2017) Effects of microplastics on sessile invertebrates in the eastern coast of Thailand: an approach to coastal zone conservation. Mar Pollut Bull 124:349–355

Tosetto L, Brown C, Williamson JE (2016) Microplastics on beaches: ingestion and behavioural consequences for beach hoppers. Mar Biol 163(10):199

Wagner M, Scherer C, Alvarez-Muñoz D, Brennholt N, Bourrain X, Buchinger S, Fries E, Grosbois C, Klasmeier J, Marti T, Rodriguez-Mozaz S, Urbatzka R, Vethaak A, Winther-Nielsen M, Reifferscheid G (2014) Microplastics in freshwater ecosystems: what we know and what we need to know. Environ Sci Eur 26:12. https://doi.org/10.1186/s12302-014-0012-7

Waller CL, Griffiths HJ, Waluda CM, Thorpe SE, Loaiza I, Moreno B, Pacheres CO, Hughes KA (2017) Microplastic in the Antarctic marine system: an emerging area of research. Sci Total Environ 598:220–227

Wessel CC, Lockridge G, Battiste D, Cebrian J (2016) Abundance and characteristics of microplastics in beach sediments: insight into microplastic accumulation in northern Gulf of Mexico estuaries. Mar Pollut Bull 109(1):178–183

Woodall LC, Gwinnett C, Packer M, Thompson RC, Robinson LF, Paterson GL (2015) Using a forensic science approach to minimize environmental contamination and to identify microfibres in marine sediments. Mar Pollut Bull 95(1):40–46

Yang CZ, Yaniger SI, Jordan VC, Klein DJ, Bittner GD (2011) Most plastic products release estrogenic chemicals: a potential health problem that can be solved. Environ Health Perspect 119:8

Zhao S, Zhu L, Li D (2016) Microplastic in three urban estuaries, China. Environ Pollut 206:597–604

Zalasiewicz J, Waters CN, Ivar D, Sul JA, Corcoran PL, Barnosky AD, Cearreta A, Edgeworth M, Gałuszka A, Jeandel C, Leinfelder R, Mcneill JR, Steffen W, Summerhayes C, Wagreich M, Williams M, Wolfe AP, Yonan Y (2016) The geological cycle of plastics and their use as a stratigraphic Indicator of the Anthropocene. Anthropocene 13:4–17

Environment Remediation Tools: Chemosensors and Biosensors

10

J. Brindha, Milind Shrinivas Dangate, and M. M. Balamurali

Abstract

Pollution is one of the major problems persisting in our environment causing an increase in the morbidity and mortality, which ultimately affects the economic growth of a nation. Therefore, environment quality sensing tools have become inevitable in everyday life. Though several tools are reported in the scientific literature and much more are available in the market, this field of research on environment sensors has always been on its priority. Simplicity, cost-effectiveness, reliability, and stability are the challenges that decide the performance of sensors. This chapter will discuss achievements made toward various chemosensors and biosensors that can pave the way toward environment remediation.

Research on environment sensors is one of the evergreen areas with contributions from transdisciplinary researchers round the world. Though technological advancements and industrial developments are happening at a daunting rate, on one hand, the hazards posed by them on the environment are increasing at a much faster rate. Therefore, it becomes necessary to develop such a tool to meet the challenges ahead in food, healthcare, and environmental sectors. With the emergence of highly infectious and/or antibiotic-resistant pathogens, advancements in biosensors are made with the need for quicker detection/treatment and ease of use in diagnostics. This chapter deals with various kinds of sensors available for the detection of pathogens and toxic chemicals, with their classifications based on the type of transducer used, along with the recent findings in each type.

J. Brindha · M. S. Dangate · M. M. Balamurali (✉)
Chemistry Division, School of Advanced Sciences, Vellore Institute of Technology, Chennai, Tamil Nadu, India

R. Prasad (ed.), *Environmental Pollution and Remediation*, Environmental and Microbial Biotechnology, https://doi.org/10.1007/978-981-15-5499-5_10

Keywords

Chemosensor · Biosensor · Pathogen detection · Piezoelectric

10.1 Introduction

In this modern era, sensors have become an inevitable device. With the explosion in technological advancements and industrial developments, the morbidity and mortality rates are also steeply rising. Therefore, it is mandatory to monitor our environment and to sustain a healthy life. Sensing involves recognition of a particular entity such as matter, energy using a specialized device termed sensor. A sensor is built with a selective sensing element coupled to a transducer which helps in converting the changes observed into a readable output signal such as electrical, magnetic, or optical, with its magnitude directly proportional to quantity sensed. Sensors are applied in human lives in various ways including medicine/health in drug/pathogen detection and also in industries like automobiles, electronics, and waste management. It is thus significant to develop sensors, and to date, several endeavors are made to move further in this area of research.

There are a number of transducers with different operating principles such as fluorescence (Li et al. 2015), surface plasmon resonance (Liu et al. 2016a; Rifat et al. 2015), colorimetric (Taneja and Tyagi 2007), cantilever (Fritz 2008)/piezoelectric crystal/quartz crystalline microbalance (Farka et al. 2013; Fawcett et al. 1988), and electrochemical mode comprising of amperometric, voltammetric and impedimetric (Chen and Shah 2013; Wan et al. 2011), and the immuno−/biochemical sensors (Menti et al. 2016; Sign and Sumana 2016). Detection of lower concentrations of target analytes such as nucleic acids or other analytes in the sample mixture has been a daunting task which is addressed by means of concentrating the sample using various methods including polymerase chain reaction and use of magnetic beads, suitable nanoparticles, or other labels. In general, a typical sensor solely relies on the selectivity of recognition element/receptor, specific recognition of an analyte, sensitivity and specificity of the detection process, enrichment of response signals, and transmission of output readable signals using the transducer.

The development in science and technology yielding incredible findings has led to increased vulnerability of all life forms to numerous toxins and infectious pathogens, which compelled the researchers to broaden the range of sensors to safeguard human lives by developing simple sensors, in particular chemosensors and biosensors targeted toward hazardous chemicals/pesticides, clinical diagnosis (Yanase et al. 2014), food industry (Scognamiglio et al. 2014), and water and environmental quality measures/monitoring (Teo and Wong 2014; Bereza-Malcolm et al. 2014; Singh et al. 2020). Since the period of the 1980s, there is an extensive growth and development of the various chemosensors, initiated by de Silva and Czarnik, known to be fathers of modern chemosensors. Chemosensors are essential for the detection of environmentally hazardous/toxic chemicals, cations, anions, neutral molecules, and physiologically important molecules/ions which are in continuous development along with new advancements in technologies for microscopic imaging. Among biosensors, the classical techniques like cell culture, cell counting, and plating can detect the presence of specific pathogens (Monis and Giglio 2006; Gracias and McKillip 2004), yet in a longer time period for completion of the detection process. It is thus certainly necessary to expand the biosensor tools with easier, sensitive, and rapid detection. In this chapter, we will be focusing on two major sensors in demand, namely, chemosensors and biosensors, along with their classifications, advantages, limitations, and interesting recent findings.

10.2 Chemosensors

There is a substantial increase in the design and development of chemosensors in the last few years. Exposure to various toxic substances, hazardous chemicals, food additives/adulterants, pesticides, industrial effluents, etc. has imposed the need for expansion of chemosensors. Parallel advancements in the field of fluorescence/colorimetric chemosensing have induced the progress of chemosensor research. Chemosensor design and development has been interdisciplinary including synthesis of organic/inorganic molecules combined with different analytical techniques. These are designed to have a high sensitivity and specificity that allow them to interact selectively with a specific target analyte present in different complex environments and are used in clinical diagnostics and detection of agricultural, industrial, or environmental pollutants, making them significant for health and safety of all life forms.

Chemosensors, like any other sensors, require the generation of a signal on binding for detection. Selection of a suitable signal transduction method is vital to achieving success in the sensing process. Commonly employed strategies in chemosensors include fluorescence, electrochemical, colorimetric, and surface plasmon resonance (SPR). Every method has its own advantages and disadvantages based on the requirements to be met, as given below with significant findings.

10.2.1 Fluorescence-Based Chemosensors

Fluorescence is one of the most frequently used signal transduction methods with promising sensing applications, as it serves as a highly sensitive technique, solely due to the emission wavelength which is usually longer compared to the excitation wavelength and the need for low analyte concentrations ($>10^{-6}$ M) for signaling. A fluorescent sensor is more like a molecular machine with the ability to signal the presence of analytes like ions/molecules. The two main parts required for designing of fluorescent sensors includes a signaling moiety or the fluorophore and a receptor that serves as a recognition molecule. The signaling molecule serves as a signal transducer, which converts the data obtained (recognition event) into an optical signal response. The recognition moiety is accountable for attachment to an analyte specifically and selectively in an efficient manner. This binding relies on the ligand

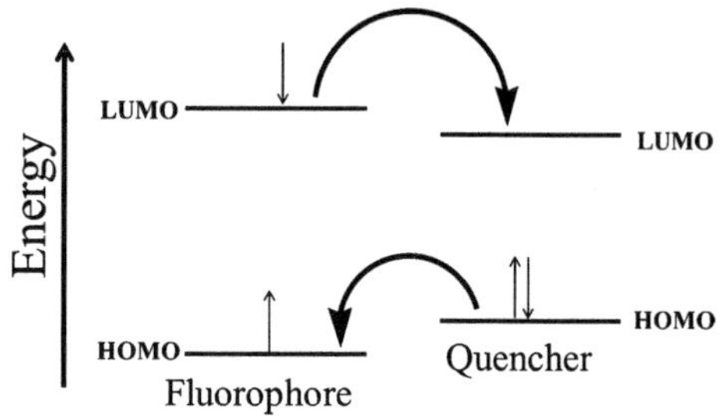

Fig. 10.1 Mechanism of energy transfer by an electron exchange process

structure, characteristics of the target analyte such as a cation or anion, or organic molecules' structure such as drug candidates. Apart from this, it is also essential to consider the nature of the solvent including polarity, pH, and ionic strength, which can affect the capability of the receptor to detect and bind the target analyte.

There are several metal ions that play a significant role in our routine life, including sodium (Na^{+}), calcium (Ca^{2+}), potassium (K^{+}), and zinc (Zn^{2+}). However, certain heavy metal ions like lead (Pb^{2+}), cadmium (Cd^{2+}), and mercury (Hg^{2+}) are highly toxic and lead to severe health and environmental issues. It is necessary to develop tools like chemosensors to detect these toxic metals/chemicals. The most typical fluorescent sensor is based on photoinduced electron transfer (PET) developed initially by de Silva et al., for sensing various cations, anions, and neutral molecules (Silva and NimaláGunaratne 1996). This is shown schematically in Fig. 10.1. Recently, a novel pyrene (Pyr1-2)-based fluorescence chemosensor for specific detection of Hg^{2+} was reported (Gao et al. 2018). These fluorescent probes Pyr1 and Pyr2 having a side chain of thioacetal moiety with carboxyl and hydroxyl group, respectively, exhibit fluorescence response specifically toward Hg^{2+}, following intramolecular charge transfer mechanism. In the presence of Hg^{2+}, the thioacetal group of Pyr1-2 probes was observed to convert into aldehyde group. Detection limits for these pyrene probes were found to be lower than 1.80 nM. For sensing or monitoring of trace amounts of Fe^{3+} in real water specimens and also for intracellular imaging, for detecting Fe^{3+} in live human breast cancer cells (MCF-7), two fluorescent chemosensors derived from pyridine and rhodamine B conjugates were developed. These were found to possess lower detection limits and lesser interference of metal cations (Song et al. 2019). A benzoindo-croconine-based colorimetric and fluorescent chemosensor (Wang et al. 2016a) was developed for the detection of metal ions, $Fe^{3+}/Cu^{2+}/Ag^{+}$ ions. It exhibited sensitivity and high selectivity to Fe^{3+}, Cu^{2+}, and Ag^{+} ions in ethanol/water (4:1, v/v). Both color change (from brown to pale yellow) and decrease in fluorescence intensity or quenching of fluorescence were reported with increase in the concentration of Fe^{3+}, with complete quenching at a concentration of 0.4 mM (Fe^{3+}). Same way, the fluorescence of the benzoindo-croconine was quenched upon increase in the concentration of Cu^{2+}/Ag^{+} ions. A new fluorescence-based chemosensor was developed for sensing organophosphorus pesticides, based on their interaction with a luminescent europium (Azab and Kamel 2016) complex by electroanalytical and fluorescent studies. These pesticides were found to have quenching effects on the characteristics emission peak for europium (III) ($\lambda = 614$ nm). Selectivity of this chemosensor was also ensured by studying the possible interferents, and also, the quenching mechanism was of dynamic type for the pesticides chlorfenvinphos, diazinon, and isofenphos, while it is static for azinphos ethyl.

Fluorescence detection in chemosensors is known for its sensitivity and wide usage, though it has limitations as well. The major one is its sensitivity to the surroundings, including the solvents, temperature, and interfering species. It needs designing of one or more fluorophores that respond with a change in fluorescence on binding, and it is important to maintain the photostability of these fluorophores used. In this regard, organic dyes are not so preferred, while the metal complexes and quantum dots possess good stability and fluorescent intensity.

10.2.2 Electrochemical Chemosensors

A commonly exploited method of detection in chemosensors is electrochemical detection. There are different ways to detect species using electrochemical methods that include ion-selective electrodes (ISEs) for sensing toxic pollutants and pH analysis in water test samples and a general research technique of cyclic voltammetry (CV) that helps in the study of thermodynamics and kinetics of electron transfer reactions apart from sensing the presence or concentration of a specific target. These methods change the physicochemical information of a specific analyte like a molecule or ion, into appropriate electrical signal possessing particular potential/current or both that could be displayed as the output according to the chemical information. The electrochemical detection is based on a chemical or physical change induced by the target analyte interacting with the electrode, on its surface, or in particular the interfacial area between the electrode and its electrolyte. These electric signals are monitored as either potential/current change or both, based on which detection is done called *potentiometry*, *amperometry* (also called *coulometry*), and *voltammetry*, respectively. Ion-selective electrodes (ISEs) are classic potentiometric sensors that help in converting the target ion's concentration into a readable output potential signal response. Membrane-based ISEs with ion-selective conducting membrane leading to an electric field generation and hence potential difference are available in market; however, most of them detect only the inorganic ions. Ion-selective membrane can also be constructed by exclusive coating of an electrode surface with an ion-selective agent. These coating-based ISEs are easy to develop and possess high selectivity and rate of response. Another new electrode includes the ion-selective (or sensitive) field-effect transistors (ISFETs), which belong to a broader class of chemically modified field-effect transistors (CHEMFETs) which are beneficial especially in equipment miniaturization and gives more data for enhanced reliability. There have been several research in the past two decades, with the aim to develop electrochemical-based molecular probes or sensors with high specificity for cationic, anionic, or neutrally charged analyte molecules (Beer 1996; Beer et al. 1999a).

Electrochemical detection is mainly used for specific target analytes that are charged, such as ions, which can undergo electron transfer reactions on an electrode. The detection of neutral molecules using electrochemical method has been a hard task. The integrated chemical processes involving a redox-responsive receptor which specifically recognizes and electrochemically detects the signaling molecule are known as electrochemical molecular recognition (Beer et al. 1999b). By means of standardized molecular imprinting procedures, a synthetic polymer receptor film specific for the cancer biomarker, neopterin, was designed and developed as a recognition unit of a potentiometric chemosensor (Sharma et al. 2016). This chemosensor worked by measuring the open circuit potential changes that occurred due to selective binding/sensing of neopterin by the polymer film. It served the purpose of detecting neopterin in serum samples with a detection limit of 22 μM. A n-type chemical doping of conducting polyaniline and the formation of Schottky barrier diode in a chemosensor (Ameen et al. 2016) paved way for electrochemical detection of hydrazine benzene chemical, with high sensitivity and detection limit of 5.11 μM. Chemosensors employing transducers coated with D- or (L-phenylalanine)-templated molecular imprinted polymer films (Iskierko et al. 2017) following extended-gate field-effect transistors were also developed for enantioselective determination of (D- or L-) phenylalanine. These were found to have a detection limit of 13 μM. Differential pulse voltammetry of two ligands made up of the diphenyl derivatives, 3-(2,4-dinitrophenoxy)phenol (L1) and 3-(2-nitrophenoxy)phenol (L2), was found to exhibit complete quenching of anodic peaks at 1.16 V for L1 and 1.34 V for L2 on adding one equivalent fluoride ions (Sharma et al. 2015). Thereby, both these ligands serve as a voltammetric chemosensor for sensing fluoride ions.

10.2.3 Colorimetric Chemosensors

Colorimetry is different from fluorescence in the working principle that they work with respect to absorption with less sensitivity relative to the emission employed in fluorescence. This method of sensing is not limited by stringent requirements of fluorophore design; instead, there are several dyes available for sensing applications. There are two common signal motifs to be considered: (i) a change in absorbance at a specific wavelength, thereby observing a color appear or fade, or (ii) monitoring the maximum absorption wavelength to change, such as initial color changes into a new second color. Sessler and Miyaji reported (Miyaji and Sessler 2001) several commercially available molecules for use as colorimetric anion sensors, including l-leucine-4-nitroanilide, 1,2-diaminoanthraquinone, 1,8-diaminoanthraquinone, 1-(4-nitrophenyl)-2-thiourea, 4-nitrophenol, 4-nitroaniline, 4-nitro-1,2-phenylenediamine, alizarin, 2,2-bi(3-hydroxy-1,4-naphthoquinone) and Direct Yellow 50.

A recent finding includes a synthesis of a coumarin derivative with benzothiazole Schiff's base structure that detects cyanide anions (CN^-) by nucleophilic addition mechanism (Fig. 10.2) (Wang et al. 2016b). This detection technique involves visible colorimetric changes, where there is a color change from reddish brown to

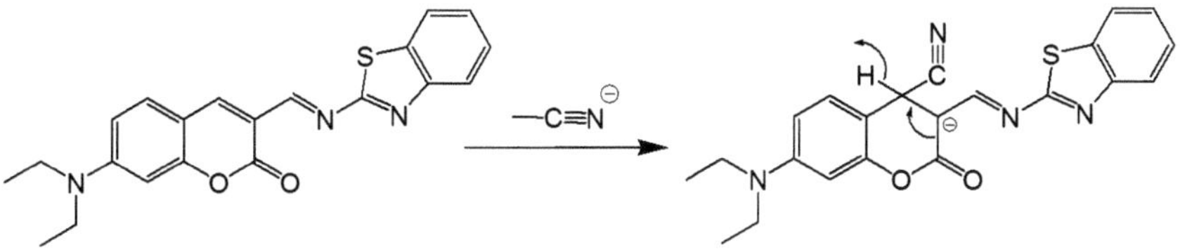

Fig. 10.2 Colorimetric sensing of cyanide ions by intramolecular charge transfer mechanism

white on the addition of cyanide in the test paper. The limit of detection of this compound for the ions, CN^-, was found to be 0.0071 μM. Also, two chemosensors have been developed from the derivatives of N-butyl-3,6-disubstituted carbazole having nitroazobenzene and nitrobenzene in each, for sensing the two strong basic anions (F^- and CN^-) where selectivity is based on polar nature of solvent and acidity and basicity of binding unit and anion, respectively (Tummachote et al. 2019). The sensing was visible in the case of both the sensors, mainly due to the intramolecular charge transfer transition with the removal of protons at the binding site. The sensor with nitrobenzene moiety was proved to have a relatively low detection limit with more stability and selectivity for the targeted anion. A simple dual chemosensor with a highly specific Schiff base derivative, an organic ligand molecule, 5-chloro-2-[(1E,2E)-3-(4-dimethylamino phenyl allylidene amino)]phenol, that acts as a colorimetric as well as surface plasmon resonance-based chemosensor for the sensing of Cu^{2+} ions (Peralta-Domínguez et al. 2016) in spite of various other possibly interfering metal ions including Cd^{2+}, Co^{2+}, Cr^{2+}, Fe^{3+}, Mg^{2+}, Ni^{2+}, Hg^{2+}, Pb^{2+}, Mn^{2+}, and Zn^{2+}. It has a lower limit of detection for copper ions of 1.25×10^{-7} M as well as a lower colorimetric limit of detection of 2×10^{-6} M through naked eyes.

10.2.4 Surface Plasmon Resonance (SPR) Chemosensor

Sensors using the optical property of the surface plasmon resonance (SPR) effect are of main interest due to their ultrahigh sensitivity and nanomolar ranges of detection limits. The mechanism of transduction of response signals based on the variation of the refractive index is an effective method to study the binding between a receptor/ligand immobilized upon a metallic surface and an analyte or the targeted solid substrate introduced into the solution in contact (Patching 2014). These sensors have several other advantages that include being label-free, the compatibility with aqueous media, ability to form miniaturized sensor tools, and integration with other probes such as fiber optics (Tabassum and Gupta 2015), electrochemical setup (Panta et al. 2009), or signal concentration with gold nanoparticles (Chang et al. 2011).

Designing highly sensitive SPR sensors depends on the formation of a gold layer adequately with the attachment of specific stable receptors possessing high affinity and selectivity for the specific analyte. The availability of ubiquitous host–guest or

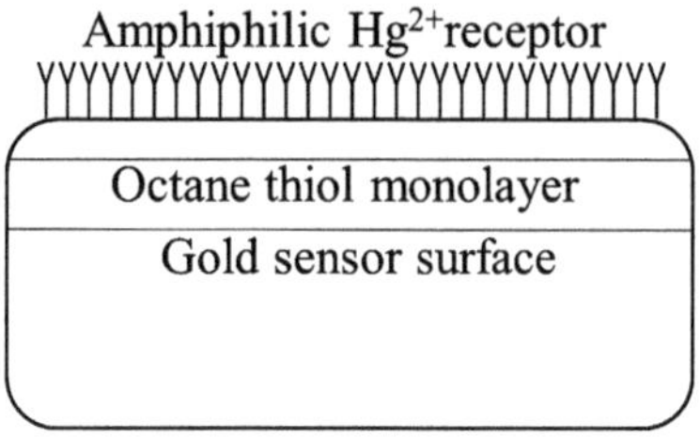

Fig. 10.3 Bilayer sensor surface for the detection of mercuric ions by surface plasmon resonance method

ligand–receptor recognition moieties such as antibody–antigen has been exploited in SPR sensors for monitoring the binding reactions among biological substrates (Couture et al. 2013). Sensing of metal ions in aqueous media is challenging. It is significant to note that SPR sensors are usually designed with a gold surface fixed and a monolayer of polymer fixed with molecular receptors that specifically bind the target (Jung et al. 2013). Similarly, electrodes with gold coatings are in use for the electrochemical detection of mercury (Martín-Yerga et al. 2013). In both cases, the detectors' efficiency is rather limited due to slow diffusion/restricted accessibility of analytes to the recognition sites. Other issues include low chemical stability, film adhesion, and high film resistivity. In this aspect, self-assembled monolayers (SAMs) of specific molecular receptors/biomolecules is emerging as an alternative (Sánchez et al. 2014). The lower chemical stability of self-assembled monolayers (SAMs), for example, thiol group oxidation and the nonspecific binding with impurities of the surface leading to defects/pinholes in the SAMs (Vericat et al. 2010). Apart from SAMs, Langmuir–Blodgett (LB) films containing large or branched amphiphilic chelators are used, which are found to have lower stability as they get leached out from the gold surface during the course of analysis (Prabhakaran et al. 2007). It was demonstrated that the combination of the Langmuir–Blodgett and self-assembled monolayers techniques (Turygin et al. 2006) resulted in sensitive and selective metal-sensing SPR chips with increased stability and homogenous orientation of the binding moieties toward the solution. This SPR bilayer methodology was applied for the direct detection of trace levels of Hg^{2+} cations. An interdigitated bilayer was developed step by step by forming self-assembly of a monolayer of octane thiol on a gold substrate. It was followed by another monolayer containing amphiphilic, highly selective mercury receptor (1,8-diamino-9,10-anthraquinone derivative) (Ermakova et al. 2013) using the Langmuir–Blodgett technique (Fig. 10.3). This resulted in a new approach for sensing metals by SPR chips. The bilayer technology used here serves the purpose of selectively quantifying mercury(II) in aqueous solutions with high sensitivity at the range of sub-nanomolar (0.01 nM) with the significant coordination properties of the amphiphilic chemosensor.

10.2.4.1 SPR Integrated Optical Fiber Approach

A novel optical chemosensor for online detection of two chemical markers, dibenzyl disulfide and furfuraldehyde, present in the insulating mineral oil of transformers, was developed (De Maria et al. 2018). It was designed based on surface plasmon resonance, fixed in a plastic optical fiber that acts as a cost-effective, optical sensing platform along with a molecularly imprinted polymer layer specific for the chemical marker, placed in contact with a thin film of gold. These specific molecular imprint polymers on the gold film form a distinct dielectric medium with optical properties like the refractive index. That would influence the sensing, upon the interaction of the target analyte to the receptor polymers imprinted on the gold film. Dibenzyl disulfide (DBDS) is known to be accountable for the corrosiveness of the mineral insulating oil used for transformer (De Maria et al. 2018). Thus, it is of paramount importance to determine its concentration in mineral oil for diagnostic purposes of the transformer. Also, furfuraldehyde, one among the major byproducts formed during degradation of the insulation paper in a transformer upon overheating. Its presence in the transformer's insulation oil could verify the status of the solid insulation in transformers, which could otherwise become an environmental hazard to many lives.

10.3 Biosensors

Biosensors have significant roles in human health care in the diagnostic procedures to find infections and to prevent epidemics by means of rapid detection techniques (Brindha et al. 2018a; Ellwanger et al. 2017; Du and Zhou 2018; Kaushik et al. 2017). They are analytical devices that work based on the recognition of biological target molecules such as microorganisms, peptide molecules, nucleic acids, proteins, or any biomolecule. The results from an ideal biosensor should be reproducible, sensitive, selective, stabile, and linear (Bhalla et al. 2016). Biosensors possess a biological targeting element that can selectively bind/interact with the target analytes and delivers an output signal depending on the extent of interaction which is then converted to a readable format by a transducer before being transferred to a reader device. The most commonly used biosensors include fluorescent label-based biosensors, surface plasmon resonance biosensors, piezoelectric biosensors, electrochemical biosensors, and biochemical/immuno-biosensors.

10.3.1 Fluorescent Label-Based Biosensor

Fluorescent labeling method (Li et al. 2015) works by energizing a fluorophore at a particular wavelength and observing the photons emitted at another wavelength in a time scale of microsecond. The common fluorophores such as calcein blue, quantum dots (Resch-Genger et al. 2008), carbon dots, and fluorescent diacetate are mostly employed as labels for signal recognition/probe molecules, when the targeted biomolecule under detection is in trace amounts in the sample. Calcein blue is one of the most commonly used blue fluorescent labels that possess the ability to detect and quantify pathogens (Sankaranarayanan et al. 2015). Here, the siderophores released by the bacteria can chelate Fe^{2+} from the iron-bound calcein blue to restore its fluorescence, which is further utilized for detecting the iron chelators quantitatively and hence the pathogenic bacterial detection in a short period of 7–8 hours. This method has been employed for detecting and quantifying Gram-positive bacteria including *Staphylococcus aureus* (Dale et al. 2004; Zawadzka et al. 2009) and Gram-negative bacteria such as *Proteus* species and *Mycobacterium tuberculosis* (Wells et al. 2013; Himpsl et al. 2010; Adler et al. 2014).

Specific biological recognition elements such as antibodies are used in conjugation fluorescent labels for pathogen detection (Yang and Li 2006). Hu et al. have employed specific antibodies labeled with fluorescein isothiocyanate to detect surface antigens of *Escherichia coli* O157:H7 with (Hu et al. 2016). The sensor surface-bound antibodies capture specific bacterial cells followed by the detachment of fluorescein. This enhances the fluorescence intensity which is then utilized to quantify bacterial cells. The detection limit using this method was found to be 3 cfu/ml. Quantum dots are yet another optical probe that can be tuned for fluorescence emission energy by modifying their sizes and chemical constituents. Simultaneous multiple-pathogen detection, of pathogens like *E. coli* O157:H7 and *Salmonella* detection, is also reported (Yang and Li 2006), using semiconductor quantum dots with multiple wavelengths: 525 nm and 705 nm as labels for their respective specific antibodies attached by means of streptavidin and biotin conjugation. Quantum dots are known to be twentyfold more intense and hundredfold relatively resistant to photobleaching as compared to other fluorescent labels/probes such as organic dyes (Vinayaka and Thakur 2010). It also has its limitations like toxicity, solubility issues, and low quantum yields (Shen et al. 2012). Like quantum dots, carbon-based nanomaterials also possess photoluminescence features (Davis et al. 1998; April et al. 2018) that are exploited for enrichment of pathogens for detection (Deng et al. 2008; Srivastava et al. 2004; Upadhyayula et al. 2009; Elkin et al. 2005). Yang et al. used anti-*S. aureus* antibody conjugated with carbon dots trapped within organosilica nanocapsules for sensitive detection of *S. aureus* (Yang et al. 2018). By using carbon dots, the fluorescence signals were enhanced 2 times as compared with other fluorescence-based immunoassays. This method resulted in the detection of *S. aureus* in the range of 1–200 cfu/ml and can be extended to other pathogens also.

10.3.1.1 Fiber Optic-Based Biosensors

Biosensors equipped with fiber optics are found to work with fluorescence-based labels using an optical transmitter for target detection and transmitting the fluorescent signals to a photodetector where it gets converted into electrical signals. Based on the waveguide patterns (Banica 2012), they can be classified as planar and cylindrical waveguides (Fig. 10.4).

Simultaneous multiple detection of *Listeria monocytogenes*, *E. coli* O157:H7, and *S. enterica* (Ohk and Bhunia 2013) was made possible with a multiplex fiber optic biosensor using waveguides coated with streptavidin and specific antibodies tagged with Alexa Fluor 647. A lower limit of detection of 10^3 cfu/ml was reported for all the above mentioned pathogens displaying negligible cross-reactivity. The detection of cell number differences of *E. coli* present in the sample (Maas et al. 2018) was determined by Maas et al. employing an optical fiber-based biosensor

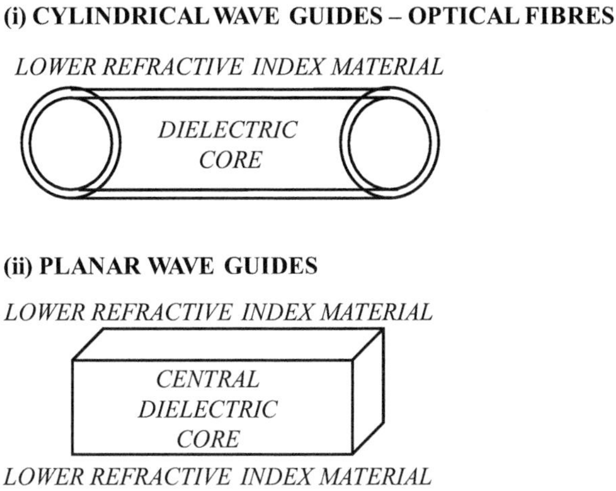

Fig. 10.4 Illustration of two types of optical waveguides based on configuration: (*i*) cylindrical waveguides and (*ii*) planar waveguides

where the polyclonal antibodies specific for *E. coli* are tagged with fluorescent secondary antibodies that are fixed to borosilicate glass fibers priorly treated with silane for the output signal. A photodiode placed at one terminal of the glass fiber monitors the optical changes taking place upon interaction with *E. coli*, the consequent changes in optical emission signals, the refractive index of glass fiber, and eventually total internal reflection. This happens to be a simple and portable biosensor that requires improvement in sensitivity.

10.3.2 Surface Plasmon Resonance Biosensors

The real-time and label-free sensing of bioanalytes was achieved with surface plasmon resonance-based biosensors. The optical phenomenon employed here is accompanied by alteration of surface plasmon wave following the changes on the sensor surface (mostly made of metals such as gold, silver, or copper) upon binding of the target analytes with specificity (Homola 2003).

In this method of sensing, a prism-coupled system is employed as depicted in Fig. 10.5 (Brindha et al. 2018b; Pi et al. 2016). The difference in signal observed in this type of sensing is attributed to the alterations in the interfacial refractive index at the top upon interaction with biomolecules.

Direct detection of *Campylobacter jejuni* was achieved with a sensor surface made of a gold chip which was coated with polyclonal antibodies specific to *C. jejuni*. This type of direct biosensing of *C. jejuni* resulted in a relatively higher limit of detection of 8×10^6 cfu/ml (Masdor et al. 2017). In order to improve the detection limits, the sandwich assay was developed that resulted in detecting as low as 4×10^4cfu/ml. This proved as an efficient and alternate method to replace the traditional enzyme-linked immunosorbent assays with a limit of detection of 10^6–10^7 cfu/ml.

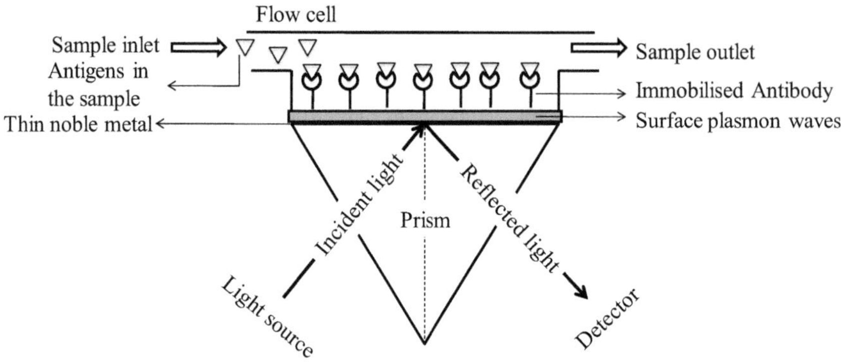

Fig. 10.5 Schematic representation on the construction and working principle of a surface plasmon resonance biosensor

10.3.2.1 Integrated Surface Plasmon Resonance Biosensors

Integration of two techniques has always yielded good results. Likewise, surface plasmon resonance accompanied with other detection techniques or signal enrichment tools including fluorophore labels, immunolabels, magnetic nanoparticles, and polymerase chain reactions are proved to enhance the sensitivity and decrease the detection limit.

Surface plasmon resonance-based biosensors with increased sensitivity were obtained by combining with immunolabels. Such sensors have been developed for detecting *Salmonella* at 10^3cfu/ml in food samples (Farka et al. 2016). It was developed by immobilizing sensor chip surface with capture antibodies with high specificity for *Salmonella* species in the test sample along with secondary antibody conjugation with horseradish peroxidase enzyme. Similarly, the use of sandwich immunoassay in surface plasmon resonance biosensors with sample enrichment using specific antibodies bound to iron oxide nanoparticles followed by magnetic isolation of these pathogenic immunocomplexes exhibited high sensitivity with a lower detection limit of 14 cfu/ml (Fig. 10.6) (Liu et al. 2016a). Here, the chip is immobilized polyclonal antibody for *Salmonella* detection using the specific immunomagnetic nanoparticles as shown in Fig. 10.6. These nanoparticles also amplify the signals by altering the refractive index differences for specific targets. DNA-immobilized surface plasmon resonance biosensors were also established to overcome the issues in the production of specific antibodies (Arya et al. 2011). A gold chip biosensor was developed with carboxylated dextran immobilized on its surface, with a complex of streptavidin/biotinylated oligonucleotide (single-stranded) probes on top (Zhang et al. 2012). The probe hybridizes with a specific and complementary, highly conserved gene of pathogen in the given sample. It had a limit of detection of 10^2 cfu/ml with detection time of 4.5 hours. The regeneration ability of the sensor surface for a minimum of 300 assay cycles makes it a cost-effective pathogen tool.

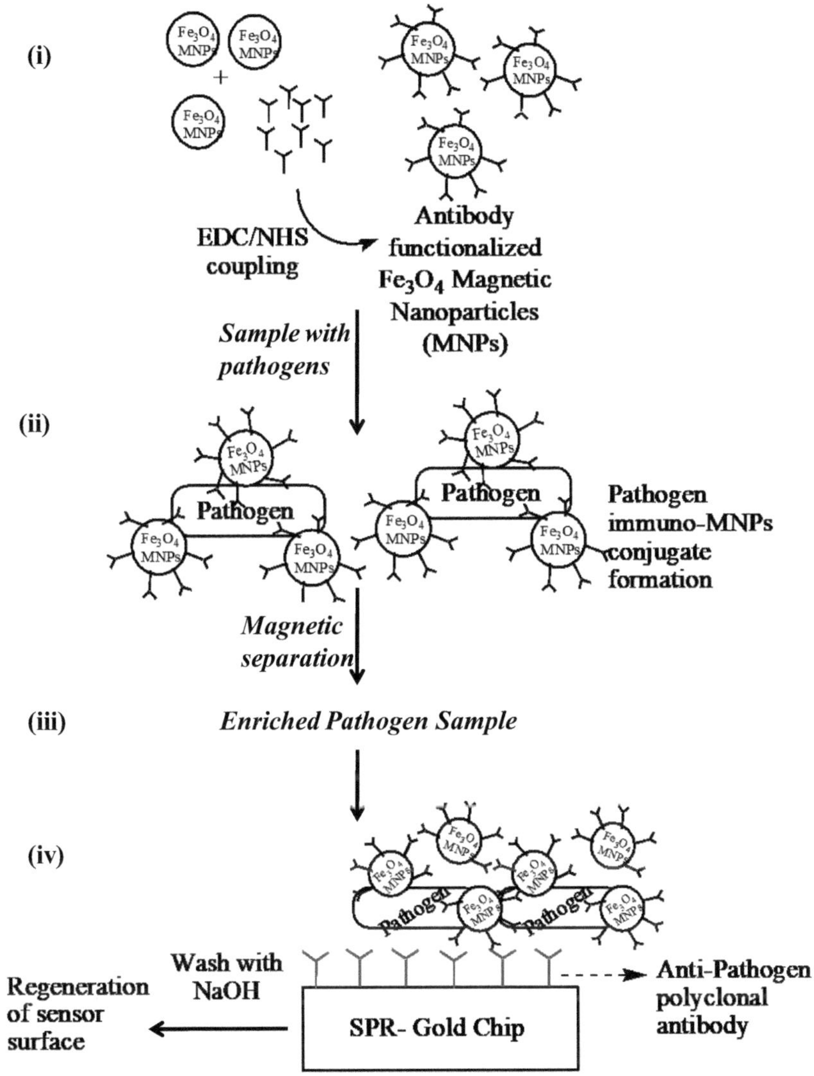

Fig. 10.6 (*i*) Antibody functionalization with magnetic iron oxide nanoparticles by EDC/NHS coupling; (*ii*) target selection by nanoparticle conjugated antibodies; (*iii*) target enrichment by magnetic separation of iron oxide nanoparticle conjugated antibodies from the sample mixture; (*iv*) target detection with an anti-pathogen polyclonal antibody. SPR, surface plasmon resonance; EDC/NHS-1, ethyl-3-(3-dimethylaminopropyl)-carbodiimide/N-hydroxysuccinimide; Fe_3O_4 MNPs, iron oxide magnetic nanoparticles

Other available pathogen-detecting sensors that require very less sample volume is the polymerase chain reaction microchip. Here, the amplified DNA is taken into the optical fiber surface plasmon resonance biosensor (Nguyen et al. 2017). This proves to be an efficient label-free method with reusability for pathogen detection, as compared to the other abovementioned methods.

10.3.3 Piezoelectric Biosensors

10.3.3.1 Piezoelectric Quartz Crystal Biosensors

Piezoelectric quartz crystal-based biosensors employ the specific attachment of a target biomolecule leading to mass change which can be detected by the equivalent variations in electrical/acoustic properties (Alder and McCallum 1983) of the piezoelectric quartz crystal. It involves a simple methodology without any sample preparation or detection labels (Arlett et al. 2011). Quartz crystal is the mainly used piezoelectric material (Deakin and Buttry 1989), for it has desirable electromechanical and chemical properties. A proportional relation exists between differences in crystal mass and its resonance which is exploited for pathogen detection in these biosensors.

There exists two modes of pathogen detection (Farka et al. 2013) in piezoelectric biosensors, namely, *active mode* or *passive mode*. *Active mode* works by wavering of the piezoelectric crystal with corresponding resonance frequency differences monitored by frequency counter (Arnau 2008), while the *passive mode piezoelectric biosensor* (Zhang et al. 2002) employs an equipment to observe the changes in mass/viscosity as a result of binding of analyte on sensor surface (Itoh and Ichihashi 2008).

A series-piezoelectric quartz crystal biosensor was developed for *S. aureus* detection (Lian et al. 2015), which uses graphene-layered interdigital gold electrodes arranged in series with piezoelectric quartz crystal, upon which aptamers/antibodies specific to the targeted pathogen were fixed as probe molecules for the detection of appropriate pathogens. Here, the targeted pathogenic DNA (i.e., *S. aureus*) hybridizes selectively to the aptamer probes, depleting the binding forces of aptamer with graphene (Shi et al. 2017). This ultimately results in variations of electrical features at the surface of the electrode and appropriate differences in oscillation frequencies yielding a detection limit of pathogen of 41 cfu/ml. Likewise, using an antimicrobial peptide probe, pleurocidin, and a transducer made of a single-walled carbon nanotubes/interdigital electrode, a multichannel series piezoelectric quartz crystal-based biosensor was developed for rapid multiple microbial detection including *S. aureus*, *Pseudomonas aeruginosa*, *Enterococcus faecalis*, *Streptococcus pneumoniae*, *Klebsiella pneumonia*, *Enterobacter cloacae*, *E. coli*, and *Candida albicans* within time period of 15 minutes (Shi et al. 2017). This biosensor finds its significance in clinical diagnosis(Jordana-Lluch et al. 2013) and food safety measures (Farahi et al. 2012). Here, the probe pleurocidin binds with target microbe causing detachment of pleurocidin from the carbon nanotubes, consequently changing the resistance of the electrodes and frequency variation of piezoelectric crystal. It is applicable for all microbes to be detected, considered as the first step for the screening of microbial blood stream infection (Gonsalves and Sakr 2010) and for testing microbial drug susceptibility. It is significant to note that enhanced sensitivity was reported for early *M. tuberculosis* detection using this piezoelectric quartz crystal-based biosensor than other methods (Zheng et al. 2007; Li et al. 2009).

10.3.3.2 Piezoelectric Cantilever Biosensors

Cantilevers denote the non-flexible structural elements that are fixed at one end of a solid firm support. Piezoelectric biosensors use biological receptors conjugated to the micro-cantilevers, that possess distinct resonance frequencies that respond on binding of target analytes in the sample, with mass increase or mechanical stress on the cantilever (Ahmed et al. 2014). This change in cantilever property is used for many pathogenic bacterial detection such as *E. coli* O157:H7 (Zhang and Hai-Feng 2004; Campbell and Mutharasan 2005a), *S. typhimurium* (Zhu et al. 2007), and *Francisella tularensis* (Ji et al. 2004) (a biowarfare agent). Detection of various bacteria like *E. coli* (Campbell and Mutharasan 2007) and *L. monocytogenes* in milk samples (Sharma and Mutharasan 2013) have been made possible using specific antibody bound to cantilevers excited by piezoelectricity. *E. coli* O157:H7 detection up to a minimum of 700 cells/ml is reported using cantilevers bound to specific monoclonal antibody of pathogen targeted (Campbell and Mutharasan 2005b). Impedance analyzers are employed to monitor the resonance frequency changes and thereby the concentration of targeted pathogen by the selective binding to specific antibody at the cantilever tips. Amplitude ratio and phase angle variations proportional to changes mass/mechanical stress were monitored with the help of an impedance analyzer (Campbell and Mutharasan 2006).

These piezoelectric cantilever biosensors are also used for specific virulence hemolysin gene, hlyA (DNA/gene)-based detection of pathogen, *L. monocytogenes*, using specific probes bound to a cantilever, detected with a fluorescent indicator and gold nanoparticles bound to secondary single-stranded DNA for hybridization signal amplification (Sankaranarayanan et al. 2015). It proves to be rapid relative to other typical pathogen detection strategies.

10.3.4 Electrochemical Biosensors

Electrochemical biosensors use the changes in electrical properties such as current, voltage, or impedance/resistance at the electrode surface on binding of biological analytes including pathogens/biomolecules in a test sample. On the basis of principle of detection and the electrical properties, they are classified as impedimetric/conductometric and amperometric/voltammetric biosensors.

It also involves another two subtypes including labeled and label-free electrochemical biosensors (Xu et al. 2017) on the basis of usage of labels such as enzymes and nanoparticles (Fei et al. 2015; Xiang et al. 2015) bound to the targeted bioanalyte in the case of labeled biosensors. However, the label-free method operates in the absence of any bound labels, based on interaction of bacterial cells at the electrode surface (Sang et al. 2016), that are discussed side by side along with other biosensor developments.

10.3.4.1 Amperometric/Voltammetric Biosensors

Biosensors detecting pathogens with respect to variations in current or potential (Monzo et al. 2015), due to processes like oxidation or reduction of the targeted biological species, where the electrode is at constant potential or current with respect to a standard reference electrode (Bard and Faulkner 2001). Cyclic sweep voltammetry is a common method to acquire data like oxidation/reduction potentials, reaction kinetics, and mechanisms (Bard and Faulkner 2001; Compton and Banks 2011). Using carbon electrodes infused with magnetic nanoparticles, the specific genomic DNA of pathogens and gold nanoparticles in a sandwich assay like format detection of *Salmonella* and *E. coli* could be performed (Blow 2015). Electroanalysis using differential pulse voltammetry and square wave voltammetry are known for better time resolution and high-frequency operation (Chen and Shah 2013). Labeled antibody-based electrochemical biosensor having an immuno-electrode made up of graphene oxide–silver nanoparticles fixed over anti-*Salmonella typhimurium* antibody, along with the help of cyclic voltammetry was proved to be used for *Salmonella* detection (Sign and Sumana 2016). Likewise, detection of other pathogens can be carried out using specific antibodies.

Amperometric sensors are known to be fast and cost-effective (Barlett 2008; Wei et al. 2009), while they also pose limitations such as poor selectivity due to other interfering constituents in the sample mixture varying the faradaic current (Monzo et al. 2015). Sensitivity in microbial/biomolecule detection enhances using semiconductor devices and field-effect transistors due to enrichment of sensor signals (Grieshaber et al. 2008; Lin et al. 2008). Recently, a DNA chip-based sensor was developed, for *S. pyogenes* detection, which works by hybridization of genomic DNA of pathogens from the sample, with the selective probe bound to the screen-printed electrode resulting in an amperometric change in current recorded using differential pulse voltammetry. It was reported to be a sensitive biosensor with a limit of detection of 130 fg DNA per 6 μl of the sample.

10.3.4.2 Impedimetric/Conductometric Biosensors

Biosensors based on impedance measurements rely on the impedance variations occurring due to voltage signal changes on the binding of targeted biomolecule/pathogen to the electrode (Bard and Faulkner 2001; Barlett 2008). The targeted pathogens interacting with the electrode are evaluated in accordance with variations in the capacitance/impedance at the electrode interface. Impedance spectroscopy offers high sensitivity and selectivity for the detection of biological analytes including pathogens (Felice et al. 1999). A carbon electrode modified with reduced graphene oxide was reported to use impedance spectroscopy (Wang et al. 2011) for the detection of methicillin-resistant *S. aureus*. Similarly, *Salmonella* detection was also successful using a label-free technique employing a combination of polypyrrole-based polymer, poly [pyrrole-co-3-carboxyl-pyrrole] copolymer, and a selective aptamer (Sheikhzadeh et al. 2016). Also, researchers have developed similar biosensors as above with gene−/immune-based (instead of aptamers) synthetically designed specific pathogenic peptides along with a record of impedance variations (Liu et al. 2016b).

More recently development of a microfluidic impedance-based biosensor is shown to be successful for *E. coli* O157:H7 (Yao et al. 2018) detection with a minimum limit of 12 cfu/ml within a period of 2 hours. Here, the impedance was monitored using the microfluidic chip, and normalization of impedance helped in the detection of *E. coli* O157:H7. Another recent patent involves the use of conductometric biosensors along with the use of nanotubes for pathogen detection (April et al. 2018). The conductometric component involving the electrodes measures the resistance variation that is proportional to target analyte concentrations specifically interacting with the recognition elements on the electrodes.

10.3.5 Biochemical Biosensors

Biochemical biosensors help in the detection of specific biochemical reactions occurring between a specific target analyte in the test sample and its specific substrate/receptor molecule on these sensors. These substrates used can be any compatible specific biomolecules such as antibodies/enzymes essential for the immuno/biochemical reactions.

10.3.5.1 Immunosensors

Immunosensors detect a biomolecule/analyte using specific immunochemical or biochemical reactions occurring between bio-recognition elements and specific receptors along with the help of transducers to convert the binding interactions into readable output signals. These immuno/biochemical reactions or the ligand–receptor binding offers high sensitivity and selectivity to the immunosensors. These sensors are equipped with signal amplification in the presence of either magnetic particles or gold nanoparticles (Wang and Alocilja 2015) for high sensitivity of pathogen detection. Developments are made in designing an integrated immunosensor, with specific antibody conjugated to gold nanoparticles employed in enzyme-linked immunosorbent assay along with enrichment of target analytes in the sample mixture, using immunomagnetic separation (Cho and Irudayaraj 2013). It was proven analytically with microtiter assay, for a highly sensitive pathogen detection including *E. coli* O157:H7 and *S. typhimurium* detection up to 3 cells/ml in buffer.

Continuous efforts are made to develop immunosensors with high sensitivity employing signal amplification strategies (Guo et al. 2013). A visible indication strategy is one of those techniques that employ color formation based on nanoparticle aggregation (Priyadarshini and Pradhan 2017). Here, the limitation is the requirement of the presence of a large number of analytes for the generation of visible colorimetric detection signal that lowers the sensitivity of the biosensor (Yoo and Lee 2016). The need for high sensitivity, real-time detection, and lower detection time remains the significant areas under research in the biosensor field, due to existing problems in the current methods. For example, the minimum limit of detection reported to date is 10^2 cfu/ml, and detection time is greater than 6 hours (Prasad and Vidyarthi 2011; Raj et al. 2015). This has encouraged researchers to develop new strategies for signal amplification such as bifunctional linkers or switchable linkers (Hahn et al. 2017; Lim et al. 2012). A bifunctional linker with biotinylated specific antibacterial antibodies is developed which binds to the specific target and leads to visible color changes due to aggregation of the target with the probe (streptavidin–gold nanoparticles), biotin–streptavidin binding reaction (Weber et al. 1989), and also the localized surface plasmon resonance phenomenon. This strategy was improved by altering the amount of streptavidin–gold nanoparticles, to result in a lower limit of detection (10 cfu/ml) of the pathogens *E. coli* and *Salmonella* (You et al. 2018). This was tested in samples of tap water, lake water, and milk samples, which are unaltered by matrix effects. Thus, it offers a speedy, highly sensitive real-time pathogen detection biosensor. Apart from this, simultaneous detection of multiple analytes is one of the major requirements fulfilled using an immunochemical method for diagnosing various infections in test samples. A unique virulence factor, for example, pyocyanin, secreted only by *P. aeruginosa* (Pastells et al. 2016), was used for its detection. By using antibodies of a specific metabolite of the virulence factor pyocyanin, 1-hydroxyphenazine, the virulence factor and hence the pathogen could be quantified/detected.

Antibiogram is a method of detection of early microbial growth. This method employed the 96-well plate format which was modified by Elavarasan et al., into a handy, polymethylmethacrylate microfluidic chip, wherein resazurin, a blue-colored water-soluble dye, was used. The response was monitored by change of color following the biochemical reactions specific for viable cells (Elavarasan et al. 2013; Kaur et al. 2013).

The reduction of water-soluble resazurin occurs in two steps when in contact with viable cells (Fig. 10.7) (Brindha et al. 2018b). The first step results in an irreversible pink colored, partially oxidized form of pink-colored resorufin. This, upon further oxidation, yields a reversible and colorless hydroresorufin (Sarker et al. 2007). This dye acts as an indicator to test cell viability, growth, and toxicity (Palomino et al. 2002). This immunoassay was carried out on a microfluidic chip not only for detecting the presence of milk pathogens but also for detecting multidrug-resistant pathogens. In this assay, as shown below in Fig. 10.7, the interpretation is based on color developed in the presence of microbial samples. When the microorganisms are susceptible to antibiotics, with no cell growth, the blue color is observed, while when it is resistant, a pink/colorless solution is observed. In the case of moderate to negligible growth of the cells, violet color is seen as a partial/complete reduction of resazurin occurs.

10.4 Conclusion and Future Prospects

All the developments of sensors in the recent past are targeted for specificity, lower detection time, lower detection limits, and enhanced sensitivity for detecting analytes. Integration of multiple labeling techniques with immuno–/enzymatic reagents or fluorophores in association with fiber optics or surface plasmon resonance is in practice, for the enhancement of selectivity/sensitivity of pathogen detection. Recently, biosensor research focuses on target analyte enrichment (Zuo

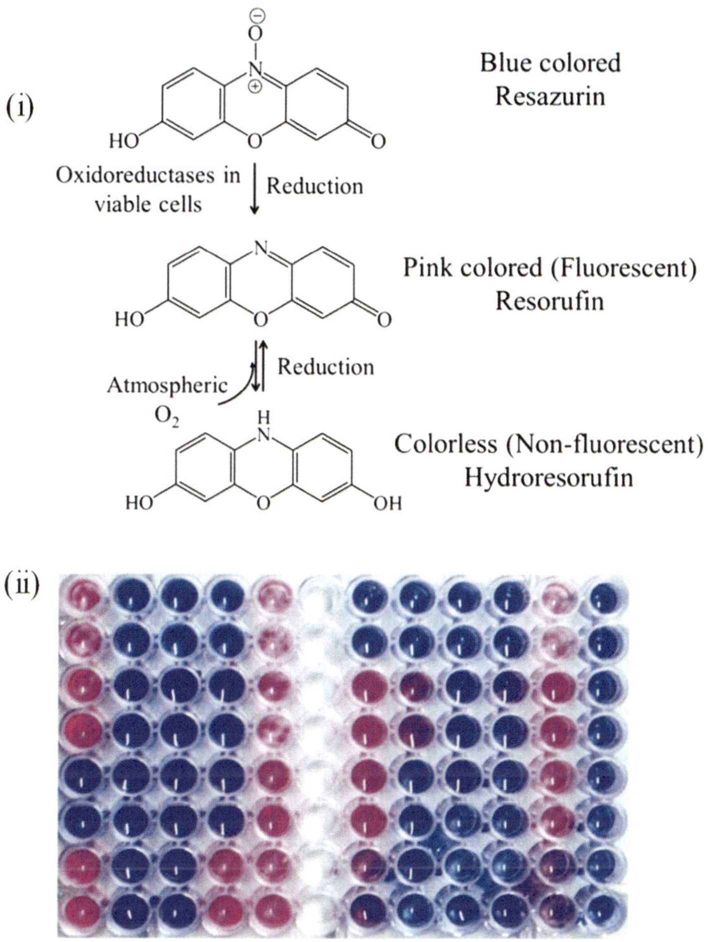

Fig. 10.7 Mechanism of immunoassay following (*i*) the reduction of resazurin to resorufin and further to hydroresorufin in the presence of viable cells, (*ii*) 96-well plate format depicting the antibiogram for a specific pathogen, blue color, blank/antibiotic susceptibility; pink/colorless, antibiotic resistance; violet, intermediate to poor growth of microbes in the sample

et al. 2013; Kim et al. 2014; Hsieh et al. 2015; Altobelli et al. 2016) and amplification of signal responses. New technologies like microfluidics, nanomaterials like nanotubes/nanoparticles, and simpler strategies such as visible indication strategy, and immuno/biochemical reactions in microchips are being developed. Most of the existing chemosensors are fluorescence based. Chemosensor development focuses on new stable fluorophore designing, integration of surface plasmon resonance with fluorophores/optical fibers, and simple miniaturized tools for sensing toxic metal ions/chemicals like pesticides in food/water or other significant environments. Future research endeavors ought to focus on bringing these user-friendly pathogen/chemical surveillance tools to the market at lower affordable prices as preventive measures against numerous existing infections or other possible epidemics outbreaks.

References

Adler C, Corbalan NS, Peralta DR, Pomares MF, de Cristóbal RE, Vincent PA (2014) The alternative role of enterobactin as an oxidative stress protector allows Escherichia coli colony development. PLoS One 9:e84734. https://doi.org/10.1371/journal.pone.0084734

Ahmed A, Rushworth JV, Hirst NA, Millner PA (2014) Biosensors for whole-cell bacterial detection. Clin Microbiol Rev 27:631–646. https://doi.org/10.1128/CMR.00120-13

Alder JF, McCallum JJ (1983) Piezoelectric crystals for mass and chemical measurements. A review. Analyst 108:1169–1189. https://doi.org/10.1039/AN9830801169

Altobelli E, Mohan R, Mach KE, Sin MLY, Anikst V, Buscarini M, Wong PK, Gau V, Banaei N, Liao JC (2016) Integrated biosensor assay for rapid uropathogen identification and phenotypic antimicrobial susceptibility testing. Eur Urol Focus. https://doi.org/10.1016/j.euf.2015.12.010

Ameen S, Akhtar MS, Shin HS (2016) Manipulating the structure of polyaniline by exploiting redox chemistry: novel p-NiO/n-polyaniline/n-Si schottky diode based chemosensor for the electrochemical detection of hydrazinobenzene. Electrochim Acta 215:200–211

April G, Yildirim N, Lee J, Cho H, Busnaina A (2018) Google patents

Arlett J, Myers E, Roukes M (2011) Comparative advantages of mechanical biosensors. Nat Nanotechnol 6:203–215. https://doi.org/10.1038/nnano.2011.44

Arnau A (2008) A review of interface electronic systems for AT-cut quartz crystal microbalance applications in liquids. Sensors 8:370–411. https://doi.org/10.3390/s8010370

Arya SK, Singh A, Naidoo R, Wu P, McDermott MT, Evoy S (2011) Chemically immobilized T4-bacteriophage for specific Escherichia coli detection using surface plasmon resonance. Analyst 136:486–492. https://doi.org/10.1039/C0AN00697A

Azab HA, Kamel RM (2016) Sensitive and selective fluorescent chemosensor for the detection of some organophosphorus pesticides using luminescent Eu (iii) complex. J Photochem Photobiol A Chem 321:33–40

Banica F-G (2012) Chemical sensors and biosensors: fundamentals and applications. Wiley

Bard AJ, Faulkner LR (2001) Fundamentals and applications. Electrochem Methods

Barlett P (2008) Bioelectrochemistry: fundamentals, experimental techniques and applications. Wiley

Beer PD (1996) Anion selective recognition and optical/electrochemical sensing by novel transition-metal receptor systems. Chem Commun:689–696

Beer PD, Gale PA, Chen GZ (1999a) Mechanisms of electrochemical recognition of cations, anions and neutral guest species by redox-active receptor molecules. Coord Chem Rev 185:3–36

Beer PD, Gale PA, Chen GZ (1999b) Electrochemical molecular recognition: pathways between complexation and signalling. J Chem Soc Dalton Trans:1897–1910

Bereza-Malcolm LT, Mann G, Franks AE (2014) Environmental sensing of heavy metals through whole cell microbial biosensors: a synthetic biology approach. ACS Synth Biol 4:535–546. https://doi.org/10.1021/sb500286r

Bhalla N, Jolly P, Formisano N, Estrela P (2016) Introduction to biosensors. Essays Biochem 60:1–8. https://doi.org/10.1042/EBC20150001

Blow M (2015) Annual meeting-International Association for Food Protection. Portland, Oregon

Brindha J, Chanda K, Balamurali M (2018a) Nanotechnology, food security and water treatment. Springer, pp 53–86

Brindha J, Chanda K, Balamurali M (2018b) Biosensors for pathogen surveillance. Environ Chem Lett 16:1325–1337

Campbell GA, Mutharasan R (2005a) Detection of pathogen Escherichia coli O157: H7 using self-excited PZT-glass microcantilevers. Biosens Bioelectron 21:462–473. https://doi.org/10.1016/j.bios.2004.11.009

Campbell GA, Mutharasan R (2005b) Escherichia coli O157: H7 detection limit of millimeter-sized PZT cantilever sensors is 700 cells/ml. Anal Sci 21:355–357. https://doi.org/10.2116/analsci.21.355

Campbell GA, Mutharasan R (2006) Detection of Bacillus anthracis spores and a model protein using PEMC sensors in a flow cell at 1 ml/min. Biosens Bioelectron 22:78–85. https://doi.org/10.1016/j.bios.2005.12.002

Campbell GA, Mutharasan R (2007) A method of measuring Escherichia coli O157: H7 at 1 cell/ml in 1 liter sample using antibody functionalized piezoelectric-excited millimeter-sized cantilever sensor. Environ Sci Technol 41:1668–1674. https://doi.org/10.1021/es061947p

Chang C-C, Lin S, Wei S-C, Chen C-Y, Lin C-W (2011) An amplified surface plasmon resonance "turn-on" sensor for mercury ion using gold nanoparticles. Biosens Bioelectron 30:235–240

Chen A, Shah B (2013) Electrochemical sensing and biosensing based on square wave voltammetry. Anal Methods 5:2158–2173. https://doi.org/10.1039/C3AY40155C

Cho IH, Irudayaraj J (2013) In-situ immuno-gold nanoparticle network ELISA biosensors for pathogen detection. Int J Food Microbiol 164:70–75. https://doi.org/10.1016/j.ijfoodmicro.2013.02.025

Compton RG, Banks CE (2011) Understanding voltammetry. World Scientific

Couture M, Zhao SS, Masson J-F (2013) Modern surface plasmon resonance for bioanalytics and biophysics. PCCP 15:11190–11216

Dale SE, Doherty-Kirby A, Lajoie G, Heinrichs DE (2004) Role of siderophore biosynthesis in virulence of staphylococcus aureus: identification and characterization of genes involved in production of a siderophore. Infect Immun 72:29–37. https://doi.org/10.1128/IAI.72.1.29-37.2004

Davis JJ, Green ML, Hill HAO, Leung YC, Sadler PJ, Sloan J, Xavier AV, Tsang SC (1998) The immobilisation of proteins in carbon nanotubes. Inorg Chim Acta 272:261–266. https://doi.org/10.1016/S0020-1693(97)05926-4

De Maria L, Cennamo N, Zeni L, Scatiggio F, Pesavento M (2018) 2018 AEIT international annual conference. IEEE, pp 1–6

Deakin MR, Buttry DA (1989) Electrochemical applications of the quartz crystal microbalance. Anal Chem 61:1147A–1154A. https://doi.org/10.1021/ac00195a001

Deng S, Upadhyayula VK, Smith GB, Mitchell MC (2008) Adsorption equilibrium and kinetics of microorganisms on single-wall carbon nanotubes. IEEE Sensors J 8:954–962. https://doi.org/10.1109/JSEN.2008.923929

Du X, Zhou J (2018) Application of biosensors to detection of epidemic diseases in animals. Res Vet Sci. https://doi.org/10.1016/j.rvsc.2018.04.011

Elavarasan T, Chhina SK, Parameswaran M, Sankaran K (2013) Resazurin reduction based colorimetric antibiogram in microfluidic plastic chip. Sensor Actuat B: Chem 176:174–180. https://doi.org/10.1016/j.snb.2012.10.011

Elkin T, Jiang X, Taylor S, Lin Y, Gu L, Yang H, Brown J, Collins S, Sun YP (2005) Immuno-carbon nanotubes and recognition of pathogens. Chembiochem 6:640–643. https://doi.org/10.1002/cbic.200400337

Ellwanger JH, Kaminski VDL, Chies JAB (2017) How to detect new viral outbreaks or epidemics? We need to survey the circulation of viruses in humans and other animals using fast, sensible, cheap, and broad-spectrum methodologies. Braz J Infect Dis 21:211–212. https://doi.org/10.1016/j.bjid.2016.12.001

Ermakova E, Michalak J, Meyer M, Arslanov V, Tsivadze A, Guilard R, Bessmertnykh-Lemeune A (2013) Colorimetric Hg2+ sensing in water: from molecules toward low-cost solid devices. Org Lett 15:662–665

Farahi R, Passian A, Tetard L, Thundat T (2012) Critical issues in sensor science to aid food and water safety. ACS Nano 6:4548–4556. https://doi.org/10.1021/nn204999j

Farka Z, Kovář D, Přibyl J, Skládal P (2013) Piezoelectric and surface plasmon resonance biosensors for bacillus atrophaeus spores. Int J Electrochem Sci 8:100–112

Farka ZK, Juřík TS, Pastucha MJ, Skládal P (2016) Enzymatic precipitation enhanced surface plasmon resonance immunosensor for the detection of salmonella in powdered milk. Anal Chem 88:11830–11836. https://doi.org/10.1021/acs.analchem.6b03511

Fawcett NC, Evans JA, Chen LC, Drozda KA, Flowers N (1988) A quartz crystal detector for DNA. Anal Lett 21:1099–1110

Fei J, Dou W, Zhao G (2015) A sandwich electrochemical immunosensor for Salmonella pullorum and Salmonella gallinarum based on a screen-printed carbon electrode modified with an ionic liquid and electrodeposited gold nanoparticles. Microchim Acta 182:2267–2275. https://doi.org/10.1007/s00604-015-1573-x

Felice CJ, Madrid RE, Olivera JM, Rotger VI, Valentinuzzi ME (1999) Impedance microbiology: quantification of bacterial content in milk by means of capacitance growth curves. J Microbiol Methods 35:37–42. https://doi.org/10.1016/S0167-7012(98)00098-0

Fritz J (2008) Cantilever biosensors. Analyst 133:855–863. https://doi.org/10.1039/B718174D

Gao Y, Ma T, Ou Z, Cai W, Yang G, Li Y, Xu M, Li Q (2018) Highly sensitive and selective turn-on fluorescent chemosensors for Hg(2+) based on thioacetal modified pyrene. Talanta 178:663–669. https://doi.org/10.1016/j.talanta.2017.09.089

Gonsalves MD, Sakr Y (2010) Early identification of sepsis. Curr Infect Dis Rep 12:329–335. https://doi.org/10.1007/s11908-010-0122-3

Gracias KS, McKillip JL (2004) A review of conventional detection and enumeration methods for pathogenic bacteria in food. Can J Microbiol 50:883–890. https://doi.org/10.1139/w04-080

Grieshaber D, MacKenzie R, Voeroes J, Reimhult E (2008) Electrochemical biosensors-sensor principles and architectures. Sensors 8:1400–1458. https://doi.org/10.3390/s80314000

Guo L, Xu Y, Ferhan AR, Chen G, Kim D-H (2013) Oriented gold nanoparticle aggregation for colorimetric sensors with surprisingly high analytical figures of merit. J Am Chem Soc 135:12338–12345. https://doi.org/10.1021/ja405371g

Hahn J, Kim E, You YS, Gunasekaran S, Lim S, Choi YJ (2017) A switchable linker-based immunoassay for ultrasensitive visible detection of Salmonella in tomatoes. J Food Sci 82:2321–2328. https://doi.org/10.1111/1750-3841.13861

Himpsl SD, Pearson MM, Arewång CJ, Nusca TD, Sherman DH, Mobley HL (2010) Proteobactin and a yersiniabactin-related siderophore mediate iron acquisition in Proteus mirabilis. Mol Microbiol 78:138–157. https://doi.org/10.1111/j.1365-2958.2010.07317.x

Homola J (2003) Present and future of surface plasmon resonance biosensors. Anal Bioanal Chem 377:528–539. https://doi.org/10.1007/s00216-003-2101-0

Hsieh K, Ferguson BS, Eisenstein M, Plaxco KW, Soh HT (2015) Integrated electrochemical microsystems for genetic detection of pathogens at the point of care. Acc Chem Res 48:911–920. https://doi.org/10.1021/ar500456w

Hu R-R, Yin Z-Z, Zeng Y-B, Zhang J, Liu H-Q, Shao Y, Ren S-B, Li L (2016) A novel biosensor for Escherichia coli O157: H7 based on fluorescein-releasable biolabels. Biosens Bioelectron 78:31–36. https://doi.org/10.1016/j.bios.2015.11.018

Iskierko Z, Checinska A, Sharma P, Golebiewska K, Noworyta K, Borowicz P, Fronc K, Bandi V, D'Souza F, Kutner W (2017) Molecularly imprinted polymer based extended-gate field-effect transistor chemosensors for phenylalanine enantioselective sensing. J Mater Chem C 5:969–977

Itoh A, Ichihashi M (2008) A frequency of the quartz crystal microbalance (QCM) that is not affected by the viscosity of a liquid. Meas Sci Technol 19:075205. https://doi.org/10.1088/0957-0233/19/7/075205

Ji HF, Yang X, Zhang J, Thundat T (2004) Molecular recognition of biowarfare agents using micromechanical sensors. Expert Rev Mol Diagn 4:859–866. https://doi.org/10.1586/14737159.4.6.859

Jordana-Lluch E, Carolan HE, Giménez M, Sampath R, Ecker DJ, Quesada MD, Mòdol JM, Arméstar F, Blyn LB, Cummins LL (2013) Rapid diagnosis of bloodstream infections with

PCR followed by mass spectrometry. PLoS One 8:e62108. https://doi.org/10.1371/journal.pone.0062108

Jung S-A, Lee T-S, Kim WM, Lee K-S, Jeong DS, Lee WS, Kim I (2013) Thickness dependence of surface plasmon resonance sensor response for metal ion detection. J Phys D Appl Phys 46:315104

Kaur N, Prasad R, Varma A (2013) Antibiotic resistance among clinical isolates of *Staphylococcus aureus* and usefulness of antibiogram. Int J Pharm Bio Sci 4(1):957–964

Kaushik A, Tiwari S, Jayant RD, Vashist A, Nikkhah-Moshaie R, El-Hage N, Nair M (2017) Electrochemical biosensors for early stage Zika diagnostics. Trends Biotechnol 35:308–317. https://doi.org/10.1016/j.tibtech.2016.10.001

Kim T-H, Park J, Kim C-J, Cho Y-K (2014) Fully integrated lab-on-a-disc for nucleic acid analysis of food-borne pathogens. Anal Chem 86:3841–3848. https://doi.org/10.1021/ac403971h

Li G, Li X, Wan J, Zhang S (2009) Dendrimers-based DNA biosensors for highly sensitive electrochemical detection of DNA hybridization using reporter probe DNA modified with au nanoparticles. Biosens Bioelectron 24:3281–3287. https://doi.org/10.1016/j.bios.2009.04.022

Li B, Yu Q, Duan Y (2015) Fluorescent labels in biosensors for pathogen detection. Crit Rev Biotechnol 35:82–93. https://doi.org/10.3109/07388551.2013.804487

Lian Y, He F, Wang H, Tong F (2015) A new aptamer/graphene interdigitated gold electrode piezoelectric sensor for rapid and specific detection of Staphylococcus aureus. Biosens Bioelectron 65:314–319. https://doi.org/10.1016/j.bios.2014.10.017

Lim S, Koo OK, You YS, Lee YE, Kim M-S, Chang P-S, Kang DH, Yu J-H, Choi YJ, Gunasekaran S (2012) Enhancing nanoparticle-based visible detection by controlling the extent of aggregation. Sci Rep 2:456. https://doi.org/10.1038/srep00456

Lin Y-H, Chen S-H, Chuang Y-C, Lu Y-C, Shen TY, Chang CA, Lin C-S (2008) Disposable amperometric immunosensing strips fabricated by au nanoparticles-modified screen-printed carbon electrodes for the detection of foodborne pathogen Escherichia coli O157: H7. Biosens Bioelectron 23:1832–1837. https://doi.org/10.1016/j.bios.2008.02.030

Liu X, Hu Y, Zheng S, Liu Y, He Z, Luo F (2016a) Surface plasmon resonance immunosensor for fast, highly sensitive, and in situ detection of the magnetic nanoparticles-enriched salmonella enteritidis. Sensor Actuat B: Chem 230:191–198. https://doi.org/10.1016/j.snb.2016.02.043

Liu X, Marrakchi M, Xu D, Dong H, Andreescu S (2016b) Biosensors based on modularly designed synthetic peptides for recognition, detection and live/dead differentiation of pathogenic bacteria. Biosens Bioelectron 80:9–16. https://doi.org/10.1016/j.snb.2016.02.043

Maas MB, Maybery GH, Perold WJ, Neveling DP, Dicks LM (2018) Borosilicate glass fiber-optic biosensor for the detection of Escherichia coli. Curr Microbiol 75:150–155. https://doi.org/10.1007/s00284-017-1359-y

Martín-Yerga D, González-García MB, Costa-García A (2013) Electrochemical determination of mercury: a review. Talanta 116:1091–1104

Masdor NA, Altintas Z, Tothill IE (2017) Surface plasmon resonance immunosensor for the detection of Campylobacter jejuni. Chemosensors 5:16. https://doi.org/10.3390/chemosensors5020016

Menti C, Henriques JA, Missell FP, Roesch-Ely M (2016) Antibody-based magneto-elastic biosensors: potential devices for detection of pathogens and associated toxins. Appl Microbiol Biotechnol 100:6149–6163. https://doi.org/10.1007/s00253-016-7624-3

Miyaji H, Sessler JL (2001) Off-the-shelf colorimetric anion sensors. Angew Chem Int Ed 40:154–157

Monis PT, Giglio S (2006) Nucleic acid amplification-based techniques for pathogen detection and identification. Infect Genet Evol 6:2–12. https://doi.org/10.1016/j.meegid.2005.08.004

Monzo J, Insua I, Fernandez-Trillo F, Rodriguez P (2015) Fundamentals, achievements and challenges in the electrochemical sensing of pathogens. Analyst 140:7116–7128. https://doi.org/10.1039/c5an01330e

Nguyen TT, Trinh KTL, Yoon WJ, Lee NY, Ju H (2017) Integration of a microfluidic polymerase chain reaction device and surface plasmon resonance fiber sensor into an inline all-in-one platform for pathogenic bacteria detection. Sensor Actuat B: Chem 242:1–8. https://doi.org/10.1016/j.snb.2016.10.137

Ohk S-H, Bhunia AK (2013) Multiplex fiber optic biosensor for detection of listeria monocytogenes, Escherichia coli O157: H7 and Salmonella enterica from ready-to-eat meat samples. Food Microbiol 33:166–171. https://doi.org/10.1016/j.fm.2012.09.013

Palomino J-C, Martin A, Camacho M, Guerra H, Swings J, Portaels F (2002) Resazurin microtiter assay plate: simple and inexpensive method for detection of drug resistance in Mycobacterium tuberculosis. Antimicrob Agents Chemother 46:2720–2722. https://doi.org/10.1016/j.ejcdt.2013.05.008

Panta YM, Liu J, Cheney MA, Joo SW, Qian S (2009) Ultrasensitive detection of mercury (ii) ions using electrochemical surface plasmon resonance with magnetohydrodynamic convection. J Colloid Interface Sci 333:485–490

Pastells C, Pascual N, Sanchez-Baeza F, Marco M-P (2016) Immunochemical determination of pyocyanin and 1-hydroxyphenazine as potential biomarkers of Pseudomonas aeruginosa infections. Anal Chem 88:1631–1638. https://doi.org/10.1021/acs.analchem.5b03490

Patching SG (2014) Surface plasmon resonance spectroscopy for characterisation of membrane protein–ligand interactions and its potential for drug discovery. Biochimica et Biophysica Acta (BBA)-Biomembranes 1838:43–55

Peralta-Domínguez D, Rodriguez M, Ramos-Ortiz G, Maldonado J, Luna-Moreno D, Ortiz-Gutierrez M, Barba V (2016) A Schiff base derivative used as sensor of copper through colorimetric and surface plasmon resonance techniques. Sensors Actuators B Chem 225:221–227

Pi S, Zeng X, Zhang N, Ji D, Chen B, Song H, Cheney A, Xu Y, Jiang S, Sun D (2016) Dielectric-grating-coupled surface plasmon resonance from the back side of the metal film for ultrasensitive sensing. IEEE Photon J 8:1–6. https://doi.org/10.1109/JPHOT.2015.2509870

Prabhakaran D, Yuehong M, Nanjo H, Matsunaga H (2007) Naked-eye cadmium sensor: using chromoionophore arrays of Langmuir–Blodgett molecular assemblies. Anal Chem 79:4056–4065

Prasad D, Vidyarthi AS (2011) Gold nanoparticles-based colorimetric assay for rapid detection of Salmonella species in food samples. World J Microbiol Biotechnol 27:2227–2230. https://doi.org/10.1007/s11274-011-0679-5

Priyadarshini E, Pradhan N (2017) Gold nanoparticles as efficient sensors in colorimetric detection of toxic metal ions: a review. Sensors Actuators B Chem 238:888–902. https://doi.org/10.1016/j.snb.2016.06.081

Raj V, Vijayan AN, Joseph K (2015) Cysteine capped gold nanoparticles for naked eye detection of E. coli bacteria in UTI patients. Sens Bio-Sens Res 5:33–36. https://doi.org/10.1016/j.sbsr.2015.05.004

Resch-Genger U, Grabolle M, Cavaliere-Jaricot S, Nitschke R, Nann T (2008) Quantum dots versus organic dyes as fluorescent labels. Nat Methods 5:763. https://doi.org/10.1038/nmeth.1248

Rifat AA, Mahdiraji GA, Sua YM, Shee YG, Ahmed R, Chow DM, Adikan FRM (2015) Surface plasmon resonance photonic crystal fiber biosensor: a practical sensing approach. IPTL 27:1628–1631. https://doi.org/10.1109/lpt.2015.2432812

Sánchez G, Curiel D, Tatkiewcz W, Ratera I, Tárraga A, Veciana J, Molina P (2014) Highly sensitive and selective detection of the pyrophosphate anion biomarker under physiological conditions. Chem Sci 5:2328–2335

Sang S, Wang Y, Feng Q, Wei Y, Ji J, Zhang W (2016) Progress of new label-free techniques for biosensors: a review. Crit Rev Biotechnol 36:465–481. https://doi.org/10.3109/07388551.2014.991270

Sankaranarayanan R, Alagumaruthanayagam A, Sankaran K (2015) A new fluorimetric method for the detection and quantification of siderophores using calcein blue, with potential as a bacterial detection tool. Appl Microbiol Biotechnol 99:2339–2349. https://doi.org/10.1007/s00253-015-6411-x

Sarker SD, Nahar L, Kumarasamy Y (2007) Microtitre plate-based antibacterial assay incorporating resazurin as an indicator of cell growth, and its application in the in vitro antibacterial screening of phytochemicals. Methods 42:321–324. https://doi.org/10.1016/j.ymeth.2007.01.006

Scognamiglio V, Arduini F, Palleschi G, Rea G (2014) Biosensing technology for sustainable food safety. TrAC Trends Anal Chem 62:1–10. https://doi.org/10.1016/j.trac.2014.07.007

Sharma H, Mutharasan R (2013) Rapid and sensitive immunodetection of Listeria monocytogenes in milk using a novel piezoelectric cantilever sensor. Biosens Bioelectron 45:158–162. https://doi.org/10.1016/j.bios.2013.01.068

Sharma R, Mittal SK, Chhibber M (2015) Voltammetric sensor for fluoride ions using diphenylether derivatives supported by NMR and theoretical studies. J Electrochem Soc 162: B248–B255

Sharma PS, Wojnarowicz A, Sosnowska M, Benincori T, Noworyta K, D'Souza F, Kutner W (2016) Potentiometric chemosensor for neopterin, a cancer biomarker, using an electrochemically synthesized molecularly imprinted polymer as the recognition unit. Biosens Bioelectron 77:565–572

Sheikhzadeh E, Chamsaz M, Turner A, Jager E, Beni V (2016) Label-free impedimetric biosensor for salmonella typhimurium detection based on poly [pyrrole-co-3-carboxyl-pyrrole] copolymer supported aptamer. Biosens Bioelectron 80:194–200. https://doi.org/10.1016/j.bios.2016.01.057

Shen J, Zhu Y, Yang X, Li C (2012) Graphene quantum dots: emergent nanolights for bioimaging, sensors, catalysis and photovoltaic devices. Chem Commun 48:3686–3699. https://doi.org/10.1039/C2CC00110A

Shi X, Zhang X, Yao Q, He F (2017) A novel method for the rapid detection of microbes in blood using pleurocidin antimicrobial peptide functionalized piezoelectric sensor. J Microbiol Methods 133:69–75. https://doi.org/10.1016/j.mimet.2016.12.005

Sign C, Sumana G (2016) Journal of Physics: Conference Series. IOP Publishing, p 012014. https://doi.org/10.1088/1742-6596/704/1/012014

Silva AP, NimaláGunaratne H (1996) Fluorescent switches with high selectivity towards sodium ions: correlation of ion-induced conformation switching with fluorescence function. Chem Commun:1967–1968

Singh S, Kumar V, Dhanjal DS, Datta S, Prasad R, Singh J (2020) Biological Biosensors for Monitoring and Diagnosis. In: Singh J, Vyas A, Wang S, Prasad R (eds) Microbial biotechnology: basic research and applications. Springer Nature Singapore, Singapore, pp 317–336

Song F, Yang C, Liu H, Gao Z, Zhu J, Bao X, Kan C (2019) Dual-binding pyridine and rhodamine b conjugate derivatives as fluorescent chemosensors for ferric ions in aqueous media and living cells. Analyst 144:3094–3102. https://doi.org/10.1039/c8an01915k

Srivastava A, Srivastava O, Talapatra S, Vajtai R, Ajayan P (2004) Carbon nanotube filters. Nat Mater 3:610–614. https://doi.org/10.1038/nmat1192

Tabassum R, Gupta BD (2015) Fiber optic manganese ions sensor using SPR and nanocomposite of ZnO–polypyrrole. Sensors Actuators B Chem 220:903–909

Taneja NK, Tyagi JS (2007) Resazurin reduction assays for screening of anti-tubercular compounds against dormant and actively growing mycobacterium tuberculosis, mycobacterium bovis BCG and mycobacterium smegmatis. J Antimicrob Chemother 60:288–293. https://doi.org/10.1093/jac/dkm207

Teo SC, Wong LS (2014) Whole cell-based biosensors for environmental heavy metals detection. Annu Res Rev Biol 4:2663–2674. https://doi.org/10.1016/S0003-2670(98)00725-9

Tummachote J, Punyain W, Thanomsak S, Sirikulkajorn A, Tomapatanaget B (2019) Colorimetric n-butyl-3,6-diamidecarbazole-based chemosensors for detection of fluoride and cyanide anions. Spectrochim Acta A Mol Biomol Spectrosc 214:384–392. https://doi.org/10.1016/j.saa.2019.02.081

Turygin DS, Subat M, Raitman OA, Arslanov VV, König B, Kalinina MA (2006) Cooperative self-assembly of adenosine and uridine nucleotides on a 2d synthetic template. Angew Chem Int Ed 45:5340–5344

Upadhyayula VK, Deng S, Smith GB, Mitchell MC (2009) Adsorption of bacillus subtilis on single-walled carbon nanotube aggregates, activated carbon and nanoceram™. Water Res 43:148–156. https://doi.org/10.1016/j.watres.2008.09.023

Vericat C, Vela M, Benitez G, Carro P, Salvarezza R (2010) Self-assembled monolayers of thiols and dithiols on gold: new challenges for a well-known system. Chem Soc Rev 39:1805–1834

Vinayaka A, Thakur M (2010) Focus on quantum dots as potential fluorescent probes for monitoring food toxicants and foodborne pathogens. Anal Bioanal Chem 397:1445–1455. https://doi.org/10.1007/s00216-010-3683-y

Wan Y, Lin Z, Zhang D, Wang Y, Hou B (2011) Impedimetric immunosensor doped with reduced graphene sheets fabricated by controllable electrodeposition for the non-labelled detection of bacteria. Biosens Bioelectron 26:1959–1964. https://doi.org/10.1016/j.bios.2010.08.008

Wang Y, Alocilja EC (2015) Gold nanoparticle-labeled biosensor for rapid and sensitive detection of bacterial pathogens. J Biol Eng 9:16. https://doi.org/10.1016/j.bios.2014.10.079

Wang Z, Zhang J, Chen P, Zhou X, Yang Y, Wu S, Niu L, Han Y, Wang L, Boey F (2011) Label-free, electrochemical detection of methicillin-resistant Staphylococcus aureus DNA with reduced graphene oxide-modified electrodes. Biosens Bioelectron 26:3881–3886. https://doi.org/10.1186/s13036-015-0014-z

Wang S, Cong T, Liang Q, Li Z, Xu S (2016a) A benzoindo-croconine based colorimetric and fluorescent chemosensor for detection Fe3+/Cu2+/Ag+ ions. Optik-Int J Light Electron Opt 127:890–895

Wang K, Feng W, Wang Y, Cao D, Guan R, Yu X, Wu Q (2016b) A coumarin derivative with benzothiazole Schiff's base structure as chemosensor for cyanide and copper ions. Inorg Chem Commun 71:102–104

Weber PC, Ohlendorf D, Wendoloski J, Salemme F (1989) Structural origins of high-affinity biotin binding to streptavidin. Science 243:85–88. https://doi.org/10.1126/science.2911722

Wei D, Bailey MJ, Andrew P, Ryhänen T (2009) Electrochemical biosensors at the nanoscale. LChip 9:2123–2131. https://doi.org/10.1039/B903118A

Wells RM, Jones CM, Xi Z, Speer A, Danilchanka O, Doornbos KS, Sun P, Wu F, Tian C, Niederweis M (2013) Discovery of a siderophore export system essential for virulence of Mycobacterium tuberculosis. PLoS Pathog 9:e1003120. https://doi.org/10.1371/journal.ppat.1003120

Xiang C, Li R, Adhikari B, She Z, Li Y, Kraatz H-B (2015) Sensitive electrochemical detection of salmonella with chitosan–gold nanoparticles composite film. Talanta 140:122–127. https://doi.org/10.1016/j.talanta.2015.03.033

Xu M, Wang R, Li Y (2017) Electrochemical biosensors for rapid detection of Escherichia coli O157:H7. Talanta 162:511–522. https://doi.org/10.1016/j.talanta.2016.10.050

Yanase Y, Hiragun T, Ishii K, Kawaguchi T, Yanase T, Kawai M, Sakamoto K, Hide M (2014) Surface plasmon resonance for cell-based clinical diagnosis. Sensors 14:4948–4959. https://doi.org/10.3390/s140304948

Yang L, Li Y (2006) Simultaneous detection of Escherichia coli O157: H7 and salmonella typhimurium using quantum dots as fluorescence labels. Analyst 131:394–401. https://doi.org/10.1039/B510888H

Yang L, Deng W, Cheng C, Tan Y, Xie Q, Yao S (2018) Fluorescent immunoassay for the detection of pathogenic bacteria at the single-cell level using carbon dots-encapsulated breakable organosilica nanocapsule as labels. ACS Appl Mater Interfaces. https://doi.org/10.1021/acsami.7b18714

Yao L, Wang L, Huang F, Cai G, Xi X, Lin J (2018) A microfluidic impedance biosensor based on immunomagnetic separation and urease catalysis for continuous-flow detection of E. coli O157: H7. Sensors Actuators B Chem 259:1013–1021. https://doi.org/10.1016/j.snb.2017.12.110

Yoo SM, Lee SY (2016) Optical biosensors for the detection of pathogenic microorganisms. Trends Biotechnol 34:7–25. https://doi.org/10.1016/j.tibtech.2015.09.012

You Y, Lim S, Hahn J, Choi YJ, Gunasekaran S (2018) Bifunctional linker-based immunosensing for rapid and visible detection of bacteria in real matrices. Biosens Bioelectron 100:389–395. https://doi.org/10.1016/j.bios.2017.09.033

Zawadzka AM, Kim Y, Maltseva N, Nichiporuk R, Fan Y, Joachimiak A, Raymond KN (2009) Characterization of a bacillus subtilis transporter for petrobactin, an anthrax stealth siderophore. Proc Natl Acad Sci 106:21854–21859. https://doi.org/10.1073/pnas.0904793106

Zhang J, Hai-Feng J (2004) An anti E. coli O157: H7 antibody-immobilized microcantilever for the detection of Escherichia coli (E. coli). Anal Sci 20:585–587. https://doi.org/10.1016/j.bios.2016.07.030

Zhang J, Su X, O'shea S (2002) Antibody/antigen affinity behavior in liquid environment with electrical impedance analysis of quartz crystal microbalances. Biophys Chem 99:31–41. https://doi.org/10.2116/analsci.20.585

Zhang D, Yan Y, Li Q, Yu T, Cheng W, Wang L, Ju H, Ding S (2012) Label-free and high-sensitive detection of salmonella using a surface plasmon resonance DNA-based biosensor. J Biotechnol 160:123–128. https://doi.org/10.1016/j.jbiotec.2012.03.024

Zheng J, Lin L, Cheng G, Wang A, Tan X, He P, Fang Y (2007) Study on an electrochemical biosensor for thrombin recognition based on aptamers and nano particles. Sci China Ser B: Chem 50:351–357. https://doi.org/10.1007/s11426-007-0062-4

Zhu Q, Shih WY, Shih W-H (2007) In situ, in-liquid, all-electrical detection of Salmonella typhimurium using lead titanate zirconate/gold-coated glass cantilevers at any dipping depth. Biosens Bioelectron 22:3132–3138. https://doi.org/10.1016/j.bios.2007.02.005

Zuo P, Li X, Dominguez DC, Ye B-C (2013) A PDMS/paper/glass hybrid microfluidic biochip integrated with aptamer-functionalized graphene oxide nano-biosensors for one-step multiplexed pathogen detection. LChip 13:3921–3928. https://doi.org/10.1039/c3lc50654a

Adsorptive Chromatography: A Sustainable Strategy for Treatment of Food and Pharmaceutical Industrial Effluents

11

Anand S. Gupta, Piyush Kumar, Soumya Pandit, and Ram Prasad

Abstract

Pollution of water bodies is the major threat to environment sustainability mainly in the industrial area. The quality of portable water bodies such as river, lakes and ponds is being tremendously compromised and deteriorated at very high rate by industrial effluents. Different techniques such as trickling-bed reactor, fluidized bed reactor (FBR), packed bed reactor (PBR), membrane process and bubble column reactor (BCR) are being used for treating effluents from different industries. The effluent is mainly composed of colours, heavy metal traces, chemicals, drugs and process-related toxins depending on the sector of industries operated. The effluents from various food and pharmaceutical industries are mainly treated by carbon as a primary treatment for colour and toxin removal, followed by membrane as a secondary treatment for various intermediates. Finally, deodorization and bacterial load reduction by chemical method are employed.

Although various techniques are being utilized, they lack sustainability and agility for effluent treatment, which can be overcome by adsorption phenomenon of chromatographic process. The thermodynamic mechanism in adsorptive chromatography can be strategically used for decolourization, toxin removal, deodorization and bacterial load reduction in a simple one-step strategic approach. The current chapter focuses on various approaches for effluent treatment of food and

A. S. Gupta (✉) · P. Kumar
Amity Institute of Biotechnology, Amity University Mumbai, Mumbai, Maharashtra, India

S. Pandit
Department of Life Sciences, School of Basic Sciences and Research, Sharda University, Greater Noida, Uttar Pradesh, India

R. Prasad
Department of Botany, Mahatma Gandhi Central University, Motihari, Bihar, India

R. Prasad (ed.), *Environmental Pollution and Remediation*, Environmental and Microbial Biotechnology, https://doi.org/10.1007/978-981-15-5499-5_11

pharmaceutical industry to maintain and sustain the environment and minimize the pollution.

Keywords

Chromatography · Membrane process · Trickling bed · Fluidized bed reactor · Packed bed reactor · Bubble column reactor

11.1 Introduction

Water bodies are being contaminated with different industrial solid waste, discharge and effluents which leads to a detrimental effect on the environment. Amongst the different industries, food and pharmaceuticals are the major contributors to environment hazards. In the food industries, the usage of various colours and additives in formulations such as confectioneries, health supplement and beverages is the key contributor (Singer et al. 2003). The pharmaceutical industry produces various active pharmaceutical ingredients (APIs) for formulations. Also, along with API, they too use varied diluents and excipient which may be colouring agents in various degrees of magnitude. Several API are used as drugs for pharmacological effect which includes antibiotics that are used therapeutically for treating infections caused by pathogenic bacteria in human, animal and birds (Cabello 2006). The discharge such as water effluents and drains from food and pharmaceutical industries contributes as a high degree of water pollutants and threats to portable water sources used for human consumptions. The quality of portable water bodies such as river, lakes and ponds is being tremendously compromised and is deteriorating at a high rate due to industrial effluents (Vignesh et al. 2011).

The major challenges of food industries are the coloured discharge in the portable water bodies which not only makes water unfit for human consumption but also adversely affects the flora and fauna of the water bodies. The phenomenon of toxic effect and accumulation into aquatic animal has predominately ruined the aquatic life system which indirectly has effect on portable water quality. Presently, the dyes and colourants used in food industries are synthetic and semisynthetic derivatives or natural origin compounds (Dotto and Pinto 2011). The dyes are chemically composed of reactive functional groups which affect the planktons present in the water bodies which are crucial for food chain of the aquatic system. Hence, these colourants may have a significant impact on the water quality irrespective of the water bodies. Similar challenge also arises for the pharmaceutical industries for dyes used for tablets, granules, lozenges and oral liquids (Hassaan and El Nemr 2017).

In the confectionery process, the colourants and dyes mainly enter in the effluent during surface washing of the lozenges for smoothening effect as a critical step for texture improvisation. The surface washing creates a high-throughput volume of effluent in food industries (Ozgun et al. 2012), whereas other steps such as spillages during colour making and addition, mould washing after process, sudden leaks and inevitable discharges through pipe fitting during the process are also contributors in the effluents. The potency of these food dyes has a high degree of risk for marine wealth indicating development of process for decolourization (Husband 2014). The consequences of these adverse effects may be hazardous on the unique species present in the microbial flora and fauna, as well as planktons in water bodies which are essentials for maintenance of chemical oxygen demand (COD) and biological oxygen demand (BOD) (Kanu and Achi 2011).

The discharge of effluents from pharmaceutical industries to water bodies introduces chemical agents such as antibiotics which are mainly responsible for potential toxicity and water contamination. These antibiotics are released into the water bodies during production of effluents from pharmaceutical industries and human discharge and disposal. The continuous exposure to antibiotics leads to development of antibiotic-resistant strains that act as a threat to public health (Kim and Aga 2007; Koch et al. 2021). Different antibiotics have been identified in crucial water resources such as groundwater, surface water and streams (Kolpin et al. 2002). Also, other environment-maintaining components such as sewage treatment plant (STP), sludge, soil and sediments are composed of antibiotic residue (Lindsey et al. 2001). Antibiotics are frequently released in large amounts along with municipal wastewater because of their incomplete metabolism by humans or because of the inappropriate disposal of unused antibiotics (Díaz-Cruz and Barceló 2005). Such unregulated activities have not only led to emergence and propagation of antibiotic resistance but also lead to issues in wastewater treatment and disposal (Kim and Carlson 2007).

Amplified exposure to these antibiotics results in development of antibiotic-resistant microorganisms that disturb the community structure by entering into a healthy individual and causing diseases (Tandukar et al. 2013; Koch et al. 2021). This has led to concerns being raised with respect to the possible effects of antibiotic residues present in the aquatic environment. Antibiotics tend to possess allergenic potential and cause side effects and toxicity. The basic classes of antibiotics majorly include beta-lactams derivatives, sulphonamides, tetracyclines, quinolones, aminoglycosides, macrolides, glycopeptides and oxazolidinones (Van Hoek et al. 2011).

11.1.1 Challenges for Effluent Treatment of Food and Pharmaceuticals Industries

The process for effluent treatment has differential aspect involving factors such as engineering and design prospect, high-throughput volume and selection of unit operation; also, the process economics and sustainability are critical consideration. The selection of different unit operations such as horizontal subsurface flow constructed wetland (HSFCW) (Rousseau et al. 2004), upflow anaerobic sludge blanket reactor (UASB) (Lettinga et al. 1980), membrane bioreactors (MB) (Stephenson et al. 2000), trickling filter (TF) (Kornaros and Lyberatos 2006), bubble column reactors (BCRs) (Smith et al. 1996), airlift reactors (ALRs) (Chan et al. 2009), packed bed bioreactors (PBRs) (Silva et al. 2002) and adsorptive

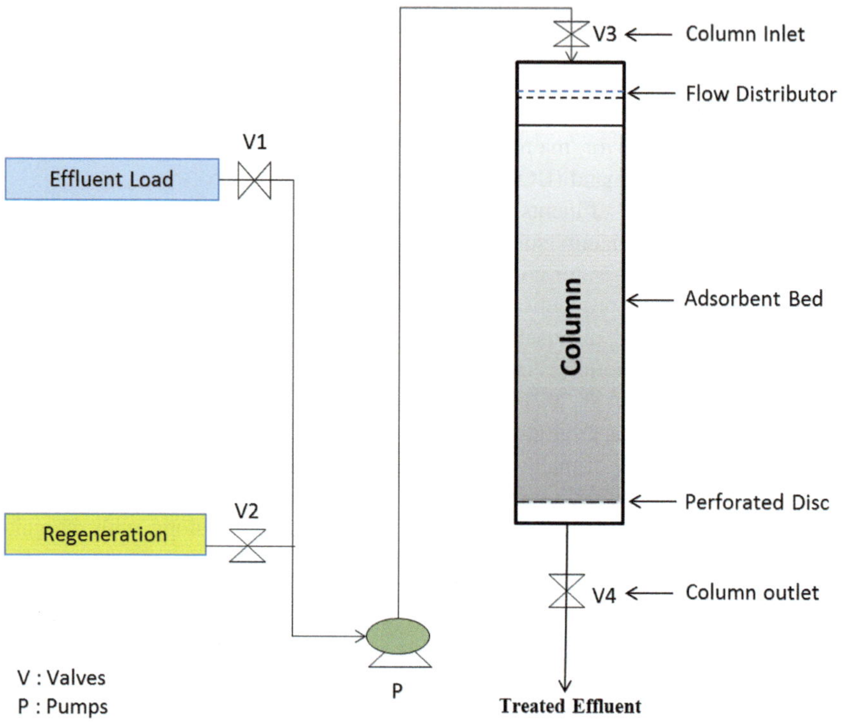

Fig. 11.1 Generalized adsorptive chromatography design for wastewater treatment from pharmaceutical and food industries

chromatography (AC) (Cooney 1998). The major challenges involved in process engineering are increasing volumetric throughput and quality output back into water bodies after treatment.

The adsorptive chromatography is performed mostly in cylindrical column with inlet and outlet control with between lies bed of adsorbent on perforated resting support at bottom (Fig. 11.1). In the adsorptive process, important concern is fluid dynamics of the chromatographic system and accurate process engineering as synergistic effect for efficacy and speed of adsorption process (Hubbuch et al. 2005). Process engineering deals with improving the transport and adsorption of target molecule such as antibiotics from pharmaceutical effluents and dyes and colours from food industries to the adsorption site. The fluid mechanic under flow condition should have reduced mass transfer resistances and axial dispersion (Chihara et al. 1978).

The current chapter focuses on various technique and approaches for effluent treatment to minimize the content of various food dyes and pharmaceutical drugs in discharges. The discussion also targets the strategic application of adsorptive chromatography as a tool for reduction and removal of toxic agent from pharmaceutical and food industry's discharge and effluents. The chapter also discusses the sustainable technique of process dovetailing, which is sequencing of different unit operations for wastewater treatment.

11.2 Techniques for Effluent Treatment from Food and Pharmaceutical Industries

11.2.1 Horizontal Subsurface Flow Constructed Wetland (HSFCW)

An HSFCW is a substantial rock- and sand-filled bowl that is planted with wetland vegetation. It is utilized for tertiary treatment of effluents such as greywater or black water. Solids are expelled in a primary treatment in a septic tank or Imhoff tank. As effluent streams on a level plane through the bowl, the sift material channels through particles and microorganisms decompose the organics. The profluent of a well-working developed wetland can be utilized for water system and aquaculture or securely be released to accepting water bodies (Wu et al. 2016). Flat stream construction wetlands are generally reasonable to construct where arrival is moderate and can be kept up by the nearby group as no cutting-edge save parts, electrical vitality or chemicals are required (Hammer 1992). The outline and execution require master information of hydrodynamics and design for treatment with continuous monitoring of BOD and COD as biochemical aspect (Tilley 2014).

11.2.2 Upflow Anaerobic Sludge Blanket Reactor (UASB)

It is a tank process involving an anaerobic treatment for removal of natural poisons and industrial toxins from effluents (Lettinga and Hulshoff Pol 1991). Wastewater enters the UASB from the bottom and flow towards upward direction. A suspended slop cover channels and regards the wastewater as it moves through it interacting with microbial flora present in the muck by anaerobic assimilation and changing it into biogas. The solid matters are held by a filtration impact of the cover. The upflow movement of effluents creates mass transfer operation without mechanical system for gas-liquid mixing. The highest point of the reactor permits gases to escape and keep a surge of the slop cover (Fig. 11.2). Just like any other oxygen-consuming treatment, UASB require a post-treatment to evacuate pathogens, yet because of a low expulsion of supplements, the water after treatment and solid at slop are utilized as a secondary stream of agribusiness (Tilley 2014).

11.2.3 Membrane Bioreactors (MB)

It is a technique which utilizes low-weight microfiltration (MF) or ultrafiltration (UF) with layer filtration gear and slime formation. The slop and films formed are utilized to partition solid and liquid. The enacted slop is utilized as auxiliary and tertiary filtration (Chang et al. 2002). The two general operations of MBR

Fig. 11.2 Schematic representation of UASB for industrial effluent and wastewater treatment

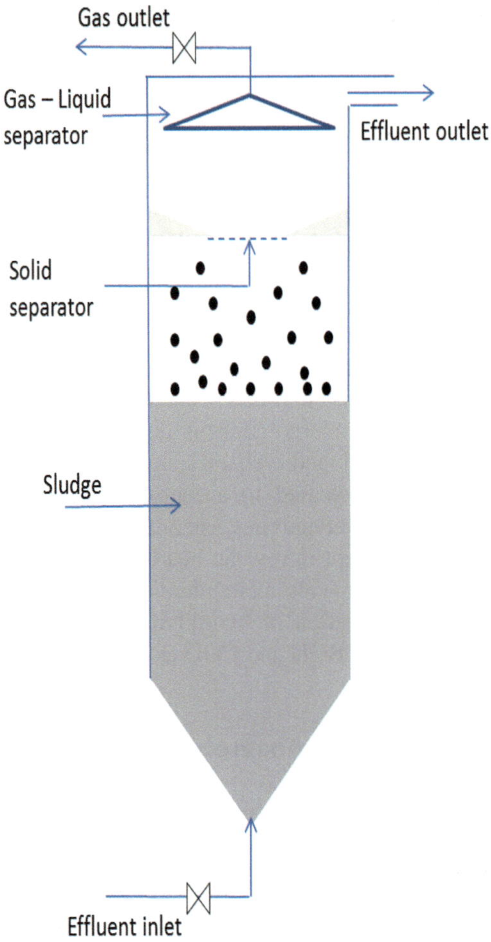

frameworks are vacuum- (or gravity-driven) and weight-driven processes. Vacuum or gravity frameworks are utilizing hollow fibre or flat sheet layers introduced in either the bioreactors or tank. Weight-driven frameworks are in-pipe cartridge frameworks (Stoquart et al. 2012). The effluent should be pretreated priorly to reduce the solid content and sediments for better efficiency of MB.

11.2.4 Trickling Filter (TF)

TF are organic reactor bed which operates by utilization of oxygen from vicinity. There is presettled wastewater that is continuously streamed over the channel. As the water travels through the pores of the channel, organic matters are utilized and decomposed by the biofilm covering the channel material (Tilley 2014). The TB is used as a reactor in which a gas-liquid mass transfer occurs concurrently in downward direction through a fixed bed of catalyst particles where reaction occurs to conversion of toxin in effluents. The term 'trickling filter' has been used because the removal of organic matter by aerobic bacterial action from wastewater streams through a bed. There are biological growths that attach themselves either to a matrix or bed of stone or any other support material over which the wastewater is made to trickle in contact with the surrounding air (Satterfield 1975).

11.2.5 Bubble Column Reactors (BCRs)

Sublation of solvents as a non-foaming effluent treatment method incorporates the advantages of bubble fractionation and fluid extraction without mixers, settlers or consequent downstream treatment (Lucas et al. 2009). Previous research on lab-scale BCRs revealed better efficiencies for removal of non-volatile and volatile organic compounds than bubble fractionation, air stripping and conventional liquid-liquid extraction (Smith et al. 1996). There are three transport mechanisms available by which pollutants mainly gases are removed. They are transport by air bubbles, water entrainment due to the rising air bubbles and molecular mass transport across the gas-water interface. After the sublation, soluble antibiotics, dye and suspended microbial debris are removed from the aqueous phase by further diversification to other unit operations, and volatile pollutants are removed from the top zone of BCR (Turhan and Turgut 2009).

In the unidirectional transport through air bubbles, the rising of air bubbles in the column leads to partition of both volatile and non-volatile compounds on the bubble surface. The extent of partitioning is determined by equilibrium relationships for bulk-phase partitioning and surface adsorption with a mass balance for the total amount of pollutant carried by a bubble of radius a. In Eq. 11.1, m is the total amount of pollutant, Γ is the surface concentration of the bubble, and C_v is the concentration of pollutant in the air bubble:

$$m = 4\pi a^2 \Gamma + \frac{4}{3}\pi a^3 C_V \tag{11.1}$$

At equilibrium, the concentrations are related to the bulk-phase water concentration through linear relationships:

$$\Gamma = K_A C_w C_v = H_c C_w \tag{11.2}$$

where K_A is the interfacial partition constant and H_c is Henry's law constant. When Eqs. 11.1 and 11.2 are combined, the following effective air concentration results:

$$C_A = \frac{m}{\frac{4}{3}\pi a^3} = \left(\frac{3}{a}K_A + H_C\right)C_w \tag{11.3}$$

The term in parentheses is the effective Henry's law constant, H. It depends inversely upon the air bubble size.

11.2.6 Airlift Reactors (ALRs)

ALRs have proved to be efficient alternative devices against conventional systems such as stirred tanks and bubble column contactors for the remediation of several antibiotic-contaminated water bodies, even more because their applications range from synthesis of chemicals, culture of plant and animal cells, production of microalgae, fuel gas and contaminated soils. These are cylindrical or rectangular vessel with a gas distributor at the inlet, usually without mechanical moving parts (Fig. 11.3) (Huang and McDonald 2009). The only energy input needed is to inject the aeration gas through a simple sparging system (Brenner et al. 1997). ALRs are considered as feasible and sustainable alternatives for stirred tank reactors (STRs), particularly for numerous bioprocesses such as plant and animal cell cultures but also for the treatment of contaminated fluid fluxes (Benyahia and Jones 1997).

The volumetric mass transfer coefficient (k_La) is the rate of gas transfer across the gas-liquid interface per unit of driving force (the driving force is the gas

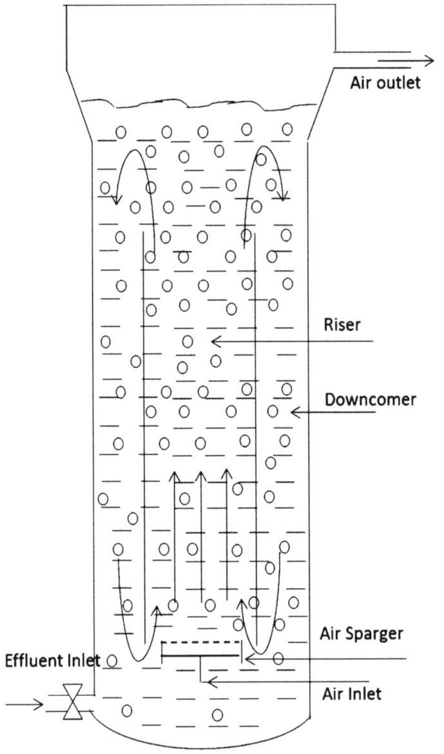

Fig. 11.3 Airlift reactor for wastewater and effluent treatment

concentration gradient between the liquid and the gas ΔC) and is defined by the following equation (Cozma and Gavrilescu 2010):

$$OTR = k_L a \, \Delta C \tag{11.4}$$

11.2.7 Fluidized Bed Reactors (FBRs)

The FBRs can be used as a high-throughput system for treatment of liquid effluent and discharges from industries. The phenomenon of fluidization leads to suspension of solid particulates in an upward-flowing stream of liquid. The liquid velocity is critically optimized to suspend the particulate matters, without expulsion from the reactor vessel. The solid particles combine to form the loosened bed quickly and also form a suspension at the top with lesser particle at the top. The fluidized material is quite often a solid, and the fluidizing medium is either a fluid or a gas. The qualities and conduct of a fluidized bed are unequivocally subject to both the solid and fluid or gas properties. Before the reactor has begun, the catalyst pellets lie on a mesh at the base of the reactor which later in process converts hazardous pollutants to non-hazardous moiety. Reactants are pumped into the reactor through a distributer consistently making the bed fluidized. The bed's conduct after introductory fluidization relies upon the condition of the reactant. In the event that it is a fluid, the bed grows consistently with expanded upward stream of the reactant (Fig. 11.4).

11.2.8 Packed Bed Bioreactors (PBRs)

The PBRs also known as fixed-bed bioreactors use particulate or immobilized biocatalysts throughout vertical tube which is completely packed beads of biocatalysts. Through this packed bed system, the effluents are introduced in through either the bottom or top of the column, and this creates the continuous liquid occupancy in PBRs. PBRs have a low attrition and impact on biocatalyst as compared to STR. PBRs are commercially used for wastewater treatment by using immobilized enzymes and cells for the production of amino acid and organic acid and also the transformation of antibiotics such as penicillins. The particles must be incompressible, and it also has to endure their own weight in PBRs without deformities or blocking the fluid flow pattern. Between the liquid medium and solid catalyst, mass transfer is carried out at high liquid flow rates through the bed, and in order to attain this, the packed beds are frequently operated with liquid recycle (Warren et al. 1976).

The mass transfer coefficient k_s should be known even before an account for external mass transfer effects can be made. k_s is based on the hydrodynamics of the reactor and properties of liquid state of matter, namely, viscosity, density and diffusivity. Values of k_s can be approximately estimated using correlations from existing literature (Moo-Young and Blanch 1981):

Fig. 11.4 Fluidized bed reactor for effluent treatment in food and pharmaceutical industries

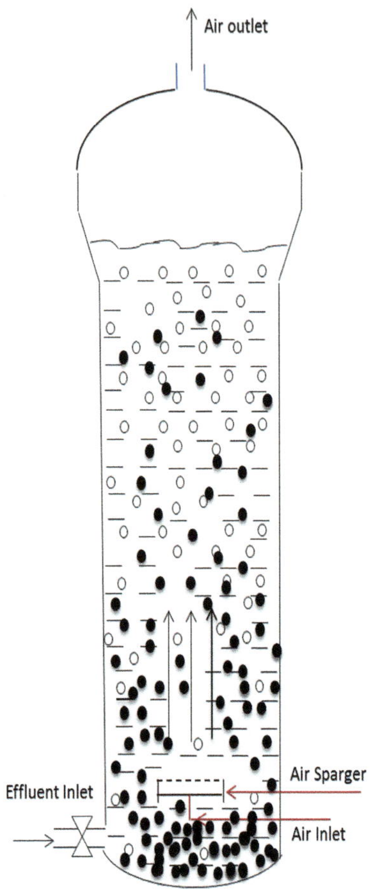

$$Re_p = \text{(particle) Reynolds number} = \frac{D_p u_{pL} \rho_L}{\mu_L}$$

$$Sc = \text{Schmidt number} = \frac{\mu_L}{\rho_L \mathcal{D}_{AL}}$$

$$Sh = \text{Sherwood number} = \frac{k_s D_p}{\mathcal{D}_{AL}}$$

$$\text{And} : Gr = \text{Grashof number} = \frac{g D_p^3 \rho_L (\rho_p - \rho_L)}{\mu_L^2}$$

where u_{pL} is the linear velocity of the particle relative to the bulk liquid, D_p is the diameter of the particle, ρ_L is the density of the liquid, μ_L is the viscosity of the liquid, $\mathcal{D}_{AL}$ is the molecular diffusivity in the liquid of component A, k_s is the liquid to solid mass transfer coefficient, g is the gravitational acceleration and lastly ρ_p is the density of the particle (Charpentier 1981; Warren et al. 1976).

The Sherwood number represents the ratio of overall as well as diffusive mass transfer rates through a boundary layer, and it contains the mass transfer coefficient. The Schmidt number, which is made up of physical properties of the system, depicts the ratio of momentum diffusivity and mass diffusivity. Sc is a constant value for Newtonian fluids at a constant temperature, pressure and composition. The Grashof number, which is important when the particles are neutrally buoyant, presents the ratio of gravitational forces to viscous forces. The form of the correlation that is used to evaluate Sh and thus k_s is based on the flow conditions, the configuration of the mass transfer system and several other factors. As per the above correlations, k_s in a packed bed is based on the liquid velocity around the particles (Warren et al. 1976). Within the range $10 < Re < 104$, the Sherwood number in packed beds has been determined as follows:

$$Sh = 0.95\, Re_p^{0.5} Sc^{0.33}$$

11.2.9 Adsorptive Chromatography (AC)

The general principle of adsorption in chromatography is being governed mainly by three important techniques such as frontal analysis, displacement method and elution method. In accordance to their application, these techniques are used in different adsorptive liquid chromatography such as analytical chromatography (HPLC) and/or preparative chromatography for capturing, isolation and purification of macromolecules (Kasai et al. 1986). The AC as a sustainable strategy for treatment of food and pharmaceutical industrial effluent involves frontal analysis method in which binding of the desired molecules can be achieved from a particular mixture stream composed of various molecules (Oka et al. 1989). The AC involves different types of thermodynamic interaction involving isotherms for particular interaction of the molecules in the effluents (Thomas 1948). The effluent is composed of multimolecular interaction, which presents a multilayered of interaction, the AC as a tool for effluent treatment involves interaction types such as hydrophobicity and ion exchange between adsorbent and molecule in effluent mixture.

The throughput of the AC depends on the binding capacity of the adsorbent for a particular molecule present in the effluent (Pereira et al. 2003). The adsorbent's adsorptive property in both the interaction is mathematically estimated by breakthrough curve (BTC). The BTC provides the information correlation between the volumes of effluent, binding capacity per mL of adsorbent at a particular flow rate. The mechanism of interaction in BTC also governs the throughput of the designed process (Thomas 1948).

11.3 Modes of Adsorptive Chromatography for Treatment of Effluents and Discharge

In AC, the modes of interaction involved are thermodynamic-driven or kinetic-driven process. The kinetic-driven process comprises the migration rate of the molecule with linear velocity (V) of the mobile phase, and least interaction between molecules and adsorbents takes place (Guiochon 2002). In thermodynamic process, involvement of high interaction, i.e. binding of the molecules to the adsorbent, is favoured; also, high capacity adds to process synergism. The thermodynamic process should have irreversible binding phenomenon as an important aspect. Both the approaches, mentioned above, have different application; e.g. kinetic-driven processes are used for purification and polishing of niche pharmaceutical and therapeutic molecules (Fig. 11.5). The thermodynamic mode can be applied for capturing and isolating desired molecules in diluted stream (Rathore and Velayudhan 2002). The capturing of different class of antibiotics from the effluents of pharmaceutical industries is a good example (Hwang et al. 1994). Also, varied class of dyes with different functional group can be captured in AC specifically (Adachi et al. 2002).

The modes of AC practised for high-throughput system in treatment of industrial effluents and discharge are thermodynamic-driven process. Thermodynamic-driven AC process can be a positive or negative chromatography. In positive chromatography, the target molecule with high interaction or affinity binds to the adsorbent, whereas negative chromatography involves binding of impurities, and target molecules remain in mobile phase. The most adopted technique for effluent treatment is positive chromatography in which the spectra of target molecules such as antibiotics, colour and dyes, toxins, proteins, peptides, sugar and organic acids are

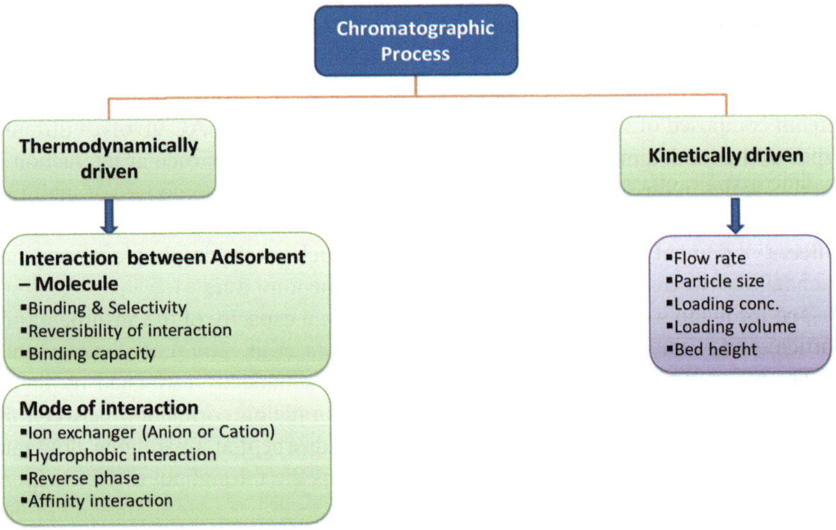

Fig. 11.5 The thermodynamic- and kinetic-driven process in adsorptive chromatography

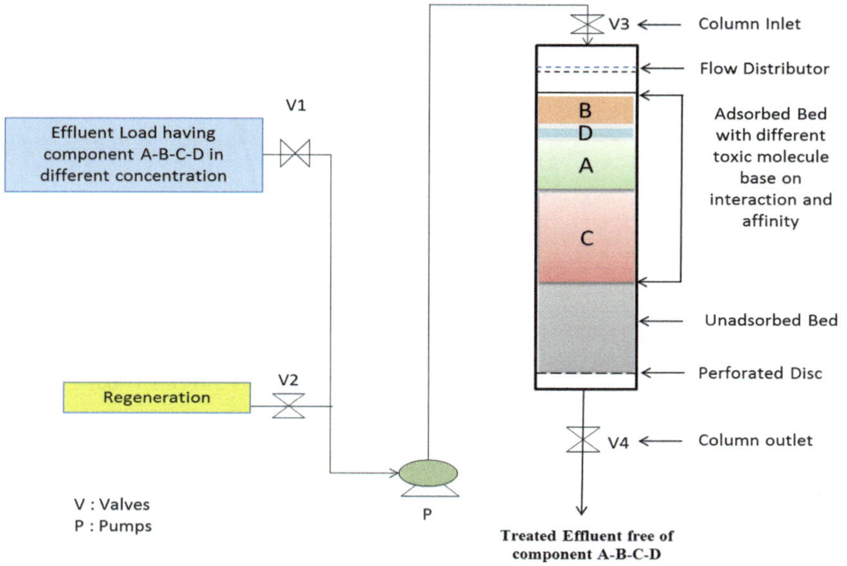

Fig. 11.6 Effluent treatment by using positive mode of AC

specifically bound to the adsorbent, and then, the treated water as effluent is introduced in the water bodies (Lee et al. 2014) (Fig. 11.6). Selection of AC as positive or negative chromatographic step is influenced by concentration of target, volume of effluent, regeneration of adsorbent after each cycle in process and mechanical stability of adsorbent at such high-throughput level of operation in the process.

Thermodynamic AC mode of operation involves steps such as equilibrium, loading, washing, elution and regeneration as a one complete cycle in the process. These steps are generally utilized for therapeutic molecule purification as AC operation. The thermodynamic mode for effluent treatment can be achieved by implicating frontal analysis technique in which the steps are curtailed to equilibrium, loading (capturing) and regeneration as one complete cycle. The ability to regenerate after each cycle makes AC as a choice of unit operation for process development and also grants techno-economic feasibility (Rathore and Velayudhan 2002).

As frontal technique is the principle underlying AC process which depends on the differential binding affinity, the molecule with high binding affinity will occupy the sites present on chromatographic bed plates from inlet zone till the bed is completely occupied (Ghorai and Pant 2005). Once the bed is saturated with molecule, the inlet and outlet concentration from the column will be equivalent. The molecule with highest affinity for the site on adsorbent will elute late, whereas molecule with low affinity will elute earlier in the process (Kasai et al. 1986). The target molecules from pharmaceutical and food industry effluents are composed of various antibiotics and colour dyes, respectively (Yoshida and Takemori 1997). The binding capacity in AC of a thermodynamic-driven process with positive mode approach is estimated by

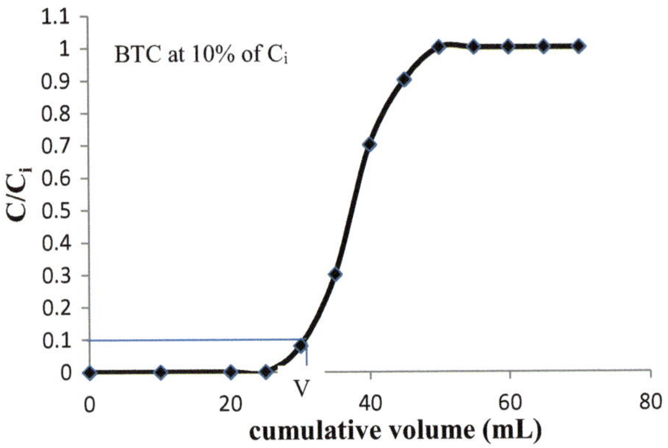

Fig. 11.7 Breakthrough curve profile of adsorptive chromatography using thermodynamic mode

BTC (Fig. 11.7). The plot is composed of C/C_i vs cumulative volume (mL) where 'C' is the concentration (mg/mL) of binding molecule at time (t) in minutes from outlet, whereas 'C_i' is the initial concentration of binding molecule in effluent load (Aksu and Gönen 2004):

$$\mathrm{BTC} = \frac{\mathrm{C_i\ X\ V}}{V_\mathrm{B}}$$

where V is the breakpoint volume at 10% of BTC and V_B is the bed volume of the column used for BTC.

The degree of binding in thermodynamic approach should have to be irreversible for reusability of the process. The quantification extent of binding depends on the number of site present on adsorbent surface and its area. The interaction on the site present on the chromatographic adsorbent is studied on the basis of the isotherms. These isotherms can be observed as linear or non-linear mechanism of binding depending on the plot of Q v/s C*, where 'Q' is the amount of molecule adsorbed per gram of adsorbent (g/g) and C* is the equilibrium concentration of the molecule in liquid phase (Fig. 11.8). The rate of these isotherms depends on their patterns predicted by observing plots as linear and non-linear. These isotherms are used to predict the mechanism of interaction such as types of diffusion in mass transfer of the high-throughput system for industrial effluent treatment (Ghorai and Pant 2005).

The mathematical models mostly for isotherm determination of adsorption phenomenon are Langmuir and Freundlich pattern for treatment of effluent from industries which is composed of multiple components and always signifies a non-linear behaviour of adsorption. The following equations are used for studying the non-linear mathematical model and their linearized form for precise estimation of slope and other variables of non-linear models (Kasai et al. 1986):

$$\text{Freundlich isotherm :} \qquad \mathrm{Q} = \mathrm{K_f.C^{1/n}} \tag{11.5}$$

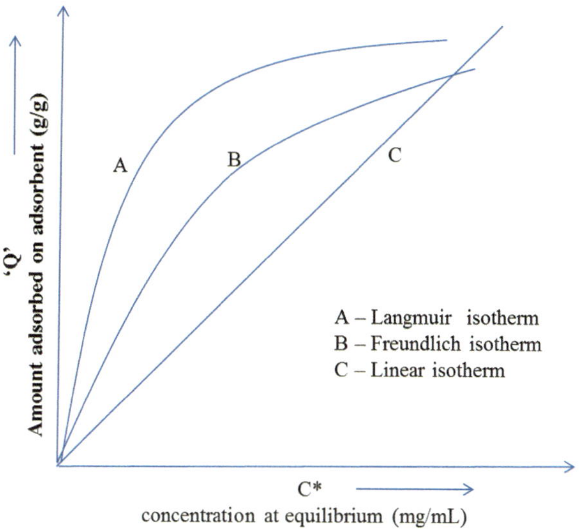

Fig. 11.8 The plot of adsorption isotherm in adsorptive chromatography

$$\text{Linearized form of Freundlich isotherm}: \log Q = \log K\text{f} + 1/\text{n} \log \text{C} \quad (11.6)$$

The plot of logQ vs logC provides the straight-line equation to estimate K_f and n which are Freundlich constant indicating the adsorption capacity and adsorption intensity, respectively:

$$\text{Langmuir isotherm}: Q = \frac{Qmax\ Kl\ C}{1 + Kl\ C} \quad (11.7)$$

$$\text{Linearized form of Langmuir isotherm}: \frac{1}{Q} = \frac{Kl}{Qmax\ C} + \frac{1}{Qmax} \quad (11.8)$$

The plot of 1/Q vs 1/C will give a straight-line equation, where Q is the quantity of solute adsorbed per unit of adsorbent, C is the concentration of protein in solution, Qmax is the maximum quantity adsorbed at high C and K_l is the dissociation constant which is a measure of strength of binding of solute to adsorbent.

In thermodynamic system for industrial effluent, treatment selection of stationary-phase and effluent-phase environment is critical for process engineering and design. The particle size of the adsorbents also has major consideration as a critical parameter for column designing for such high-throughput system for treatment of industrial effluent. It is technically wise to choose particle size with larger particle size so as to reduce the pressure drop across the operating column at process scale-up from bench level to process scale. The binding capacity is the important factor for selection of adsorbent in any mode of AC process. It is always a target to achieve optimum loading (g of molecule/g of adsorbent) and also to enhance loading on the column by increasing the aspect ratio (D/h) (Rathore and Velayudhan 2002).

11.4 Dovetailing of Unit Operations for Sustainable Technique

The emerging trend to treat effluent from industrial sectors and maintain the ecological balance in the water bodies as a mandate from different officials has led to process dovetailing. The dovetailing involves the correct sequencing of various unit operations to get the resultant outcome with recyclable quality for daily reuse or re-entering in water bodies (Bijan and Mohseni 2005). The general theme practised traditionally is alone natural-based operations such as HSFCW, UASB and TF which depends on the bed formation or wet land generation with plantation (Champagne et al. 2017). The single-unit operation cannot serve the objective of removal of antibiotics, colours, dyes, drug traces, toxic protein and peptides, fats and carbohydrates (Høibye et al. 2008). The dovetailing strategies by combining with correct sequence such as ALRs-AC, AC-TF, FBRs-AC, TF-MB-AC, FBRs-ALRs and AC-AC2 as a tandem approach (Qiu et al. 2013). The dovetailing of AC with other unit operation has not been practised widely, whereas the tandem approach of AC can be used for pharmaceutical and food industry effluent and discharge (Gennaro et al. 1989) (Fig. 11.9).

The AC1 and AC2 can be selected based on the molecular descriptors such as molecular weight, size and structure, functional group charge, hydrophobicity, conductivity and density for capturing and isolation of the hazardous and toxic molecule in the effluents. The combination of AC1 and AC2 can be followed for ideal sequencing of unit operations (Table 11.1).

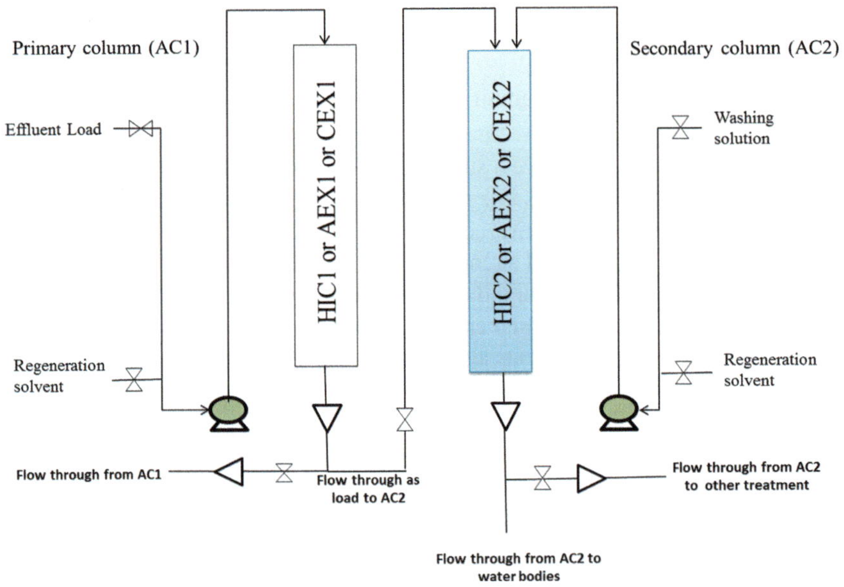

Fig. 11.9 Treatment of effluents by tandem approach involving AC with different types

Table 11.1 The combination of various AC types in dovetailing process for industrial effluent

AC1	AEX	CEX	AEX	HIC	CEX	HIC	AEX1	CEX1	HIC1
AC2	CEX	AEX	HIC	AEX	HIC	CEX	AEX2	CEX2	HIC2
AC1-AC2	AEX-CEX	CEX-AEX	AEX-HIC	HIC-AEX	CEX-HIC	HIC-CEX	AEX1-AEX2	CEX1-CEX2	HIC1-HIC2

11.5 Conclusions and Future Prospects

The conventional methods of effluent management and treatment such as UASB, TF and HSFCW are predominantly dependent on wetland generation and require substantial time lapse and high cost for their construction. Thus, long-term maintenance of such methods is a challenge for sustainability. The adsorption phenomenon in chromatography as strategic approach can be used as an agile and simple and technically feasible method for effluent treatment. The mechanism of AC provides high throughput and sustainability. The model of construction for AC process can be moulded into any stream and outlet with accurate engineering design. The aspect of adsorbent regeneration in AC for higher number of cycle makes it highly economical compared to other techniques. The wide application of AC for specific removal of toxic molecules such as antibiotics, colour and dyes, proteins, peptides, sugar and organic acids from effluent has resulted into a better technique compared to ALRs, PBRs and BCRs. The mass transfer coefficient (kLa) in AC can be easily predicted and understood by isotherm data which provide the engineers an accurate vision for design and process development at various scale.

The concept of dovetailing by hyphenating two different techniques or similar techniques with different types or modes should be explored for effluent treatment from industries of different sectors. The sequencing in the dovetailing process for effluent stream treatment should be correctly performed prior to scale-up models for high-throughput system. The advancements in AC based on types and modes with high binding and site should be utilized for high-volume throughput and scalability, which may provide more efficient and economical process. The correct hyphenation sequence involved in high mass transfer technique such as ALRs, BCRs and FBRs should be utilized amongst or along with AC as strategic approach for rendering water back into mainstream water bodies with enhanced portability. The bacterial and organic load can be easily reduced by MBs and TF as conventional approach for pretreatment of effluents which later can be introduced to AC for enhancing its portability.

The strategic and continuous processes such as tandem approach along with simulation of AC for high-throughput and sustainable technique can be achieved as advancement. Further, utilization of high-end continuous AC technique such as simulated moving bed (SMB) can be used at various scales of process. Liquid-solid continuous fluidized bed (LSCFB), another approach, can also be used for continuous effluent treatment at low rate of method for small-scale industry discharge.

Acknowledgement We thank Dr. Neetin Desai, Director, Amity Institute of Biotechnology, Amity University Mumbai, for infrastructural support. We also acknowledge graduate students Denice Peter and Anuskha Nair for their help in literature survey. The authors also thank Dr. Nilesh Wagh for his editorial help in writing the chapter.

References

Adachi T, Ando S, Watanabe J (2002) Characterization of synthetic adsorbents with fine particle sizes for preparative-scale chromatographic separation. J Chromatogr A 944:41–59

Aksu Z, Gönen F (2004) Biosorption of phenol by immobilized activated sludge in a continuous packed bed: prediction of breakthrough curves. Process Biochem 39:599–613

Benyahia F, Jones L (1997) Scale effects on hydrodynamic and mass transfer characteristics of external loop airlift reactors. J Chem Technol Biotechnol 69:301–308

Bijan L, Mohseni M (2005) Integrated ozone and biotreatment of pulp mill effluent and changes in biodegradability and molecular weight distribution of organic compounds. Water Res 39:3763–3772

Brenner A, Ben-Shushan N, Siegel M, Merchuk J (1997) Pilot plant performance and model calibration of a sequencing batch air-lift reactor. Water Sci Technol 35:121–127

Cabello FC (2006) Heavy use of prophylactic antibiotics in aquaculture: a growing problem for human and animal health and for the environment. Environ Microbiol 8:1137–1144

Champagne P, Liu L, Howell M (2017) Aerobic treatment in cold-climate countries. In: Current developments in biotechnology and bioengineering. Elsevier, Amsterdam, pp 161–201

Chan YJ, Chong MF, Law CL, Hassell D (2009) A review on anaerobic–aerobic treatment of industrial and municipal wastewater. Chem Eng J 155:1–18

Chang I-S, Le Clech P, Jefferson B, Judd S (2002) Membrane fouling in membrane bioreactors for wastewater treatment. J Environ Eng 128:1018–1029

Charpentier J-C (1981) Mass-transfer rates in gas-liquid absorbers and reactors. In: Advances in chemical engineering, vol 11. Elsevier, pp 1–133

Chihara K, Suzuki M, Kawazoe K (1978) Adsorption rate on molecular sieving carbon by chromatography. AICHE J 24:237–246

Cooney DO (1998) Adsorption design for wastewater treatment. CRC press

Cozma P, Gavrilescu M (2010) Airlift reactors: hydrodynamics, mass transfer and applications in environmental remediation. Environ Eng Manag J 9:681–702

Díaz-Cruz MS, Barceló D (2005) LC–MS2 trace analysis of antimicrobials in water, sediment and soil. TrAC Trends Anal Chem 24:645–657

Dotto G, Pinto L (2011) Adsorption of food dyes acid blue 9 and food yellow 3 onto chitosan: stirring rate effect in kinetics and mechanism. J Hazard Mater 187:164–170

Gennaro R, Skerlavaj B, Romeo D (1989) Purification, composition, and activity of two bactenecins, antibacterial peptides of bovine neutrophils. Infect Immun 57:3142–3146

Ghorai S, Pant K (2005) Equilibrium, kinetics and breakthrough studies for adsorption of fluoride on activated alumina. Sep Purif Technol 42:265–271

Guiochon G (2002) Preparative liquid chromatography. J Chromatogr A 965:129–161

Hammer DA (1992) Designing constructed wetlands systems to treat agricultural nonpoint source pollution. Ecol Eng 1:49–82

Hassaan MA, El Nemr A (2017) Health and environmental impacts of dyes: mini review. Am J Environ Sci Eng 1:64–67

Høibye L, Clauson-Kaas J, Wenzel H, Larsen HF, Jacobsen BN, Dalgaard O (2008) Sustainability assessment of advanced wastewater treatment technologies. Water Sci Technol 58:963–968

Huang T-K, McDonald KA (2009) Bioreactor engineering for recombinant protein production in plant cell suspension cultures. Biochem Eng J 45:168–184

Hubbuch J, Thömmes J, Kula M-R (2005) Biochemical engineering aspects of expanded bed adsorption. In: Technology transfer in biotechnology. Springer, pp 101–123

Husband T (2014) The sweet science of candymaking. ChemMatters 10:5–8
Hwang BK, Ahn SJ, Moon SS (1994) Production, purification, and antifungal activity of the antibiotic nucleoside, tubercidin, produced by Streptomyces violaceoniger. Can J Bot 72:480–485
Kanu I, Achi O (2011) Industrial effluents and their impact on water quality of receiving rivers in Nigeria. J Appl Technol Environ Sanit 1:75–86
Kasai K-I, Oda Y, Nishikata M, Ishii S-I (1986) Frontal affinity chromatography: theory for its application to studies on specific interactions of biomolecules. J Chromatogr B Biomed Sci Appl 376:33–47
Kim S, Aga DS (2007) Potential ecological and human health impacts of antibiotics and antibiotic-resistant bacteria from wastewater treatment plants. J Toxicol Environ Health Part B 10:559–573
Kim S-C, Carlson K (2007) Temporal and spatial trends in the occurrence of human and veterinary antibiotics in aqueous and river sediment matrices. Environ Sci Technol 41:50–57
Koch N, Islam NF, Sonowal S, Prasad R, Sarma H (2021) Environmental antibiotics and resistance genes as emerging contaminants: methods of detection and bioremediation. Curr Res Microbial Sci. https://doi.org/10.1016/j.crmicr.2021.100027
Kolpin DW, Furlong ET, Meyer MT, Thurman EM, Zaugg SD, Barber LB, Buxton HT (2002) Pharmaceuticals, hormones, and other organic wastewater contaminants in US streams, 1999–2000: a national reconnaissance. Environ Sci Technol 36:1202–1211
Kornaros M, Lyberatos G (2006) Biological treatment of wastewaters from a dye manufacturing company using a trickling filter. J Hazard Mater 136:95–102
Lee MFX, Chan ES, Tey BT (2014) Negative chromatography: progress, applications and future perspectives. Process Biochem 49:1005–1011
Lettinga G, Hulshoff Pol L (1991) UASB-process design for various types of wastewaters. Water Sci Technol 24:87–107
Lettinga G, Van Velsen A, Sd H, De Zeeuw W, Klapwijk A (1980) Use of the upflow sludge blanket (USB) reactor concept for biological wastewater treatment, especially for anaerobic treatment. Biotechnol Bioeng 22:699–734
Lindsey ME, Meyer M, Thurman EM (2001) Analysis of trace levels of sulfonamide and tetracycline antimicrobials in groundwater and surface water using solid-phase extraction and liquid chromatography/mass spectrometry. Anal Chem 73:4640–4646
Lucas MS, Peres JA, Lan BY, Puma GL (2009) Ozonation kinetics of winery wastewater in a pilot-scale bubble column reactor. Water Res 43:1523–1532
Moo-Young M, Blanch H (1981) Design of biochemical reactors mass transfer criteria for simple and complex systems. In: Reactors and reactions. Springer, pp 1–69
Oka H, Ikai Y, Kawamura N, Yamada M, Harada K-I, Yamazaki Y, Suzuki M (1989) Improvement of chemical analysis of antibiotics: XV. Isocratic high-performance liquid chromatographic methods for the analysis and preparative separation of the components of bacitracin. J Chromatogr A 462:315–322
Ozgun H et al (2012) Confectionery industry: a case study on treatability-based effluent characterization and treatment system performance. Water Sci Technol 66:15–20
Pereira MFR, Soares SF, Órfão JJ, Figueiredo JL (2003) Adsorption of dyes on activated carbons: influence of surface chemical groups. Carbon 41:811–821
Qiu G, Song Y-h, Zeng P, Duan L, Xiao S (2013) Characterization of bacterial communities in hybrid upflow anaerobic sludge blanket (UASB)–membrane bioreactor (MBR) process for berberine antibiotic wastewater treatment. Bioresour Technol 142:52–62
Rathore A, Velayudhan A (2002) Scale-up and optimization in preparative chromatography: principles and biopharmaceutical applications, vol 88. CRC Press
Rousseau DP, Vanrolleghem PA, De Pauw N (2004) Model-based design of horizontal subsurface flow constructed treatment wetlands: a review. Water Res 38:1484–1493
Satterfield CN (1975) Trickle-bed reactors. AICHE J 21:209–228

Silva AJ, Varesche M, Foresti E, Zaiat M (2002) Sulphate removal from industrial wastewater using a packed-bed anaerobic reactor. Process Biochem 37:927–935

Singer RS, Finch R, Wegener HC, Bywater R, Walters J, Lipsitch M (2003) Antibiotic resistance—the interplay between antibiotic use in animals and human beings. Lancet Infect Dis 3:47–51

Smith JS, Valsaraj KT, Thibodeaux LJ (1996) Bubble column reactors for wastewater treatment. 1. Theory and modeling of continuous countercurrent solvent sublation. Ind Eng Chem Res 35:1688–1699

Stephenson T, Brindle K, Judd S, Jefferson B (2000) Membrane bioreactors for wastewater treatment. IWA publishing

Stoquart C, Servais P, Bérubé PR, Barbeau B (2012) Hybrid membrane processes using activated carbon treatment for drinking water: a review. J Membr Sci 411:1–12

Tandukar M, Oh S, Tezel U, Konstantinidis KT, Pavlostathis SG (2013) Long-term exposure to benzalkonium chloride disinfectants results in change of microbial community structure and increased antimicrobial resistance. Environ Sci Technol 47:9730–9738

Thomas HC (1948) Chromatography: a problem in kinetics. Ann N Y Acad Sci 49:161–182

Tilley E (2014) Compendium of sanitation systems and technologies. Eawag

Turhan K, Turgut Z (2009) Decolorization of direct dye in textile wastewater by ozonization in a semi-batch bubble column reactor. Desalination 242:256–263

Van Hoek AH, Mevius D, Guerra B, Mullany P, Roberts AP, Aarts HJ (2011) Acquired antibiotic resistance genes: an overview. Front Microbiol 2:203

Vignesh R, Karthikeyan B, Periyasamy N, Devanathan K (2011) Antibiotics in aquaculture: an overview. South Asian J Exp Biol 1:114–120

Warren L, Warren L, McCabe W, Smith JC (1976) Unit operations of chemical engineering. Mc Graw Hill Company

Wu J, Feng Y, Dai Y, Cui N, Anderson B, Cheng S (2016) Biological mechanisms associated with triazophos (TAP) removal by horizontal subsurface flow constructed wetlands (HSFCW). Sci Total Environ 553:13–19

Yoshida H, Takemori T (1997) Adsorption of direct dye on cross-linked chitosan fiber: breakthrough curve. Water Sci Technol 35:29–37

Remediation of Heavy Metals Through Genetically Engineered Microorganism

12

Neerja Srivastava

Abstract

Bioremediation of soil, groundwater, and activated sludge sites can be done by genetically engineered microorganisms (GEMs) with increased breakdown capacity for a number of pollutants. However, very few studies related to GEM applications in environmental ecosystems have been done. Unfortunately, the single approach to fully exploit the potential of GEMs in bioremediation is through continuous field release studies. Therefore, to get accurate knowledge for assessing the whole efficiency and dangers of GEM entrance into natural ecosystems, field studies are necessary. These days, remediation of a large number of heavy metals through the use of GE bacteria is in the limelight because of its environmental friendly nature and reduced damage to human health in contrast to various physico-chemical methods. For successfully reclaiming heavy metal-polluted sites by in situ bioremediation utilizing engineered bacteria, a mixture of microbiological and ecological information, biochemical procedures and field engineering methods are necessary components. While developing and using GE bacteria to enhance bioremediation, different important research problems have been recognized which prepared for possible future research. Genetic engineering of native microflora, adapted for local environmental surroundings, may remediate polluted sites more efficiently and will prove it much easier and more eco-friendly strategy. However, various issues related to the release of genetically designed bacteria in field settings have to be dealt with. There are potential dangers related to the application of GE bacteria in the field especially the modes by which molecular genetics could add to the risk mitigation. Environmental and public health issues should be dealt with by the molecular biologists.

N. Srivastava (✉)
Department of Biochemistry, IBSBT, CSJM University, Kanpur, India

R. Prasad (ed.), *Environmental Pollution and Remediation*, Environmental and Microbial Biotechnology, https://doi.org/10.1007/978-981-15-5499-5_12

Keywords

Genetically engineered microorganism · Heavy metals · Bioremediation · Biosorption · Siderophores · Microbial remediation

12.1 Introduction

Microorganisms and metals are present together with a long period of time in history. This is represented through the presence of a vast variety of divalent and transition metals in the active centres of different enzymes. Various chemical properties of the metal are utilized for maintenance of the structure of protein as well as catalysis of key reactions in the metabolic pathways. Metals are needed in minuscule quantities for performing usual cell metabolism, while their intake is dependent on complex homeostatic mechanisms that guarantee enough but not excess increase. Various other metals have no biologically important function and cause damage, mostly because of their desire for the sulfhydryl groups in proteins, which they obstruct and make inactive. Therefore, physiologically, metals are of three main types: (i) those which are necessary and are not harmful like Ca and Mg, (ii) necessary but damaging at an elevated level (usually, Fe, Mn, Zn, Cu, Co, Ni and Mo) and (iii) harmful like Hg or Cd. The chemical identity of the specific element is responsible for interactions with metals. The need of living systems to both gain and discard metals has selected a variety of mechanisms of interaction which make microorganisms adapted to an altering and often unfriendly environment. The appearance of metal tolerance determinants is a very early event in evolution (Silver 1998). In recent times, mobilization from metal ores by man has formed new, metal-rich places with higher selective pressure for metal tolerance. Various prokaryotes have particular resistance determinants which can withstand different concentration of various elements through a number of mechanisms. Different types of bacteria have the capability to withstand toxic concentrations of metals which pollute the environment. This information formed the basis for new approaches to remediate heavy metal pollution which creates more damage than all other types of pollution in combination (Nriagu and Pacyna 1988). Microorganisms are successfully used in the treatment of heavy metal pollution in a number of approaches. Various upcoming technologies are dependent on increase in the biosorption of metals into biomass or the precipitation of ions by using some metal-related aspect of bacterial metabolism. Due to their specific physico-chemical characters, mercury and metalloids (As) can occasionally be taken away from contaminate sites through microbes that change them into volatile species, which is an alternative to the immobilization/precipitation strategy (Valls and de Lorenzo 2002).

12.2 Heavy Metals

Due to the development of various industries, heavy metals are delivered untreated in the environment with wastes which results in serious environmental pollution and endanger human survival. Heavy metals like As, Cr, Pb, Hg, Cd and U are always present in the industrial effluents, which is a major environmental issue because of their toxic and non-biodegradable character and the extended biological half-lives making their removal from biological tissues a difficult task (Aiking et al. 1984).

Those ions which have partially or completely filled d-orbitals and atomic weights ranging from 63.5 to 200.6 with a specific gravity greater than 5 are known as heavy metals. Living organisms need small amounts of certain heavy metals like cobalt, copper, iron, manganese, molybdenum and vanadium. However, a high concentration of these essential metals is toxic to the living organism. Some other heavy metals like cadmium, chromium, mercury, lead, arsenic and antimony are especially concern with surface water systems and often transported by runoff water and pollute water sources downstream from the industrial site. These heavy metals can bind to the surface of microorganisms and may even go inside the cells. Inside the microorganism, the heavy metals are chemically transformed as the microorganism utilizes a chemical reaction to digest food.

These metals pose health problems to those who inhabit these polluted sites. Metal waste is commonly found in soil, sediments and water. Most probably, metal contaminants make their entries through breathing, eating, drinking and skin interaction. Metals like mercury, lead and arsenic are harmful to the kidneys, diminish mental capabilities and cause weakness, headaches, abdominal cramps, diarrhoea and anaemia (Kristanti et al. 2011). Severe contact with these pollutants can harm kidney and brain permanently (Kristanti et al. 2011). But metals like cadmium are very harmful and induce DNA breakage. Industrial sewages contaminated with metals are discharged into sewage treatment plants which ultimately increase metal concentrations in the activated sludge. However, arsenic, copper, iron, nickel, etc. are important in less concentration but cause damage at high concentration (Baker 1999). Heavy metal contamination causes many health problems like headache, irritability, abdominal pain and various symptoms associated with nervous system, anxiety, bladder and kidney cancer (Baker 1981) through the transfer of important nutritional minerals from their original place, thereby blocking their biological function or accumulating to disrupt the function of important organs and glands like heart, brain, kidneys, bone and liver (Balasubramanian et al. 2008). The microbial world is diverse, because of which microorganisms can change themselves to toxic concentrations of heavy metals and turn into "metal resistant" (Bogacka 2011; Gupta et al. 2016).

12.2.1 Traditional Methods for Treating Environments Contaminated by Heavy Metals

Environments contaminated by heavy metals are removed through means of conventional technologies based on physico-chemical principles and are ineffective and costly. In one of the processes used to eliminate metals from aqueous solutions to increase pH, reagents are added, which change metals from a soluble to insoluble form (hydroxides) resulting in their precipitation. This method generates lots of mud in the final wastewater with high concentrations of metals which is difficult to remove. While other complicated methods have single or several stages like (1) precipitation with hydroxides, carbonates or sulphides; (2) redox chemistry; (3) sorption (adsorption with activated carbon/ion exchange); (4) utilization of membranes (ultrafiltration, electrodialysis and RO, i.e. reverse osmosis); (5) recovery of electrolytes; (6) evaporation; (7) liquid-liquid extraction and (8) electrodeposition. But, these days, a new technology bioremediation has emerged as a substitute because of its simplicity, effectiveness and low cost (Goyal et al. 2003; Tabak et al. 2005; Hameed 2006; Machado et al. 2008; Wang and Chen 2009; Elen Aquino Perpetuo et al. 2011).

Bioremediation utilizes plants or microorganisms, viable or non-viable, natural or genetically engineered to remediate environments contaminated with organic molecules which are difficult to degrade (xenobiotics) and to lessen the damage of harmful heavy metals, by changing them into non-toxic or less toxic elements, thereby forming harmless products (Cunningham and Berti 1993; Davis et al. 2003; Prasad and Aranda 2018). Various strategies can be applied to improve the bioremediation which can be dependent on the nature of the contaminated environment. One such method, biostimulation, encourages the growth of indigenous microorganisms through the addition of nutrients in the contaminated site. As a result, the rate of biodegradation/bioremediation can be increased. In another method, i.e. bioaugmentation or bio addition, microbial populations are added to local, outside or genetically modified organisms (GMO), at places with shortage of native microbes or where they are not able for competition.

Microbes have developed various methods to overcome the damages of metals and metalloids, through storage and resistance or by limiting their bioavailability or toxicity through biomethylation as well as transformation (Thakare et al. 2021). The high level of waterborne heavy metals is a big environmental problem in aquatic ecosystems globally. Many of these heavy metals move to groundwater, and others are stored in seafood or in plants and are the main toxic source for humans. The rhizosphere has increased microbial activity which increases storage, transformation, degradation and biomethylation of Se with other trace elements. Microbes residing in the rhizosphere easily remove harmful heavy metals or metalloids coming through wastewaters by biosorption, sulphide precipitation, and biotransformation (reduction, volatilization).

A lot of microorganisms like bacteria, fungi, yeasts and algae have been studied for utilizing in bioremediation processes, and some of them are used as heavy metal biosorbents (Deeb and Altalhi 2009; Dobson and Burgess 2007; Dowdy and Volk

1983). The main advantage of biosorption over conventional treatment approach is its cost-effectiveness, great efficiency, reduction of chemical as well as biological sludge, sensitivity towards particular metals, no extra nutrient necessity, renewal of the biosorbent and the possibility of metal recovery (Fan et al. 2007; Garbisu and Alkorta 2001). Various prokaryotes as well as eukaryotes remove harmful heavy metal ions, because of different natural abilities to biosorb, which provides them differing tolerance at 10–20 mg/L of metal ions because of their mobility, solubility and bioavailability potential (Gadd and White 1993; Gawali et al. 2014; Gomes et al. 2013; Goyal et al. 2003). A new technique biosorption has been explored for the elimination of toxic metals from contaminated sites which is dependent on the metal binding capability of various biological materials.

There is a need to isolate bacterial strains with new metabolic capabilities to start both biochemical and genetical degradation pathways. *Bacillus* (Hamer 1986; Hameed 2006), *Pseudomonas* (Huckle et al. 1993; Huston et al. 2002) and *Streptomyces* (Infante et al. 2014; Jain et al. 2012) are powerful metal sorbent bacteria. Biosorption in mixtures that change qualitatively and quantitatively is based on two methods – active and passive, which are metabolism-independent and metabolism-dependent (Jain and Bhatt 2014), respectively – and depends on biomass variety, its origin, possibility and processing type. In addition to biosorption, certain other advantageous processes like ion exchange, complexation, precipitation, adsorption, siderophores, biosurfactants, oxidation-reduction (redox), biomethylation, metal-binding cysteine-rich peptides, metallothioneins (MTs), glutathione (GSH), natural phytochelatins (PCs) and synthetic phytochelatin (EC20) as well as the "cell-surface display" system remediate heavy metals. There is a requirement to improve these techniques for the remediating heavy metals for the welfare of mankind (Gupta et al. 2016; Thakare et al. 2021)

12.3 Microbial Diversity and Bioremediation

Bioremediation employ microbes to eliminate, counteract, or detoxify contaminants from the polluted environment. There are various studies concerning the characterization of the bacterial population, their reaction to contaminants, recognition of genes required for degradation, and a lot of others (Das 2014). According to various reports, the presence of a large number of unidentified microbes assisting in bioremediation of polluted environment can be detected only through the strategies which do not depend upon culture (Marzorati et al. 2010). Gene analysis of 16S rRNA has changed the way of detecting the microbial variety in the natural environment through both culture-dependent and culture-independent strategies (Das et al. 2014). No standard method is available for evaluating environmental microflora for application in bioremediation practices, although polyphasic strategies are employed widely for studying microbial variety in the environment. Molecular biology techniques are widely used in studying microbial ecology. In fact, a clear understanding of the function of metal-resistant genes from the different microbial population is required before their use in polluted environments (Dash and Das

2012). For the purpose of bioremediation, new culture-independent applications obtain assistance of sequencing as well as in silico strategies for screening of both sequence as well as function-dependent genes (Khan et al. 2013). By knowing the physico-chemical parameters of the environmental conditions, the environmental DNA and functional RNA can be manipulated additionally to devise the species relations in a community structure. Actually, the complex nature and variation in biological organization at various stages is known as microbial diversity which incorporates the magnitude and allocation of genetic information within microbial species in microbial communities, their difference in community structure, complications associated with interactions, number of trophic levels and number of guilds (Hinojosa et al. 2010). Microbes consist of de novo capability for decomposing all naturally occurring compounds by the principle of microbial infallibility (Dash et al. 2013). Microorganisms degrade organic contaminants through oxidizing them to carbon dioxide, while for toxic metals, microorganisms can only change the species of metal and their mobility (Lovley and Coates 1997). Presently, microbial diversity with metabolic capacity partially has been searched, and further use of the unnoticed genetic resources will increase the remediation practices of toxic metals (Naser 2013). Additionally, newly developed molecular genetics of bioremediation and knowledge-based strategies of balanced protein modification will give more light in the production designer biocatalysts for environmental bioremediation (Pieper and Reineke 2000; Paul et al. 2005; Das et al. 2016).

12.4 Utilization of the Natural Biodiversity

12.4.1 Microbial Remediation of Heavy Metals Polluted Soils

Many bacteria such as *Bacillus subtilis*, *Pseudomonas putida*, and *Enterobacter cloacae* are employed effectively for decreasing high toxicity of chromium into the less toxic Cr (VI–III) (Ajmal et al. 1996). *Bacillus thuringiensis* has removed more Cd and Zn from soil polluted with Cd as well as the effluent of the metal industry. It is thought that siderophore (Fe-complexing molecules) are produced by bacteria which may be responsible for the removal of these metals from the soil which ultimately affects their bioavailability (Khan 2005). Sulphate-reducing bacteria like *Desulfovibrio desulfuricans* indirectly bioremediate by bioprecipitation which changes sulphate to hydrogen sulphate and ultimately combines with heavy metals such as Cd and Zn to form insoluble metal sulphides. There is a lot of debate concerning genetically modified organisms and also the reality that the heavy metal remains in the soil in this strategy of bioremediation which are prime restrictions in implementation. One approach in bioremediating polluted soils is to make the soil favourable for soil microbes. This is known as biostimulation in which nutrients are added as manure or other organic amendments for carbon source to the microorganisms present in the soil. These additional nutrients will boost the growth and activities of microorganisms implicated in the remediation procedure and also raises the effectiveness of bioremediation. Biochar is presently being used for its capability of managing heavy metal polluted soils (Namgay et al. 2010). When the contaminated soil was remediated with biochar, less availability of heavy metals is observed which decreased plant absorption of the metals. Biochar has the capability to increase soil pH unlike other organic amendments because properties of biochar differ generally on the basis of its production process and on the feedstock employed in its making. The influence of various biochar amendments depends on the accessibility of heavy metals occurring in soil. Additional research is required for understanding the influence of biochar on soil microorganisms as well as effects of the interaction between biochar and soil microbes on remediation of heavy metal polluted soils as these studies are very few (Singh et al. 2015).

12.5 Unique Mechanisms in Bacteria: Natural Development of Resistance Towards Toxic Elements

Bacteria are known to flourish in every type of environment, covering from frigid poles to the hot deserts, to the wet swamps, to the aquatic systems. They have some special properties like tiny in size, having large surface-area-to-volume ratio, capable to move genetic characters and their flexibility. Bacteria generally degrade organic pollutants into non-hazardous products by using them as a carbon source. But, for inorganic metal contamination, microorganisms employ three different tolerance mechanism, like flushing of the contaminant metals out of the cell through transporters, conversion of the metals into harmless forms and biosorption. Mostly, biosorption and the enzymatic change of metal into another form are joined, meaning once a metal is absorbed inside the cell, enzymes will precipitate it in the form of salt (Williams et al. 2012). The primary line of defence against heavy metals is efflux, which includes heavy metal efflux resistance nodulation cell division (HME-RND) protein superfamily members together with efflux pumps (Nies 2003). The second process through which bacteria are known to tolerate as well as get rid of harmful metals is biosorption (Vieira and Volesky 2010), which is actually the binding of contaminants at the surface of the cell. It includes the biosorbent, a biological material which is a solid phase and solvent (liquid phase) with the dissolved species which has to be absorbed (sorbate, metal ions). The affinity of sorbent to sorbate is retained until equilibrium is reached between the sum of solid-bound sorbate species and its residual part in the solution, till the sorbate is removed. In bacteria, metallothioneins (MTs), a group of proteins that are cysteine rich having a molecular weight of 0.5–14 kDa, mediate sorption of metals (Blindauer et al. 2002). In certain cases, they may also contain histidine residues. It is already known that microbial metallothioneins behave like a storehouse for zinc also guard the cells from cadmium toxicity (Klaassen and Liu 1999). They may also work as free radical scavengers and bind with toxic superoxide and hydroxide ions. Cysteine is changed into cystine by oxidation and release bound metals in the environment. The biosorption mechanism regenerates biosorbent perfectly and has very efficient possibilities of metal recovery (Kratchovil and Volesky 1998). But, sometimes, early saturation of metal-interactive sites is also problematic. On the surface of the bacterial cell wall, many functional groups are present containing a vast range of polysaccharides and proteins performing as active sites for binding various metal ions. The binding is based on the living or non-living biomass status, a variety of biomaterials, metal solution chemistry characteristics, and surrounding environmental setting (Gee and Dudeney 1988). Bacteria can detoxify metals through reduction to a different oxidation state. General reduction mechanisms change Hg^{2+} to Hg^0, Cr^{6+} to Cr^{3+} and $AsO_4{}^{3-}$ to $AsO_3{}^{3-}$. Metal precipitation can be done through metallic phosphate precipitation as a result of dissimilarity reduction or secondary outcome of metabolic mechanisms concerned with the transformed metals (Valls and de Lorenzo 2002). The process of chemical modification of compounds through biological agents is known as biotransformation, while formation of CO_2, $NH_4{}^+$ and H_2O is called mineralization. Reduction of inorganic mercury (Hg^{2+}) to elemental mercury (Hg^0) mediated through mercuric ion reductase is one example of this mechanism (Dash and Das 2012). Some other similar reactions are the change of arsenate to arsenite and chromium (VI) to its less harmful form chromium (III). But, unsafe waste sites are usually polluted with organic compounds and metals. Media composition, soil type and pH decide bioavailability of metals and controls the degree to which metals oppose biodegradation. In an environment where a lot of different pollutants are present, the non-biodegradable metal component is removed or stabilized by different processes like mobilization, separation and collection, off-site transfer and dumping. Metals hinder the mechanism of organic pollutant degradation through specific degradation enzymes as well as enzymes concerned with general metabolism. The metal bioavailability is controlled through their interaction with organic compounds. Rosner and Aumercier (1990) reported that common intermediate of aromatic hydrocarbons, salicylate enhanced uptake of cadmium and toxicity in *Escherichia coli*. Biodegradation rate of compounds is decreased, and the acclimation period is increased by the metals which are present as soluble complexed species and ionic solutes in an environment (Sandrin and Maier 2003). In an environment which is co-contaminated, the energy supplies to preserve simultaneous metal resistance and organic degradation are elevated; thus, the instantaneous activity of bioremediation via microbes is required to execute both the activities under environmental conditions. Therefore, organic contaminant-degrading metal-tolerant bacteria with biosorption and biotransformation activities are necessary for eliminating both metal toxicity and organic pollution (Das et al. 2016).

12.5.1 Microorganisms as Biosorbents of Heavy Metals

Various reports show that several prokaryotes and eukaryotes consist of diverse natural capabilities to biosorb harmful heavy metal ions which provide them dissimilar built-in tolerance especially in diluted solutions (between 10 and 20 mg/L^{-1}) because of their mobility, solubility and bioavailability powers of these metal ions (Malik 2004; Tabak et al. 2005; Kim et al. 2007; Chen and Wang 2008). Being eukaryotic or prokaryotic decides the way by which microorganisms interact with heavy metal ions. Eukaryotes are extra susceptible to the harmful effects of metal toxicity than bacteria. Various tolerance processes are activated in the presence of damaging metal concentrations, such as the induction of peptides of the metal-binding protein family, like metallothioneins (MTs); the control of the intracellular level of metals from the cytoplasm to the inside of vacuoles, induction of protein transporters of ligand metal complexes and entry of metal ions from ion channels existing in the cell wall. In bacteria, all these processes are often coded by plasmids, which help their dispersal from cell to cell (Valls and de Lorenzo 2002). For bioremediating heavy metals industrially, low-budget biomaterials should be used which can be a derivative or waste product with more elimination power because less price of this biomass is vital for the economical viability of the process (Volesky and Holan 1995; Zouboulis et al. 2001). A lot of work has been done to improve the tolerance and/or the microorganism capacity to collect heavy metal ions as well as those with different parameters like pH (Naeem et al. 2006), temperature, different metal concentrations and biomass (Soares et al. 2003; Kim et al. 2005), competition between ions of diverse elements, microorganism and metal contact period (Kotrba and Ruml 2000), components of the culture medium (Ghosh et al. 2006); bioaugmentation/biostimulation (Roane and Pepper 2001; Silva et al. 2004), tolerance to harmful effects of heavy metals of Gram-positive/Gram-negative bacteria (Samuelson et al. 2002); intracellular/extracellular bioaccumulation, viable/non-viable cells, free/immobilized cells and biological mechanisms through aerobic/anaerobic microorganisms (Dias et al. 2002; Liu et al. 2005; Tabak et al. 2005; Wang and Chen 2006). New techniques for removing toxic metals from polluted sites are mainly centred upon biosorption, which is dependent on the metal binding potential of different biological materials. The capability of biological materials to store heavy metals from wastewater by either through metabolic pathways or through physico-chemical methods is known as biosorption (Fourest and Roux 1992). Known important metal absorbents are algae, bacteria, fungi and yeasts (Volesky 1987). Many local organisms isolated from heavy metals polluted sites are resistant to heavy metal toxicity and are natural initiating point for all biotechnological applications. It is therefore essential to isolate bacterial strains with new metabolic potential and to set up degradation pathways both biochemically and genetically. Most effective metal biosorbent bacteria are *Bacillus* (Nakajima and Tsuruta 2004; Tunali et al. 2006), *Pseudomonas* (Chang et al. 1997; Uslu and Tanyol 2006) and *Streptomyces* (Mameri et al. 1999; Selatnia et al. 2004). In the biosorption solid-phase sorbent or biosorbent, a biological material and a liquid phase solvent usually water with dissolved sorbate metal ions are present. The sorbate is favoured because sorbent has a greater affinity for the sorbate species and attached itself there through various processes till equilibrium is achieved in between the total solid-bound sorbate species and its remaining part in the solution. The extent with which sorbent binds to the sorbate defines its distribution between solid and liquid phases.

12.5.2 Biosorbent Material

Some microorganisms show very powerful biosorbent behaviour towards metallic ions due to the chemical composition of the microbial cells. This kind of biosorbent contains dead and metabolically non-active cells. Certain varieties of biosorbents most likely have a wide range which bind and accumulate most of the heavy metals without any special preference, while some of them are particular for specific metals. Some of the labs use easily obtainable biomass, while others are separated particular strains of microorganisms and some processed existing unprocessed biomass to a certain level to increase its biosorption properties. Current biosorption research has concentrated on waste products, which are derivatives or remnants of big-scale industrial processes like mycelia waste obtained from fermentation procedure, solid waste of olive oil-making plants (Pagnanelli et al. 2002), activated sludge of sewage treatment plants (Hammaini 2003), biosolids (Norton et al. 2004) and aquatic macrophytes (Keskinkan et al. 2003). The biosorption procedure is intricate which consist of mainly ion exchange, chelation, adsorption through physical forces, entrapment in intra- and interfibrillar capillaries as well as spaces of the structural polysaccharide network due to the concentration gradient and diffusion by cell walls and membranes. There are many chemical groups that are supposed to draw and sequester metals in biomass like acetamido groups of chitin; structural polysaccharides in fungi; amino and phosphate groups in nucleic acids; amido, amino, sulphydryl and carboxyl groups in proteins; hydroxyls in polysaccharide and basically carboxyls and sulphates in polysaccharides of marine algae of divisions Phaeophyta, Rhodophyta and Chlorophyta. But, the occurrence of a specific functional group never guarantees biosorption maybe because of steric, conformational, and other barriers.

12.5.3 The Choice of Metal for Biosorption Process

Suitable metals for biosorption investigations are chosen on the basis of point of interest and the effect of different metals. Thus, they can be categorized into four major classes: (a) harmful heavy metals, (b) strategic metals, (c) valuable metals, and (d) radionuclides. Of these, categories (a) and (d) are of prime consideration environmentally with respect to their elimination from the environment as well as from point source effluent discharges. Besides the toxicological point of view, the attention to particular metals is dependent on the representation of its behaviour in relation to ultimate generalization of outcome about their biosorbent uptake. Chromium, arsenic and selenium are suitable choices for studying the harmful effects as well as attractive solution chemistry of elements. Strategic and valuable metals are significantly considered because of their recovery value, although they are not environmentally alarming. Studies on biosorption were started in the 1980s (Volesky and Holan 1995; Volesky 2001). In the past, studies in this area employed bioremediation with microorganisms only, to decompose organic compounds (Lovley and Coates 1997). A lot of work has been done since then, to explore the binding characteristics of heavy metals to various kind of biomass (Chen and Wang 2008). Usually, biosorption is a metal ion physico-chemical interaction with microorganisms's particular cell surface groups that may increase or prohibit intracellular transfer or affect the transformation mechanism as well as biomineralization which is extracellular precipitation (Davis et al. 2003; Gavrilescu 2004). Biosorption is dependent on metabolism-independent passive or metabolism-dependent active storage methods (Wang and Chen 2006), in mixtures that vary qualitatively as well as quantitatively and are based on a variety of biomass, its source, practicability and type of handing out (Veglio and Beolchini 1997). Metal accumulation processes can be categorized on the basis of their dependence on cellular metabolism or according to the position of the metal inside the cell.

12.5.4 Biosorption Mechanisms

The intricate structure of microorganisms indicates that microbial cells have several routes for metal uptake. There are several biosorption processes which are still not yet fully resolved. They can be categorized on the basis of several facts. Intracellular storage of metals takes place through the metal transport across the cell membrane depending upon cell metabolism meaning that this type of biosorption is possible only in viable cells. Mostly, it is linked with an active microorganism's defence system, which responds to toxic metal presence. While in non-metabolism-based biosorption, metal is taken up by physico-chemical reactions of metal with that of functional groups of the microbial cell surface. This relies on physical adsorption, ion transfer and chemical sorption and is independent of cell metabolism. Mainly, polysaccharides, proteins and lipids are present in microbial cell walls with several metal-binding groups like carboxyl, sulphate, phosphate and amino groups. This non-metabolism-dependent biosorption is comparatively fast that can be reversed back (Kuyucak and Volesky 1988). During precipitation, metals are taken up by the solution as well as the cell surface (Ercole et al. 1994). In addition, it can be cell metabolism-dependent if the microorganism generates compounds due to the existence of toxic metals favouring the precipitation process. Precipitation may not be cell metabolism-dependent, if it is happening after a chemical reaction of the metal with the cell surface. Because of biomaterial complexity, several of these mechanisms may happen together in scales on the basis of biosorbent and environmental surroundings (Kefala et al. 1999; Volesky 2001; Gavrilescu 2004). Wang and Chen (2006) have centred biosorption studies on the choice of best possible biomass varieties whose prospects by sorption will be based on three areas, i.e. biology, chemistry and engineering (Kefala et al. 1999). Biosorption potential is influenced principally by three reasons: (1) properties of the metal ion like atomic weight, ionic ray and valence; (2) environmental settings like pH, temperature, ionic strength, contact period and biomass concentration; and (3) the biosorbent character which possibly decides variations in choice as well as metal ions affinity because the variety and microorganisms species, growth conditions, physiological state and age influence the binding procedure of heavy metals (Wang and Chen 2006; Chen and Wang 2008). Various approaches are developed through recombinant DNA technology to get genetically superior strains to employ in the biosorption mechanism. The bacterial cell wall is geared up by these various approaches to perform as anchors through metal ion-binding polypeptides. Of these, one investigation was in connection with the fusion protein of the α-domain of IgA protease of *Neisseria gonorrhoeae* and metallothionein (MT) from rats (Valls et al. 2000) and lpp-ompA-various sizes of peptides (EC20) (Bae et al. 2000). The major biosorption processes developed by microorganisms are also regarded as resistance or tolerance mechanisms towards heavy metals.

12.5.4.1 Transport Across Cell Membranes

Transfer of heavy metals through microbial cell membranes is probably conducted by similar processes which are employed to pass on metabolically significant ions like potassium, magnesium and sodium. The metal transfer systems may get puzzled because of the existence of similarly charged heavy metal ions as well as ionic radius linked with necessary ions. This type of process is not related to metabolic activity. Principally, biosorption by living organisms includes two stages: the first one is metabolism-independent, where metals are attached to the cell walls, and the second one is intracellular intake, which is metabolism-dependent and metal ions are transferred through the cell membrane (Costa and Leite 1991; Gadd 1988; Huang et al. 1990; Nourbakhsh et al. 1994).

12.5.4.2 Ion Exchange

The cell wall of a microorganism possesses several types of polysaccharides and has a substitution of bivalent metal ions through the counter ions of the polysaccharides as alginates in marine algae are found as K^{+}, Na^{+}, Ca^{2+} and Mg^{2+} salts. Heavy metal biosorption can be done by replacement through counter ions like CO^{2+}, Cu^{2+}, Cd^{2+} and Zn^{2+}(Kuyucak and Volesky 1988). Copper biosorption through *Ganoderma lucidium* (Muraleedharan and Venkobachar 1990) and *Aspergillus niger* fungi is one of the examples of this phenomena.

12.5.4.3 Complexation

Electrostatic interaction of metallic ion chelating agent with a polymer produced by a viable or unviable microorganism results in an extracellular complexation or coordination. This could be due to the biosurfactants, polysaccharides, proteins and nucleic acids. Electrostatic attraction is because of electron pairs of chelating agents, and no electron transfer will take place if they adhere to the metallic ions. The final composition has the electric charge which is the total of all individual charges of every member of the complex (Veglio and Beolchini 1997; Davis et al. 2003). "Oxidative stress" in the cell will happen if this detoxification/complexation system is filled to its full capacity (Gavrilescu 2004). A complex is formed over the cell surface by the interaction of metals with active groups which removes metal from the solution. According to Aksu (1992), *Chlorella vulgaris* and *Zoogloea ramigera* carry out biosorption of copper by adsorption as well as by creating coordination bonds amongst metals and amino and carboxylic groups present in polysaccharides of the cell wall. Complexation is the sole process which is accountable for calcium, magnesium, cadmium, zinc, copper and mercury storage through *P. syringae*. Production of various organic acids like citric acid, oxalic acid, gluonic acid, fumaric acid, lactic acid and malic acid by microbes may chelate harmful metals and produce metallo-organic molecules. These organic acids assist in solubilizing metal compounds and discharging through their surfaces. Carboxyl groups present in polysaccharides and other polymers of microbes may biosorb or complex metals.

12.5.4.4 Precipitation

Precipitation can be of two types, cellular metabolism-dependent and cellular metabolism-independent. In the first case, metal are eliminated from solutions usually through the active defence system of the microorganisms. They respond to the presence of those compounds which form harmful metal and supports the precipitation procedure. Chemical interaction of metal with the cell surface results in cellular metabolism-independent precipitation. These different biosorption procedures can happen at the same time.

12.5.4.5 Physical Adsorption

Metal ion of solutions attaches itself to microbial cell wall polyelectrolytes with the help of various interactions like electrostatic, Van der Waals, covalent bonding, redox and biomineralization to attain electroneutrality during physical adsorption. This method is independent of metabolism as well as reversible and has great potential due to many advantages, particularly treatment of huge level of wastewater having low amount of contaminants (Ahluwalia and Goyal 2007; Kuroda and Ueda 2010; Nishitani et al. 2010). Cell wall potential negatively attracts the metal ions and depends on pH in physical adsorption. pH affects metal storage in yeasts, algae and bacteria similarly. As in yeast, at $pH < 2$, the metal storage is practically zero due to the linkage of cell wall active sites to protons at low pH, which limits the metal cation strategy causing repulsion. Due to this, pH rises, and a greater number of groups such as acetamide chitin, structural polysaccharides of fungi, phosphate and amino groups of nucleic acids, amino and carboxyl groups of proteins and hydroxyl groups of polysaccharides are substituted by negative charges which further raises the attraction between metallic cations and cell surface adsorption. Due to this, there is a decline in metallic ions solubility which ultimately decreases bioavailability, and precipitation takes place (Esposito et al. 2002; Chen and Wang 2008; Nishitani et al. 2010; Kuroda and Ueda 2011). Hence, pH of 4–8 is usually termed as "good" for heavy metal biosorption for roughly every kind of biomass (Borro and Fein 2005; Wang and Chen 2006; Machado et al. 2010). The function of extracellular polymeric substances (EPS) excreted by bacteria for eliminating heavy metal ions through adsorption has also been investigated (Gupta et al. 2000). According to Kuyucak and Volesky (1988), electrostatic forces of metal ions of solutions and microbial cell walls are responsible for biosorption of various metals like uranium, cadmium, zinc, copper and cobalt by dead biomass of algae, fungi and yeasts. Electrostatic attractions are also performing copper biosorption by *Z. ramigera* bacterium and the *C. vulgaris* alga (Aksu 1992) and chromium biosorption through *G. lucidum* and *A. niger* fungi.

12.5.4.6 Siderophores

Certain microorganisms are having low molecular weight similar to heavy metal chelating agents and are known as siderophores. Certain particular iron chelators siderophores are formed in the medium, when microorganisms are cultured in the iron-lacking medium. They have a significant function in the complexation of toxic metals as well as radionuclides through increase in their solubility. Siderophores are of low molecular weight and consist of catecholate, phenolate or hydroxamate for attachment which are produced by various bacteria like *Actinomycetes Azotobacter* and genus *Pseudomonas* to confine the iron ions needed for their metabolic activity as well as for biosorption (Pattus and Abdallah 2000; Gazsó 2001; Das et al. 2007).

12.5.4.7 Biosurfactants

Most surfactants utilized for bioremediation are manufactured industrially from petroleum, but microorganisms can also produce them. Naturally produced biosurfactants are saponins made in plants, glycolipids in microorganism and bile salts in the bodies of organisms. They contain various advantages in comparison to industrially produced surfactants, like less toxicity towards decomposing microorganisms and low recalcitrance in the environment, more varieties of chemical structures and ability to work on a wider setting at diverse temperatures and pHs (Bognolo 1999). Biosurfactants are metabolically produced by aerobic microbes like bacteria, fungi and yeasts and are released into the medium. The hydrophilic part is made from amino acids, peptides or saccharides, while the hydrophobic fraction generally contains saturated or unsaturated fatty acids (Inamuddin et al. 2021). Biosurfactants have the capability to form numerous structures like micelles, vesicles, sphere-shaped or irregular lamellar structures, amongst others (Champion et al. 1995; Mulligan 2005; Li and Li 2011). When challenged by insoluble organic as well as inorganic pollutants like hydrocarbons, oil, pesticides and heavy metals such as uranium, cadmium and lead, microorganisms remove their shortcomings naturally through the excretion of these structures in culture medium attached to their cell walls. This helps in transporting and translocating insoluble substrates which ultimately assist in biosorption. In heavy metal ion bioremediation, a complicated harmless biosurfactant/metallic ion structure is produced which emulsify and solubilize ions and ultimately compartmentalize them physically. Rhamnolipids formed by *P. aeruginosa* is amongst highly employed natural biosurfactants in bioremediation (Bognolo 1999; Champion et al. 1995; Mulligan 2005; Tabak et al. 2005; Zhang et al. 2005).

12.5.4.8 Oxidation-Reduction (Redox)

Metal ions, metalloid and organometal compounds can be mobilized or immobilized by microorganisms which ultimately promotes redox processes. Prokaryotes alone can gain energy from oxidation of metals Mn^{2+}, Fe^{2+}, Co^{2+}, Cu, AsO^{2-}, Se0 or SeO_3^{2-} or by reducing Mn^{4+}, Fe^{3+}, Co^{3+}, AsO_4 $^{2-}$, SeO_4^{2-} or SeO_3^2 (Gavrilescu

2004). Following the reduction of Fe^{3+} ion to Fe^{2+} or the Mn^{4+} ion to Mn^{2+}, solubility is enhanced. Microbial species have the capacity to immobilize heavy metals perfectly by reducing heavy metal ions to a lower oxidation state, to produce less bioactive metallic elements (load zero) (Valls and de Lorenzo 2002; Gadd 2004).

12.5.4.9 Biomethylation

Metal ions can be changed from highly toxic to less harmful form by the microorganisms through biomethylation process. In aerobic as well as anaerobic environment, different varieties of bacteria, filamentous fungi and yeasts can methylate Hg, As, Cd, Se, Sn, Te and Pb ions making them highly immobile and prepare them for participating in those processes which lessen their toxicities. In this enzymatic process, a methyl group (CH_3) is transferred to metals and metalloids. The resultant methylated compounds are highly diverse in solubility and volatility as well as damaging (Roane and Pepper 2001; Gadd 2004), such as intermediates of processing for Hg^0, methyl and dimethyl mercury, which are more harmful in comparison to inorganic Hg ions. Inorganic forms of As are more damaging in comparison to methylated species (acids and methyl-As dimethyl-As), while the methylated and inorganic forms of Se and Cd are more harmful (Roane and Pepper 2001; Tabak et al. 2005).

12.5.4.10 Metal-Binding Cysteine-Rich Peptides

Cells produce cysteine-rich peptides, metallothioneins (MTs), glutathione (GSH) or phytochelatin (PCs) when exposed to toxic levels of heavy metals. All of them are low-molecular-weight and non-enzymatic compounds which can withstand thermo-coagulation as well as acid precipitation. They make complexes with divalent metals and metal-thiols which are significant metabolites for fighting ROS (Bae et al. 2000, 2001).

Metallothioneins (MTs)

A group of well-preserved structures of proteins which behave like antioxidants is known as metallothioneins (MTs) and are found in all living beings. They are low molecular weight as well as cysteine rich. Thiol groups (SH) of cysteine chemical structure have the capability to arrest metal ions like Cd^{2+}, Fe^{2+}, Hg^{2+}, Cu^{2+} and Zn^{2+}. MTs possess two different domains: one in the N-terminal region is known as β domain, while the other present in the C-terminal region is called as α domain. The β domain has nine cysteine residues which bind three divalent ions, while the α domain already consists of eleven cysteine residues which bind four ions; therefore, a single molecule binds total seven ions (Cobbett and Goldsbrough 2002; Thirumoorthy et al. 2007). The MTs play various roles like heavy metal detoxification and defence to the presence of ROS. Therefore, MTs are accountable for decreasing the influence of oxidative stress due to these ions, but they also retain homeostatic cellular redox balance. According to these features, only metals can induce protein synthesis (Cobbett and Goldsbrough 2002; Smith et al. 2007).

Glutathione (GSH)

All living beings have glutathione (GSH), L-glutamyl-L-cysteinyl-glycine, an antioxidant which is soluble as well as highly significant non-protein thiol. Although it has (Glu-Cys-Gly) amino acids, its biochemical features are due to a cysteine thiol group which is found in its active site (Bae and Mehra 1997; Penninckx 2000, 2002; Mendoza-Cózatl et al. 2005). The GSH regulates its own synthesis and takes part in different mechanisms such as it regulates intracellular redox state, inactivates ROS, transfers GSH linked amino acid as well as other molecules and builds up sulphur and cysteine. In mammals, it is present in liver at higher levels and has same biosynthesis in plants, yeast and protists. The mitochondria and the nucleus are not damaged by the action of ROS because of their own GSH reservation which is critical or helpful in protecting these structures (Penninckx 2002; Inouhe 2005; Mendoza-Cózatl et al. 2005). In yeast, the cell defence system against Cd^{2+} toxicity works through the making of a GSH-Cd^{2+} complex which reduces the lipid peroxidation of the cell membrane and permits the transfer of GSH-Cd^{2+} conjugate inside the vacuole. This causes reduction in the levels of toxic metals of cytosol which ultimately assists in decreasing oxidative stress (Penninckx 2000; Adamis et al. 2004; Kobayashi et al. 2006; Preveral et al. 2006).

Natural Phytochelatins (PCs) and Synthetic Phytochelatin (EC20)

Some small peptides rich in cysteine with general structure (Glu-Cys) nGly ($n = 2–11$) are known as phytochelatins (Grill et al. 1985; Cobbett 2000). PC synthase catalyzes glutathione (GSH) in different stages to produce PCs (Grill et al. 1985; Gupta et al. 2005). Because of them, ions become capable of ligating heavy metal ions by their thiol as well as carboxyl residues (Kobayashi et al. 2006; Inouhe 2005). Plants, fungi, nematodes, parasites and algae along with cyanobacteria possess PCs. These PCs have better capability to bind heavy metal ions (1 atom per cysteine) in comparison to MTs, although they have been placed under MT-III class. The initial efforts of inducing recombinant PCs *E.coli* strongly opposed by type γ chemical bonds formed between Glu-Cys units as a result of multienzyme mechanisms. Type α bonds found in amino acid chains of all proteins are not like these bindings (Bae et al. 2001; Cobbett 2000; Penninckx 2000; Gupta et al. 2005; Inouhe 2005; Hirata et al. 2005; Mendoza-Cózatl et al. 2005; Wu et al. 2006). Another option to this was to produce an in vitro gene which encodes proteins analogous to PCs with the structure (Glu-Cys) nGly (ECs) in which all amino acids are chemically linked through type α. Therefore, 20 units of Glu-Cys (EC) repeats are present in the synthetic phytochelatin EC20 which has higher capability to bind heavy metal ions in comparison to natural PCs. Production of recombinant bacterial strains is a routine process now (Bae et al. 2000, 2001; Xu et al. 2002; Lee et al. 2002; Wu et al. 2006), but not much work is done in yeast (Schmitt et al. 2006). Induction of EC20 can be referred to identify microorganisms with better capability to bind heavy metal ions to use in bioremediation. MTS, PCs and ECs along with other genetic engineering approaches have been searched to produce new recombinant microbial strains for getting these proteins to stay attached to the external surface cell ("cell-surface display") (Deng et al. 2003; Merle et al. 2003; Kim et al. 2005; Nishitani et al. 2010). The objective was to raise the potential of these unique microorganisms in adsorption procedures concerning heavy metal ions than non-recombinant microorganisms (Bae et al. 2000, 2001; Kuroda et al. 2002; Kuroda and Ueda 2003; Jiang et al. 2007).

12.5.4.11 The "Cell-Ssurface Display" System

Cell surface proteins are highly significant biomolecules as they are present on the interface of cell and its environment. The cells consist of various methods to bond specific proteins on the surface and restrict them to specific areas. A variety of these methods are being employed in bacteria as well as in *S. cerevisiae* (Kondo and Ueda 2004). The appearance of heterologous peptides on the surface of the cell ("cell-surface display") is a great procedure which is very commonly practised in the biotechnology in different procedures like in manufacturing recombinant vaccines, antigens, antibodies, enzymes and library peptides (Kuroda et al. 2001; Chen and Georgiou 2002; Samuelson et al. 2002; Rutherford and Mourez 2006; Wang et al. 2007; Kuroda and Ueda 2010, 2011; Nishitani et al. 2010).

12.6 Factors Affecting Biosorption

The study of the efficiency of metal uptake through microbial biomass is necessary for the industrial implementation of biosorption to acquire knowledge for the equilibrium of the procedure needed to plan the machinery for using. Metal uptake is generally calculated through parameter "q", which denotes as the milligrams of metal collected per gram of biosorbent material, while "qH" is described as a function of the metal stored, the sorbent material utilized and the working environ ment. The following characteristics influence the biosorption method:

1. Temperature – It does not appear to affect the biosorption performance in the range of 20–35°C (Aksu 1992);
2. pH – It appears to be the most significant factor in the biosorption process as it influences the metals solution chemistry and the functional group activity of biomass and fights amongst metallic ions (Machado et al. 2010);
3. Biomass concentration in solution – It appears to affect specific uptake as with lower biomass concentrations specific uptake rises (Fourest and Roux 1992; Gadd and De Rome 1988). According to Gadd and De Rome (1988), higher biomass concentration interferes with binding sites. Fourest and Roux (1992) nullified this theory by making low metal concentration in the solution accountable for fall in specific uptake. Therefore, for using microbial biomass like biosorbent, this fact has to be considered.
4. Biosorption – It is basically utilized to remediate wastewater if there is the presence of more than one metal ion and the occurrence of the other metal ions affects the elimination of a specific metal. For instance, uranium is taken up by bacterium, fungus and yeast biomass and is unaffected by the presence of manganese, cobalt, copper, cadmium, mercury and lead in the solution

(Sakaguchi and Nakajima 1991). Quite the opposite, the occurrence of Fe^{2+} and Zn^{2+} affects the uptake of uranium through *Rhizopus arrhizus* (Tsezos and Volesky 1982), while cobalt uptake by various microorganisms is totally banned by the presence of uranium, lead, mercury and copper (Sakaguchi and Nakajima 1991; Elen Aquino Perpetuo et al. 2011).

12.7 Diversity of Metal-Resistant Genes and Biotechniques in Metal-Resistant Bacteria for Bioremediation

Microorganisms have developed methods to tolerate harmful natural compounds in the polluted environment and remediate them for their own profit (Guo et al. 2010). The capacity of bacteria to oppose toxic metals is because of highly modified genetic systems, through which they synthesize proteins and make it possible to flourish in the presence of these elements. Bacteria continue to exist due to induction of various metal-tolerant genes for harmful metals like cadmium, chromium, copper, lead, mercury and nickel. They can live in an extremely toxic environment because of these tolerant genes which are employed further for bioremediation. The objective of the selecting metal is the comparative plentiness of these metal pollutants in the surroundings and their toxicity level as specified by USEPA and the US Department of Labor (https://www.osha.gov/SLTC/metalsheavy/). Bacteria have their own defence mechanisms which develop a tolerance for these harmful metals and may be utilized to clean-up the polluted environments. Cadmium is an extremely toxic non-essential metal which can be very dangerous even in very small concentrations. According to the first report on cadmium genetics, no special cadmium resistance method is present in bacteria (Silver and Misra 1984). However, it was extensively opposed by Trevors et al. (1986) as he investigated cadmium transport, tolerance and toxicity in bacteria, algae and fungi. Tolerance to cadmium is due to cad operon of plasmid found in *Staphylococcus* spp. as well as by czc operon at 3972-bp element of *P. aeruginosa* (Crupper et al. 1999; Chakraborty and Das 2014). cadA and cadB operons are parts of cad operon (Zhang et al. 2015), while two genes, cadA and cadC, are constituent of cadA operon present in plasmid pI258 (Nucifora et al. 1989). CadA protein has homology with ArsB protein of the ars operon and forms an energy-dependent ATPase which effluxes cadmium from the bacterial cell to give protection. CadC protein is the transcriptional regulator of the operon (Hsieh et al. 2010). A 204-residue polypeptide is encoded by the cadB operon whose mechanism of action is still not totally known. It is presumed that cadB, present on the plasmid pII147 cellularly, binds cadmium, most probably at the plasma membrane (Smith and Novick 1972). According to Crupper et al. (1999), another novel cadmium tolerance process is present in the plasmid pRW001 of *Staphylococcus aureus* which is due to the presence of cadD system with its two genes, cadD and cadX*. cadD is similar to the gene cadB, while cadX* is encoding an inactive transcription regulator. However, this operon imparts only low cadmium tolerance. According to Chaouni et al. (1996), another cadmium-tolerant operon is present in the plasmid pLUG10 of *S. lugdunensis* which had a cadB-like gene for tolerance and another gene cadX which is a transcription regulator gene. High cadmium tolerance is due to both of these two genes. cadX has 40% similarity with cadC sequence, while the cadX* gene of pRW001 plasmid is also having similarity with the first 78 nucleotides of cadX gene sequence. After this, it is truncated due to which its transcription regulation capacity is lost. The cadX protein also has around 30% homology with ArsR protein of the ars operon (Yoon and Silver 1991). These are various genetic adjustments through which bacteria acquire resistance in the cadmium-polluted environment. Practically, degradation of metals cannot be done; thus, nearly all biological metal remediation strategies are dependent on detoxification as well as immobilization of metal to decrease its biological toxicity and also obstruct metal transfer. Remediation of cadmium through bacteria primarily is done by binding metal to the bacterial cell wall. Near to cadmium, calcium ions as well as protons are discharged which shows the cell wall's competitive binding behaviour. According to Plette et al. (1996), when bivalent ions surround the cell wall, it becomes positively charged depending upon pH-dependent charging along with metal binding. Carboxylic as well as phosphatic sites of the bacterial cell wall are employed for powerful coordination with the cadmium ions which assist in their site remediation. Peptidoglycan, phospholipids and lipopolysaccharides are constituent of the cell wall of Gram-negative bacteria. Several processes such as ion exchange, complexation, coordination, adsorption, electrostatic interaction, chelation and microprecipitation bind metals on the cell coat due to higher anionic character and charged behaviour of lipopolysaccharides (Vijayaraghavan and Yun 2008). Intracellular cadmium (36%) sequestration is done by *Pseudomonas* strain H1, and *Bacillus* strain H9 to (Roane and Pepper 2001) lessen cadmium toxicity. These strains produce exopolymers (EPS) which store cadmium and lessen soluble cadmium by 22% and 11%, respectively. Another option is to use microbial biomass for cadmium bioremediation which is cheap and eco-friendly. This process is performed by independent extracellular adsorption through surface complexation, ion exchange or electrostatic interaction resulting in intracellular storage by the surface of the cell (Vargas-García et al. 2012). According to Khan et al. (2015), 18.8%, 37%, and 56% Cd^{2+} are removed from the aqueous medium after 48, 96, and 144 h, respectively, while Cd^{2+} level enhanced in the medium after 192 h and intracellular Cd^{2+} storage is reduced. It is because of the activity of efflux system in *E. coli* for its existence. This shows that metal accumulation and efflux system in bacteria are interdependent. The CdS nanoparticle utilization is another approach cadmium remediation which is produced by functionalized EPS of *P. aeruginosa* JP-11 to remove 88.66% of cadmium from aqueous solutions (Raj et al. 2016).

12.7.1 Chromium

Seventh highly abundant metal on the earth is chromium which is found in the environment in two stable states, i.e. trivalent Cr^{3+} and hexavalent Cr^{6+}. Chromium is responsible for oxidative damage and inhibition of sulphate membrane transport in bacteria. Microbes have developed two processes to tolerate chromium. One of them is the efflux of chromate from the cells, and the second one reduces toxic Cr^{6+} to less toxic Cr^{3+} through enzymatic processes. chrA gene is encoding chromate efflux protein whose homologs are present in eubacteria, archaea and even eukaryotes. Nies et al. (1998) reported the presence of two chromate efflux pumps which have six transmembrane division, while Diaz-Magana et al. (2009) reported the formation of 12 segmented single heterodimers by these two pumps instead of separate units in *E. coli*. In *Ochrobactrum tritici*, a different report illustrated the existence of a chromium resistance operon which is tolerating Cr^{6+} till 50 mM (Branco et al. 2008). This operon is situated on the 7189-bp transposable element TnOtChr which has four genes in chrBACF which were not continuous and interrupted by the other genes. chrB is operon regulator and induced its expression in the presence of chromium (Chihomvu et al. 2015). A chromate ion transporter is encoded by chrA which is sensitive to Cr^{6+} and insensitive to Cr^{3+}. ChrA possesses a motif GGX12VX4WX16PGPX9/8G (X = any amino acid) having homology with several other species also. ChrC has 202-amino acid protein which is having similarity with iron/manganese superoxide dismutase, although its exact role is still not well known (Morais et al. 2011). ChrF is also similar to putative superoxide dismutase proteins (Branco et al. 2008), but its function is also not understood. There is also no influence on the level of chromate resistance by the deletion of chrF2 (Juhnke et al. 2002). Gonzalez et al. (2005) recognized ChrR as chromate reductase which has the signature sequence LFVTPEYNXXXXXX-LKNAIDXXS and can also give extra protection towards H_2O_2. Usually, chromate transport reaction requires the family of chromate ion transporters. ChrJ, chrK, and chrL genes were also found in *Arthrobacter* sp. strain FB24 (Henne et al. 2009). According to Viti et al. (2013), chrJ is encoding a putative malate/quinone reductase protein, chrK, which is specifying a protein with YVTN beta-propeller repeat, while chrL is creating a probably conserved lipoprotein of the LppY/LpqO family. An additional chromate detoxification mechanism is the reduction of Cr^{6+} to Cr^{3+} by the enzymes (Batool et al. 2012). In chromium-resistant *Lactobacillus* strain, the existence of chromate-reducing enzyme as well as rapid reduction of Cr^{6+} is found to be useful for bioremediation of Cr^{6+} in the polluted environment (Mishra et al. 2012). According to Cervantes and Campos-García (2007), Cr^{6+} reduction can be categorized by three processes like aerobic reduction in NADH or NADPH presence through reductases, Cr^{6+} utilization as an electron acceptor in the electron transport system, or reaction of Cr^{6+} with organic compounds, while, in anaerobic reduction, Cr^{6+} behaves like an electron acceptor and the catalyzing enzymes usually exhibit a flavin oxidoreductase activity. One of such enzymes, ferric reductase from *Paracoccus denitrificans*, reduces Fe^{3+} as well as Cr^{6+}. Another example is YieF Cr^{6+} reductase from *E. coli* which has a structural similarity with the ChrR of *P. putida* (Ackerley et al. 2004). Many bacterial strains like *Enterobacter* sp. and *Pseudomonas* sp. are also isolated and used in Cr^{6+} reduction in anaerobic conditions employing chromate reductase (Kamaludeen et al. 2003). The YieF reaction transfers four electrons, of which three are reducing Cr^{6+} to Cr^{3+} and only one combines with O_2 to produce reactive oxygen species (ROS) (Viti et al. 2013). As these ROS are damaging to the cell, superoxide dismutases operate and change these ROS to oxygen or water. In the same way,

ArsH of *Synechocystis* sp. PCC 6803 has the capacity to reduce Cr^{6+} (Xue et al. 2014). Actually, these are various genetic adjustment processes present in bacteria for tolerating chromium toxicity.

12.7.2 Copper

Copper is the third oldest commonly utilized metal which has vast uses in wires, motors, architecture and medicines. Copperedius results from the extra amount of copper in the body, which produces ROS and can damage protein, lipids, DNA. Copper has two oxidation states, Cu(I) and Cu(II), which can dislodge iron (Fe) from available Fe–S clusters of dehydratases as well as from other iron–sulphur proteins (Macomber and Imlay 2009). According to the study of Tetaz and Luke (1983), a plasmid pRJ1004 confers copper resistance in bacteria. Afterwards, Bender and Cooksey (1986) reported indigenous pPT23D plasmids in *P. syringae pv. tomato* helping from copper toxicity. Later on, Mellano and Cooksey (1988) confirmed the existence of a copper inducible cop operon in the 35-kb pPT23D plasmid. It has a lot of similarity to the copper resistance genes of *P. cichorii* and *P. fluorescens* (Cooksey et al. 1990). Wunderli-Ye and Solioz (1999) illustrated the process of copper tolerance in the bacterium *Enterococcus hirae*, which has four genes: copYABZ. CopY codes for a repressor in response to copper, while copZ codes for chaperone protein. It was suggested that copA coding for a protein under restricted environment to assist in the uptake of copper and CopA as well as CopB are found to be P-type ATPases. Incidentally *Streptomyces* sp. AB2A retained copper in the beginning and later on removed harmful metals which indicated the activity of extracellular cupric reductase in the organism (Albarracín et al. 2008). In *E. coli*, copper tolerance has also been investigated in detail. It was observed that *E. coli* possess a dual regulatory procedure for copper tolerance. The first one is the cus (copper sensing) locus which is a two-component system (Munson et al. 2000). It consists of two regulator genes, namely, cusR and cuss to make a regulator–sensor pair to control cusCFBA expression. The CusCBA proteins are like cation/proton antiporter complexes to remove metal ions from the cell with an influx of H^+. CusF binds with copper in the periplasmic space to increase copper storage in the cells (Yu et al. 2014). A small 10-kDa protein CusF is present in periplasmic space which is responsible for linking single copper in each polypeptide. It is associated with copper tolerance and has many methionine, aspartate and histidine residues needed for linking copper. However, the participation of methionine residues for CusF linkage with copper was recognized by site-directed mutagenesis. CusF sequence analysis exposed an N-terminal leader sequence as well as possible cleavage location of signal peptidase which indicated the periplasmic position of CusF (Franke et al. 2003). The second system is cue or copper efflux system, which is involved with copper homoeostasis. copA and cueO, the two genes, are under the control of the regulatory gene cueR. CopA is found to be a P-type ATPase, while multicopper oxidase is coded by cueO to oxidize Cu^+. Oxidation of Cu(I) in less harmful Cu (II) by CueO reduces dioxygen to water in four different single-electron transfer stages (Djoko et al. 2010). Type 1 (T1), type 2 (T2) and two type 3 (T3), three different kinds of copper atoms, are present in multicopper oxidases. Of them, T1 is in the protein's interior and oxidizes the substrate, while T2 and two T3 make a trinuclear centre (TNC) where reduction of dioxygen takes place. But, there is an additional methionine-rich helix in cueO which inhibits the solvent access to the T1 position and makes an extra copper-binding location for storage of copper (Singh et al. 2004). An additional regulatory gene CsoR with Cu^{+} inhibits repression of copper tolerance genes (Chang et al. 2014). It was observed that copper transport operon copYAZ is present in a *Streptococcus* strain which has the heavy metal-binding proteins named copY and copZ (Vats and Lee 2001). *Pseudomonas fluorescens* possess copRSCD operon (Hu et al. 2009), while opposite to this, *Helicobacter pylori* possess two different operons hpcopA and hpcopP to export and import copper (Ge and Taylor 1996). One more copper regulatory system is present in *B. subtilis* which is mediated through YcnJ and controlled by YcnK and CsoR (Chillappagari et al. 2009). These genes collectively maintain copper homoeostasis inside the cell. But, the primary method to remove cytoplasmic is an ATPase-driven copper efflux system. Periplasmic copper management, multicopper oxidases, metallochaperones and RND systems are implicated in this mechanism (Bondarczuk and Piotrowska-Seget 2013). According to Grass et al. (2004), in *E. coli*, not only multicopper oxidase cueO is induced in the company of copper to oxidize Cu(I) in the form of Cu(II) in the periplasm, but also enterobactin, a catecholate consisting ligands, is also produced. Like siderophore, enetrobactin has capability to reduce Cu(II), and it was anticipated that enterobactin oxidation through CueO has an accompanying role in Cu tolerance by not producing highly toxic Cu(I) ions. Furthermore, one intermediate of enterobactin biosynthesis, 2,3-dihydroxybenzoic acid (DHB), strongly combines with Cu ions, behaving like a Cu sink. Additionally, yersiniabactin of uropathogenic *E. coli* sequester Cu (II) outside bacterial cells protects the bacteria from intracellular killing and inhibits its catechol-mediated reduction in the form of Cu(I) (Chaturvedi et al. 2012). Moreover, the Cu(II)–yersiniabactin complex possesses superoxide dismutase activity to safeguard bacteria by oxidative stress inside phagocytic vesicles (Chaturvedi et al. 2014).

12.7.3 Lead

Lead (Pb) is a very harmful, unrelating and non-essential metal pollutant. Many bacteria like *Arthrobacter* spp., *B. megaterium*, *P. marginalis*, *Citrobacter freundii*, *S. aureus*, and *E. coli* are lead tolerant. Endogeneous pMOL30 megaplasmid present in the bacterium *Cupriavidus* (*Ralstonia*) *metallidurans* CH34 is probably one of the most investigated lead resistance operons (Borremans et al. 2001). This pbr operon has several structural genes as well as one regulatory gene (pbrR). Of these, pbrT gene is coding for Pb(II) uptake protein, pbrA is coding for P-type Pb(II) efflux ATPase, and pbrB is coding for expected integral membrane protein whose role is not known, while pbrC is coding for expected prolipoprotein signal peptidase. It was thought that pbrA exports Pb(II) from the cytoplasm, which is changed to a phosphate salt through inorganic phosphate formed by PbrB. Hynninen et al. (2009) reported that pbrB is coding for an undecaprenyl pyrophosphate phosphatase. A Pb (II)-binding protein was coded by pbrD which is present downstream of pbrC and is critical for lead sequestration. Under the lead's presence, the regulator pbrR activates the transcription of pbrABCD operon. Pbr operon is controlled through pbrR, which is having similarity to the family of heavy metal ion-sensing regulatory genes merR. pbrR activated the induction of pbrABCD like single transcriptional unit under Pb^{2+}presence. pbrABC combination is thought to have some role in lead efflux, while pbrD is believed to have role in Pb^{2+} storage (Borremans et al. 2001). PbrD consists of a possible cysteine-rich metal-binding motif (Cys-7XCys-Cys-7X-Cys-7X-His-14X-Cys) and also possesses many proline and serine residues (Jarosławiecka and Piotrowska-Seget 2014). Monchy et al. (2007) illustrated that gene pbrU is also present and is induced under the presence of Pb^{2+}. However, pbrU encodes a permease which belongs to the main facilitator superfamily (MFS) present at the inner membrane of *C. metallidurans* (Taghavi et al. 2009). *Cupriavidus* metallidurans was identified as *Ralstonia metallidurans*, and the name of some of the genes in the Deutsche Sammlung von Mikroorganismen und Zellkulturen (DSMZ) database are still starts with BRme meaning *R. metallidurans* (von Rozycki and Nies 2009). Roane (1999) studied the intracellular compartmentalization of 0.6 mM lead through metallothionein-like proteins in *B. megaterium*. Bacterial metallothionein (BmtA) is present in *P. aeruginosa* strain WI-1 of Mandovi estuary which stored intracellularly 26.5 mg lead/g dry weight of cells to decrease the harmful influences of lead (Naik et al. 2012a). *Salmonella choleraesuis* and *Proteus penneri* stored 19 and 22 mg lead, respectively, as an expression of metallothionein protein of smtAB gene was enhanced in them (Naik et al. 2012b). Thus, the occurrence of metallothioneins in bacteria may be exploited for bioremedaiting lead at polluted sites. Metal immobilization through extracellular sequestration is critical for the regulation of metal toxicity. Extracellular polymeric substances (EPS) have different substances like polysaccharides, proteins, nucleic acids, humic substances and lipids which possess numerous functional groups such as hydroxyl, carboxyl, amides and phosphoryls having a strong attraction for heavy metals with greater specificity and affinity (Bhaskar and Bhosle 2006; Bramhachari et al. 2007). De et al. (2007) also studied the binding of lead through negatively charged compounds of EPS in *P. aeruginosa* CH07. *Pseudomonas marginalis* is capable of tolerating up to 2.5 mM lead through sequestration of lead in an exopolymer (Roane 1999). In the same way, in *Paenibacillus jamilae*, biosorption of 303.03 mg lead/g EPS through EPS from lead solution takes place (Morillo et al. 2008). There are various enzymatic activities in the bacterial EPS which help in transforming harmful metals through chemical reaction, precipitation or entrapment (Pal and Paul 2008). Thus, lead-resistant bacterial strains which generate EPS may function as a potential bioremediative agent, i.e. as lead biosorbent in those places which are polluted by lead. Another approach to decrease metal bioavailability as well as toxicity is to make insoluble complex through bioprecipitation of dangerous metals. Precipitation of lead into lead sulphide (PbS) was observed in *B. iodonium* GP13 and

B. pumilus S3 (De et al. 2008), while immobilization of lead into pyromorphite which is an insoluble lead phosphate mineral makes phosphate-solubilizing bacterium *E. cloacae* lead tolerant (Park et al. 2011). This lead reclamation process is an efficient, environmental friendly lead bioremediation strategy. Another approach for detoxification is the induction of siderophores under lead stress. There is a production of siderophores like pyochelin and pyoverdine in lead-resistant P. *aeruginosa* strain 4EA through activation by lead (Naik and Dubey 2011). Thus, genetic modifications providing lead resistance followed by several biotransformation procedures for remediating lead may be utilized for microbial remediation of lead.

12.7.4 Mercury

Mercury is one of the highly dangerous elements in the universe and has severe health consequences compared to other harmful metal contaminants. It becomes concentrated in sediments, soils, atmosphere and water. The most probable source of mercury ingestion is fish eating. Mercury also gets an entry in the environment as industrial wastes. Two separate operons provide mercury resistance in bacteria. One of them is a narrow-spectrum mer operon, and the other one is the broadspectrum mer operon (Silver and Phung 2013). Transposons, Tn5037 (Kalyaeva et al. 2001) and Tn5070 (Mindlin et al. 2001) possess most simple mer operons, while the most complicated mer operon is found present at Tn5718 (Schneiker et al. 2001). Genes merR, merT, merC, merF, merP and merD are present on the narrow-spectrum mer operon (Dash and Das 2012). This operon is induced by inorganic mercury (Hg^{2+}) and gives tolerance just for inorganic mercury salts. merR behaves like the positive transcriptional regulator of the operon and is transcribed independently. Irrespective of the existence of extracellular mercury or lack of intracellular mercury, it combines with the promoter/operator portion of the operon for controlling the induction of the functional genes positively as well as negatively. MerP is the outermost protein of the operon and found situated in the periplasmic space. It has a β-α-β-β-α-β fold and has 72 amino acids. The two α helices are overlaying the four-strand antiparallel β sheet (Eriksson and Sahlman 1993). The mercury-binding site has the GMTCAAC consensus sequence. MerP functions as a scavenger of inorganic mercury ions and brings them on the MerT protein (Hamlett et al. 1992). MerT is coded through the transposon Tn501 and is a 116-amino acid protein, which accepts the inorganic mercury from MerP on the plasma membrane and transfers it in the cell by combining with cysteine residues over its transmembrane helices. Mercuric resistance will be lost by eliminating or mutating these residues. The precise method of MerC is unclear. But, it was observed that MerC has a role in mercury transportation and accumulation (Kiyono et al. 2013). Induction of merC in *Arabidopsis thaliana* and *Nicotiana tabacum* has doubled their capacity of storing mercury (Sasaki et al. 2006). On the contrary, MerF is a 8.7-kDa protein which possesses two transmembrane helices. It also behaves like a broad-spectrum mercury transporter. Mercuric ion reductase is coded by a 1600-nucleotide-long merA gene (Moore et al. 1990). This dimeric enzyme's role is to decrease the level of Hg^{2+} in the cell for producing volatile Hg^0 for easily discharging it from the cell. MerA protein is a flavoprotein, which needs NADPH for donating electrons to carry out the reaction. MerD produces a secondary regulator protein which combines weakly with the same promoter/operator region as MerR and controls the operon negatively (Nascimento and Chartone-Souza 2003). merH is recognized as a mercuric ion transporter (Schue et al. 2009) which is situated upstream of the merA gene and transports Hg^{2+} ions through cysteine residues and induced simultaneously with merA itself. merH is assumed to work like a metal-trafficking protein to merR, which subsequently induces the mer operon to volatilize Hg^{2+} through merA (Schelert et al. 2013). merI is also recognized just downstream to merA gene (Schelert et al. 2006). Although its exact role is still not known, a wide variety of patterns and order of mer operon can be investigated further for bioremediation application (Rebello et al. 2013). The broad-spectrum mer operons have similarity in their genes with the narrow-spectrum operons. Besides the already existing genes, some extra genes such as merE, merG and merB are also found. The broad-spectrum operons defend against organic mercury (Barkay et al. 2003). These compounds are very harmful because they can go effortlessly inside the cell without the help of transporter molecules. Organic mercury (R-Hg) diffuses passively inside the cell or is moved inside through MerE or MerG. merE was primarily found in the transposon Tn21 and can assist in the translocation of methyl mercury (CH_3Hg^+) as well as inorganic mercury (Boyd and Barkay 2012). Tolerance to phenylmercury is because of MerG which is a 20 kD protein produced by a 654-bp gene (Kiyono and Pan-Hou 1999). It was found to be situated in between merA and merB on the operon and give protection to the cell from phenylmercury through prohibiting its entry inside the cell (Schneiker et al. 2001). Another enzymatic molecule of the mer operon merB encodes the enzyme organmercurial lyase. MerB protein is catalyzing the protonolysis of the carbon–mercury bond which produces ionic mercury as well as a reduced hydrocarbon. The mercuric ion reductase reduces the ionic mercury to the elemental form Hg^0 (Dash and Das 2012) which is volatilized out of the bacterial cell because of its elevated vapour pressure. Microbes bioremediate mercury through several enzymatic transformations such as reduction of Hg^{2+} to Hg^0, organomercurial compound breakdown, Hg^{2+}methylation and oxidation of Hg^0 into Hg^{2+}. A transgenic bacterium *B. cereus* BW- 03(pPW-05) was constructed through the transformation of the plasmid containing mer operon of marine bacterium *B. thuringiensis* PW-05 found in the Bay of Bengal (Dash et al. 2014) into a novel marine bacterium *B. cereus* BW-03 by Dash and Das (2015) having the capacity for biosorption of mercury to resist mercury. It has eliminated (>99%) mercury supplement in laboratory condition through parallel volatilization (>53%) as well as biosorption (>40%). Many mercury-tolerant mer gene-containing bacterial isolates were separated which could volatilize and reduce Hg^{2+} to Hg^0 (De et al. 2008). Proteins of the mer operon reduce mercury as well as breakdown organomercurial compounds. Additionally, the electrochemical potential of Hg^{2+}/Hg^0 at pH 7 is +430 mV which shows that $Hg2^+$ can be reduced to the elemental form by the living cell which is harmless for bacteria. As the melting point/boiling point of mercury is low (−39/357 °C), metallic mercury can diffuse passively from the cell and can be volatilized into the air or can precipitate because of its low solubility in water eliminates harmful Hg^{2+} (Wagner-Dobler 2003). Microbial as well as abiotic transformations can regulate the production of methyl mercury in the environment. Hg^{2+} methylation and MeHg degradation are involved in direct transformation. Reduction of Hg^{2+} to Hg^0 as well as its cyclic oxidation influences formation of MeHg indirectly through the regulation of Hg^{2+} which is the substrate for methylation (Barkay and Wagner-Dobler 2005). In fact, these days, an eco-friendly and cheaper treatment technology is employed for cleaning mercury-polluted wastewater through mercury-tolerant microbes. This was proved through the manufacturing of pilot plants to eliminate mercury which remediates 100 m^3 of 50% wastewater having mercury per day (Wagner-Dobler 2003). In addition to enzymatic transformation, metallothioneins and polyphosphates can also be employed for bioremediating mercury through sequestration of mercury ions in a form which is biologically inactive. Polyphosphate kinase is coded by ppk gene and participates in polyphosphate biosynthesis. These are orthophosphate polymers with negative charges and can bind with mercury ions (Kornberg 1995). ppk genes expressed in several transgenic bacteria express ppk genes to tolerate as well as store about 16 μM of mercury from solutions (Pan-Hou et al. 2002). An additional strategy is the production of mercuric sulphide (HgS) through directly reacting Hg^{2+} with H_2S generated anaerobically in *Clostridium cochlearium* (Pan-Hou and Imura 1981). It was observed in the same type of studies that *Klebsiella* aerogenes NCTC418 generated HgS in continuous aerobic culture with mercury chloride. Biological elimination of mercury was done through a mercury-reducing biofilm having natural as well as engineered mercuric reductase (Brunke et al. 1993). Sulphate-reducing bacteria is also utilized as a source of H_2S which precipitates metal as sulphides. The HgS generates high-solubility products which eliminate mercury through H_2S (Hakansson et al. 2008).

12.7.5 Nickel

Nickel is present in five stable isotopes: ^{58}Ni, ^{60}Ni, ^{61}Ni, ^{62}Ni, and ^{64}Ni. Of these, ^{58}Ni is highly abundant metal in the environment. Nickel plays a very significant function in the biochemistry of microbes as well as plants. The enzyme urease consists of nickel. Other enzymes such as hydrogenases, superoxide dismutase and glyoxalase enzyme possess Ni-Fe clusters or employ nickel like a co-factor. Nickel is extremely harmful to animals and humans because of its ability to cross the placenta and influence the fetus. Nickel tolerance in bacteria is usually mediated through efflux pumps. One such process of resistance has been investigated by Grass et al. (2000) in *C.* (*Ralstonia*) *metallidurans* CH34. They observed that cnrYHXCBAT gene system codes cnrCBA efflux pump. Transcription begins at the cnr promoter through cnrY and cnrC after nickel gets an entry into the periplasm. The products of the three genes cnrYXH control the induction of complete gene cluster. cnrH is actually an extracytoplasmic function (ECF) sigma factor (Grass et al. 2000) that continuously expresses cnrCBA. Membrane-bound proteins cnrX and cnrY are present in the periplasm and probably work like antisigma factors. The cnrCBA codes for an extremely proficient pump which can be induced in the presence of just micromolar quantity of nickel. These gene products produce an efflux pump which removes an extra amount of nickel from the cell. Grass et al. (2005) also described the presence of nre (nickel resistance) operons in *Achromobacter xylosoxidans* 31A. They observed the nre locus on the plasmid pTOM9 and found that just a single gene, nreB, was accountable to provide the whole tolerance for nickel. However, just like cnr operon, this operon is not detoxifying Ni^{2+}, but removing it from the cell. Previously, Schmidt and Schlegel (1994) found one other operon ncc operon in the same pTOM9 plasmid, which is giving joint tolerance for nickel, cobalt and cadmium. However, these processes of nickel resistance do not participate much in nickel bioremediation. Seven ORFs were investigated and named as nccYXHCBAN. Nucleotide sequences have a strong similarity to the cnr as well as czc operons of *Alcaligenes eutrophus* CH34 (Tibazarwa et al. 2000). The NccX protein is having 76 amino acids with several His residues which shows that it might be working like a metal-binding protein used for nickel remediation (Trepreau et al. 2011). Rodrigue et al. (2005) studied the yohM gene in *E. coli* and observed that it is coding a membrane-bound polypeptide with the capacity to tolerate nickel and cobalt. More copies of yohM resulted in a decreased level of intracellular nickel. yohM, therefore, designated now as rcnA, was the first nickel efflux system found in *E. coli*. One more efflux pump for cadmium–zinc–nickel was found in *Helicobacter pylori* and designated as cznABC (Stahler et al. 2006). Of these three genes, cznA and cznC were responsible for nickel homoeostasis. *E. coli* JM109 was genetically manipulated to simultaneously induce nickel transport system and metallothionein gene which can eliminate as well as recover Ni^{2+} from aqueous solution. There is a six-time enhancement in the nickel-binding ability in comparison to host cells after the linearized Langmuir isotherm (Deng et al. 2003). In another investigation, the nickel/cobalt transferase gene, NiCoT of *S. aureus* ATCC6538, was amplified and ligated into vector pET-3c which bioaccumulated 11.33 mg/g nickel, three times in comparison to the original *E. coli* BL21 strain (Zhang et al. 2007). RcnA nickel (Ni) and cobalt (Co) efflux systems were repressed genomically for genetically manipulating *E. coli* to combine the induction of a particular metallic transporter, NiCoT of *Novosphingobium aromaticivorans* which was under plasmid regulation. This Ni/Co buster strain is accountable for increased nickel (II) and cobalt (II) uptake as well as storage of metal about 6 mg/g bacterial dry weight in first 10 min of treatment. An extra synthetic adherence operon was incorporated into the plasmid with the capacity to manufacture bulky biofilm structures in nickel (II) as well as cobalt (II) presence. Thus, genetic manipulation shows enhanced metal sequestration and biofilm production through *E. coli* which can be employed for biofiltration of nickel and cobalt industrially through immobilized cells (Duprey et al. 2014). Nickel can be transported to enzyme precursors by protein–protein interactions in a complicated serial mechanism through nickel metallochaperones. To bind as well as to eliminate nickel, these metallochaperones can be engineered. Actually, it was reported that *P. cepacia* 120S (living and dead biomass) eliminates about 234.4 μg/ml nickel, while *B. subtilis* 117S eliminates 117.2 and 351.6 μg/ml nickel through living and dead biomass, respectively (Abdel-Monem et al. 2010). It was observed that Ni-resistant *B. thuringiensis* residing in the soil can withstand up to 10 mM Ni and eliminates about 82% of it through biosorption from the medium (Das et al. 2014). Recently, in another strategy, nickel oxide nanoparticles synthesized from *Microbacterium* sp. MRS-1 were used for treating nickel electroplating industrial effluent and removed about 95% nickel (Sathyavathi et al. 2014).

12.8 Discovery of Novel Metal-Resistant Genes Involved in Bioremediation

Bacteria have developed certain remarkable processes for their existence integrated into their genome under heavy metal stress. They produce a variety of enzymes and proteins to get assistance in countering these adverse conditions (Johnsen et al. 2005). Bacteria can be manipulated genetically or metabolically to generate products which provide unique characters to the host cells. At present, options of novel adaptation mechanisms in microorganisms for eliminating and ultimately detoxifying contaminants are being explored by workers. Genes which code for the enzymes that can alter the oxidation state of heavy metal from more harmful to less harmful forms can be introduced like for coding bacterial merA and mercuric reductase incorporation into other bacteria to improve bioremediation (Dash and Das 2015). Therefore, altered genes can be incorporated which assists in acquiring new detoxification methods for heavy metals (Arora et al. 2010). One more strategy which can be employed is the utilization of in silico procedures. Currently, due to the introduction of computers and software, it is now feasible to obtain knowledge on any subject from a single source. Several databases are available which give knowledge about the harmful effects of compounds as well as their location, features and degradation pathways. Certain significant databases are USEPA (http://www.epa.gov/), ATSDR (http://www.atsdr.cdc.gov/) and KEGG PATHWAY Database (http://www.genome.jp/). Khan et al. (2013) have compared the utilization of these computational sources to perform bioremediation virtually prior to test it on the location. Eleven software tools were listed by them to forecast the harmful effects of compounds. Ten databases containing knowledge about harmful effects of several compounds were also explained by them. Additionally, 15 programs were also employed to test the environmental degradability of compounds. Biodegradability Evaluation and Simulation System (BESS) and Biochemical Network Integrated Computational Explorer (BNICE) are two of the highly successful programs. Nowadays, due to this information with the assistance of biological pathway prediction software like Scansite 2.0 (Obenauer et al. 2003), BioCyc (Karp et al. 2005), SMART5 (Letunic et al. 2006), STRING7 (Von Mering et al. 2007) and KEGG, exhaustive studies on interactions of different biomolecules in silico are possible, and new proteins and genes can also be deduced which may be utilized in bioremediation (Kanehisa and Goto 2000; Das et al. 2016).

12.9 Genetically Engineered Microorganisms

For effectively performing in situ bioremediation, collective ecological and microbiological information as well as biochemical processes and field engineering plan are necessary. Not only ethical issues but also several other issues obstruct a successful production of recombinant strain. Due to biotic as well as abiotic factors, two prominent barriers are: competition of engineered microorganism with other natural population for nutrition and other resources and selection pressure (Singh et al. 2011). Thus, to get the desirable results, it is necessary to opt for correct bacterial strain in terms of growth potential as well as nutrient response. For degrading different contaminants, several recombinant bacterial systems have been developed. Chakrabarty and group in 1971 successfully developed the first genetically modified microbe, which was patented in 1980 in the US Supreme Court. It was modified from the genus *Pseudomonas*, which have the ability to degrade crude oil constituents (Ezezika and Singer 2010; Kumar et al. 2013).

These GEMs are produced through the detection and modifications of some specific genetic sequences. GEMs show increased degradability for a broad variety of xenobiotics from various environmental sources and have the capability to bioremediate them. GEM designing is dependent on the knowledge of microbes and xenobiotic interaction and its genetic base, biochemical processes, the structure of operon, molecular biology and its ecological application.

Various strategies are available for the production of GEMs for bioremediation. In the first strategy, organisms appropriate for modification with the relevant genes are recognized. For example, microorganisms suitable to thrive in soil environment cannot adapt to the aquatic environment and therefore cannot be employed effectively. Thus, aquatic microbes can be employed for producing GEMs to bioremediate aquatic sources. The supplementation of nutrients to the inoculated environment is not required in the application of these organisms which decrease the costs as well as maintenance. *Anabaena* sp. and *Nostoc ellipsosporum* are generated through inserting linA of *P. paucimobilus* and fcbABC of *Arthrobacter globiformis*, respectively. The gene linA regulates the biodegradation of lindane (-γ-hexachlorocyclohexane), while fcbABC provides the capacity to biodegrade halobenzoates, which can be employed to remediate these contaminants of water sources.

In the second strategy, pathway construction, extension and regulation is done. GEMs are produced through improvement in already present catabolic pathways or extension of these pathways for degrading those compounds which wild strains cannot degrade. The entire catabolic pathway is coded either by a single microorganism or through a group of microorganisms executing single or many stages in the bioremediation of xenobiotics. In this manner, manufactured GEMs have degradation capacities of diverse microbial communities because of modification in gene sequences which ultimately improve the effectiveness and value of the catabolic pathways.

Enzyme specificity and affinity are modified in the third strategy. Transcription and translation of specific genes produce enzymes which catalyze different steps of the metabolic pathway. GEMs are generated through hybrid gene clusters which modify activities of enzymes as well as enzyme substrate specificities. These gene clusters code for the enzyme with better transforming capacity. By genetically modifying *E. coli* strain, a hybrid gene cluster for degrading trichloroethylene (TCE) is expressed.

In the fourth strategy, bioprocess development, its observation and regulation as well as bioaffinity, bioreporter, sensor usage for chemical sensing, decrease in toxicity, and end-point analysis is performed. There are fewer reports on this strategy. However, a lux gene-based system has been developed which has many benefits to monitor bioremediation methods. Bioluminescence is very easy to detect and has no need for costly instruments, additional chemicals or co-factors from outside. GEMs also have chemical sensors which permit the observation of bioavailability of pollutants rather than just their presence. GEMs generated by bioluminescence assist in gaining knowledge about the microbes present in the region of contamination as well as of the last stage of bioremediation.

To overcome the restrictions of usage of wild-type microbes, various developments in GEMs have been done. Microbes cannot be used in anaerobic conditions as they are limited to aerobic catabolic as well as co-metabolic pathways. To utilize them in an anaerobic environment, production of GEMs through the incorporation of genes for oxygenases is done. Before using microbes for bioremediation, each and every xenobiotic existing in a multi-contaminated environment should be considered. Because of incorporating various genes from different microbes in a single microorganism, GEMs attain diverse characters which can be utilized effectively for bioremediation. But, it is highly important to take precautions for the safety of the environment as well as human health during the application of GEMs for remediating contaminants. Those microbes or GEMs that utilize xenobiotics as a source of carbon, energy and/or nitrogen can obtain nutrients and flourish there. In fact, they may scatter in an uninhibited way which may prove harmful. Scientists have developed a new approach for "suicidal genetically engineered microorganisms (SGEMs)" through searching antisense RNA-controlled plasmid dependence, proteic plasmid dependence and inducible bacterial degradative operons. For designing new S-GEM, information on killer–anti-killer genes is required to create microbes prone to programmed cell death following xenobiotic degradation. This technology assists in eliminating microbes just after bioremediation through autolysis and thus decrease the dangers for human beings and environment.

Usage of GEM-dependent xenobiotic remediation is in the front because it is an environmental and human-friendly strategy. But, production of GEMs is not much because of the restricted knowledge about the genes. A second major hurdle in the implementation of GEM is its controlling measures and related risks. Although this problem can be solved through the production and utilization of suicidal genetically engineered microorganisms. In future, this can become a highly effective technique if we acquire the knowledge about microbes as bioremediating devices, their genomes and biochemical mechanism exercises. This will permit the usage of

GEMs in the polluted environment from its present restricted use in laboratory research only (Kulshreshtha 2013).

12.10 Manipulation of Bacterial Genetic System for Enhanced Bioremediation

In the environment, the majority of contaminants have higher persistence principally because of suboptimal degradative pathways (Timmis and Pieper 1999). Most of the organic contaminants are degraded through biological pathways, while harmful inorganic metals are altered through several enzymes. Various intracellular and extracellular events are present in bacteria which bioremediate harmful contaminants from the environment. Under the presence of harmful metals in the environment, tolerant bacteria produce various intracellular as well as extracellular enzymes to eliminate or degrade the toxic type of metals to non-toxic/less toxic types. Every enzymatic pathway possesses a step which is rate-limiting. To enhance the bioremediation capacity, manipulation of this rate-limiting step could be a solution. There are various techniques which involve the manipulation of bacterial systems which are given below. The current genetic engineering techniques permit the designing of microorganisms with the ability to alter particular metal pollutants. The chances of making artificial permutation of genes that are not found in nature give enormous results for these GMOs which have to be used for in situ removal of metal contaminants. The major regular techniques comprise engineering with a single gene or operon, modification of existing gene sequences and pathway switching.

12.10.1 Engineering Single Gene or Gene Cluster/Operon

Microorganisms widespread in the high metal-contaminated sites adapt themselves to tolerate the harmful effects through modifying several genetic processes. These strains due to their intrinsic capability to live in a contaminated environment are favoured for the purpose of bioremediation. Therefore, through the incorporation of a single gene or group of genes in the intrinsic bacterium, a bacterium is produced which because of its newly acquired adaptation of these strains is appropriate for bioremediation purposes. Sandaa et al. (1999) observed that majority of Gram-positive and α- proteobacteria are existing in heavy metal polluted soils. As most indigenous bacteria have the capability to live in this unfavourable environment, genetically altered recombinant strains may be utilized for increasing bioremediation. Genetic modification of local bacterial population is prefered, but the main crisis here is that cloned genes are unstable and their transfer to the next coming generations is also difficult. According to Lorenzo et al. (1998), minitransposons of naturally existing Tn5 and Tn10 transposons can be utilized. These minitransposons are unique because only the functional part of a DNA can be cut off and cloned in the Tn5 vector to subsequently insert it into the chromosome of Gram-negative bacteria. Ruiz et al. (2011) have done the transformation of *E. coli* JM109 with vectors which causes increased expression of metallothionein (mt1) as well as polyphosphate kinase (ppk) genes. Metallothioneins are proteins which are membrane-bound and bind with metals subsequently reducing them. This modified bacterium stores Hg over 100 μM. Radiation-tolerant *Deinococcus radiodurans* has been transformed through the mercuric ion reductase gene (merA) of *E. coli* BL308 host by Brim et al. (2000). This mutant strain flourishes under both high radiation and mercury. They have also volatilized inorganic Hg (Hg^{2+}) into elemental Hg (Hg^0). Dash and Das (2015) have produced transgenic strain *B. cereus* BW- 03(pPW-05) in the same manner which contains plasmid of a wild strain of *B. thuringiensis* PW-05 containing mer operon. Hg volatilization and Hg biosorption, the two processes of mercury tolerance, are present in the resultant transgenic strain which can be commonly employed for in situ elimination of mercury from the polluted environments. To tolerate mercury, genetic alteration of heavy metal-tolerant bacterium *C. metallidurans* strain MSR33 was done (Rojas et al. 2011). This bacterium can resist inorganic mercury as well as methyl mercury both with copper and chromate. According to Chaturvedi and Archana (2014), two metal-binding peptides were expressed in *D. radiodurans* R1, which is an attractive approach to develop metal resistance. A synthetic gene (EC20) was created through overlap extension which was expressed in DR1 under the native groESL promoter and codes for phytochelatin analogue as well as cyanobacterial metallothionein (MT) gene, smtA. This recombinant strain is 2.5-fold more tolerant to Cd^{2+} and stored 1.21-fold more Cd^{2+}. Thus, integrating desirable genes inside compatible bacteria provides improved strategies to remediate the environment.

12.10.2 Modification of Intrinsic Genes

There are several drawbacks for utilizing a genetically modified microorganism in various environmental conditions. Some of the incorporated genes may be unstable in different environments to perform the desired role (Dixit et al. 2015). In some cases, under the presence of pollutants, the expression level of the foreign genes is influenced so much that it has a negative impact on bioremediation applications (Kiyono and Pan-Hou 1999). Therefore, as an alternative to these foreign genes, the native genes may be attacked or modified. This method permits the microorganisms growth in their natural surroundings besides performing the desired roles in bioremediating metals. There are several investigations which focus on changing/modifying the already existing group of genes of native microflora for bioremediation. Three cadmium-tolerant *P. aeruginosa* strains were isolated from industrial sludge by Kermani et al. (2010). Upon exposure to the dyes acridine orange and acriflavine, mutations in these strains could tolerate 7 mM of Cd^{2+}. In *E. coli*, cloning and overexpression of phytochelatin synthase of *Schizosaccharomyces pombe* which genetically altered *E. coli* strain to store 25 times more Cd^{2+} than control strain (Kang et al. 2007). According to Wu et al. (2006), in the soil bacterium *P. putida* 06909 cadmium, the binding ability was increased three times more through overexpression of a synthetic phytochelatin EC20 which also doubled the growth of the cell. Brim et al. (2003) have generated the close relative of the radiation-tolerant *D. radiodurans* as well as *D. geothermalis*. It was observed that *D. geothermalis* was decreasing Hg^{2+} as well as Fe^{3+}, U^{6+} and even Cr^{6+} through the assistance of plasmids. In *E. coli* overexpression of nfsA reduced 1.5-fold more chromate (Ackerley et al. 2004). Overproduction of ChrR in *P. putida* reduces 24 times more Cr^{6+} (Gonzalez et al. 2005).

12.10.3 Pathway Switching

In pathway switching, manufacturing, expansion and control of some new genetic processes are incorporated for bioremediation purposes. For total bioremediation of toxic metals, an efficient method is to create a group of microbes for every particular stage. Not much work has been done till now to improve the bacterial strains to eliminate metals through making new consortium. The strategy of pathway switching may be also used for eliminating metals utilizing GMOs. Because Cu–Mb (copper–methanobactin) uptake through methanotrophic bacteria is quite obvious, a process for the uptake of Cu–Mb complexes at the same time can be developed rather than separation of Cu from Mb before the uptake (Balasubramanian et al. 2011). Further research in this area can be done by detecting an appropriate pathway to acquire metals easily from the surroundings through the knowledge of the processes of transport machineries. GMO usage for bioremediation is in front position because of their effectiveness and low-cost strategies. But total knowledge about the genes is needed which can be gained through microarray as well as fluorescent in situ hybridization techniques. The risks and the control measures linked with these organisms are certain other hurdles. Future of bacterial bioremediation depends on the production of suicidal genetically engineered microorganisms (S-GEMs) which will increase the use of these GMOs for on-site implementations. These S-GEMs are manufactured by utilizing a killer gene as well as controlling circuit which regulates the expression of the killer gene against the presence or absence of environmental indicators (Paul et al. 2005). Therefore, the manufactured S-GEMs will have programmed cell death because of the existence of killer–anti-killer genes following the elimination of harmful materials for their safe disposal in the environment. (Das et al. 2016).

12.11 Biodegradation of Heavy Metals

Heavy metals like mercury (Hg), arsenic (As), cadmium (Cd), nickel (Ni), cobalt (Co) and lead (Pb) are stored in the environment because of very fast industrialization and result in serious harm to the important organs in living beings (Singh et al. 2011). Detoxification mechanism of genetically modified bacteria is helpful in bioremediation of metals. Against high Hg toxicity, the bacterial world has developed an amazing collection of tolerance processes. A group of Hg tolerance genes (*mer*) in an operon promote the bacteria to transform harmful Hg^{2+} into volatile Hg with the help of Hg reductase (Barkay et al. 2003). The narrow Hg tolerant *mer* operon breaks down mercury in the sequence of three successive stages, which are transfer of Hg^{2+} into the cell, NADPH-dependent enzymatic transformation of the ionic mercury into less toxic elementary mercury (Hg^0) and at last the control of those functional genes which are involved in the transport and transformation (Singh et al. 2011). This operon possesses several genes which are functionally different like *merA* and *merB* involved in the reduction, *merT* and *merP* working like transporters, *mer R* and *merD* used in the regulation and newly discovered *merE* and *merH* used for membrane transport (Ruiz and Daniell 2009; Kiyono et al. 2009; Schue et al. 2009). Genetically modified *E. coli* possess simultaneously *merT-merP* and MT genes which have the potential to eliminate Hg^{2+} efficiently from electrolytic wastewater (Deng and Wilson 2001). *Deinococcus radiodurans*, the highst radiation-tolerant bacterium, was altered through the expression of *merA* gene from *E. coli* BL308. The resultant recombinant was able to grow in radiations and high levels of ionic mercury. It has efficiently reduced Hg ions to less harmful volatile elemental mercury (Brim et al. 2006). Metal-binding peptides phytochelatins as well as metallothioneins increase the ability to bind metals in microorganisms (Bae et al. 2001). Cloning of the genes of these proteins has been done in *E. coli* from plants and fungi. Recombinant bacteria, having increased bioaccumulation ability and more affinity for chosen metal, select the metal ions for storing from multicomponent contaminants (Sauge-Merle et al. 2003). For bioaccumulating Cd^{2+} from the heavy metal contaminated place, Deng et al. (2007) developed a genetically modified *E. coli* JM109. At the same time, by expressing fusion proteins for phytochelatin (PC), metal-binding peptide (EC20), and RsaA-6His, respectively, several genetically modified bacteria like *Mesorhizobium huakuii, P. putida* and *Caulobacter crescentus* have been developed for the storage of Cd^{2+} (Sriprang et al. 2003; Wu et al. 2010; Patel et al. 2010). Arsenic (As) is highly dangerous for animal health as it is stored in the edible portion of plants, soil and polluted water. From here, this inorganic compound enters the food chain. Because of this, to urgently eliminate As from the polluted location, arsenite S-adenosylmethionine methyltransferase gene (*arsM*) was isolated from *Rhodopseudomonas palustris* and cloned in *E. coli* (Qin et al. 2006), which methylated inorganic As into less harmful and volatile trimethylarsine (TMA). Recombinant bacteria modified through *arsM* gene eliminate As from the polluted soil perfectly by volatilizing As (Liu et al. 2011). These recombinant bacteria stored more As than control strain. Expression of another metalloregulatory protein ArsR in *E. coli* provided more affinity to eliminate As from the polluted sites (Kostal et al. 2004). In several organisms, more affinity for membrane transport of Ni^{2+} is provided because of the metallothionein protein (MT), glutathione S-transferase fusion protein (GST-MT) and *nixA*- encoded membrane transport protein (Singh et al. 2011). It is thought that higher levels of Ni^{2+} could be stored in the engineered bacteria due to enhanced expression of these transport proteins (Kumar et al. 2013).

12.12 Risk Mitigation in Genetically Modified Bacteria

12.12.1 Monitoring of Recombinant Strains

The monitoring of the future of environmentally released recombinant bacteria is required which can be done by genetically selecting them through easily detectable markers (Wilson and Lindow 1993). Earlier, lacZ, xylE and gusA genes were used in the investigations. But, in recent times, they have been substituted with gfp genes for green fluorescent protein (March et al. 2003) as well as lux genes of bioluminescence (Ripp et al. 2000), which are absent in soil bacteria and can be revealed with more sensitivity. It is very difficult to do genuine bioremediation field trials through the use of live genetically altered bacteria due to the compliance of environmental regulations in the United States as well as Europe. In one specific remarkable trial which is done in large lysimeters, genetically modified *P. fluorescens* with the capacity to degrade naphthalene was utilized. The existence of lux genes, a bioluminescent signal on the degradation of naphthalene and salicylate because of the existence of lux genes, is a very unique character in designing of this bacterium which allowed on-line observation of the degradation mechanism (Sayler and Ripp 2000).

12.12.2 Horizontal Transfer of Recombinant Genes to Other Microorganisms

One of the most strong oppositions of this approach is that once discharged into the environment, recombinant genes may be transferred from their host to other bacteria. The horizontal transfer through transformation, transduction and conjugation of several genes such as antibiotic resistance, heavy metal resistance, symbiotic as well as degradative in different environmental conditions has been analyzed (Davison 1999; Koch et al. 2021). It was concluded that interspecific horizontal genes transfer for various bacteria for several genes and in different environmental conditions. Therefore, it is quite possible that recombinant genes entered into the environment may extend into other related bacteria if extra precautions are not taken. Specially assessments of low, or undetectable, DNA transfer in laboratory investigations (e.g. $<10^{-9}$ cell^{-1} generation^{-1}) may not be relevant because of large numbers of bacteria that could be discharged as well as multiplied in longer periods. On the contrary, several techniques have been developed to decrease horizontal gene transfer to other bacteria.

12.12.3 Use of Defective Transposons

Genetic modifications normally depend upon plasmid vectors, but they are not appropriate for environmental discharge because of their susceptibility to horizontal transfer (Davison 1999, 2002a, b; De Lorenzo 1994; Torres et al. 2003a). Actually, non-conjugative plasmids like RSF1010, generally employed in genetic modification with *Pseudomonas*, keeps the oriT region which allows their perfect transfer under the presence of self-transmissible plasmids like RP4. The plasmid can still be transferred through the making of cointegrate which utilizes homology of common DNA sequences, even if the oriT region is not present. Various efforts to make vectors with the focus on the biodegradation field-release targets on defective transposons which can be incorporated only once into the chromosome and cannot be used again because of the absence of the transposase gene. Genes incorporated in the minitransposons exhibited a very low degree of horizontal transfer ($<10^{-9}$ cell^{-1} generation^{-1} (Torres et al. 2003b).

12.12.4 Removal of Antibiotic Resistance Genes

It is an exceptional event; therefore, transposition vectors need very powerful selective marker generally antibiotic resistance. But, on biosafety grounds for releasing into the environment, an antibiotic marker is not required. Several transposons with resistance for zinc, mercury or tellurite have also been planned (Davison 2002a, b), but they can permit selective advantage in certain conditions. As the selective markers are utilized only to discover the presence of transposition, the easiest solution is to cut it afterwards through a site-specific recombination event (Davison 2002b; Schweizer 2003). Such contemporary procedures permit the cutting of DNA sequences present in between two site-specific recombination sites like loxP or FRT after transient expression of the recombinase (cre or FLP). In one instance, an antibiotic resistance marker was removed through the site-specific ResA resolvase; therefore, recombinant bacteria possess only the lux genes essential to monitor the environment. Besides improving biosafety concerns, this technique permits reuse of selectable markers for sequential strain productions.

12.12.5 Suicide Mechanisms Preventing Escape of Recombinant Bacteria

Even implementation of defective transposons has the possibility of horizontal transfer through recombination of conjugative plasmid or through bacterial conjugation. Therefore, several vectors are created which possess extra safety measures through the use of a large number of suicide processes (Davison 2002a). In the most simple way, suicide may be inhibited through an environmental signal such as the pollutant which has to be degraded, permitting cell survival. The absence of this signal causes induction of the suicide gene, which results in cell death. In one of the best-studied instances, the lethal gef gene carrying a mini-Tn5 transposon was regulated by the LacI repressor. Therefore, cell survival is dependent on the continuous existence of this negative control factor. On the contrary, transcription of the lacI gene itself is dependent upon the positive-activator XylS of the meta-cleavage pathway which is active under the presence of 3-methylbenzoate. Therefore, the cells are viable only in the presence of 3-methylbenzoate. Cells not having 3-methylbenzoate would not survive, either because of its degradation or because of its escape to the environment, because of Gef expression. The escape rate from killing in the lab was approximately 10^{-8} $cell^{-1}$ $generation^{-1}$. The testing of the system was done in fields with bulk soil or rhizosphere soil and was very efficient. There was no proof that it spread out of the experimental area (Torres et al. 2003a). To evade lethal function loss because of mutation, double strains with extra function depending upon the asd gene were made for diaminopimelic acid synthesis. This gene is compulsory needed in cell wall synthesis, and strains without it need totally external diaminopimelic acid, which is not present in the soil environments. A *P. putida* strain with asd gene deletions has an alternative asd gene which is under positive regulation by the XylS activator. This same XylS activator negatively regulates gef expression too through transcription of the lacI gene as explained earlier. Therefore, a strain devoid of 3-methylbenzoate would not survive in an environment because of killing by Gef protein and also because of the absence of diaminopimelic acid. The survival level of this strain due to the lack of 3-methylbenzoate was under the detection range ($<10^{-9}$ $cell^{-1}$ $generation^{-1}$ (Ronchel and Ramos 2001).

12.12.6 Suicide Mechanisms Preventing Horizontal Gene Transfer

The suicide scheme is created to allow reliable killing of recombinant strains from being discharged into the wild. The same type of schemes could stop the transfer of recombinant genes into other bacteria through conjugation, transduction or transformation. One diverse conditional lethal scheme was dependent on the colicin E3 (colE3) gene which kills several diverse bacteria through breaking of 16S ribosomal RNA (Torres et al. 2003b). Under natural condition, this lethal function is opposed through an immunity function (encoded by the immE3 gene) providing a poison/antidote effect. The colE3 gene was kept inside a plasmid, and the immE3 gene was put on the chromosome of *E. coli* or *P. putida* to test whether the horizontal transfer of plasmid DNA to other bacteria could be stopped by using the killing function. Therefore, horizontal transfer of the plasmid to other bacteria would cause cell death in the recipient because the colE3 gene would be transferred without chromosomal immE3 gene. This scheme has the capability to stop horizontal transfer into a broad range of Gram-negative bacteria (Torres et al. 2003a). The main trouble with this kind of control system is that the killing process can simply be overcome through a mutation which prohibits the killing function. One of the solutions for this is to utilize two autonomous killing schemes, so that mutation of one scheme does not overcome the other. This type of twofold scheme utilizes a plasmid with two lethal genes encoding colE3 toxin as well as EcoRI restriction endonuclease. They are killed through totally diverse processes, i.e. prohibition of protein synthesis and degradation of DNA. In a true degradation experiment, these functions will lie side by side with some genes like biodegradation genes, so that a deletion eliminating colE3 as well as ecoRIR genes would also eliminate the gene to be regulated. The respective "antidote" genes immE3 and ecoRM (EcoRI methylase) were put into the bacterial chromosome so that they are not transferred with the plasmid. This dual control system exhibited higher control of gene transfer than the individual lethality functions alone (Torres et al. 2003a; Davison 2005).

12.13 Future Directions

Detection of new genes and proteins which have the capacity for environmental friendly cleaning will be of great help to attain increased bioremediation. Random mutations may occur in those genes which are responsible for providing heavy metal tolerance to microbes. These mutations generally have dangerous influences on the organism but sometimes also have positive influences which create strains of higher detoxification capacity. This event is known as gain-of-function mutation (Arora et al. 2010). Complete genome analysis must be done through a microarray technique to recognize novel genes which may be induced under the presence of a specific contaminant. Investigation of the expression of a variety of genes immediately is made possible through this technique by which it may become possible for scientists to decide whether only a single gene or a group of genes is causing the detoxification of a specific contaminant. Another simple procedure which can be applied for investigating the presence of some target genes is fluorescence in situ hybridization (FISH) (Pernthaler et al. 2002). FISH is usually utilized in medical science to identify the existence of pathogens if infection occurs. But, its range may be extended to other studies as well. Metagenomic analyses make possible to detect and investigate several other unknown genomes which have the capacity to control pollution. At last, once particular genes are detected, they may be moved to other powerful strains and firmly induced for creating microbes which are more efficient candidates for doing bioremediation. Microbes consist of several special traits like production of biofilm, generation of biosurfactant, production of secondary metabolites and various others to tolerate the stress environment. These characteristics of metal resistant bacteria may be employed for their increased use in bioremediation. Multispecies biofilm communities have been investigated for their metal resistance as well as biomineralization characters recently (Golby et al. 2014) which exhibited the potential of using microbial population for bioremediation of metals and require further investigations (Das et al. 2016).

References

Abdel-Monem MO, Al-Zubeiry AH, Al-Gheethi AA (2010) Biosorption of nickel by Pseudomonas cepacia 120S and Bacillus subtilis 117S. Water Sci Technol 61(12):2994–3007

Ackerley DF, Gonzalez CF, Keyhan M, Blake R, Matin A (2004) Mechanism of chromate reduction by the Escherichia coli protein, NfsA, and the role of different chromate reductases in minimizing oxidative stress during chromate reduction. Environ Microbiol 6:851–860

Adamis PD, Gomes DS, Pinto ML, Panek AD, Eleutherio EC (2004) The role of glutathione transferases in cadmium stress. Toxicol Lett 154(1):281–288

Ahluwalia SS, Goyal D (2007) Microbial and plant derived biomass for removal of heavy metals from wastewater. Bioresour Technol 98(12):2243–2257

Aiking H, Stijnman AC, Garderen V, Heerikhuizen HV, Riet JVT (1984) Inorganic phosphate accumulation and cadmium detoxification in *Klebsiella aerogenes* NCTC 418 growing in continuous culture. Appl Environ Microbiol 47:374–377

Ajmal M, Rafaqa AK, Bilquees AS (1996) Studies on removal and recovery of Cr (VI) from electroplating wastes. Water Res 30:1478–1482

Aksu Z (1992) The biosorption of Cu (II) by C. vulgaris and Z. ramigera. Environ Technol 13 (1):579–586

Albarracin VH, Avila AL, Amoroso MJ, Abate CM (2008) Copper removal ability by Streptomyces strains with dissimilar growth patterns and endowed with cupric reductase activity. FEMS Microbiol Lett 288:141–148

Arora PK, Srivastava A, Singh VP (2010) Application of monooxygenases in dehalogenation, desulphurization, denitrification and hydroxylation of aromatic compounds. J Bioremed Biodegr 1:112. https://doi.org/10.4172/2155-6199.1000112

Bae W, Mehra RK (1997) Metal-binding characteristics of a phytochelatin analog (Glu- Cys)2Gly. J Inorg Biochem (Elsevier Sci) 68:201–210

Bae W, Chen W, Mulchandani A, Mehra RK (2000) Enhanced bioaccumulation of heavy metals by bacterial cells displaying synthetic phytochelatins. Biotechnol Bioeng 70:518–524

Bae W, Mehra R, Mulchandani A, Chen W (2001) Genetic engineering of *Escherichia coli* for enhanced bioaccumulation of mercury. Appl Environ Microbiol 67:5335–5338

Baker AJM (1981) Accumulators and excluders – strategies in the response of plants to heavy metals. J Plant Nutr 3:643–654

Baker AJ (1999) Metal hyperaccumulator plants: a review of the biological resource for possible exploitation in the phytoremediation of metal-polluted soils. In: Phytoremediation of contaminated soil and water. CRC Press, Boca Raton

Balasubramanian S, Bentley DR, Swerdlow HP, Smith GP, Milton J et al (2008) Accurate whole human genome sequencing using reversible terminator chemistry. Nature 456:53–59

Balasubramanian R, Kenney GE, Rosenzweig AC (2011) Dual pathways for copper uptake by methanotrophic bacteria. J Biol Chem 286:37313–37319

Barkay T, Wagner-Dobler I (2005) Microbial transformations of mercury: potentials, challenges, and achievements in controlling mercury toxicity in the environment. Adv Appl Microbiol 57:1–52

Barkay T, Miller SM, Summers AO (2003) Bacterial mercury resistance from atoms to ecosystems. FEMS Microbiol Rev 27:355–384

Batool R, Yrjala K, Hasnain S (2012) Hexavalent chromium reduction by bacteria from tannery effluent. J Microbiol Biotechnol 22:547–554

Bender CL, Cooksey DA (1986) Indigenous plasmids in Pseudomonas syringae pv. tomato: conjugative transfer and role in copper resistance. J Bacteriol 165:534–541

Bhaskar PV, Bhosle NB (2006) Bacterial extracellular polymeric substances (EPS) a carrier of heavy metals in the marine food-chain. Environ Int 32:192–198

Blindauer CA, Harrison MD, Robinson AK, Parkinson JA, Bowness PW, Sadler PJ, Robinson NJ (2002) Multiple bacteria encode metallothioneins and SmtA-like zinc fingers. Mol Microbiol 45 (5):1421–1432

Bogacka EK (2011) Surface properties of yeast cells during heavy metal biosorption. Eur J Chem 9:348–351

Bognolo G (1999) Biosurfactants as emulsifying agents for hydrocarbons. Colloids Surf A Physicochem Eng Asp 152:41–52

Bondarczuk K, Piotrowska-Seget Z (2013) Molecular basis of active copper resistance mechanisms in gram-negative bacteria. Cell Biol Toxicol 29:397–405

Borremans B, Hobman JL, Provoost A, Brown NL, van Der Lelie D (2001) Cloning and functional analysis of the pbr lead resistance determinant of Ralstonia metallidurans CH34. J Bacteriol 183:5651–5658

Borro DM, Fein JB (2005) The impact of ionic strength on the adsorption of protons, Pb, Cd, and Sr onto the surfaces of gram negative bacteria: testing non-eletrostatic, diffuse, and triple-layer models. J Colloid Interface Sci 286:110–126

Boyd ES, Barkay T (2012) The mercury resistance operon: from an origin in a geothermal environment to an efficient detoxification machine. Front Microbiol 3. https://doi.org/10.3389/fmicb.2012.00349

Bramhachari PV, Kavi Kishor PB, Ramadevi R, Kumar R, Rao BR, Dubey SK (2007) Isolation and characterization of mucous exopolysaccharide produced by Vibrio furnissii VB0S3. J Microbiol Biotechnol 17:44–51

Branco R, Chung AP, Johnston T, Gurel V, Morais P, Zhitkovich A (2008) The chromate-inducible chrBACF operon from the transposable element TnOtChr confers resistance to chromium (VI) and superoxide. J Bacteriol 190(21):6996–7003

Brim H, McFarlan SC, Fredrickson JK, Minton KW, Zhai M, Wackett LP, Daly MJ (2000) Engineering Deinococcus radiodurans for metal remediation in radioactive mixed waste environments. Nat Biotechnol 18:85–90

Brim H, Venkateswaran A, Kostandarithes HM, Fredrickson JK, Daly MJ (2003) Engineering Deinococcus geothermal is for bioremediation of high-temperature radioactive waste environments. Appl Environ Microbiol 69:4575–4582

Brim H, Osborne JP, Kostandarithes HM, Fredrickson JK, Wackett LP, Daly MJ (2006) Deinococcus radiodurans engineered for complete toluene degradation facilitates Cr (VI) reduction. Microbiology 152:2469–2477

Brunke M, Deckwer WD, Frischmuth A, Horn JM, Lünsdorf H, Rhode M, Röhricht M, Timmis KN, Weppen P (1993) Microbial retention of mercury from waste streams in a laboratory column containing merA gene bacteria. FEMS Microbiol Rev 11(1–3):145–152

Cervantes C, Campos-García J (2007) Reduction and efflux of chromate by bacteria. In: Molecular microbiology of heavy metals. Springer, Berlin/Heidelberg, pp 407–419

Chakraborty J, Das S (2014) Characterization and cadmium-resistant gene expression of biofilm-forming marine bacterium Pseudomonas aeruginosa JP-11. Environ Sci Pollut Res. https://doi.org/10.1007/s11356-014-3308-7

Champion JT, Gilkey CJ, Lamparski H, Retterer J, Miller RM (1995) Electron microscopy of rhamnolipid (biosurfactant) morphology: effects of pH, cadmium, and octadecane. J Colloid Interface Sci 170:569–574

Chang L, Meier J, Smikth M (1997) Application of plant and earthworm bioassays to evaluate remediation of a lead-contaminated soil. Arch Environ Contam Toxicol 32:166–171

Chang FM, Coyne HJ, Cubillas C, Vinuesa P, Fang X, Ma Z, Ma D, Helmann JD, García-de los Santos A, Wang YX, Dann CE 3rd, Giedroc DP (2014) Cu(I)-mediated allosteric switching in a copper-sensing operon repressor (CsoR). J Biol Chem 289(27):19204–19217

Chaouni LBA, Etienne J, Greenland T, Vandenesch F (1996) Nucleic acid sequence and affiliation of pLUG10, a novel cadmium resistance plasmid from Staphylococcus lugdunensis. Plasmid 36:1–8

Chaturvedi R, Archana G (2014) Cytosolic expression of synthetic phytochelatin and bacterial metallothionein genes in Deinococcus radiodurans R1 for enhanced tolerance and bioaccumulation of cadmium. Biometals 27(3):471–482

Chaturvedi KS, Hung CS, Crowley JR, Stapleton AE, Henderson JP (2012) The siderophore yersiniabactin binds copper to protect pathogens during infection. Nat Chem Biol 8:731–736

Chaturvedi KS, Hung CS, Giblin DE, Urushidani S, Austin AM, Dinauer MC, Henderson JP (2014) Cupric yersiniabactin is a virulence-associated superoxide dismutase mimic. ACS Chem Biol 9:551–561

Chen W, Georgiou G (2002) Cell-surface display of heterologous proteins: from highthroughput screening to environmental applications. Biotechnol Bioeng 79:496–503

Chen C, Wang J (2008) Removal of Pb2+, Ag +, Cs+, and Sr 2+ from aqueous solution by brewery waste biomass. J Hazard Mater 151:65–70

Chihomvu P, Stegmann P, Pillay M (2015) Characterization and structure prediction of partial length protein sequences of pcoA, pcoR and chrB genes from heavy metal resistant bacteria from the Klip River, South Africa. Int J Mol Sci 16:7352–7374

Chillappagari S, Miethke M, Trip H, Kuipers OP, Marahiel MA (2009) Copper acquisition is mediated by YcnJ and regulated by YcnK and CsoR in Bacillus subtilis. J Bacteriol 191:2362–2370

Cobbett CS (2000) Phytochelatins and their roles in heavy metal detoxification. Plant Physiol 123:825–832

Cobbett C, Goldsbrough P (2002) Phytochelatins and Metallothioneins: roles in heavy metal detoxification and homeostasis. Annu Rev Plant Biol 53:159–182

Cooksey DA, Azad HR, Cha JS, Lim CK (1990) Copper resistance gene homologs in pathogenic and saprophytic bacterial species from tomato. Appl Environ Microbiol 56:431–435

Costa ACA, Leite SGF (1991) Metals biosorption by sodium alginate immobilized Chiarella homosphaera cells. Biotechnol Left 13:555–562

Crupper SS, Worrell V, Stewart GC, Iandolo JJ (1999) Cloning and expression of cadD, a new cadmium resistance gene of Staphylococcus aureus. J Bacteriol 181:4071–4075

Cunningham SD, Berti WR (1993) Remediation of contaminated soils with green plants: an overview. In Vitro Cell Dev Biol Plant 29:207. https://doi.org/10.1007/BF02632036

Das A, Prasad R, Srivastava A, Giang PH, Bhatnagar K, Varma A (2007) Fungal siderophores: structure, functions and regulations. In: Varma A, Chincholkar SB (eds) Microbial siderophores, vol 12. Springer-Verlag, Berlin, pp 1–42

Das S (ed) (2014) Microbial biodegradation and bioremediation. Elsevier Inc

Das P, Sinha S, Mukherjee SK (2014) Nickel bioremediation potential of Bacillus thuringiensis KUNi1 and some environmental factors in nickel removal. Biorem J 18(2):169–177

Das S, Dash HR, Chakraborty J (2016) Genetic basis and importance of metal resistant genes in bacteria for bioremediation of contaminated environments with toxic metal pollutants. Appl Microbiol Biotechnol 100:2967–2984. https://doi.org/10.1007/s00253-016-7364-4

Dash HR, Das S (2012) Bioremediation of mercury and the importance of bacterial mer genes. Int Biodet Biodeg 75:207–213

Dash HR, Das S (2015) Enhanced bioremediation of inorganic mercury through simultaneous volatilization and biosorption by transgenic marine bacterium Bacillus cereus BW-03(pPW-05). Int Biodeterior Biodegrad 103:179–185

Dash HR, Mangwani N, Chakraborty J, Kumari S, Das S (2013) Marine bacteria: potential candidates for enhanced bioremediation. Appl Microbiol Biotechnol 97:561–571

Dash HR, Mangwani N, Das S (2014) Characterization and potential application in mercury bioremediation of highly mercury-resistant marine bacterium Bacillus thuringiensis PW-05. Environ Sci Pollut Res 21(4):2642–2653

Davison J (2005) Risk mitigation of genetically modified bacteria and plants designed for bioremediation. J Ind Microbiol Biotechnol 32(11–12):639–650

Davis TA, Volesky B, Mucci AA (2003) Review of the biochemistry of heavy metal biosorption by brown algae. Water Res 37(18):4311–4330

Davison J (1999) Genetic exchange between bacteria in the environment. Plasmid 42:73–91

Davison J (2002a) Genetic tools for pseudomonads, rhizobia and other gram-negative bacteria. BioTechniques 32:386–401

Davison J (2002b) Towards safer vectors for field release of recombinant bacteria. Environ Biosaf Res 1:9–18

De Lorenzo V (1994) Designing microbial systems for gene expression in the field. Trends Biotechnol 12:365–371

De J, Ramaiah N, Bhosle NB, Garg A, Vardanyan L, Nagle VL, Fukami K (2007) Potential of mercury resistant marine bacteria for detoxification of chemicals of environmental concern. Microbes Environ 22:336–345

De J, Ramaiah N, Vardanyan L (2008) Detoxification of toxic heavy metals by marine bacteria highly resistant to mercury. Mar Biotechnol 10:471–477

Deeb BE, Altalhi AD (2009) Degradative plasmid and heavy metal resistance plasmid naturally coexist in phenol and cyanide assimilating bacteria. Am J Biochem Biotechnol 5(2):84–93

Deng X, Wilson DB (2001) Bioaccumulation of mercury from wastewater by genetically engineered *Escherichia coli*. Appl Microbiol Biotechnol 56:276–279

Deng X, Li QB, Lu YH, Sun DH, Huang YL, Chen XR (2003) Bioaccumulation of nickel from aqueous solutions by genetically engineered E. coli. Water Res 37:2505–2511

Deng X, Yi XE, Liu G (2007) Cadmium removal from aqueous solution by gene modi fi ed *Escherichia coli* JM109. J Hazard Mater 139:340–344

Dias MA, Lacerda ICA, Pimentel PF, Castro HF, Rosa CA (2002) Removal of heavy metals by an Aspergillus terreus strain immobilized in a polyurethane matrix. Lett Appl Microbiol 34:46–50

Diaz-Magana A, Aguilar-Barajas E, Moreno-Sánchez R, Ramírez-Díaz MI, Riveros-Rosas H, Vargas E, Cervantes C (2009) Short-chain chromate ion transporter proteins from Bacillus subtilis confer chromate resistance in Escherichia coli. J Bacteriol 191:5441–5445

Dixit R, Malaviya D, Pandiyan K, Singh UB, Sahu A, Shukla R, Singh BP, Rai JP, Sharma PK, Lade H, Paul D (2015) Bioremediation of heavy metals from soil and aquatic environment: an overview of principles and criteria of fundamental processes. Sustainability 7(2):2189–2212

Djoko KY, Chong LX, Wedd AG, Xiao Z (2010) Reaction mechanisms of the multicopper oxidase CueO from Escherichia coli support its functional role as a cuprous oxidase. J Am Chem Soc 132:2005–2015

Dobson RS, Burgess JE (2007) Biological treatment of precious metal refinery wastewater: a review. Miner Eng 20(6):519–532

Dowdy RH, Volk VV (1983) Movement of heavy metals in soils. Chem Mobil Reactiv Soil Syst (Chemicalmobilit) 4:229–240

Duprey A, Chansavang V, Frémion F, Gonthier C, Louis Y, Lejeune P, Dorel C (2014) BNiCo Buster^: engineering E. coli for fast and efficient capture of cobalt and nickel. J Biol Eng 8 (1):1–11

Elen Aquino Perpetuo, Cleide Barbieri Souza, Claudio Augusto Oller Nascimento (2011) Engineering bacteria for bioremediation, progress. In: Angelo Carpi (ed) Molecular and environmental bioengineering – from analysis and modeling to technology applications. InTech, Reijka. ISBN: 978-953-307-268-5. Available from: http://www.intechopen.com/books/progress-in-molecular-and-environmental-bioengineering-from-analysis-and-modeling-to-technology-applications/engineering-bacteria-for-bioremediation

Ercole C, Veglio F, Toro L, Ficara G, Lepidi A (1994) Immobilisation of microbial cells for metal adsorption and desorption. In: Mineral bioprocessing II. Snowboard, Utah

Eriksson PO, Sahlman L (1993) 1H NMR studies of the mercuric ion binding protein MerP: sequential assignment, secondary structure and global fold of oxidized MerP. J Biomol NMR 3:613–626

Esposito A, PAgnanelli FE, Veglio F (2002) pH-related equilibria models for biosorption in single metal systems. Chem Eng Sci 57(3):307–313

Ezezika OC, Singer PA (2010) Genetically engineered oil-eating microbes for bioremediation: prospects and regulatory challenges. Technol Soc 32:331–335

Fan Q, He J, Xue H et al (2007) Competitive adsorption, release and speciation of heavy metals in the Yellow River sediments, China. Environ Geol 53:239. https://doi.org/10.1007/s00254-007-0638-5

Fourest E, Roux JC (1992) Heavy metal adsorption by fungal mycelial by-products: influence of pH. Appl Microbiol Biotechnol 37:399–403

Franke S, Grass G, Rensing C, Nies DH (2003) Molecular analysis of the copper-transporting efflux system CusCFBA of Escherichia coli. J Bacteriol 185(13):3804–3812

Gadd GM (1988) Heavy metal and radionuclide by fungi and yeasts. In: Norris PR, Kelly DP (eds) Biohydrometallurgy. A. Rowe, Chippenham/Wilts

Gadd GM (2004) Microbial influence on metal mobility and application for bioremediation. Geoderma 122(2–4):109–119

Gadd GM, De Rome L (1988) Biosorption of copper by fungal melanine. Appl Microbiol Biotechnol 29(6):610–617

Gadd GM, White C (1993) Microbial treatment of metal pollution – a working biotechnology? Trends Biotechnol 11:353–359

Garbisu C, Alkorta I (2001) Phytoextraction: a cost-effective plant-based technology for the removal of metals from the environment. Bioresour Technol 77:229–236

Gavrilescu M (2004) Removal of heavy metals from the environment by biosorption. Eng Life Sci 4 (3):219–232

Gawali AA, Nanoty VD, Bhalekar UK (2014) Biosorption of heavy metals from aqueous solution using bacterial EPS. Int J Life Sci 2:373–377

Gazsó LG (2001) The key microbial processes in the removal of toxic metals and radionuclides from the environment. Mini ReVol CEJOEM 7:178–185

Ge Z, Taylor DE (1996) Helicobacter pylori genes hpcopA and hpcopP constitute a cop operon involved in copper export. FEMS Microbiol Lett 145:181–188

Gee AR, Dudeney AWL (1988) Adsorption and crystallization of gold at biological surfaces. Biohydrometallurgy. Science & Technology Letters, London, UK, pp 437–451

Ghosh SK, Chaudhuri R, Gachhui R, Mandal A, Ghosh S (2006) Effect of mercury and organomercurials on cellular glucose utilization: a study using resting mercury-resistant yeast cells. J Appl Microbiol 102:375–383

Golby S, Ceri H, Marques LLR, Turner RJ (2014) Mixed-species biofilms cultured from an oil sand tailings pond can biomineralize metals. Microb Ecol 68:70–80

Gomes KM, Rebello RC, Duarte RS, Rachid CT (2013) Diversity of mercury resistant *Escherichia coli* strains isolated from aquatic systems in Rio de Janeiro, Brazil. Int J Biodivers 6:1–8

Gonzalez CF, Ackerley DF, Lynch SV, Matin A (2005) ChrR, a soluble quinone reductase of Pseudomonas putida that defends against H2O2. J Biol Chem 280:22590–22595

Goyal N, Jain SC, Banerjee UC (2003) Comparative studies on the microbial adsorption of heavy metals. Adv Environ Res 7:311–319

Grass G, Große C, Nies DH (2000) Regulation of the cnr cobalt and nickel resistance determinant from Ralstonia sp. strain CH34. J Bacteriol 182:1390–1398

Grass G, Thakali K, Klebba PE, Thieme D, Muller A, Wildner GF, Rensing C (2004) Linkage between catecholate siderophores and the multicopper oxidase CueO in Escherichia coli. J Bacteriol 186:5826–5833

Grass G, Fricke B, Nies DH (2005) Control of expression of a periplasmic nickel efflux pump by periplasmic nickel concentrations. Biometals 18:437–448

Grill E, Winnacker EL, Zenk MH (1985) Phytochelatins: the principal heavy-metal complexing peptides of higher plants. Science 230:674–676

Guo H, Luoa S, Chen L, Xiao X, Xi Q, Wei W, Zeng G, Liu C, Wan Y, Chen J, He Y (2010) Bioremediation of heavy metals by growing hyperaccumulaor endophytic bacterium Bacillus sp. L14. Bioresour Technol 101:8599–8606

Gupta R, Ahuja P, Khan S, Saxena RK, Mohapatra H (2000) Microbial biosorbents: meeting challenges of heavy metal pollution in aqueous solutions. Curr Sci 78(8):967–973

Gupta DK, Tohoyama H, Joho M, Inouhe M (2005) Changes in the levels of phytochelatins and related metal-binding peptides in chickpea seedlings exposed to arsenic and different heavy metal ions. J Plant Res 17:253–256

Gupta A, Joia J, Sood A, Sood R, Sidhu C et al (2016) Microbes as potential tool for remediation of heavy metals: a review. J Microb Biochem Technol 8:364–372

Hakansson T, Suer P, Mattiasson B, Allard B (2008) Sulphate reducing bacteria to precipitate mercury after electrokinetic soil remediation. Int J Environ Sci Technol 5(2):267–274

Hameed MSA (2006) Continuous removal and recovery of lead by alginate beads free and alginate-immobilized *Chlorella vulgaris*. Afr J Biotechnol 5:1819–1823

Hamer DH (1986) Metallothionein. Annu Rev Biochem 55:913–995

Hamlett NV, Landale EC, Davis BH, Summers AO (1992) Roles of the Tn21 merT, merP, andmerC gene products in mercury resistance and mercury binding. J Bacteriol 174(20):6377–6385

Hammaini A (2003) Simultaneous uptake of metals by activated sludge. Miner Eng 16:723–729
Henne KL, Nakatsu CH, Thompson DK, Konopka AE (2009) High-level chromate resistance in Arthrobacter sp. strain FB24 requires previously uncharacterized accessory genes. BMC Microbiol 9:199
Hinojosa BM, Garcia-Ruiz R, Carreira JA (2010) Utilizing microbial community structure and function to evaluate the health of heavy metal polluted soils. In: Sherameti I, Varma A (eds) Soil heavy metals. Springer, Berlin/Heidelberg
Hirata K, Tsuji N, Miyamoto K (2005) Biosynthetic regulation of phytochelatins, heavy metal-binding peptides. J Biosci Bioeng 100(6):593–599
Hsieh PF, Lin HH, Lin TL, Wang JT (2010) CadC regulates cad and tdc operons in response to gastrointestinal stresses and enhances intestinal colonization of Klebsiella pneumoniae. J Infect Dis 202:52–64
Hu YH, Wang HL, Zhang M, Sun L (2009) Molecular analysis of the copper-responsive CopRSCD of a pathogenic Pseudomonas fluorescens strain. J Microbiol 47:277–286
Huang C, Huang C, Morehart AL (1990) The removal of copper from dilute aqueous solutions by Saccharomyces cerevisiae. Water Res 24:433–439
Huckle JW, Morby AP, Turner JS, Robinson NJ (1993) Isolation of a prokaryotic metallothionein locus and analysis of transcriptional control by trace metal ions. Mol Microbiol 7:177–187
Huston WM, Jennings MP, McEwan AG (2002) The multi-copper oxidase of *Pseudomonas aeruginosa* is a ferroxidase with a central role in iron acquisition. Mol Microbiol 45:1741–1750
Hynninen A, Touze T, Pitkänen L, Mengin-Lecreulx D, Virta M (2009) An efflux transporter PbrA and a phosphatase PbrB cooperate in a lead-resistance mechanism in bacteria. Mol Microbiol 74:384–394
Inamuddin, Ahamed MI, Prasad R (2021) Microbial biosurfactants: preparation, properties and applications. Springer Singapore. ISBN 978–981–15-6606-6. https://www.springer.com/gp/book/9789811566066
Infante JC, De Arco RD, Angulo ME (2014) Removal of lead, mercury and nickel using the yeast *Saccharomyces cerevisiae*. Revista MVZ Córdoba 19:4141–4149
Inouhe M (2005) Phytochelatins. Braz J Plant Physiol 17(1):650–678
Jain S, Bhatt A (2014) Molecular and *in situ* characterization of cadmium-resistant diversified extremophilic strains of Pseudomonas for their bioremediation potential. Biotech 4:297–304
Jain AN, Udayashankara TH, Lokesh KS (2012) Review on bioremediation of heavy metals with-microbial isolates and amendments on soil residue. Int J Sci Res 6:2319–7064
Jarosławiecka A, Piotrowska-Seget Z (2014) Lead resistance in microorganisms. Microbiology 160:12–25
Jiang ZB, Song HT, Gupta N, Ma LX, Wu ZB (2007) Cell surface display of functionally active lipases from Yarrowia lipolytica in Pichia pastoris. Protein Expr Purif 56:35–39
Johnsen AR, Wick LY, Harms H (2005) Principles of microbial PAHdegradation in soil. Environ Pollut 133:71–84
Juhnke S, Peitzsch N, Hübener N, Grobe C, Nies DH (2002) New genes involved in chromate resistance in Ralstonia metallidurans strain CH34. Arch Microbiol 179:15–25
Kalyaeva ES, Kholodii GY, Bass IA, Gorlenko ZM, Yurieva OV, Nikiforov VG (2001) Tn5037, a Tn21-like mercury resistance transposon from Thiobacillus ferrooxidans. Russ J Genet 37:972–975
Kamaludeen SPB, Arunkumar KR, Avudainayagam SA, Ramasamy K (2003) Bioremediation of chromium contaminated environments. Indian J Exp Biol 41:972–985
Kanehisa M, Goto S (2000) KEGG: Kyoto encyclopedia of genes and genomes. Nucleic Acids Res 28:27–30
Kang SH, Singh S, Kim JY, Lee W, Mulchandani A, Chen W (2007) Bacteria metabolically engineered for enhanced phytochelatin production and cadmium accumulation. Appl Environ Microbiol 73:6317–6320

Karp PD, Ouzounis CA, Moore-Kochlacs C, Goldovsky L, Kaipa P, Ahrén D, Ópez-Bigas N (2005) Expansion of the BioCyc collection of pathway/genome databases to 160 genomes. Nucleic Acids Res 33:6083–6089

Kefala MI, Zouboulis AIE, Matis KA (1999) Biosorption of cadmium ions by Actinomycetes and separation by flotation. Environ Pollut 104(2):283–293

Kermani AJN, Ghasemi MF, Khosravan A, Farahmand A, Shakibaie MR (2010) Cadmium bioremediation by metal-resistant mutated bacteria isolated from active sludge of industrial effluent. Iran J Environ Health Sci Eng 7(4):279–286

Keskinkan O, Goksu MZL, Yuceer A, Basibuyuk M, Forster CF (2003) Heavy metal adsorption characteristics of a submerged aquatic plant (Myriophyllum spicatum). Process Biochem 39 (2):179–183

Khan AG (2005) Role of soil microbes in the rhizospheres of plants growing on trace metal contaminated soils in phyto-remediation. J Trace Elem Med Biol 18(4):355–364

Khan F, Sajid M, Cameotra SS (2013) In silico approach for the bioremediation of toxic pollutants. J Pet Environ Biotechnol 4:2

Khan Z, Nisar MA, Hussain SZ, Arshad MN, Rehman A (2015) Cadmium resistance mechanism in Escherichia coli P4 and its potential use to bioremediate environmental cadmium. Appl Microbiol Biotechnol 99(24):10745–10757

Kim SK, Lee BS, Wilson DB, Kim EK (2005) Selective cadmium accumulation using recombinant E. coli. J Biosci Bioeng 99(2):109–114

Kim SU, Cheong YH, Seo DC, Hu JS, Heo JS, Cho JS (2007) Characterisation of heavy metal tolerance and biosorption capacity of bacterium strain CPB4 (Bacillus spp.). Water Sci Technol 55(1–2):105–111

Kiyono M, Pan-Hou H (1999) The merG gene product is involved in phenylmercury resistance in Pseudomonas strain K-62. J Bacteriol 181:726–730

Kiyono M, Sone Y, Nakamura R, Pan-Hou H, Sakabe K (2009) The Mer E protein encoded by transposon Tn21 is a broad mercury transporter in *Escherichia coli*. FEBS Lett 583:1127–1131

Kiyono M, Oka Y, Sone Y, Nakamura R, Sato MH, Sakabe K, Pan-Hou H (2013) Bacterial heavy metal transporter MerC increases mercury accumulation in Arabidopsis thaliana. Biochem Eng J 71:19–24

Klaassen CD, Liu SCJ (1999) Metallothionein: an intracellular protein to protect against cadmium toxicity. Annu Rev Pharmacol Toxicol 39:267–294

Kobayashi I, Fujiwara S, Saegusa H, Inohe M, Metsumoto H, Tsuzuki M (2006) Relief of arsenate toxicity by Cd-stimulated phytochelatin synthesis in the green alga Chlamydomonas reinhardtii. Mar Biotechnol 8:94–101

Koch N, Islam NF, Sonowal S, Prasad R, Sarma H (2021) Environmental antibiotics and resistance genes as emerging contaminants: methods of detection and bioremediation. Curr Res Microbial Sci. https://doi.org/10.1016/j.crmicr.2021.100027

Kondo A, Ueda M (2004) Yeast cell-surface display-applications of molecular display. Appl Microbiol Biotechnol 64:28–40

Kornberg A (1995) Inorganic polyphosphate: toward making a forgotten polymer unforgettable. J Bacteriol 177:491–496

Kostal JRY, Wu CH, Mulchandani A, Chen W (2004) Enhanced arsenic accumulation in engineered bacterial cells expressing ArsR. Appl Environ Microbiol 70:4582–4587

Kotrba P, Ruml T (2000) Bioremediation of heavy metal pollution exploiting constituents, metabolites and metabolic pathway of livings. Appl Environ Microbiol 65(8):1205–1247

Kratchovil D, Volesky B (1998) Advances in the biosorption of heavy metals. Trends Biotechnol 16:291–300

Kristanti RA, Hadibarata T, Toyama T, Tanaka Y, Mori K (2011) Bioremediation of crude oil by white rot fungi Polyporus sp. S133. J Microbiol Biotechnol 21:995–1000

Kulshreshtha S (2013) Genetically engineered microorganisms: a problem solving approach for bioremediation. J Bioremed Biodegr 4:e133. https://doi.org/10.4172/2155-6199.1000e133

Kumar S, Dagar VK, Khasa YP, Kuhad RC (2013) Genetically modified microorganisms (GMOs) for bioremediation. In: Kuhad RC, Singh A (eds) Biotechnology for environmental management 191 and resource recovery. Springer India. https://doi.org/10.1007/978-81-322-0876-1_11

Kuroda K, Ueda M (2003) Bioadsorption of cadmium ion by cell surface-engineered yeasts displaying metallothionein and hexa-His. Appl Microbiol Biotechnol 63:182–186

Kuroda K, Ueda M (2010) Engineering of microorganisms towards recovery of rare metal ions: mini review. Appl Microbiol Biotechnol 87:53–60

Kuroda K, Ueda M (2011) Yeast biosorption and recycling of metal ions by cell surface engineering. Microbial Biosorption Metals 10:235–247. https://doi.org/10.1007/978-94-007-0443/5-10

Kuroda K, Shibasaki S, Ueda M, Tanaka A (2001) Cell surface engineered yeast displaying a histidine oligopeptide (hexa-His) has enhanced adsorption of and tolerance to heavy metal ions. Appl Microbiol Biotechnol 57:697–701

Kuroda K, Ueda M, Shibasaki A, Tanaka A (2002) Cell surface-engineered yeast with ability to bind, and self-aggregate in response tom copper ion. Appl Microbiol Biotechnol 59:259–264

Kuyucak N, Volesky B (1988) Desorption of cobalt-laden algal biosorbent. Biotechnol Bioeng 33:815–822

Lee W, Wood T, Chen W (2002) Engineering TCE-degrading Rhizobacteria for heavy metal accumulation and enhanced TCE degradation. Biotechnol Bioeng:399–403. https://doi.org/10.1002/bit.20950

Letunic I, Copley RR, Pils B, Pinkert S, Schultz J, Bork P (2006) SMART 5: domains in the context of genomes and networks. Nucleic Acids Res 34:D257–D260

Li Y, Li B (2011) Study on fungi-bacteria consortium bioremediation of petroleum contaminated mangrove sediments amended with mixed biosurfactants. Adv Mater Res 183–185:1163–1167

Liu J, Zhang Y, Huang D, Song G (2005) Cadmium induced MTs synthesis via oxidative stress in yeast Saccharomyces cerevisiae. Mol Cell Biochem 280(1–2):139–145

Liu S, Zhang F, Chen J, Sun GX (2011) Arsenic removal from contaminated soil via biovolatilization by genetically engineered bacteria under laboratory conditions. J Environ Sci 23(9):1544–1550

Lorenzo V, Herrero M, Sánchez JM, Timmis KN (1998) Minitransposons in microbial ecology and environmental biotechnology. FEMS Microbiol Ecol 27:211–224

Lovley DR, Coates JD (1997) Bioremediation of metal contamination. Curr Opin Biotechnol 8:285–289

Machado MD, Santos MSF, Gouveia C, Soares HMVM, Soares EV (2008) Removal of heavy metal using a brewer's yeast strain of Saccharomyces cerevisiae: the flocculation as a separation process. Bioresour Technol 99:2107–2115

Machado MD, Soares EV, Helena MVM, Soares HMVM (2010) Removal of heavy metals using a brewer's yeast strain of Saccharomyces cerevisiae: chemical speciation as a tool in the prediction and improving of treatment efficiency of real electroplating effluents. J Hazard Mater 180:347–353

Macomber L, Imlay JA (2009) The iron-sulfur clusters of dehydratases are primary intracellular targets of copper toxicity. Proc Natl Acad Sci U S A 106:8344–8349

Malik A (2004) Metal bioremediation through growing cells. Environ Int 30:261–278

Mameri N, Boudries N, Addour L, Belhocine D, Lounici H, Grib H, Pauss A (1999) Batch zinc biosorption by a bacterial nonliving Streptomyces rimosus biomass. Water Res 33:1347–1354

March JC, Rao G, Bentley WE (2003) Biotechnological applications of green fluorescent protein. Appl Microbiol Biotechnol 62:303–315

Marzorati M, Balloi A, De Ferra F, Daffonchio D (2010) Identification of molecular markers to follow up the bioremediation of sites contaminated with chlorinated compounds. Methods Mol Biol 668:219–134

Mellano MA, Cooksey DA (1988) Induction of the copper resistance operon from Pseudomonas syringae. J Bacteriol 170:4399–4401

Mendoza-Cózatl D, Tavera HL, Navarro AH, Sánchez RM (2005) Sulfur assimilation and glutathione metabolism under cadmium stress in yeast, protists and plants. FEMS Microbiol 29 (4):653–671

Merle SS, Cuiné S, Carrier P, Pradines CL, Luu DT, Peltier G (2003) Enhanced toxic metal accumulation in engineered bacterial cells expressing Arabidopsis thaliana phytochelatin synthase. Appl Environ Microbiol 69(1):490–494

Mindlin S, Kholodii G, Gorlenko Z, Minakhina S, Minakhin L, Kalyaeva E, Nikiforov V (2001) Mercury resistance transposons of gramnegative environmental bacteria and their classification. Res Microbiol 152:811–822

Mishra R, Sinha V, Kannan A, Upreti RK (2012) Reduction of chromium-VI by chromium resistant Lactobacilli: a prospective bacterium for bioremediation. Toxicol Int 19:25–30

Monchy S, Benotmane MA, Janssen P, Vallaeys T, Taghavi S, van der Lelie D, Mergeay M (2007) Plasmids pMOL28 and pMOL30 of Cupriavidus metallidurans are specialized in the maximal viable response to heavy metals. J Bacteriol 189(20):7417–7425

Moore MJ, Distefano MD, Zydowsky LD, Cummings RT, Walsh CT (1990) Organomercurial lyase and mercuric ion reductase: nature's mercury detoxification catalysts. Acc Chem Res 23:301–308

Morais PV, Branco R, Francisco R (2011) Chromiumresistance strategies and toxicity: what makes Ochrobactrum tritici 5bvl1 a strain highly resistant. Biometals 24:401–410

Morillo JA, Garcia-Ribera R, Quesada T, Aguilera M, Ramos-Cormenzana A, Monteoliva-Sanchez M (2008) Biosorption of heavy metals by the exopolysaccharide produced by Paenibacillus jamilae. World J Microbiol Biotechnol 24:2699–2704

Mulligan CN (2005) Environmental applications for biosurfactants. Environ Pollut 133:183–198

Munson GP, Lam DL, Outten FW, O'Halloran TV (2000) Identification of a copper-responsive two-component system on the chromosome of Escherichia coli K-12. J Bacteriol 182:5864–5871

Muraleedharan TR, Venkobachar C (1990) Mechanism of biosorption of copper (II) by Ganoderma lucidum. Biotechnol Bioeng 35:320–325

Naeem A, Woertz JR, Fein JB (2006) Experimental measurement of proton, Cd, Pb, Sr, and Zn adsorption onto the fungal species Saccharomyces cerevisiae. Environ Sci Technol 40:5724–5729

Naik MM, Dubey SK (2011) Lead-enhanced siderophore production and alteration in cell morphology in a Pb-resistant Pseudomonas aeruginosa strain 4EA. Curr Microbiol 62:409–414

Naik MM, Pandey A, Dubey SK (2012a) Pseudomonas aeruginosa strain WI-1 from Mandovi estuary possesses metallothionein to alleviate lead toxicity and promotes plant growth. Ecotoxicol Environ Saf 79:129–133

Naik MM, Shamim K, Dubey SK (2012b) Biological characterization of lead resistant bacteria to explore role of bacterial metallothioneinin lead resistance. Curr Sci 103:1–3

Nakajima A, Tsuruta T (2004) Competitive biosorption of thorium and uranium by Micrococcus luteus. J Radioanal Nucl Chem 260:13–18

Namgay T, Singh B, Singh BP (2010) Influence of biochar application to soil on the availability of As, Cd, Cu, Pb, and Zn to maize (Zea maize L.). Soil Res 48(6–7):638–647

Nascimento AM, Chartone-Souza E (2003) Operon mer: bacterial resistance to mercury and potential for bioremediation of contaminated environments. Genet Mol Res 2(1):92–101

Naser HA (2013) Assessment and management of heavy metal pollution in the marine environment of the Arabian Gulf: a review. Mar Pollut Bull 72:6–13

Nies DH (2003) Efflux-mediated heavy metal resistance in prokaryotes. FEMS Microbiol Rev 27:313–339

Nies DH, Koch S, Wachi S, Peitzsch N, Saier MH (1998) CHR, a novel family of prokaryotic proton motive force-driven transporters probably containing chromate/sulfate antiporters. J Bacteriol 180:5799–5802

Nishitani T, Shimada M, Kuroda K, Ueda M (2010) Molecular design of yeast cell surface for adsorption and recovery of molybdenum, one of rare metals. Appl Microbiol Biotechnol 86:641–648

Norton L, Baskaran K, McKenzie T (2004) Biosorption of zinc from aqueous solutions using biosolids. Adv Environ Res 8(3–4):629–635

Nourbakhsh M, Sag Y, Ozer D, Aksu Z, Kustal T, Caglar A (1994) A comparative study of various biosorbents for removal of chromium (VI) ions from industrial waste waters. Process Biochem 29:1–5

Nriagu JO, Pacyna JM (1988) Quantitative assessment of worldwide contamination of air, water and soils by trace metals. Nature 333:134–139

Nucifora G, Chu L, Misra TK, Silver S (1989) Cadmium resistance from Staphylococcus aureus plasmid pI258 cadA gene results from a cadmium-efflux ATPase. Proc Natl Acad Sci U S A 86:3544–3548

Obenauer JC, Cantley LC, Yaffe MB (2003) Scansite 2.0: proteome-wide prediction of cell signalling interactions using short sequencemotifs. Nucleic Acids Res 31:3635–3641

Pagnanelli F, Esposito A, Vegliò F (2002) Multi-metallic modelling for biosorption of binary systems. Water Res 36:4095–4105

Pal A, Paul AK (2008) Microbial extracellular polymeric substances: central elements in heavy metal bioremediation. Indian J Microbiol 48:49–64

Pan-Hou HS, Imura N (1981) Role of hydrogen sulfide in mercury resistance determined by plasmid of Clostridium cochlearium T-2. Arch Microbiol 129(1):49–52

Pan-Hou H, Kiyono M, Omura T, Endo G (2002) Polyphosphate produced in recombinant Escherichia coli confers mercury resistance. FEMS Microbiol Lett 207:159–164

Park JH, Bolan N, Meghraj M, Naidu N (2011) Concomitant rock phosphate dissolution and lead immobilization by phosphate solubilising bacteria (Enterobacter sp.). J Environ Manag 92:1115–1120

Patel J, Zhang Q, Michael R, McKay L, Vincent R, Xu Z (2010) Genetic engineering of *Caulobacter crescentus* for removal of cadmium from water. Appl Biochem Biotechnol 160:232–243

Pattus F, Abdallah M (2000) Siderophores and iron-transport in microorganisms: review. J Chin Chem Soc 47:1–20

Paul D, Pandey G, Pandey J, Jain RK (2005) Accessing microbial diversity for bioremediation and environmental restoration. Trends Biotechnol 23:135–142

Penninckx MA (2000) A short review on the role of glutathione in the response of yeasts to nutritional, environment, and oxidative stresses. Enzyme Microbiol Technol 26:737–742

Penninckx MJ (2002) An overview on glutathione in Saccharomyces versus nonconventional yeasts. FEMS Yeast Res 2:295–305

Pernthaler A, Pernthaler J, Amann R (2002) Fluorescence in situ hybridization and catalyzed reporter deposition for the identification of marine bacteria. Appl Environ Microbiol 68:3094–3101

Pieper DH, Reineke W (2000) Engineering bacteria for bioremediation. Curr Opin Biotechnol 11:262–270

Plette AC, Benedetti MF, van Riemsdijk WH (1996) Competitive binding of protons, calcium, cadmium, and zinc to isolated cell walls of a gram-positive soil bacterium. Environ Sci Technol 30(6):1902–1910

Prasad R, Aranda E (2018) Approaches in bioremediation: the new Era of environmental microbiology and nanobiotechnology. Springer International Publishing (978–3–030-02369-0). https://www.springer.com/gp/book/9783030023683

Preveral S, Ansoborlo E, Mari S, Vavasseur A, Forestier C (2006) Metal(loid)s and radionuclides cytotoxicity in Saccharomyces cerevisiae. Role of YCF1, glutathione and effect of buthionine sulfoximine. Biochimie 88(11):1651–1663

Qin J, Rosen BP, Zhang Y, Wang GJ, Franke S, Rensing C (2006) Arsenic detoxification and evolution of trimethylarsine gas by a microbial arsenite S adenosylmethionine methyltransferase. Proc Natl Acad Sci U S A 103:2075–2080

Raj R, Dalei K, Chakraborty J, Das S (2016) Extracellular polymeric substances of a marine bacterium mediated synthesis of CdS nanoparticles for removal of cadmium from aqueous solution. J Colloid Interface Sci 462:166–175

Rebello RCL, Gomes KM, Duarte RS, Rachid CTCC, Rosado AS, Regua-Mangia AH (2013) Diversity of mercury resistant Escherichia coli strains isolated from aquatic systems in Rio de Janeiro, Brazil. Int J Biodivers. https://doi.org/10.1155/2013/265356

Ripp S, Nivens DE, Werner C, Sayler GS (2000) Bioluminescent most-probable-number monitoring of a genetically engineered bacterium during a long-term contained field release. Appl Microbiol Biotechnol 53:736–741

Roane TM (1999) Lead resistance in two bacterial isolates from heavy metal-contaminated soils. Microb Ecol 37:218–224

Roane TM, Pepper IL (2001) Environmental microbiology. In: Roane TM, Pepper IL (ed) Microorganisms and metal pollutants. Academic, Vol 17, pp 403–423

Rodrigue A, Effantin G, Mandrand-Berthelot MA (2005) Identification of rcnA (yohM), a nickel and cobalt resistance gene in Escherichia coli. J Bacteriol 187:2912–2916

Rojas LA, Yáñez C, González M, Lobos S, Smalla K, Seeger M (2011) Characterization of the metabolicallymodified heavy metal-resistant Cupriavidus metallidurans strain MSR33 generated for mercury bioremediation. PLoS One 6:e17555

Ronchel MC, Ramos JL (2001) Dual system to reinforce biological containment of recombinant bacteria designed for rhizoremediation. Appl Environ Microbiol 67:2649–2656

Rosner JL, Aumercier M (1990) Potentiation by salicylate and salicyl alcohol of cadmium toxicity and accumulation in Escherichia coli. Appl Environ Microbiol 34:2402–2406

Ruiz NO, Daniell H (2009) Genetic engineering to enhance mercury phytoremediation. Curr Opin Biotechnol 20:1–7

Ruiz ON, Alvarez D, Gonzalez-Ruiz G, Torres C (2011) Characterization of mercury bioremediation by transgenic bacteria expressing metallothionein and polyphosphate kinase. BMC Biotechnol 11:82

Rutherford N, Mourez M (2006) Surface display of proteins by gram negative bacterial autotransporters. Microb Cell Fact 5(1):22

Sakaguchi T, Nakajima A (1991) Accumulation of heavy metals such as uranium and thorium by microorganisms. In: Smith RW, Misra M (eds) Mineral bioprocessing. The minerals, metals and materials society, Pennsylvania

Samuelson P, Gunneriusson E, Nygren PPA, Stahl S (2002) Display of proteins on bacteria. J Biotechnol 96(2):129–154

Sandaa RA, Torsvik V, Enger O, Daae FL, Castberg T, Hahn D (1999) Analysis of bacterial communities in heavy metal-contaminated soils at different levels of resolution. FEMS Microbiol Ecol 30:237–251

Sandrin TR, Maier RM (2003) Impact of metals on the biodegradation of organic pollutants. Environ Health Perspect 111(8):1093

Sasaki Y, Hayakawa T, Inoue C, Miyazaki A, Silver S, Kusano T (2006) Generation of mercury-hyperaccumulating plants through transgenic expression of the bacterial mercury membrane transport protein MerC. Transgenic Res 15:615–625

Sathyavathi S, Manjula A, Rajendhran J, Gunasekaran P (2014) Extracellular synthesis and characterization of nickel oxide nanoparticles from Microbacterium sp. MRS-1 towards bioremediation of nickel electroplating industrial effluent. Bioresour Technol 165:270–273

Sauge-Merle S, Cuine S, Carrier P, Lecomte-Pradines C, Luu DT, Peltier G (2003) Enhanced toxic metal accumulation in engineered bacterial cells expressing *Arabidopsis thaliana* phytochelatin synthase. Appl Environ Microbiol 69:490–494

Sayler GS, Ripp S (2000) Field applications of genetically engineered microorganisms for bioremediation processes. Curr Opin Biotechnol 11:286–289

Schelert J, Drozda M, Dixit V, Dillman A, Blum P (2006) Regulation of mercury resistance in the crenarchaeote Sulfolobus solfataricus. J Bacteriol 188:7141–7150
Schelert J, Rudrappa D, Johnson T, Blum P (2013) Role of MerH in mercury resistance in the archaeon Sulfolobus solfataricus. Microbiology 159:1198–1208
Schmidt T, Schlegel HG (1994) Combined nickel-cobalt-cadmium resistance encoded by the ncc locus of Alcaligenes xylosoxidans 31A. J Bacteriol 176:7045–7054
Schmitt M, Schwanewilm P, Ludwig J, Fraté LH (2006) Use of PMA1 as a housekeeping biomarker for assessment of toxicant- induced stress in Saccharomyces cerevisiae. Appl Environ Microbiol 72(2):1515–1522
Schneiker S, Keller M, Dröge M, Lanka E, Pühler A, Selbitschka W (2001) The genetic organization and evolution of the broad host range mercury resistance plasmid pSB102 isolated from a microbial population residing in the rhizosphere of alfalfa. Nucleic Acids Res 29:5169–5181
Schue M, Dover LG, Besra GS, Parkhill J, Brown NL (2009) Sequence and analysis of a plasmid-encoded mercury resistance operon from Mycobacterium marinum identifies MerH, a newmercuric ion transporter. J Bacteriol 191:439–444
Schweizer HP (2003) Applications of the Saccharomyces cerevisiae Flp-FRT system in bacterial genetics. J Mol Microbiol Biotechnol 5:67–77
Selatnia A, Boukazoula A, Kechid BN, Bakhti MZ, Chergui A, Kerchich Y (2004) Biosorption of lead (II) from aqueous solution by a bacterial dead Streptomyces rimosus biomass. J Eng Biochem 19(2):127–135
Silva E, Fialho AM, Sá-Correia I, Burns RG, Shaw LJ (2004) Combined bioaugmentation and biostimulation to cleanup soil contaminateed with high concentrations of atrazine. Environ Sci Technol 15–38(2):632–637
Silver S (1998) Genes for all metals – a bacterial view of the periodic table. The 1996 Thom Award Lecture. J Ind Microbiol Biotechnol 20:1–12
Silver S, Misra TK (1984) Bacterial transformations of and resistances to heavy metals. In: Genetic control of environmental pollutants. Springer, Boston, pp 23–46
Silver S, Phung LT (2013) Bacterial mercury resistance proteins. In: Encyclopedia of Metalloproteins. Springer, New York, pp 209–217
Singh SK, Grass G, Rensing C, Montfort WR (2004) Cuprous oxidase activity of CueO from Escherichia coli. J Bacteriol 86:7815–7817
Singh JS, Abhilash PC, Singh HB, Singh RP, Singh DP (2011) Genetically engineered bacteria: an emerging tool for environmental remediation and future perspectives. Gene 480(1–2):1–9, 36851
Singh SN, Goyal SK, Singh SR (2015) Bioremediation of heavy metals polluted soils and their effect on plants. Agriways 3(1):19–24
Smith K, Novick RP (1972) Genetic studies on plasmid-linked cadmium resistance in Staphylococcus aureus. J Bacteriol 112:761–772
Smith MC, Sumner ER, Avery SV (2007) Glutathione and Gts1p drive beneficial variability in the cadmium resistances of individual yeast cells. Mol Microbiol 66(3):699–712
Soares EV, Hebbelinck K, Soares HM (2003) Toxic effects caused by heavy metals in the yeast Saccharomyces cerevisiae: a comparative study. J Microbiol 49(5):336–343
Sriprang R, Hayashi M, Ono H, Takagi M, Hirata K, Murooka Y (2003) Enhanced accumulation of Cd2+ by a *Mesorhizobium* sp. transformed with a gene from *Arabidopsis thaliana* coding for phytochelatin synthase. Appl Environ Microbiol 69:179–796
Stahler FN, Odenbreit S, Haas R, Wilrich J, Van Vliet AH, Kusters JG, Kist M, Bereswill S (2006) The novel Helicobacter pylori CznABC metal efflux pump is required for cadmium, zinc, and nickel resistance, ureasemodulation, and gastric colonization. Infect Immun 74:3845–3852
Tabak HH, Lens P, Hullebusch EDV, Dejonghe W (2005) Developments in bioremediation of soil and sediments polluted with metals and radionuclides–1. Microbiolal processes and mechanisms affecting bioremediation of metal contamination and influencing meal toxicity. Rev Environ Sci Biotechnol 4:115–156

Taghavi S, Lesaulnier C, Monchy S, Wattiez R, Mergeay M, van der Lelie D (2009) Lead (II) resistance in Cupriavidus metallidurans CH34: interplay between plasmid and chromosomally-located functions. Antonie Van Leeuwenhoek 96:171–182

Tetaz TJ, Luke RK (1983) Plasmid-controlled resistance to copper in Escherichia coli. J Bacteriol 154:1263–1268

Thakare M, Sarma H, Datar S, Roy A, Pawar P, Gupta K, Pandit S, Prasad R (2021) Understanding the holistic approach to plant-microbe remediation technologies for removing heavy metals and radionuclides from soil. Curr Res Biotechnol. https://doi.org/10.1016/j.crbiot.2021.02.004

Thirumoorthy N, Kumar KTM, Sundar AS, Panayappan L, Chatterjee M (2007) Metallothionein: an overview. World J Gastroenterol 13(7):993–996

Tibazarwa C, Wuertz S, Mergeay M, Wyns L, van Der Lelie D (2000) Regulation of the cnr cobalt and nickel resistance determinant of Ralstonia eutropha (Alcaligenes eutrophus) CH34. J Bacteriol 182(5):1399–1409

Timmis KN, Pieper DH (1999) Bacteria designed for bioremediation. Trends Biotechnol 17:201–204

Torres B, Garcia J, Diaz E (2003a) Plasmids as tools for containment. In: Funnell B, Phillips G (eds) Plasmid biology. ASM, Washington, DC, pp 589–601

Torres B, Jaenecke S, Timmis KN, Garcia JL, Diaz E (2003b) A dual lethal system to enhance containment of recombinant micro-organisms. Microbiology 149:3595–3601

Trepreau J, de Rosny E, Duboc C, Sarret G, Petit-Hartlein I, Maillard AP, Covès J (2011) Spectroscopic characterization of the metal-binding sites in the periplasmic metal-sensor domain of CnrX from Cupriavidus metallidurans CH34. Biochemist 50:9036–9045

Trevors JT, Stratton GW, Gadd GM (1986) Cadmium transport, resistance, and toxicity in bacteria, algae, and fungi. Can J Microbiol 32:447–464

Tsezos M, Volesky B (1982) The mechanism of uranium biosorption by Rhizopus arrhizus. Biotechnol Bioeng 24(2):385–401

Tunali S, Çabuk A, Akar T (2006) Removal of lead and copper ions from aqueous solutions by bacterial strain isolated from soil. Chem Eng J 115:203–211

Uslu G, Tanyol M (2006) Equilibrium and thermodynamic parameters of single and binary mixture biosorption of lead(II) and copper(II) ions onto Pseudomonas putida: effect of temperature. J Hazard Mater 135:87–93

Valls M, de Lorenzo V (2002) Exploiting the genetic and biochemical capacities of bacteria for the remediation of heavy metal pollution. FEMS Microbiol Rev 26:327–338

Valls M, Atrian S, Lorenzo V, Fernadéz LA (2000) Engineering a mouse metallothionein on the surface of Ralstonia eutropha CH34 for immobilization of heavy metals in soil. Nat Biotechnol 18:661–665

Vargas-García MC, López MJ, Suárez-Estrella F, Moreno J (2012) Compost as a source of microbial isolates for the bioremediation of heavy metals: in vitro selection. Sci Total Environ 431:62–67

Vats N, Lee SF (2001) Characterization of a copper-transport operon, copYAZ, from Streptococcus mutans. Microbiology 147:653–662

Veglio F, Beolchini F (1997) Removal of metals by biosorption: a review. Chem Hydrometall 44:301–316

Vieira RH, Volesky B (2010) Biosorption: a solution to pollution? Int Microbiol 3:17–24

Vijayaraghavan K, Yun YS (2008) Bacterial biosorbents and biosorption. Biotechnol Adv 26:266–291

Viti C, Marchi E, Decorosi F, Giovannetti L (2013) Molecular mechanisms of Cr (VI) resistance in bacteria and fungi. FEMS Microbiol Rev 38:633–659

Volesky B (1987) Biosorbents for metal recovery. Trends Biotechnol 5(4):96–101

Volesky B (2001) Detoxification of metal-bearing effluents: biosorption for the next century. Hydrometallurgy 59(2–3):203–216

Volesky B, Holan ZR (1995) Biosorption of heavy metal. Biotechnol Prog 11:235–250

Von Mering C, Jensen CJ, Kuhn M, Chaffron S, Doerks M, Krüger B, Bork P (2007) STRING 7-recent developments in the integration and prediction of protein interactions. Nucleic Acids Res 35:D358–D362

von Rozycki T, Nies DH (2009) Cupriavidus metallidurans: evolution of a metal-resistant bacterium. Antonie Van Leeuwenhoek 96(2):115–139

Wagner-Dobler I (2003) Pilot plant for bioremediation of mercurycontaining industrial wastewater. Appl Microbiol Biotechnol 62(2–3):124–133

Wang J, Chen C (2006) Biosorption of heavy metal by Saccharomyces cerevisiae. Biotechnol Adv 24:427–451

Wang J, Chen C (2009) Biosorbents for heavy metals removal and their future. Biotechnol Adv 27:195–226

Wang Q, Li L, Chen M, Qi Q, Wang PG (2007) Construction of a novel system for cell surface display of heterologous proteins on Pichia pastoris. J Biotechnol Lett 29:1561–1566

Williams GP, Gnanadesigan M, Ravikumar S (2012) Biosorption and biokinetic studies of halobacterial strains against Ni^{2+}, Al^{3+} and Hg^{2+} metal ions. Bioresour Technol 107:526–529

Wilson M, Lindow SE (1993) Release of recombinant microorganisms. Annu Rev Microbiol 47:913–944

Wu CH, Wood TK, Mulchandani A, Chen W (2006) Engineering plant-microbe symbiosis for rhizoremediation of heavy metal. Appl Environ Microbiol 72(2):1129–1134

Wu G, Kang H, Zhang X, Shao H, Chu L, Ruan C (2010) A critical review on the bio-removal of hazardous heavy metals from contaminated soils: issues, progress, eco-environmental concerns and opportunities. J Hazard Mater 14:1–8

Wunderli-Ye H, Solioz M (1999) Copper homeostasis in Enterococcus hirae. In: Copper transport and its disorders. Springer, Boston, pp 255–264

Xu Z, Bae W, Nulchandani A, Mehra RK, Chen W (2002) Heavy metal removal by novel CBD-EC20 sorbents immobilized on cellulose. Biomacromolecules 3:462–465

Xue XM, Yan Y, Xu HJ, Wang N, Zhang X, Ye J (2014) ArsH from Synechocystis sp. PCC 6803 reduces chromate and ferric iron. FEMS Microbiol Lett 356:105–112

Yoon KP, Silver S (1991) A second gene in the Staphylococcus aureus cadA cadmium resistance determinant of plasmid pI258. J Bacteriol 173:7636–7642

Yu P, Yuan J, Deng X, Ma M, Zhang H (2014) Subcellular targeting of bacterial CusF enhances Cu accumulation and alters root to shoot Cu translocation in Arabidopsis. Plant Cell Physiol 55 (9):1568–1581

Zhang GL, Wu YZ, Qian XP, Meng Q (2005) Biodegradation of crude oil by Pseudomonas aeruginosa in the presence of rhamnolipids. J Zhejiang Univ Sci 6(8):725–730

Zhang YM, Yin H, Ye JS, Peng H, Zhang N, Qin HM, Yang F, He BY (2007) Cloning and expression of the nickel/cobalt transferase gene in E. coli BL21 and bioaccumulation of nickel ion by genetically engineered strain. Huan Jing Ke Xue 28(4):918–923

Zhang H, Zhou Y, Bao H, Zhang L, Wang R, Zhou X (2015) Plasmidborne cadmium resistant determinants are associated with the susceptibility of Listeria monocytogenes to bacteriophage. Microbiol Res 172:1–6

Zouboulis AI, Matis KA, Lazaridis NK (2001) Removal of metal ions from simulated wastewater by Saccharomyces yeast biomass: combining biosorption and flotation processes. Sep Sci Technol 36:349–365

Heavy Metal Removal Processes by Sulfate-Reducing Bacteria

13

María Isabel Neria-González and Ricardo Aguilar-López

Abstract

Actually, natural phenomena and anthropogenic activities have led to large environmental pollution in all kinds of ecosystems. From the above exist the needed to implement technological actions in order to diminish the pollution problems. Among these technological actions, the biotechnology processes have the great advantage of high specificity for the removal of chemical compounds as heavy metals with low energy consumption. The microorganisms are the biological agents most used in the treatments of polluting compounds, due to their degradation capacities of organic and inorganic pollutants, but the metals are not degraded but only can be modified in their redox state, converting them to less toxic forms. Precipitation and biosorption are the most employed process for metal ion removal from water. Microalgae, fungi and bacteria have been used successfully in the removal of metals, but in the last decade, sulfate-reducing bacteria (SRB) have taken great relevance in the processes of metal removal from wastewater. The removal of heavy metals is promoted when hydrogen sulfide is produced; it reacts with metal ion and forms metal sulfides, which are insoluble and tend to precipitate; it is not the only mechanism by which metal ions are removed, biosorption mechanisms can also be carried out (with biomass and production of exo-polysaccharides), immobilization and enzymatic reduction of the metal ion to less toxic and insoluble forms, but this will depend on the type of bacteria and their tolerance to metals.

M. I. Neria-González (✉)
División de Ingeniería Química y Bioquímica, Tecnológico de Estudios Superiores de Ecatepec, Ecatepec de Morelos, Edo. México, Mexico
e-mail: mineriag@tese.edu.mx

R. Aguilar-López
Departamento de Biotecnología y Bioingeniería, CINVESTAV-IPN, Ciudad de México, Mexico

R. Prasad (ed.), *Environmental Pollution and Remediation*, Environmental and Microbial Biotechnology, https://doi.org/10.1007/978-981-15-5499-5_13

Specifically, the species of the genus *Desulfovibrio* have been the most studied for metal removal with a high efficiency, where the main mechanism is the precipitation of metal sulfides among others. *Desulfovibrio alaskensis* 6SR exhibits a high metallic resistance with respect to other sulfate-reducing bacteria, including other microorganisms; since it shows strong resistance to Cr(VI), Cd (II), Pb(II) and Zn(II), some of these are considered as extremely toxic to biota. *Desulfovibrio alaskensis* strain 6SR is capable to remove more of the 98% Cr (VI) Cd(II), Pb(II) and Zn(II) in solution. The main mechanism of removal is the precipitation of the corresponding metallic sulfides, followed by adsorption of these by the produced EPS, and transmission electron micrographs show a slight metal accumulation at the intracellular level and periplasmic space. Also, the chromium reducing for the hydrogen sulfide has been analyzed by the sulfate reduction in independent reactors, as well as in culture per batch of *D. alaskensis*. The results indicate that the bacterium is able to grow up to a concentration of 18 mg/L of Cr(VI), and contrary to *D. vulgaris*, the reduction of sulfate does not interrupt at any time of the chromium reducing. Finally, a molecular analysis with respect to cadmium and chromium resistance mechanisms demonstrated the presence of *cad*A and *chr*A genes. Both genes are induced by Cd, Zn, Pb and Cr; the codified proteins by these genes are involucre to abate the oxidative stress provoked by heavy metal non-essentials.

Keywords

Heavy metals · Sulfate-reducing bacteria · Wastewater · Microbial reduction · *Desulfovibrio alaskensis*

13.1 Introduction

Ever since that the human is done sedentary, the usage of metals was incrementing; now they are used for everything, from construction to improve our health. Some metals are also necessary for the vital processes of any organisms. But where do we obtain the metals? In answer to this question, the Earth's crust is the main source of metals. So, gold, silver, platinum and others are found as the uncombined elements or native or free state, known as non-reactive metals too. In general, most metals are found combined with other elements to form compounds; on both cases, these are in rocks named ores. Most metals are extracted from ores by different extraction methods that depend upon the metal's position in the reactivity series. In principle, any metal could be extracted from its compound using electrolysis, but the using of large amounts of electrical energy results expensive, and other types of extraction methods are required, but this is another story. The fact is that the high demand of metals for various anthropogenic activities as the manufacture of steel, foundries, electroplating, auto parts, fuel production, manufacture of electronic devices, manufacture of agrochemicals and manufacture of batteries, among others, is the main source of metallic contamination (Haferburg and Kothe 2010). The metals are released to the environment as solutes or particles that can reach high concentrations, especially near the discharge site (Krishna and Govil 2007). In general, heavy metals accumulate in aquatic environments, mainly in sediments, debris and organic matter, where they may be consumed by fish, which tend to accumulate metals in the gills and intestines magnifying their concentration to a toxic level. This generates serious problems in health and the trophic chain of organisms affecting all kinds of ecosystems.

In actuality exist the needed to implement technological actions in order to diminish the pollution problems. Among these technological actions, the biotechnology processes have the great advantage of high specificity for the removal of chemical compounds as heavy metals with low energy consumption. The microorganisms are the biological agents most used on the treatments of polluting compounds, due to their degradation capacities of organic and inorganic pollutants, but the metals are not degraded but only can be modified in their redox state, converting them to less toxic forms (Wood and Wang 1983). Precipitation and biosorption are the most employed process for metal ion removal from water. Microalgae, fungi and bacteria have been used successfully in the removal of metals, but in the last decade, sulfate-reducing bacteria (SRB) have taken great relevance in the processes of metal removal from wastewater (López-Pérez et al. 2016). The removal of heavy metals is promoted when hydrogen sulfide is produced; it reacts with metal ions and forms metal sulfides, which are insoluble and tend to precipitate; however, it is not the only mechanism by which metal ions are removed, but biosorption mechanisms can also be carried out (with biomass and production of exo-polysaccharides), immobilization and enzymatic reduction of the metal ion to less toxic and insoluble forms, but this will depend on the type of bacteria and their tolerance to metals (Li et al. 2018). Specifically, the species of the genus *Desulfovibrio* have been the most studied for metal removal with a high efficiency, where the main mechanism is the precipitation of metal sulfides between others. In particular, *D. alaskensis* 6SR exhibits a high metallic resistance with respect to other sulfate-reducing bacteria, including other microorganisms. Therefore, this strain is considered as a model for the removing of metals under anaerobic conditions.

13.2 Heavy Metal Removal

Heavy metals are chemical elements; in principle, these are the simpler substances of the matter. The chemical elements are found ordered and classified in the periodic table. Currently, this contains 118 chemical elements; the most were discovered and some synthesized. Metals are the most abundant chemical elements, and these are classified in the periodic table as alkali metals, alkaline earth metals, transition metals, lanthanides and actinides.

However, the term "heavy metals" has been widely used to refer to any metallic or semimetallic chemical element. In the scientific literature, an authoritative definition is not found. But the authors have considered the density to refer to the term "heavy"; some of them have proposed densities among 3.5–7.5 g/cm. Therefore, a heavy metal is a metal or metalloid with a relatively high density and associated with pollution and toxicity (Duffus 2002). Other criteria as the atomic weight or mass were used to give consistency to the term "heavy", but not prospered. With respect to pollution and toxicity, heavy metals are highly toxic to very low concentrations. The heavy metals tend to accumulate with respect to time, exceeding their permissible concentration in the environment; the metallic accumulation unchains damages on the biological systems (human, animals, microorganisms and plants), provoking important problems of toxicity for environmental health and safety (Velea et al. 2009). Also, the term "heavy metal" in the legal aspects implies that the pure metal and all its compounds have the same physicochemical, biological and toxicological properties, which is false, as will mention below.

Heavy metals can be classified as (1) toxic metals, (2) essential metals for living organisms and (3) radionuclides such as uranium (Gadd 2010; Wood and Wang 1983). (1) Toxic metals – the toxicity of the heavy metals can define as the ability of a metal to cause negative effects on living organisms and depends on the bioavailability of the metals, and it is aggravated by their long-term persistence in the environment. Arsenic, fluorine, cadmium, mercury, chromium and lead are some examples of extremely toxic elements to biota, even at very low concentrations. Lead is one metal with the most retention time in soil (150–5000 years). (2) Essential metals, of all elements in the periodic table, 30 are required for microbial life, although not all are necessary for the growth and cell division of every microbial species. Among them, the carbon, nitrogen, hydrogen and oxygen are the bulk elements, and 26 of them are required in intermediate to trace amounts. Twenty-two of the 26 elements are found to be essential for life in higher organisms; see Fig. 13.1. An overabundance of any of these elements can cause build-up to an intracellular toxic level, which can result in death (Wood and Wang 1983). (3) Radionuclides, the radioactive elements, are formed by chemical elements whose atomic nuclei are unstable. As a consequence of this instability, its atoms emit subatomic particles intermittently and randomly. The imbalance is corrected by the release of excess neutrons or protons, in the form of α particles that are really helium nuclei and β particles that can be electrons or positrons. Among the radioactive elements are polonium, astatus, radon, francium, radio, actinium, thorium, protoactinium, uranium, neptunium, plutonium, americium, lawrencio, curio, berkelio, california, einsteinium, fermium, mendelevian, and nobel.

The contamination for heavy metals is caused by natural phenomena and human activities. The human activities demand the development of new chemicals, materials and enormous quantities of energy and exploit natural resources and discharge of wastewater from metal-related industry, which result in environmental pollution, mainly of water-body (Haferburg and Kothe 2010; Thakare et al. 2021). Generally, the heavy metals are present in form of soluble salts in water, that these cannot be separated by ordinary physical separation systems. Physicochemical processes such as chemical precipitation, chemical oxidation or reduction, electrochemical treatment, evaporative recovery, filtration, ion exchange and membrane technologies have been widely used to remove metallic ions from industrial wastewater (Das et al. 2008; Li et al. 2018). But these processes may be ineffective or

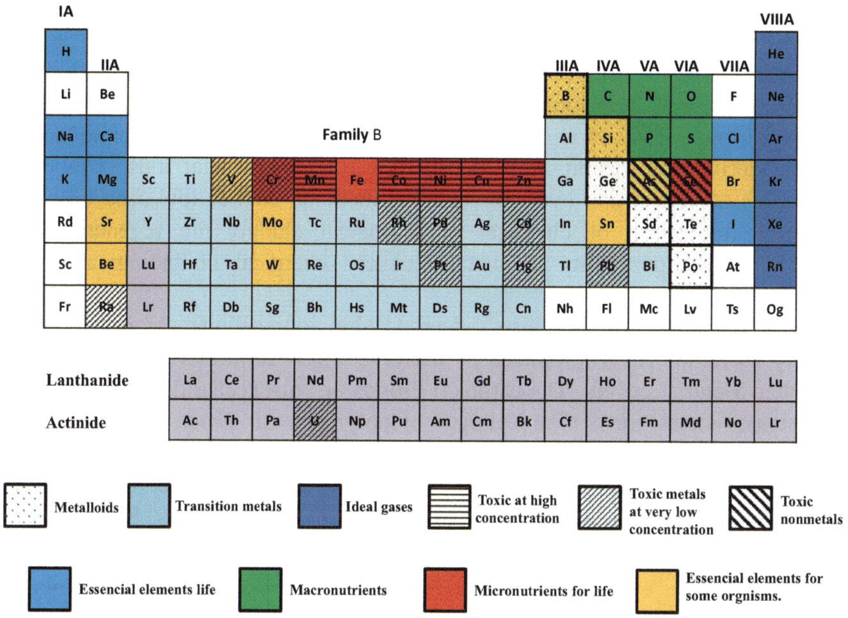

Fig. 13.1 Periodic table of the elements. This table illustrates the essential elements of life. Most of the transition metals are considered as toxic and cannot be processed by living organisms, but some are essential for life

expensive by the high energy requirement, especially when the solutions contain of 1–100 mg metal per litre (Das et al. 2008). Therefore, the technical applicability, cost-effectiveness and plant simplicity are the key factors in selecting the most suitable treatment method to remove heavy metals. However, the biotechnology offers alternative biological methods to the removal of heavy metals (Joo et al. 2015). Although the inorganic elements cannot be destroyed, microorganisms can alter their redox state. Oxidation-reduction reactions mediated by microorganisms help in the conversion of a highly toxic, soluble, and mobile species into a species less soluble and toxic, e.g. the biological reduction of Cr(VI) to Cr(III). When some metals as Cu, Zn, Ni, Cd, As, U, Np and Tc combines with the sulfide, hydroxide or carbonate anion form salts whose solubility is low and precipitate in aqueous solution.

13.3 Main Mechanisms of Biological Removal

13.3.1 Biosorption and Bioaccumulation

Biosorption is the most employed technique for the removing of metals in solution, it is a form of passive uptake of metallic ions by a sorption material, and it can be an alternative to the conventional technologies. Microbial biomass and agricultural waste are biomaterials most used (Ahluwalia and Goyal 2007; Singh and Goyal

2007). Bacteria of the genera *Bacillus* and *Streptomyces* and yeasts possess the capacity of adsorbing high amounts of metals from solution, due to the adsorption of the metallic ions by ionizable groups of the cell surface constituents, such as peptidoglycan, cellular membrane and capsule (carboxyl, amino, phosphate and hydroxyl groups) (Das et al. 2008; Ahluwalia and Goyal 2007). Macroalgae and alginate derivatives exhibit high affinity towards many metal ions. The major advantages of biosorption over conventional methods include low cost, high efficiency, minimization of chemical or biological sludge, the regeneration of biosorbents and possibility of metal recovery chemical or biological sludge (Gavrilescu 2004).

Heavy metal ions are also adsorbed by extracellular biopolymers or exopolysaccharide (EPS) produced by bacteria, which form the bacterial capsules and other bacterial covers. The carboxyl groups of polysaccharides hold back and accumulate metal ions. Extracellular biopolymers of *Enterobacter cloacae*, *Marinobacter* sp., *Klebsiella aerogenes* and *Acinetobacter* sp. have shown metallic accumulation. However, a considerable accumulation of ions of copper, lead and zinc was demonstrated on *Pseudomonas aeruginosa* cell biofilms. EPS acts as a barrier of protecting bacterial cells inside of biofilm of the toxicity of metal ions (Ianieva 2009).

The accumulation of intracellular metal ions responds to changes in cell membrane permeability. Thus, some metallic ions can enter the cell via the systems responsible for the uptake of essential elements, as in the case of *Ralstonia metallidurans*, the ions of cadmium, zinc, cobalt, nickel and manganese enter at the cell using systems of magnesium transport. Another example is chromate using sulfate transport system (Ianieva 2009; Gadd 2010). The passage of metal ions into the cell causes response detoxification of toxic metals, which is based on the expression of proteins capable of being complex with metallic ion. These metallic complexes are out from the cell, and these accumulate in cell membrane surface or in the periplasmic space or are deposited in internal vacuoles called inclusion bodies. These are called resistance mechanisms and are associated with metallothioneins and metallohistins, molecules of intracellular storage and detoxification able to bind heavy metals (Haferburg and Kothe 2010; Thakare et al. 2021). This may provide an excellent source of applicable mechanisms in environment biotechnological decontamination.

13.3.2 Precipitation

In the precipitation produced chemically stable forms of metal and only use to reducible metals. Metabolic activity of the microorganisms contributes to the removing indirect or direct of metals, e.g. metal precipitation by secreted phosphate generated from polyphosphate hydrolysis. In this context, the precipitation by phosphates and sulfides has been investigated due to the low solubility of their metal compounds. This mechanism is suggested to remove metals and actinides from wastewater. Therefore, the selective precipitation of metals with the hydrogen sulfide produced by sulfate-reducing bacteria (SRB) is the most studded (Joo et al.

2015; Li et al. 2018). The main process of removal of heavy metals followed by SRB is definitely the precipitation of metal sulfides. The generation of sulfide diminishes the acidity promoting the precipitation of metals as insoluble metal sulfides that can be easily separated. The process consists of two stages: (1) the production of H_2S by SRB and (2) the precipitation of metals by the biologically produced H_2S; it reacts with metal ions and produces insoluble metal sulfides that can easily separate from a solution.

$$\underset{(e^- \text{ donor})}{\text{Lactate}} + SO_4^{2-} \rightarrow \underset{(e^- \text{ acceptor})}{2CH_3COO^-} + HS^- + HCO_3^- \tag{13.1}$$

$$Me^{2+} + HS^- \rightarrow MeS \downarrow + H^+ \tag{13.2}$$

13.3.3 Microbial Reduction of Metallic Ions

Microbial reduction of some metals and metalloids as Cr(VI), Mn (IV), Tc(VII), U (VI) and Se(VI) has been proposed like a bioremediation strategy; particularly, their reduced forms [Cr(III), Mn (II), Tc(IV), Se(0) and U(IV) Se(0)] are insoluble and less toxic precipitates. Uranium reduction can carry out under aerobic and anaerobic conditions. The reduction of U(VI) under anaerobic conditions forms uraninite (U (IV)), which is an insoluble mineral. It oxidized to U(VI) with nitrate acting as the electron acceptor; this could provide a strategy for solubilizing and extracting microbial U(IV) precipitates from the subsurface (Finneran et al. 2002; Silver and Phung 2005). Generally, the microbial metallic reduction utilizes electron donors such as ethanol and acetate and the metallic ions as electron acceptors under anaerobic conditions, sulfate-reducing bacteria being an excellent alternative (Cabrera et al. 2006).

Finally, the using of microorganisms in the removal of metals is due to the metallic resistance capacity of each strain, which are enveloped as an evaluative characteristic to survive in a hostile habitat. But it is clear that such resistance is due to resistance mechanisms that microorganisms induce in the presence of toxic heavy metals. The main mechanisms of metal resistance studied in bacteria are related to (1) cellular components that capture metal ions for neutralizing their toxicity, (2) enzymes that modify the redox state of metals or metalloids to less toxic forms and (c) transporters of the membrane that eject harmful species outside cellular cytoplasm (Ramírez et al. 2008). Resistance mechanisms of transporters of the membrane are efflux systems, which contain proteins belonging to three families: resistance, nodulation, and cellular division (RND), cation diffusion facilitator (CDF) and P-type ATPases. In Gram-negative bacteria, both P-type ATPases and CDF proteins are predominant; proteins transport specific substrates through the plasma membrane into the periplasm. P-type ATPases predominantly transfer metal ions with high affinity for sulfhydryl groups [Cu(I)/Ag(I), Zn(II)/Cd(II)/Pb(II)], while CDF proteins specifically interact with ions of divalent metals [Zn(II), Co

(II), Ni(II), Cd(II) and Fe(II)]. The most relevant resistance mechanism in bacteria is related to the P1B-type ATPases and chemiosmotic systems (Naghma et al. 2005; Silver and Phung 2005). Genomic studies of the interrelationships to metal-induced proteome and metabolome changes allow in silico searches for genes encoding metal-responsive proteins (Haferburg and Kothe 2010; Chance et al. 2004). The proteins encoded could be either involved in metal homeostasis, thus being of interest for improving metal resistance of strains for bioremediation.

13.4 Sulfate-Reducing Bacteria

Reducing sulfates constitute a group of obligate anaerobic prokaryotes (bacteria and archaea); they present a morphological and physiological diversity. These microorganisms live in anoxic habitats and have great ecological importance in the carbon and sulfur cycles, because they mineralize the organic matter of the anaerobic environments. In marine sediments, 50% of organic matter is oxidized by sulfate reduction with an equivalent or higher yield than in an aerobic process (Barton 1995). Within this group of prokaryotes, the sulfate-reducing bacteria (SRB) are the most abundant and are widespread in natural habitats such as marine sediments, lakes and saltwater lagoons and oil fields, as well as in the gastrointestinal tracts of many animals, including humans. In environments with low sulfate levels, such as bodies of freshwater, they have relevance in the mineralization of organic matter (Muyzer and Stams 2008). Some SRB are able to survive in the presence of oxygen, but no growth has been observed (Fournier et al. 2004). Also, SRB have industrial (biocorrosion), environmental (bioremediation) and health (inflammation of the intestine) implications (Bartosch et al. 2004); consequently, the SRB have been studied extensively.

Sulfate-reducing bacteria were discovered by Martinus Willem Beijerinck in 1895 and described as the use of enrichment cultures for "sulfur fermentation". With his cultivation technique, he was able to isolate colonies surrounded by a black precipitate, ferrous sulfide. Beijerinck had isolated the first sulfate-reducing bacteria from the Dutch city canal in Delft. The morphological description of the bacterium corresponded to curved bacilli with movement, which is the reason why it was named *Spirillum desulfuricans*. Beijerinck also suggested studying other terminal electrons acceptors, besides sulfate, and studying the distribution of these bacteria in marine environments and soil (Voordouw 1995). SRB are chemolithotrophic microorganisms, capable of using sulfate as the final electron acceptor in the degradation of organic matter, a process called sulfate reduction, where hydrogen sulfide is generated. The dissimilatory reduction of sulfate is a large-scale process limited to SRB; however, they can also reduce other oxidized forms of sulfur such as sulfite and thiosulfate and other inorganic compounds as nitrite or nitrate. Some SRB are able to integrate elemental sulfur as a substrate in the respiration, and other SRB can even respire with oxygen. SRB can grow in a sulfate-dependent manner using hydrogen and a wide range of organic compounds, but polymeric compounds as polysaccharides and proteins are not typically used by them. Therefore, SRB are very versatile with respect to the electron donors and acceptors for their growth.

Sulfate-reducing bacteria most studied are *Desulfovibrio* spp., which are curved bacilli, are spiroidal bacilli and occasionally are straight bacilli, and present mobility by polar flagella. Under stress conditions, some bacteria are polymorphic, such as *D. africanus*, *D. salexigens* and *D. gigas*. Nutritionally, the cultures are easily enriched with lactate, a reducing agent for their growth and in some cases of vitamins. Lactate is partially oxidized to acetate for most of the species. Some strains are capable of growing in the presence of H_2 as a source of energy and acetate plus CO_2 or yeast extract as a source of carbon, while carbohydrates are a source unusual of carbon. Very few members carry out a fermentative metabolism. Others are moderate halophiles by their requirement of NaCl (20–30 g/L). Not only is sulfate the electron acceptor, but also the sulfite and thiosulfate and in some cases nitrate are final electron acceptors (Devereux et al. 1990; Voordouw 1995). Some bacteria have the capacity to reduce iron (III), uranium (VI), chromium (VI), pertechnetate (VII), selenite (VI) and arsenate (VI), but these reduction processes are not coupled to growth (Cabrera et al. 2006; Muyzer and Stams 2008).

13.4.1 Classification and Phylogeny

During the first six decades of the last century, the knowledge of new SRB with similar cellular characteristics but different morphologies generated serious problems in the classification of this bacterial group. In 1936, Kluyver and Niel gave the prospects for a natural system of classification of bacteria, as an attempt to devise a bacterial natural classification of their own, based primarily on morphological criteria, but attaching great weight to physiological characters. Based on the morphological and physiological criteria, the genus *Desulfovibrio* was established, and "*Spirillum desulfuricans*", discovered by Beijerinck, was the first species, but with a novel name *Desulfovibrio desulfuricans*. Later, Campbelly and Postgate proposed the genus *Desulfotomaculum* (1965). Both continued studding the Gram-negative SRB and discovered the presence of cytochrome c_3, indicated by an absorption band at 552–554 mμ, forming a "pyridine hemochromogen" of the hematoheme class, observed by other investigators too. This generated a greater understanding of Gram-negative SRB that the genus *Desulfovibrio* was improved (Postgate and Campbell 1966). For 25 years, other SRB were isolated, and other genera that appeared such as *Desufobacter*, *Desulfobulbus*, *Desulfococcus*, *Desulfonema*, *Desulfobotulus*, *Desulfobacterium*, *Desulfomonile*, and *Desulfoarculus* were described (Widdel and Pfennig 1984). Thermophilic Gram-negative bacteria able to reduce sulfate were classified inner of the genera *Thermodesulfomicrobium* and *Thermodesulfobacterium*, and thermophilic Gram-positive sulfate reducers were placed in *Desulfotomaculum*. On the other hand, two isolate sulfate reducers from anaerobic submarine hydrothermal areas were denominated as archaeal sulfate reducers, which were classified as *Archaeoglobus fulgidus* and *A. profundus*. Then, three basic cell groups of SRB were proposed:

gram-negative eubacteria, gram-positive eubacteria and archaebacteria (Barton 1995).

In the 1990s of the last century, advances in molecular biology and DNA technology, as well as the classification of organisms proposed by Woese, based on ribosomal gen analysis, had a great impact on the traditional classification, particularly of microorganisms (Woese et al. 1990). The phylogenetic analysis based on the comparative analysis of the sequences of the 16S rRNA gene establishes phylogenetic relationships among microorganisms up to a genus and species approach. Therefore, the importance generated around SRB and with the development of rRNA phylogenetic analysis gave a great step for the taxonomy and phylogeny of the sulfate-reducing prokaryotes. The phylogenetic analysis of the 16S rRNA gene showed that the sulfate- reducing prokaryotes are divided into four phylogenetic lineages: mesophilic Gram-negative SRB, Gram-positive spore-forming SRB, thermophilic sulfate-reducing archaea (SRA) and SRB (Castro et al. 2000). Gram-negative SRB are located in the Deltaproteobacteria subdivision, where there are other non-sulfate-reducing bacteria, but within this division, there are two relevant families of SRB: Desulfovibrionaceae and Desulfobacteriaceae. The family Desulfovibrionaceae is basically constituted by species of the genus *Desulfovibrio* (Voordouw 1995). The main genera of the family Desulfobacteriaceae are *Desulfobacter*, *Desulfococcus*, *Desulfobulbus*, *Desulfobacterium*, *Desulfosarcina*, *Desulfoneum* and *Desulfotulus*, among others. The genus *Desulfotomaculum* is the most representative of Gram-positive SRB, and their species form spores. The thermophilic reducing sulfate prokaryotes are divided into bacterial and archaea species (Castro et al. 2000).

In the first decade of this century, Muyzer and Stams reported that due to the discovery and isolation of new species (Muyzer and Stams 2008) and the increasing of the number of 16S rRNA gene sequences reported in the Genebank, this group of microorganisms is divided in seven phylogenetic lineages: five within the Bacteria and two within the Archaea. Most SRB belong to the class Deltaproteobacteria with approximately 23 genera, and the Gram-positive bacteria within the *Clostridia*, followed by lineages *Nitrospirae* (genus *Thermodesulfovibrio), Thermodesulfobacteria* and *Thermodesulfobiaceae* (only thermophilic sulfate reducers). The lineages to SRA belong to genus *Archaeoglobus* in the Euryarchaeota and the genera *Thermocladium* and *Caldivirga* in the Crenarchaeota.

In the next decades up to today, using of molecular technique and functional markers as *dsr* or *asr*, functional genes that codify dissimilatory sulfite reductase and anaerobic sulfite reductase, respectively, has increased the knowledge of novel sulfate/sulfite-reducing microorganisms (Jiang et al. 2009; Meyer and Kuever 2007; Wagner et al. 2005). In parallel, different studies revealed the capacity sulfate or sulfite-reducing in at least 13 additional bacterial and archaeal lineages, which at the moment were not associated with the metabolism dissimilatory of sulfate or sulfite. These include the phyla Acidobacteria, Planctomycetes, Verrucomicrobia and Armatimonadetes, among others. Besides, 8 of these 13 are candidate phyla without isolated representatives, which only represent uncultured microorganisms (Müller et al. 2015; Wörner and Pester 2019).

Currently, the application of next-generation sequencing methods of DNA in studies of metagenomic and metatranscriptomic has revealed the existence of novel sulfate reducers not yet cultivated in different environments, mainly marine habitats, and they are classified in others lineages to those already known (Anantharaman et al. 2018). Therefore, the knowledge of phylogenetic relationships of this particular microbial group allows a better understanding of their energy and nutritional demands, as well as a perception of their way of life in a given environment, and their possible biotechnological application.

13.4.2 Biotechnological Implications

Biotechnology offers an alternative to develop and to innovate methods using as tools the knowledge biochemistry and the organism's manipulation, with the goal of obtaining valuable products or improving an industrial process while maintaining the natural environment. In this sense, environmental pollution and waste treatment are mainly treated by microorganisms (Mani and Kumar 2014). In this way, SRB have great implications in the context of environmental biotechnology.

Sulfate-reducing bacteria can cause serious problems for industries, the production of sulfide, a highly reactive and corrosive compound, provoking corrosion of iron and steel, which affects mainly to oil, hydroelectric and metal-mechanic industry (Gadd 2010; Hernández-Gayoso et al. 2004). Metallic corrosion is one of the main damages causing great economic losses in pipeline systems of the petroleum industry. In particular, *D. desulfuricans*, *D. vulgaris*, *D. vietnamensis* and *D. alaskensis* play an important role in the biocorrosion of metallic surfaces due to H_2S production, which modifies the environmental pH and leads to the formation of the corrosion product FeS. SRB also promote the development of biofilms in aquatic environments, which are associated with located corrosion of the metal surfaces of oil pipelines. The production of hydrogen sulfide by SRB sours oil and gas; this reduces the quality and the cost of them (Dinh et al. 2004; Muyzer and Stams 2008: Neria-González et al. 2006). Also, SRB have health implications since some species of *Desulfovibrio* can act as opportunistic pathogens; they are associated with primary bacteremia and abdominal infections, such as abscesses and cholecystitis (Urata et al. 2008).

Due to the demand of sulfuric acid on the manufacture of fertilizer and its use in applications in oil refining, pigment production, steel treatment and non-ferrous metal extraction and manufacture of explosives, detergents, plastics and fibres, it results in the occurrence of sulfate in wastewater. The presence of sulfate has an impact on the biological treatment in the wastewaters. In nature, sulfate reducers coexist with other microorganisms, which drifts in metabolic interactions, i.e. the sulfate reducers compete with the methanogens and acetogens for common substrates, such as hydrogen and acetate, but in the absence of sulfate, sulfate reducers grow acetogenically in syntrophic with the methanogens. These microbial interactions are important in the treatment of wastewater. The heavy metals are other pollutants very common of the wastewater; its presence is a very serious threat to the environment owing to its toxicity, even at a low concentration, and bioaccumulation potential (Velea et al. 2009). Thus, SRB are an alternative method for sulfate and heavy metal removal from wastewater, as well as oxidized sulfur compounds from gas and water (Klonowska et al. 2008). Comparing the biological sulfate reduction with conventional physical and chemical methods, it is an effective technique to treat and reduce both heavy metals and sulfate from contaminated systems.

On the other hand, the toxicity of heavy metals in the microorganisms, including SRB, deactivates enzymes because these react with the functional groups, provoking the denaturation of proteins; also, there is a competition with essential cations. In both cases, they can reduce bacterial metabolic activity or cause death. The concentration of the metal in solution is an important variable, since this depends on the bacterial ability to immobilize the heavy metals. Several studies have determined the capacity of tolerance towards the heavy metals on different cultures of SRB, since their high metal tolerance and removal capabilities are particularly attractive in heavy metal removal processes. The toxic concentration of heavy metals for SRB can reach up to 100 mg/L but in a mixed culture of SRB for some metals is as follows: Zn (25–40 mg/L), Pb (75–80 mg/L), Cu (4–20 mg/L), Cd ([4–20 mg/L), Ni (10–20 mg/L) and Cr (60 mg/L) (Cabrera et al. 2006; Utgikar et al. 2002). In marine SRB, a high tolerance level (500 μM) towards the most toxic metals, Hg(II) and Cd (II), has been reported (El-Naggar 2009; Joo et al. 2015). The application of SRB as an alternate technology on removal metal offers advantages such as high metal removal at low pH, stable sludge, very low operation costs, and minimal energy consumption (Ayangbenro et al. 2018).

13.5 Removing of Heavy Metal by *Desulfovibrio*

13.5.1 *Desulfovibrio alaskensis* as a Model of Removing of Metals

The species of the genus *Desulfovibrio* are the most studied in this field (Joo et al. 2015; Qian et al. 2016). The bacterial response to heavy metals depends on the concentration and availability of metals, and on the response depends the mechanisms of action towards the metals, which include precipitation, reduced uptake, formation and sequestration of heavy metals in insoluble complexes, enzymatic oxidation or reduction to less toxic species, efflux from the cell, metabolic bypass and repair (Gadd 2010; Haferburg and Kothe 2010). Such mechanisms can be evaluated, particularly, in *D. alaskensis* strain 6SR showing high efficiency in the removal of different metallic ions as Cd, Zn, Pb and Cr (López-Pérez et al. (2015); Neria-González et al. 2011; Peña-Caballero et al. 2016).

Desulfovibrio alaskensis was isolated from a soured oil well in Prudhoe Bay, Alaska, and its morphological description corresponds to gram-negative, mobile curved bacilli with a single polar flagellum. The optimal growth conditions of the bacterium in Postgate medium C correspond at a pH 7, 2.5% (w/v) NaCl and temperature 37 °C, using lactate as the main source of carbon and energy (Postgate 1984). Therefore, it is considered as a mesophilic and moderated halophilic

Table 13.1 Minimum inhibitory concentrations of metallic ions for different species of *Desulfovibrio*

Species	Cr (VI)	Cd (II)	Zn (II)	Pb (II)	References
Desulfovibrio alaskensis strain 6SR	18	200	130	200	López-Pérez et al. (2015), Peña-Caballero et al. (2016), and Morón-Vázquez (2015)
Desulfovibrio vulgaris	5	20	20	80	Klonowska et al. (2008) and Cabrera et al. (2006)
Desulfovibrio magneticus	–	1.3	–	–	Arakaki et al. (2002)
Desulfovibrio desulfuricans	–	50	50	1.2	Li et al. (2018) and Muyzer and Stams (2008)
Desulfovibrio sp.[a]	15		15	–	Cabrera et al. (2006)

[a]The strain was evaluated with Cr(III)

bacterium (Feio et al. 2004). Otherwise, the 6SR strain was isolated from a biofilm formed within an oil pipeline from the southeast of Mexico; the phylogenetic characterization showed strong homology with *D. alaskensis* (Neria-González et al. 2006). Subsequent researches indicate that *D. alaskensis* is an SRB frequently found in biofilms developed in the oil pipelines, and these are associated to metallic biocorrosion of pipelines of the oil industry (Hernández-Gayoso et al. 2004; Neria-González et al. 2006).

Desulfovibrio alaskensis strain 6SR grows in Postgate medium C with a concentration of 3.0% NaCl, pH 7 at 45 °C. The growth of the bacterium and the generation of -SH favour the production of EPS. In a natural environment, the EPS induces the development of biofilms that allows microorganisms to survive under stress conditions; also, it offers protection to environmental changes and toxic substances. But, this situation is not very different in pure cultures because when hydrogen sulfide accumulates, the EPS is produced as a protection mechanism of the toxicity of hydrogen sulfide, since despite the sulfate-reducing nature of *D. alaskensis* strain 6SR, its growth is inhibited by the accumulation of this (Barton 1995; Neria-González et al. 2011). On the other hand, the EPS have an important role in the adsorption of heavy metals for their chelating properties, an important factor in the removal of metals in solution. The production of EPS by *D. alaskensis* strain 6SR under optimal growth conditions in Postgate medium C is estimated in 780 mg/L, highest compared with the ones in cultures of *Desulfovibrio* H0407 (239 mg/L) and *Desulfovibrio* LM1 (169 mg/L), obtaining a yield of 6.14 mg EPS/mg cellular protein (Neria-González et al. 2011).

Desulfovibrio alaskensis strain 6SR is able to grow in the presence of Cr, Cd, Pb and Zn; some of these are considered as extremely toxic to biota. The minimal inhibitory concentration (MIC) shows that the strain 6SR has a strong resistance to these metals, and in comparison with other SRB, *D. alaskensis* strain 6SR is the most resistant; see Table 13.1.

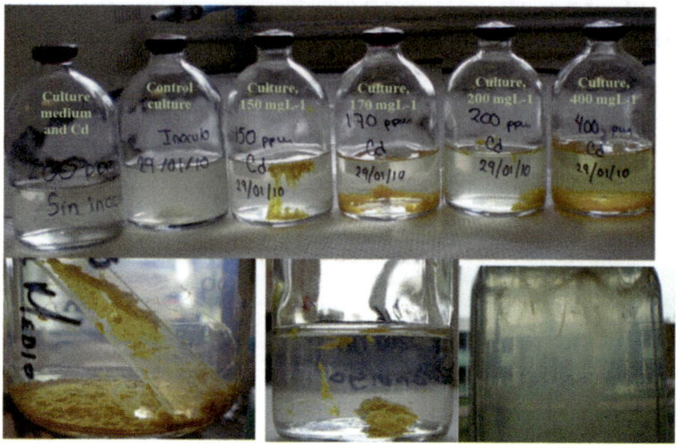

Fig. 13.2 Growth of *Desulfovibrio alaskensis* strain 6SR in Postgate medium C with a high cadmium concentration (0–400 mg/L). Images show biofilm with the precipitated cadmium sulfide and the control culture. *D. alaskensis* strain 6SR was up to 200 mg/L Cd(II)

Based on to their metallic resistance, the removal of Cr(VI), Cd(II), Pb(II) and Zn (II) has been studied in cultures of *D. alaskensis* strain 6SR. Such studies showed a removal rate of 78.3% for 18 mg/L Cr(VI), and for 150 mg/L Cd(II), Pb(II) and Zn (II), removal rates were greater than 98%. The main mechanism of removal is the precipitation of the corresponding metallic sulfides, followed by adsorption of these by the produced EPS and a slight metal accumulation at the level of the periplasmic space, as shown in the image obtained by transmission electron microscopy (Avilés Trejo and Salazar López 2014; López-Pérez et al. 2013b, 2015; Peña-Caballero 2016). The removal mechanisms for zinc, lead and cadmium is related to the production of H_2S, generated from the anaerobic respiration of the bacterium, which reacts with zinc, lead and cadmium, forming the respective metallic sulfide, allowing simple recovery methods. Specifically, for cadmium, *D. alaskensis* 6SR is able to grow under high cadmium concentrations, of the order of 200 mg/L Cd (II) without affecting the sulfate-reducing metabolism, and the production EPS is considerable (López-Pérez et al. 2013a, b); see Fig. 13.2.

Precipitation is the main mechanism of cadmium removal, cadmium sulfide being a yellow precipitate, followed by its adsorption in EPS. The highest amount of cadmium was contained in the EPS or biofilm (99.4%); see Fig. 13.3. The accumulated cadmium in the biofilm could be associated with a quick production of EPS. In the free biomass, only 0.47% Cd (0.8 mg/L) was detected. Electronic micrographs show a very low intracellular and periplasmic accumulation of cadmium in the cells (López-Pérez et al. 2015). The capacity to remove cadmium by cultures of *D. alaskensis* strain 6SR overtake the capacity of other sulfate-reducing systems, including other bacterial species; see Fig. 13.4 and Table 13.2. This confirms that the capacity of removing toxic metals by *D. alaskensis* strain 6SR is higher than some physicochemical methods and other microorganisms, which require special conditions, i.e. the microalgae require illumination conditions and large areas, and the fungi generate a large amount of biomass that demands an aeration special system and mixing. Also, species as *D. vulgaris* (11 mg/L), *D. magneticus* (1.3 mg/L), *D. desulfuricans* (56 mg/L) and *Desulfovibrio* sp. (20 mg/L) have a lower cadmium removal capacity (Fig. 13.3). Further, we

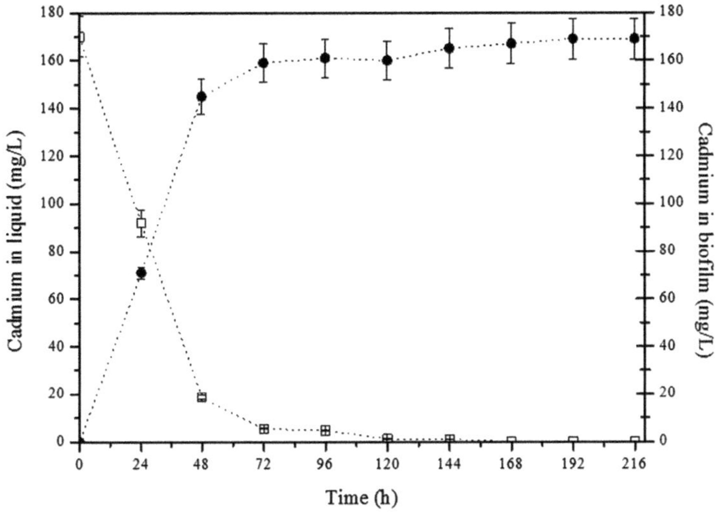

Fig. 13.3 Cadmium removal by *Desulfovibrio alaskensis* strain 6SR. Symbols: (□) concentration of cadmium in the liquid phase and (•) concentration of cadmium adsorbed in the biofilm

think that the anaerobic of culture and handling conditions of SRB could be a disadvantage in the removal metals but it is quite the opposite. Bacterial growth under anaerobic conditions produces less biomass, and its simple removal metal mechanism allows easy recovery of the metallic sulfide, and some of them have an added value in the manufacture of solar cells (López-Pérez et al. 2015; Rangel-Chávez et al. 2015).

In the case of hexavalent chromium, *D. alaskensis* strain 6SR is able to grow up to a concentration of 18 mg/L Cr(VI); see Figs. 13.5 and 13.6. However, some species as *D. vulgaris* Hildenborough have the capacity to reduce Cr(VI) to Cr(III) by the action of dehydrogenase and the cytochrome c_3, but cells are unable to use Cr (VI) like a terminal electron acceptor linked to growth (Klonowska et al. 2008; Franco et al. 2018). This has been demonstrated using a fresh culture medium with sulfate and chromium, which was inoculated with cells harvested in the exponential phase and washed. The results indicated that chromium decoupled lactate consumption from sulfate reduction, while chromium reduction is carried out. This explains that electron flow from cytochrome c_3 is a non-specific process that can be diverted from sulfate by other redox partners including Cr(VI). Then, the reduction of chromium in culture for the batch of *D. alaskensis* strain 6SR was studied under the same methodology and conditions followed by Klonowska et al. (2008). The results revealed that the reduction of Cr(VI) is due to the production of H_2S, since the sulfate-reducing process was not inhibited by chromium hexavalent, such as demonstrated to *D. vulgaris* Hildenborough; see Fig. 13.7 (Avilés-Trejo and Salazar-López 2014; Peña-Caballero et al. 2016). Therefore, *Desulfovibrio alaskensis* strain 6SR is not able to reduce Cr(VI) via enzymatic reduction.

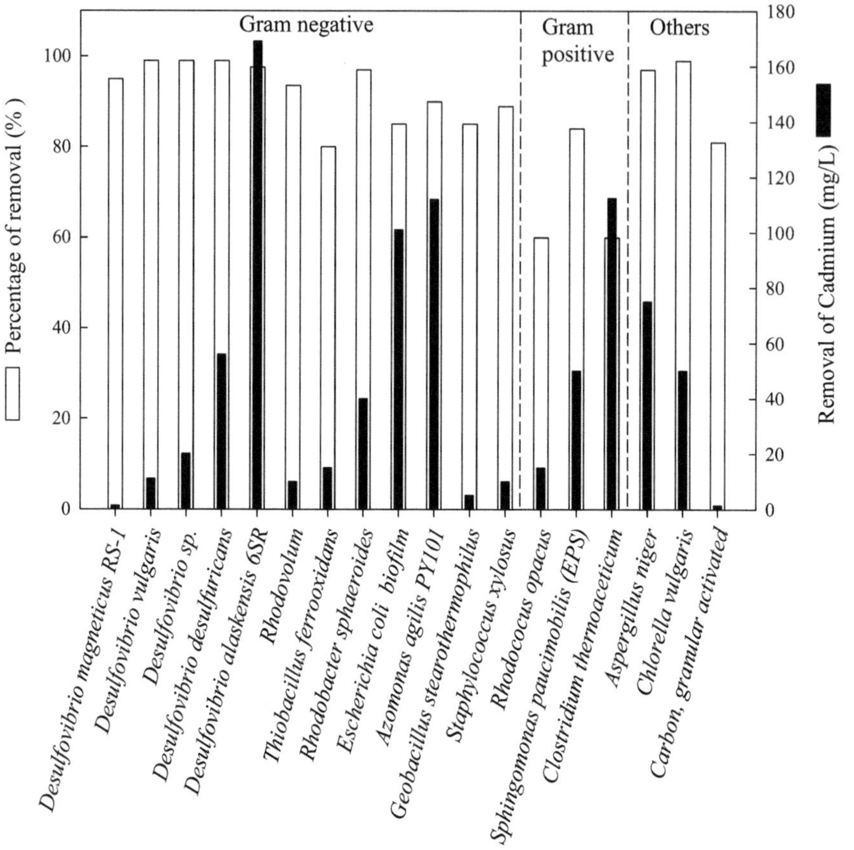

Fig. 13.4 Removal of cadmium obtained among *Desulfovibrio alaskensis* 6SR and other microorganisms. The graphic was made using data previously reported (Arakaki et al. 2002; Cunningham and Lundie 1993; Yun-guo et al. 2006; Quintelas et al. 2009; Selatnia et al. 2004; Sinha and Mukherjee 2009; Vásquez et al. 2007; Watanabe et al. 2003; White and Gadd 1998; Ziagova et al. 2007)

The chromium reduction carried out between 16 and 24 hours under the sulfate-reducing process. The measuring of chromium in liquid phase and biofilm indicated a deficit of chromium added to culture medium; see Fig. 13.8. The remaining chromium was localized in the free biomass, this was reveled by a TEM analysis when evaluate the accumulation of chromium in the cells. The micrographs obtained from *D. alaskensis* strain 6SR were not stained when chromium was present, whereas cells not exposed to the metal were stained with lead citrate. Comparison of both micrographs showed an accumulation of chromium on extern and cellular membrane and periplasmic space in the cell; see Fig. 13.9. For this reason, the chromium removal follows a precipitation by chromium sulfide and bioaccumulation in the biofilm and cellular surface.

Table 13.2 Cadmium removal by different microorganisms

Kind of microorganisms	Cd (mg/L)	Adsorbent (g)	Uptake capacity (mg/g)	Method	References
E. coli biofilm	85	6.54	13	Batch experiments – Biosorption performance	Quintelas et al. (2009)
Pseudomonas aeruginosa	11	0.26	43	Immobilized on granular activated charcoal	Sinha (2009)
Pseudomonas sp.	10	0.03	278	Biosorption	Ziagova et al. (2007)
Rhodobacter sphaeroides (S)	20	0.87	23	Batch experiments – Biosorption performance	Watanabe et al. (2003)
Rhodovulum sp. (PS88)	20	0.54	37	Batch experiments – Biosorption performance	Watanabe et al. (2003)
Rhodococcus opacus	9	2.31	4	Batch-scale basis	Vásquez et al. (2007)
Staphylococcus xylosus	9	0.04	250	Biosorption	Ziagova (2007)
Streptomyces rimosus	210	3.32	63	Batch experiments – Biosorption performance	Selatnia et al. (2004)
Clostridium thernoaceticum	110	–	–	Precipitation of cadmium	Cunningham and Lundie (1993)
Mixed-culture SRB (*Desulfotomaculum*)	7	–	–	Batch experiments	White and Gadd (1998)
Desulfovibrio magneticus RS-1	1.3	–	–	Batch experiments	Arakaki et al. (2002)
Desulfovibrio desulfuricans	100	–		Batch experiments	Joo et al. (2015)
Desulfovibrio alaskensis 6SR	170	2.03[a]	83	Batch and continue experiments	López-Pérez et al. (2013b, 2015)
Aspergillus niger	75	4.84	16	Agitation rate on the biosorption	Yun-guo et al. (2006)
Chlorella vulgaris	150	1.73	87	Batch stirred system	Aksu and Donmez (2006)
Spirulina platensis	2	0.03	48	Batch experiments	Murugesan et al. (2008)
Dead algae, marine	252	3.15	80	Batch experiments	Herrero et al. (2006)

[a]Based on the dry mass of biofilm

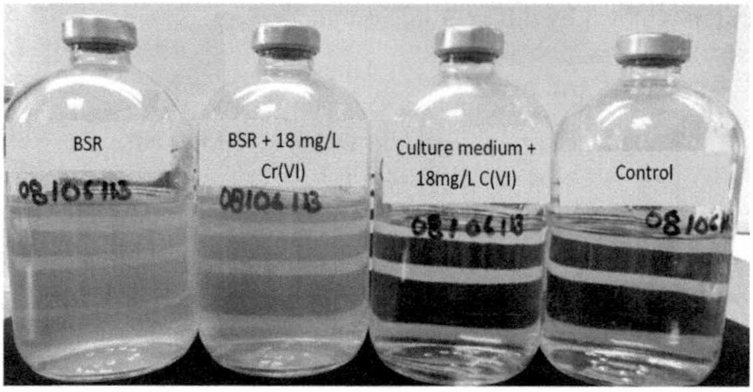

Fig. 13.5 Growth of *Desulfovibrio alaskensis* strain 6SR. Postgate medium C supplemented with 2.5% NaCl, pH 7, and 18 mg/L Cr(VI). A control with culture medium and Cr (VI) indicates that the culture medium does not affect chromium speciation or precipitate it. Postgate medium C is the negative control

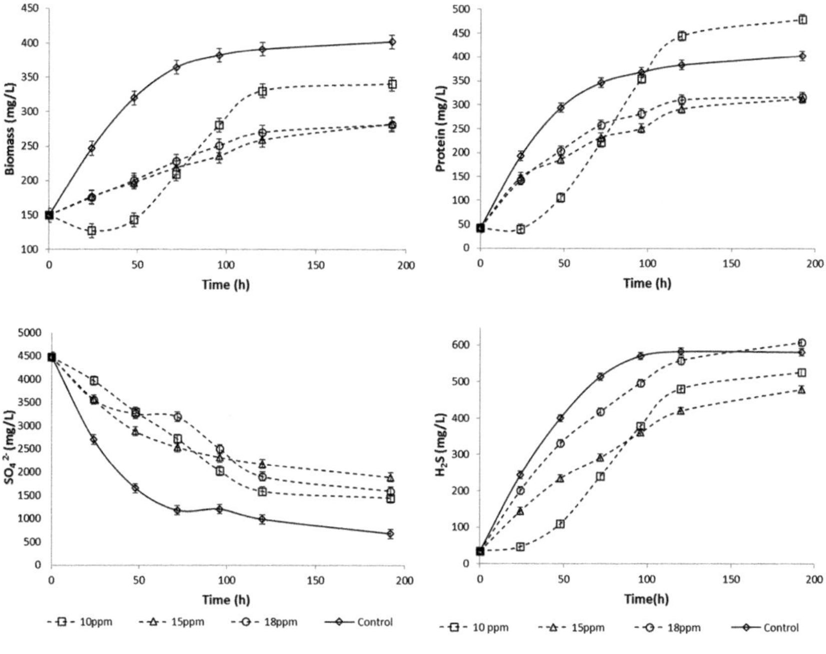

Fig. 13.6 Growth of *Desulfovibrio alaskensis* strain 6SR in Postgate medium C with 0, 10, 15 and 18 mg/L Cr(VI)

Due to the industrial interest of chromium, *D. alaskensis* is employed for hexavalent chromium and sulfate reduction. Chromium is reduced by hydrogen sulfide produced by the sulfate reduction in independent reactors; this alternative process was studied in the electrochemical treatment of Cr(VI) from wastewater (Peña-Caballero et al. 2016), as shown in Fig. 13.10.

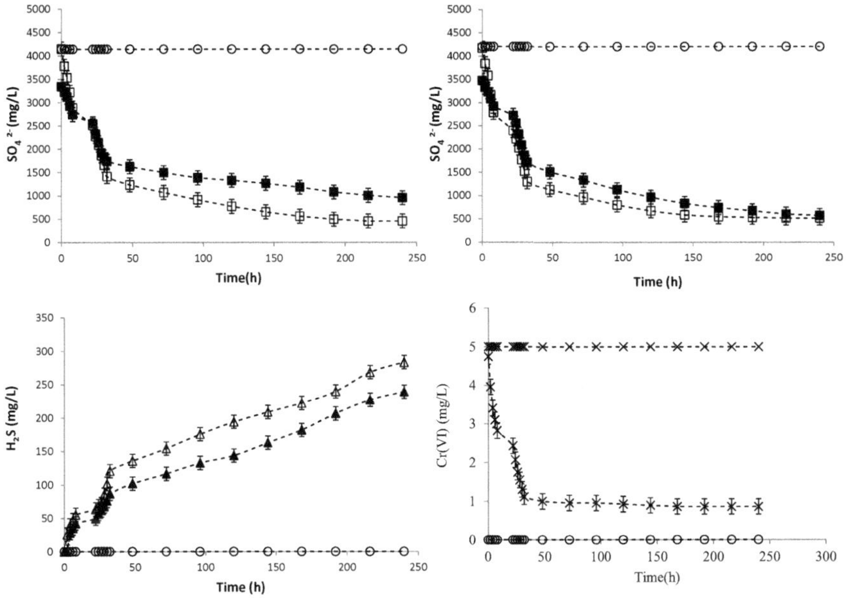

Fig. 13.7 Growth of *Desulfovibrio alaskensis* strain 6SR in Postgate medium C and 5 mg/L Cr (VI), under the same conditions of growth followed by Klonowska et al. 2008. The sulfate reduction process was not inhibited. Full markers represent the culture with Cr(VI). Empty markers represent control culture

Metallic removal studies in *D. alaskensis* strain 6SR showed that the bacterium is a strong candidate for the development of metallic removal processes of contaminated wastewater. *Desulfovibrio alaskensis* strain 6SR showed that there is no inhibition of biomass growth or sulfate reduction in the presence of high concentrations of Cd(II), Pb (II) and Zn(II) (150 mg/L), contrary to other SRB capable of growing in concentrations below 100 mg/L of metal ions. But even more, the metallic removal observed in *D. alaskensis* strain 6SR cultures was almost 100%, in the particular case of cadmium; the removal was very similar to concentrations higher than 150 mg/L Cd(II), (López-Pérez et al. 2015), the removal efficiency of 5 and 18 mg/L Cr(VI) was around 80% to both concentrations (Fig. 13.8b). Studies on the harmful effects of these heavy metals have revealed that cadmium and divalent metals are able to replace the essential ions to cell, provoking a build-up on cellular structure and block functional groups of macromolecules, promoting damage to the integrity of the cellular membrane and inactivation of cellular enzymes (Hossain et al. 2012); likewise, zinc at 40 mg/L initial concentration and above inhibits the cellular growth (Sani et al. 2001). However, the molecular studies showed that *D. alaskensis* 6SR has a resistance mechanism relationship to ATPase type P1B system, which responds to cadmium, lead and zinc (Morón-Vázquez 2015). The main mechanisms of metal resistance studied in Gram-negative bacteria are related to the transporters of the membrane

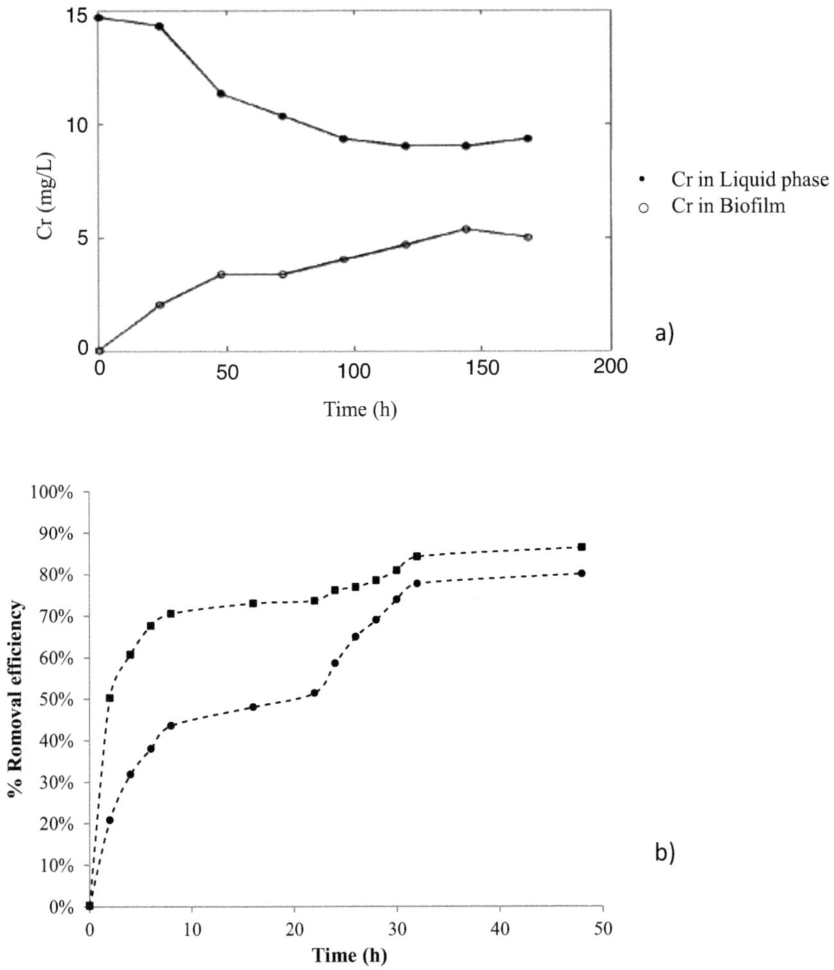

Fig. 13.8 Chromium removal. (**a**) Balance for chromium in liquid phase and biofilm (EPS). Chromium was assayed by atomic absorption spectrophotometry (AAS) (Atomic Absorption Spectrometer Specter AA-20 plus, Varian). (**b**) Efficiency of chromium removal: • 18 mg/Cr and ◆ 5 mg/L

that throw out harmful species of cellular cytoplasm such as ATPases type P1B (Nies 1999; Ramírez et al. 2008). Besides, the most relevant resistance mechanism in bacteria is related to the P1B-type ATPases and chemiosmotic systems. Metallomic specific researches have showed the interrelationships of metal-induced proteome and metabolome changes (Lara-Chavero et al. 2018; Metallomics 2012). The increase in genomic sequencing opens the knowledge of the genes that are related to the mechanisms of metallic resistance. Furthermore, the influence of metal ions on gene expression is of great interest in understanding metal resistance in bacteria. Some genes as *cad*A encode for such resistance mechanisms, which are

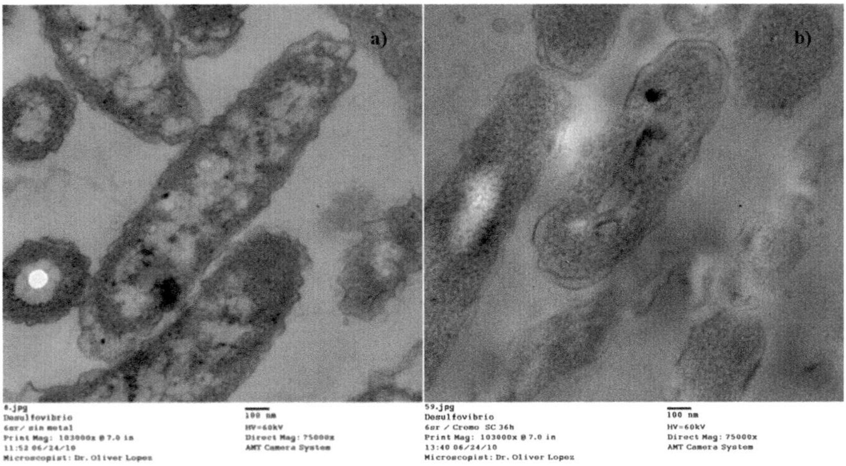

Fig. 13.9 Transmission electron micrographs of *Desulfovibrio alaskensis* strain 6SR. (**a**) Cells of *D. alaskensis* strain 6SR stained with lead acetate. (**b**) *D. alaskensis* strain 6SR grown in the presence of 18 mg/L Cr(VI) (72 h incubation)

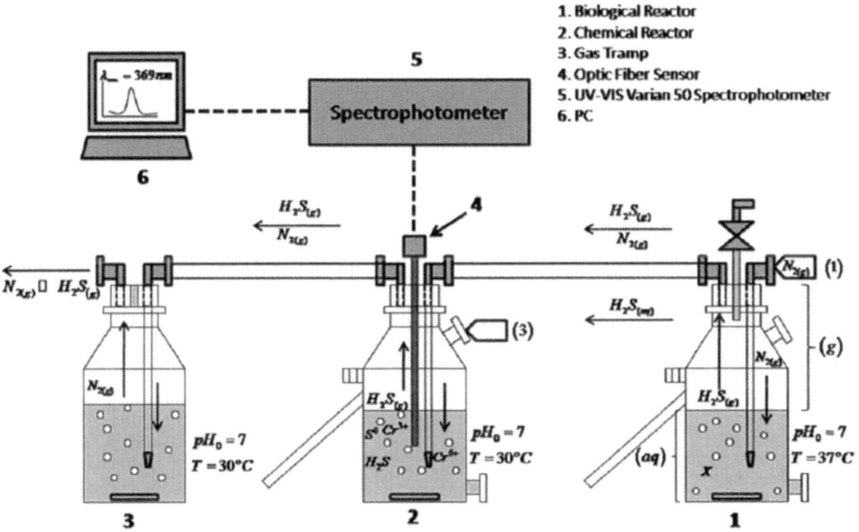

Fig. 13.10 Schematic diagram of the chromium removal from wastewater by biological sulfide. (Reported by Peña-Caballero et al. 2016)

induced by cadmium and zinc, and *chr*A gene is induced by hexavalent chromium (Naghma et al. 2005; Ding et al. 2005). The molecular analysis with respect to cadmium and chromium resistance mechanisms demonstrated the presence of *cad*A and *chr*A genes, when the partial sequences were analysed phylogenetically; see Figs. 13.11 and 13.12. Both genes are induced by Cd, Zn, Pb and Cr; the codified

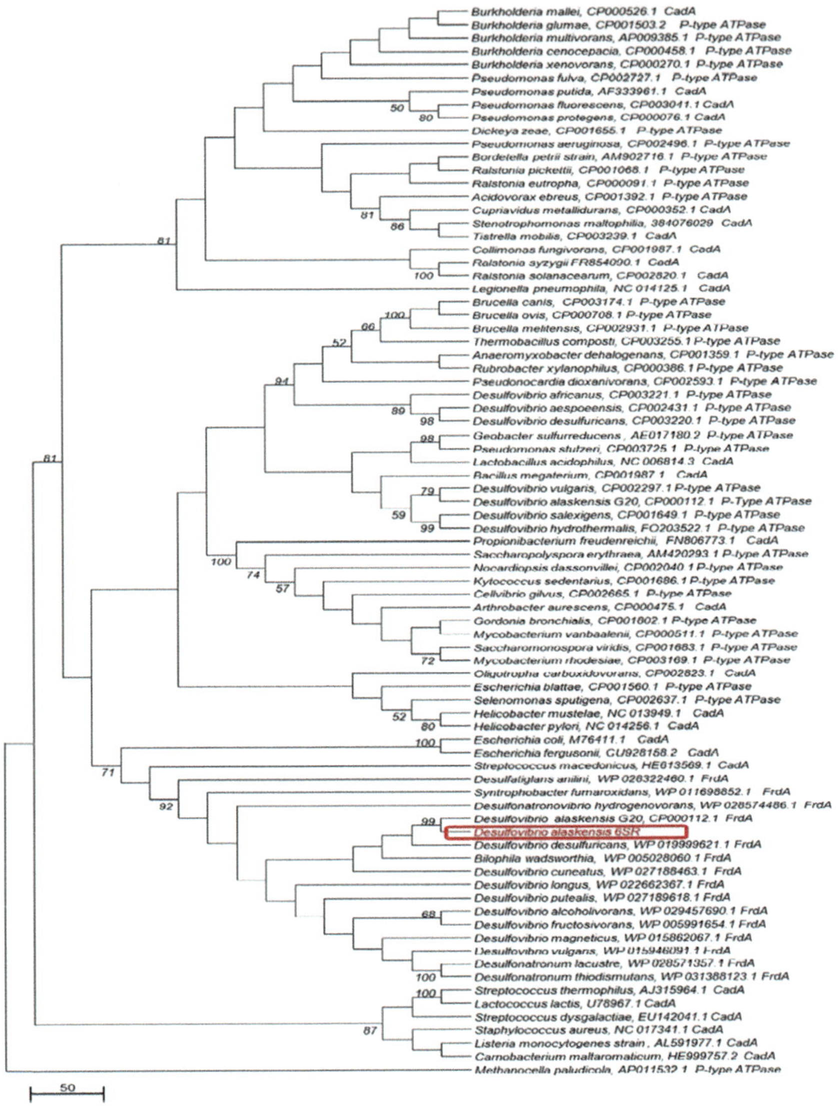

Fig. 13.11 Phylogenetic tree for the sequence obtained of *cad*A gene. The analysis was carried out with 550 aligned amino acids with related sequence to species that exhibit resistance to cadmium. The tree was built using a maximum likelihood algorithm. The scale bar represents 50 nucleotide substitutions per 100 amino acid. Bootstrap values, expressed as a percentage of 1000 repetitions, are shown at branching points. Only values greater than 50% are displayed

protein by these genes is involucre to abate the oxidative stress provoked by heavy metal nonessentials. The presence of thioredoxin codified by *chr*A gene has been reported in enzymatic chromium reduction processes. The capacity to reduce U

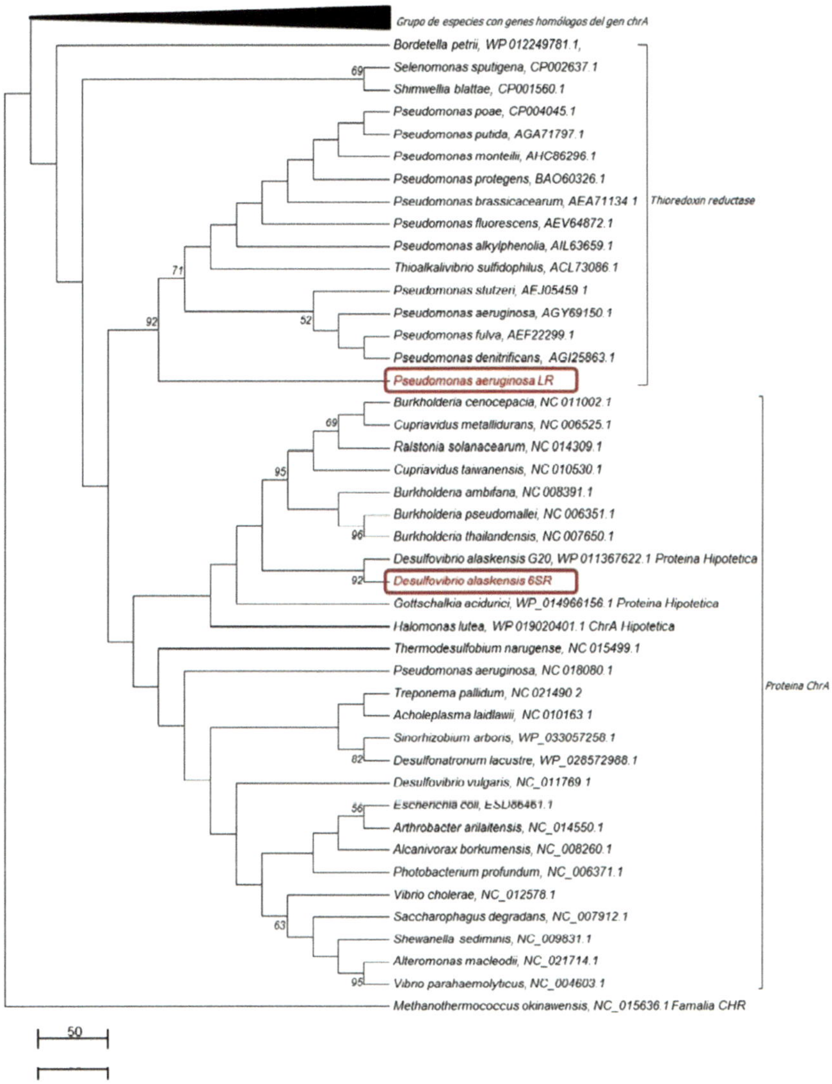

Fig. 13.12 Phylogenetic tree for the sequence obtained of *chr*A gene. The analysis was carried out with 550 aligned amino acids with related sequence to species that exhibit resistance to cadmium. The tree was built using a maximum likelihood algorithm. The scale bar represents 50 nucleotide substitutions per 100 amino acid. Bootstrap values, expressed as a percentage of 1000 repetitions, are shown at branching points. Only values greater than 50% are displayed

(VI) and Cr(VI) of *D. desulfuricans* G20 is related to the action of the thioredoxin reductase and NADPH (Li and Lee 2009). This suggests that the strain 6SR could have a resistance mechanism involving enzymatic reduction of Cr(VI), like a response to the oxidative stress generated by metallic ion.

References

Ahluwalia SS, Goyal D (2007) Microbial and plant derived biomass for removal of heavy metals from wastewater. Bioresour Technol 98:2243–2257. https://doi.org/10.1016/j.biortech.2005.12.006

Aksu Z, Donmez G (2006) Binary biosorption of cadmium (II) and nickel (II) onto dried *Chlorella vulgaris*: co-ion effect on mono-component isotherm parameters. Process Biochem 41:860–868. https://doi.org/10.1016/j.procbio.2005.10.025

Anantharaman K, Hausmann B, Jungbluth SP et al (2018) Expanded diversity of microbial groups that shape the dissimilatory sulfur cycle. ISME J 12:1715–1728. https://doi.org/10.1038/s41396-018-0078-0

Arakaki A, Takeyama H, Tanaka T, Matsunaga T (2002) Cadmium recovery by a sulfate-reducing magnetotactic bacterium, *Desulfovibrio magneticus* RS-1, using magnetic separation. Appl Biochem Biotechnol 98–100:833–840. https://doi.org/10.1385/abab:98-100:1-9:833

Avilés Trejo M, Salazar López M d C (2014) Reduction of Cr (VI) to Cr (III) via cellular, using Desulfovibrio alaskensis 6SR as a biological model. Tesis Ingeniería Bioquímica, Tecnológico de Estudios Superiores de Ecatepec. (in Spanish)

Ayangbenro AS, Olanrewaju OS, Babalola OO (2018) Sulfate-reducing bacteria as an effective tool for sustainable acid mine bioremediation. Front Microbiol 9:1986. https://doi.org/10.3389/fmicb.2018.01986

Barton L (1995) Biotechnology handbooks. Sulfate-reducing bacteria. Springer

Bartosch S, Fite A, Macfarlene GT, McMurdo MET (2004) Characterization of bacterial community in faces in healthy elderly volunteers and hospitalized elderly patients by using Real-Time PCR and effects of antibiotic treatment on the fecal microbiota. Appl Environ Microbiol 70:3575–3581. https://doi.org/10.1128/AEM.70.6.3575-3581.2004

Cabrera G, Perez R, Gomez JM, Abalos A, Cantero D (2006) Toxic effects of dissolved heavy metals on *Desulfovibrio vulgaris* and *Desulfovibrio* sp. strains. J Hazard Mater 135:40–46. https://doi.org/10.1016/j.jhazmat.2005.11.058

Campbell LL, Postgate JR (1965) Classification of the spore-forming sulfate reducing bacteria. Bacteriol Rev 29:359–363

Castro HF, Williams NH, Ogram A (2000) Phylogeny of sulfate-reducing bacteria. FEMS Microbiol Ecol 31:1–9. https://doi.org/10.1111/j.1574-6941.2000.tb00665.x

Chance MR, Fiser A, Sali A, Pieper U, Eswar N, Xu G, Fajardo JE, Radhakannan T, Marinkovic N (2004) High-throughput computational and experimental techniques in structural genomics. Genome Res 14:2145–2154. https://doi.org/10.1101/gr.2537904

Cunningham DP, Lundie LL Jr (1993) Precipitation of cadmium by *Clostridium thermoaceticum*. Appl Environ Microbiol 59:7–14

Das N, Vimala R, Kartika P (2008) Biosorption of heavy metals – an overview. Indian J Biotechnol 7:159–169

Devereux R, He S-H, Doyle C et al (1990) Diversity and origin of Desulfovibrio species: phylogenetic definition of a family. J Bacteriol 172:3609–3619. https://doi.org/10.1128/jb.172.7.3609-3619.1990

Ding H, Harrison K, Lu J (2005) Thioredoxin reductase system mediates iron binding in IscA and iron delivery for the iron-sulfur cluster assembly in IscU. J Biol Chem 280:30432–30437. https://doi.org/10.1074/jbc.M504638200

Dinh HY, Kuever J, MuBmann M, Hassel AW, Stratmann M, Widdel F (2004) Iron corrosion by novel anaerobic microorganisms. Nature 427:829–832. https://doi.org/10.1038/nature02321

Duffus JH (2002) "Heavy metals" – a meaningless term? (IUPAC technical report). Pure Appl Chem 74:793–807. https://doi.org/10.1351/pac200274050793

El-Naggar MM (2009) Heavy metals removal by marine sulfate reducing bacteria. Egypt J Exp Biol (Bot) 5:215–219. http://www.egyptseb.org

Feio MJ, Zinkevich V, Beech IB, Llobet-Brossa E, Eaton P et al (2004) *Desulfovibrio alaskensis* sp. nov., a sulphate-reducing bacterium from a soured oil reservoir. Int J Syst Evol Microbiol 54:1747–1752. https://doi.org/10.1099/ijs.0.63118-0

Finneran KT, Anderson RT, Nevin KP, Lovley DR (2002) Bioremediation of uranium-contaminated aquifers with microbial U(VI) reduction. Soil Sediment Contam 11:339–357. https://doi.org/10.1080/20025891106781

Fournier M, Dermoun ZM-C, Dolla A (2004) A new function of the *Desulfovibrio vulgaris* Hildenborough [Fe] hydrogenase in the protection against oxidative stress. J Biol Chem 279:1787–1793. https://doi.org/10.1074/jbc.M307965200

Franco LC, Steinbeisser S, Zane GM, Wall JD, Fields MW (2018) Cr(VI) reduction and physiological toxicity are impacted by resource ratio in *Desulfovibrio vulgaris*. Appl Microbiol Biotechnol 102:2839–2850. https://doi.org/10.1007/s00253-017-8724-4

Gadd GM (2010) Metals, minerals and microbes: geomicrobiology and bioremediation. Microbiology 156:609–643. https://doi.org/10.1099/mic.0.037143-0

Gavrilescu M (2004) Removal of heavy metals from the environment by biosorption. Eng Life Sci 4:219–232. https://doi.org/10.1002/elsc.200420026

Haferburg G, Kothe E (2010) Metallomics: lessons for metalliferous soil remediation. Appl Microbiol Biotechnol 87:1271–1280. https://doi.org/10.1007/s00253-010-2695-z

Hernández-Gayoso MJ, Zavala-Olivares G, Ruiz-Ordaz N et al (2004) Microbial consortium influence upon steel corrosion rate, using polarization resistance and electrochemical noise techniques. Electrochim Acta 49:4295–4302. https://doi.org/10.1016/j.electacta.2004.03.038

Herrero R, Cordero B, Lodeiro P, Rey-Castro C, Sastre de Vicente ME (2006) Interactions of cadmium(II) and protons with dead biomass of marine algae *Fucus* sp. Mar Chem 99:106–116. https://doi.org/10.1016/j.marchem.2005.05.009

Hossain MA, Piyatida P, Teixeira da Silva JA, Fujita M (2012) Molecular mechanism of heavy metal toxicity and tolerance in plants: central role of glutathione in detoxification of reactive oxygen species and methylglyoxal and in heavy metal chelation. J Bot 872875:1–37. https://doi.org/10.1155/2012/87287

Ianieva OD (2009) Mechanisms of bacteria resistance to heavy metals. Mikrobiol Z 71:54–65. (Article in Russian)

Jiang L, Zheng Y, Peng X et al (2009) Vertical distribution and diversity of sulfate-reducing prokaryotes in the Pearl River estuarine sediments, Southern China. FEMS Microbiol Ecol 70:249–262. https://doi.org/10.1111/j.1574-6941.2009.00758.x

Joo JO, Choi J-H, Kim IH, Kim YK, Oh BK (2015) Effective bioremediation of Cadmium (II), Nickel (II), and Chromium (VI) in a marine environment by using *Desulfovibrio desulfuricans*. Biotechnol Bioprocess Eng 20:937–941. https://doi.org/10.1007/s12257-015-0287-6

Klonowska A, Clark ME, Thieman SB, Giles BJ, Wall JD, Fields MW (2008) Hexavalent chromium reduction in *Desulfovibrio vulgaris* Hildenborough causes transitory inhibition of sulfate reduction and cell growth. Appl Microbiol Biotechnol 78(6):1007–1016. https://doi.org/10.1007/s00253-008-1381-x

Kluyver AJ, van Niel CB (1936) Prospects for a natural system of classification of bacteria. Zentr Bakteriol Parasitenk Abt II 94:369–403

Krishna AK, Govil PK (2007) Soil contamination due to heavy metals from an industrial area of Surat, Gujarat, Western India. Environ Monit Assess 124:263–275. https://doi.org/10.1007/s10661-006-9224-7

Lara-Chavero ZZ, Neria-González MI, Aguilar-López R (2018) Chromium-resistant *Pseudomonas stutzeri* LRO and its possible resistance mechanism. In: X biochemical engineering international congress. J Bioeng Biomed Res, pp 55–57. XVI biomedicine and molecular biotechnology scientific meetings proceedings, Merida, Yucatán, México

Li X, Lee KR (2009) Thioredoxin is involved in U (VI) and Cr (VI) reduction in *Desulfovibrio desulfuricans* G20. J Bacteriol 191:4924–4933. https://doi.org/10.1128/JB.00197-09

Li X, Lan SM, Zhu ZP, Zhang C et al (2018) The bioenergetics mechanisms and applications of sulfate-reducing bacteria in remediation of pollutants in drainage: a review. Ecotoxicol Environ Saf 30:162–170. https://doi.org/10.1016/j.ecoenv.2018.04.025

López-Pérez PA, Neria-González MI, Aguilar López R (2013a) Cadmium concentration stabilization in a continuous sulfate reducing bioreactor via sulfide concentration control. Chem Pap 67:326. https://doi.org/10.2478/s11696-012-0274-8

López-Pérez PA, Neria-González MI, Flores-Cotera LB, Aguilar-López R (2013b) A mathematical model for cadmium removal using a sulfate reducing bacterium: *Desulfovibrio alaskensis* 6SR. Int J Environ Res 7:501–512. https://doi.org/10.22059/IJER.2013.630

López-Pérez PA, Aguilar-López R, Neria-González MI (2015) Cadmium removal at high concentration in aqueous medium: mediated by *Desulfovibrio alaskensis*. Int J Environ Sci Technol 12:1975–1986. https://doi.org/10.1007/s13762-014-0601-4

López-Pérez PA, Neria González MI, Pérez García MR, Aguilar López R (2016) Concentrations monitoring via software sensor for bioreactors under model parametric uncertainty: application to cadmium removal in an anaerobic process. Alex Eng J 55:1893–1902. https://doi.org/10.1016/j.aej.2016.03.013

Mani D, Kumar K (2014) Biotechnological advances in bioremediation of heavy metals contaminated ecosystems: an overview with special reference to phytoremediation. Int. J. Environ Sci Technol 11:843–872. https://doi.org/10.1007/s13762-013-0299-8

Metallomics: whence and whither (2012) Whence: the inorganic side of life. (Editorial) Metallomics 4: 1017–1019. https://doi.org/10.1039/C2MT90041F

Meyer B, Kuever J (2007) Molecular analysis of the diversity of sulfate-reducing and sulfur-oxidizing prokaryotes in the environment, using the *apr*A as functional marker gene. Appl Environ Microbiol 73:7664–7679. https://doi.org/10.1128/AEM.01272-07

Morón-Vázquez MA (2015) Identification of metal resistance genes in the sulfate reducing bacteria *Desulfovibrio alaskensis* 6SR. Tesis Maestría en Ciencias, Tecnológico de Estudios Superiores de Ecatepec. (in Spanish)

Müller AL, Kjeldsen KU, Rattei T, Pester M, Loy A (2015) Phylogenetic and environmental diversity of DsrAB-type dissimilatory (bi)sulfite reductases. ISME J 9:1152–1165. https://doi.org/10.1038/ismej.2014.208

Murugesan AG, Maheswari S, Bagirath G (2008) Biosorption of cadmium by live and immobilized cells of *Spirulina platensis*. Int J Environ Res 2:307–312. https://doi.org/10.22059/IJER.2010.209

Muyzer G, Stams AJM (2008) The ecology and biotechnology of sulphate-reducing bacteria. Review. Nat Rev Microbiol 6:441–454. https://doi.org/10.1038/nrmicro1892

Naghma N, Hilary K, Nuzhat A, Geoffrey M (2005) Cadmium accumulation and DNA homology with metal resistance genes in sulfate-reducing bacteria. Appl Environ Microbiol 71:4610–4618. https://doi.org/10.1128/AEM.71-4618.2005

Neria-González I, Wang ET, Ramírez F, Romero JM, Hernández-Rodríguez C (2006) Characterization of bacterial community associated to biofilms of corroded oil pipelines from the Southeast of Mexico. Anaerobe 12:122–133. https://doi.org/10.1016/j.anaerobe.2006.02.001

Neria-González MI, Figueroa-Estrada JC, Cruz-Diaz MR, Aguilar-López R (2011) Adaptive smooth observer design for state estimation 4 in *Desulfovibrio alaskensis* 6SR cultures. Rev Mex Ing Quím 10:137–146

Nies DH (1999) Microbial heavy-metal resistance. Appl Microbiol Biotechnol 51:730–750. https://doi.org/10.1007/s002530051457

Peña-Caballero V, Aguilar-López R, López-Pérez PA, Neria-González MI (2016) Reduction of Cr (VI) utilizing biogenic sulfide: an experimental and mathematical modeling approach. Desalin Water Treat 57:13056–13065. https://doi.org/10.1080/19443994.2015.1055811

Postgate JR (1984) The sulphate reducing bacteria, 2nd edn. Cambridge University Press, New York

Postgate JR, Campbell IL (1966) Classification of *Desulfovibrio* species, the nonsporulating sulfate-reducing bacteria. Bacteriol Rev 30:732–738

Qian J, Wei L, Liu R et al (2016) An exploratory study on the pathways of Cr (VI) reduction in sulfate-reducing up-flow anaerobic sludge bed (UASB) reactor. Sci Rep 6:23694. https://doi.org/10.1038/srep23694

Quintelas C, Rocha Z, Silva B et al (2009) Biosorptive performance of an *Escherichia coli* biofilm supported on zeolite NaY for the removal of Cr(VI), Cd(II), Fe(III) and Ni(II). Chem Eng J 152:110–115. https://doi.org/10.1016/j.cej.2009.03.039

Ramírez M, Díaz C, Vargas E, Riveros H, Campos J, Cervantes C (2008) Mechanisms of bacterial resistance to chromium compounds. Biometals 21:321–332. https://doi.org/10.1007/s10534-007-9121-8

Rangel-Chávez RG, Neria-González MI, Márquez-Herrera A et al (2015) Synthesis of CdS nanocrystals by employing the by-products of the anaerobic respiratory process of *Desulfovibrio alaskensis* 6SR bacteria. J Nanomater 2015:Article ID 60397. https://doi.org/10.1155/2015/260397

Sani RK, Peyton BM, Brown LT (2001) Copper-induced inhibition of growth of *Desulfovibrio desulfuricans* G20: assessment of its toxicity and correlation with those of zinc and lead. Appl Environ Microbiol 67:4765–4772. https://doi.org/10.1128/AEM.67.10.4765-4772.2001

Selatnia MZ, Bakhti A, Madani L, Kertous Y, Mansouri K (2004) Biosorption of Cd^{2+} from aqueous solution by a NaOH-treated bacterial dead *Streptomyces rimosus* biomass. Hydrometallurgy 75:11–24. https://doi.org/10.1016/j.hydromet.2004.06.005

Silver S, Phung LT (2005) A bacterial view of the periodic table: genes and proteins for toxic inorganic ions. J Ind Microbiol Biotechnol 32:587. https://doi.org/10.1007/s10295-005-0019-6

Singh SA, Goyal D (2007) Microbial and plant derived biomass for removal of heavy metals from wastewater. Bioresour Technol 98:2243–2257. https://doi.org/10.1016/j.biortech.2005.12.006

Sinha S, Mukherjee SK (2009) *Pseudomonas aeruginosa* KUCD1, a possible candidate for cadmium bioremediation. Braz J Microbiol 40:655–662. https://doi.org/10.1590/S1517-838220090003000030

Thakare M, Sarma H, Datar S, Roy A, Pawar P, Gupta K, Pandit S, Prasad R (2021) Understanding the holistic approach to plant-microbe remediation technologies for removing heavy metals and radionuclides from soil. Curr Res Biotechnol. https://doi.org/10.1016/j.crbiot.2021.02.004

Urata T, Kikuchi M, Yusuke TH, Kiyoko Y et al (2008) Bacteremia caused by *Desulfovibrio fairfieldensis*. J Infect Chemother 14:368–370. https://doi.org/10.1007/s10156-008-0629-9

Utgikar VP, Harmon SM, Chaudhary N et al (2002) Inhibition of sulfate-reducing bacteria by metal sulfide formation in bioremediation of acid mine drainage. Environ Toxicol 17:40–48. https://doi.org/10.1002/tox.10031

Vásquez TGP, Botero AEC, de Mesquita LMS, Torem ML (2007) Biosorptive removal of Cd and Zn from liquid streams with a *Rhodococcus opacus* strain. Miner Eng 20:939–944. https://doi.org/10.1016/j.mineng.2007.03.014

Velea T, Gherghe L, Predica V, Krebs R (2009) Heavy metal contamination in the vicinity of an industrial area near Bucharest. Environ Sci Pollut Res 16:S27–S32. https://doi.org/10.1007/s11356-008-0073-5

Voordouw G (1995) The genus *Desulfovibrio*: the centennial. Appl Environ Microbiol 61:2813–2819

Wagner M, Loy A, Klein M et al (2005) Genes for identification of sulfate-reducing prokaryotes. Methods Enzymol 397:469–489. https://doi.org/10.1016/S0076-6879(05)97029-8

Watanabe M, Kawahara K, Sasaki KN (2003) Noparatnaraporn, biosorption of cadmium ions using a photosynthetic bacterium, *Rhodobacter sphaeroides S* and a marine photosynthetic bacterium, *Rhodovulum* sp. and their biosorption kinetics. J Biosci Bioeng 95:374–378. https://doi.org/10.1016/81389-1723(03)80070-1

White C, Gadd GM (1998) Accumulation and effects of cadmium on sulphate-reducing bacterial biofilms. Microbiology 144:1407–1415. https://doi.org/10.1099/00221287-144-5-1407

Widdel F, Pfennig N (1984) Section 7. Dissimilatory sulfate- or sulfur reducing bacteria. In: Krieg NR, Holt JG (eds) Bergey's manual of systematic bacteriology, vol 1. Williams & Wilkins, Baltimore, pp 663–679

Woese C, Kandlert O, Wheelis M (1990) Towards a natural system of organisms: proposal for the domains Archaea, Bacteria, and Eucarya. Proc Natl Acad Sci 87:4576–4579. https://doi.org/10.1073/pnas.87.12.4576

Wood JM, Wang H-K (1983) Microbial resistance to heavy metals some microorganisms have developed "strategies" for combating effects of toxic inorganics, and several may prove useful for their removal from wastewater. Environ Sci Technol 17:12. https://doi.org/10.1021/es00118a002

Wörner S, Pester M (2019) The active sulfate-reducing microbial community in littoral sediment of oligotrophic lake constance. Front Microbiol 10:1–12. https://doi.org/10.3389/fmicb.2019.00247

Yun-guo L, Ting F, Guang-ming Z, Xin L, Qing T, Fe Y, Ming Z, Wei-hua X, Yu-e H (2006) Removal of cadmium and zinc ions fi-om aqueous solution by living *Aspergillus niger*. Trans Nonferrous Met Soc China 16:681–686. https://doi.org/10.1016/S1003-6326(06)60121-0

Ziagova Z, Dimitriadis G, Aslanidou D et al (2007) Comparative study of Cd(II) and Cr (VI) biosorption on *Staphylococcus xylosus* and *Pseudomonas* sp. in single and binary mixtures. Bioresour Technol 98:2859–2865. https://doi.org/10.1016/j.biortech.2006.09.043

Biological Implications of Dioxins/Furans Bioaccumulation in Ecosystems

14

Jose Gnanaleela Aswin Jeno, Ravichandran Rathna, and Ekambaram Nakkeeran

Abstract

Dioxins and furans are a class of persistent, bioaccumulative and toxic organic pollutants in the environment that have garnered attention within the scientific community. This chapter encompasses the bioaccumulation of dioxins/furans in the ecosystem. Further, information on the source and exposure of dioxins/furans is discussed. In addition, the presence of dioxins/furans in the global environment and its regulatory responses are also presented. Persistent dioxins/furans undergo bioaccumulative and biomagnification in the food chain and led to metabolic transformations in human. A detailed compilation of health effects in humans is also elaborated.

Keywords

Bioaccumulation · Biomagnification · Dioxins/furans · Global environment · Health effects

14.1 Introduction

The continuous expansion of civilizations around the globe has resulted in several scientific developments on one aspect, whereas the other side causes deleterious environmental damages. Rapid urbanization, deforestation, population explosion, industrialization and pollution accelerate the accumulation of persistent organic pollutants in the environment. Persistent organic pollutants (POP) are a peculiar type of pollutants that have garnered huge attention within the scientific community

J. G. A. Jeno · R. Rathna · E. Nakkeeran (✉)
Department of Biotechnology, Sri Venkateswara College of Engineering (Autonomous), Sriperumbudur Tk, Tamil Nadu, India
e-mail: nakkeeran@svce.ac.in

R. Prasad (ed.), *Environmental Pollution and Remediation*, Environmental and Microbial Biotechnology, https://doi.org/10.1007/978-981-15-5499-5_14

owing to their persistency and toxic accumulation in the ecosystem. This class of persistent compounds include polychlorinated biphenyls (PCBs), polychlorinated dibenzo-p-dioxins (PCDDs), polychlorinated dibenzofurans (PCDFs), benzo(a)-pyrene (BaP), hexachlorobenzene (HCB), polychlorinated naphthalene, octachlorostyrene, 1,1,1-trichloro-2,2-diethane, alkyl-lead pesticides such as mirex, dieldrin, aldrin, chlordane and toxaphene. Of these POPs, PCDDs, PCDFs and PCBs (dioxin-like compounds) are the three prime categories of dioxins and furans that have gained importance in the scientific community.

PCDDs/PCDFs and PCBs are the most dangerous anthropogenic toxicants since these chemicals have a widely reported environmental damage history over the past few decades and also highlighted as "the most toxic man-made chemical" ever made. PCDDs and PCDFs are two classes of planar polycyclic halogenated aromatic hydrocarbons with a total of 410 different compounds, only 30 of which are considered to be significant congeners. The characteristic of the congeners is certain to depend on the nature, position and number of halogen atoms bound to the aromatic carbon rings; one such classic example is 2,3,7,8-tetrachlorodibenzo-p-dioxin (2,3,7,8-TCDD), chlorine substituted in the 2, 3, 7, and 8 positions of the aromatic benzene ring, which is the most toxic congeners. 2,3,7,8-TCDDs are known for its lowest lethal dose median (LD_{50}) among the dioxin family in male guinea pigs which is nearly 1 μg/kg of body. Dioxins/furans were not produced or created intentionally for market value but were the by-products of several industrial processes and combustion techniques. An example of such a phenomenon is the burning of plastic which contains chlorinated phenolic compounds that release dioxins and are termed to be accidental contaminants. Due to the accumulation of members of dioxins family in the fat tissues and thereby entering the food chain, efforts are being made to minimize the exposure of such chemicals into the environment.

Since the nineteenth century, there has been an incidence of dioxin and furan pollution from industries. German industrial workers were overexposed to 2,3,7,8-TCDD or dioxin that resulted in chloracne, oozing skin lesions. During the Vietnam War (1965–1971), United States (US) military sprayed herbicide dubbed Agent Orange as a defoliant, which was contaminated with 2,3,7,8-TCDD and showed serious health effects. Dioxin pollution and its ill effects came to public attention after the explosion of Industrie Chimiche Meda Societa Anonima (ICMESA) chemical factory in Seveso, Italy, in 1976. Boda et al. (2018) demonstrated the associations of dioxins in breast milk and cord blood sex hormones in Vietnamese newborn infants residing near the former US Air Force base. TCDD has the highest degree of toxicity rated 1 (reference value), while others are less toxic. In 2005, the World Health Organization (WHO) re-evaluated the toxic equivalency factor (TEF) values defined on the 1998 WHO-TEF value for human risk assessment (Van den Berg et al. 2006). According to Ren and Zheng (2009), the differences between dioxin and furan estimation with each unit should not exceed 28%. Reduction in levels of dietary dioxins/furans in food sources between 1998 and 2018 was observed in the general population of Spain (González et al. 2018).

This chapter encompassed the current status of dioxin/furan accumulation and its biological implication in the ecosystem. In addition, detailed information on the sources and exposure risk of dioxins/furans are explained. This chapter also discusses the presence of dioxins and furans in the global environment and related regulatory measures. A detailed compilation of bioaccumulation and biomagnification of dioxins/furans in the food chain and its adverse side effects in human is also elaborated. For brevity, PCDDs/PCDFs are termed as dioxins/furans unless specified in the chapter.

14.2 Nature of Dioxins and Furans

Dioxins/furans are volatile to an extent, slightly soluble in water and readily soluble in lipids. Their lipid solubility aids directly to cross the cell membranes and accumulates in the fat tissues of several organisms. These compounds have high toxic properties causing severe abnormalities, developmental problems and interference with regulatory hormonal functions. These are non-polar, poorly water-soluble, lipophilic compounds generally resistant to chemical breakdown through an acid-base reaction, redox and hydrolysis. The nature of dioxins and furans is summarized in Table 14.1. These physicochemical properties of dioxins and furans favour their long-range transport and have been detected in arctic organisms. In fact, PCDDs/PCDFs are a serious threat to public health due to their persistence in the environment as illustrated by the half-life of 2,3,7,8-TCDD in the soil of 10–12 years (Mukerjee 1998).

14.3 Sources and Exposure of Dioxins/Furans

Dioxins/furans are formed as the result of undesirable by-products from anthropogenic activities including certain industrial applications and process. Figure 14.1 illustrates the various sources of dioxins and furans responsible for its release into the environment.

Table 14.1 Nature of dioxins and furans

Nature	Dioxins	Furans
Chemical formula	$C_4H_4O_2$	C_4H_4O
Physical state and appearance	Solid	Liquid
Molecular weight (g/mol)	321.96	68.07
Melting range (°F)	581–617	−122.8
Boiling range (°F)	789.8–834.8	89.6
Solubility in water (g/L)	Insoluble	Very slightly soluble in cold water
Vapour pressure (at 20 °C)	7.4×10^{-8} mm Hg	493 mm Hg
Specific gravity (water = 1)	1.827	0.94

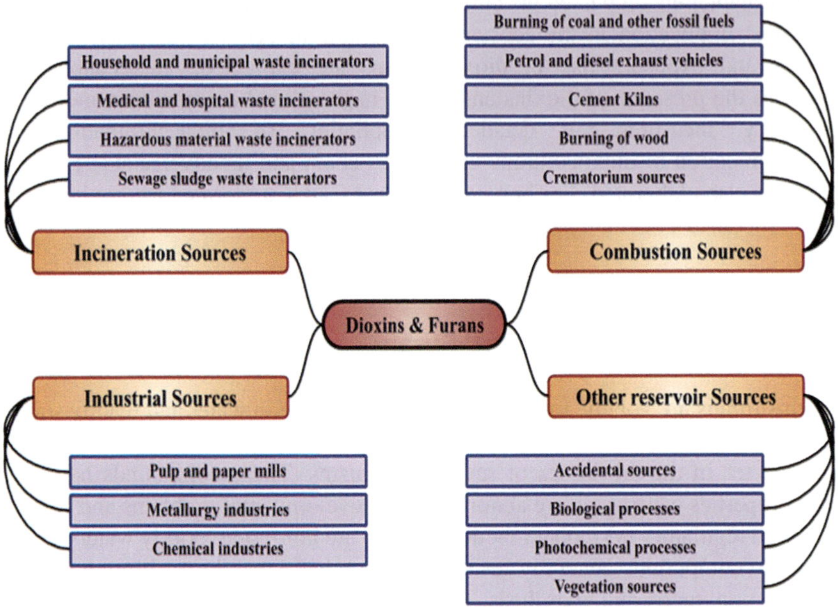

Fig. 14.1 Sources of dioxins and furans

14.3.1 Incineration

Incineration is a thermal waste reduction method that involves combustion process and one of the major anthropogenic sources for the emission dioxins and furans. However, incineration of contaminated wastes in rotary kilns is usual practice in many industrialized countries, as they are best suited to accept solid, pasty or liquid waste. Municipal waste incineration, hazardous waste incineration, medical waste incineration, light-fraction shredder waste incineration, waste wood and waste biomass incineration and sewage sludge incineration are the important waste incineration categories in a nation. The major incineration categories that emit dioxins and furans are municipal waste incineration, hazardous waste incineration and medical waste incineration, while the rest are least studied. Medical waste incinerators in China reported emitting 1.18 kg I-TEQ of PCDDs/PCDFs in the year 2004 (11.5% total PCDD/PCDF emission). In order to minimize the emission of PCDDs/PCDFs from medical waste incinerators, an advanced gas cleaning and stable combustion system should be enforced (Li et al. 2010). Municipal solid waste incinerators contribute to the emission of dioxins, furans and derivatives of dioxins through stack flue gases and ashes (Wang et al. 2010). Therefore, appropriate control strategies should be installed to regulate the emission of dioxins and furans.

In India, municipal waste incineration is absent and around 94% of the municipal solid waste simply dumped, while the rest is either composted (4%) or recycled

(2%). In case of hazardous waste incineration in India, there are around 120 hazardous waste incinerators and 11 hazardous waste landfills and releases 5505.6 g TEQ/a in air and residue. The release of dioxins and furans from medical waste incineration in India is 272.37 g TEQ/a in air and residue. An amount of 14.6 g TEQ/a dioxins and furans is released in Portugal from waste incineration (Quina et al. 2011). In the United States, the annual emission in the year 2000 is less than 12 g TEQ/a dioxins and furans which are released from municipal solid waste (Psomopoulos et al. 2009). The release of dioxins and furans from incineration sources in the developed nation is less than the developing nation due to abatement strategies and sophisticated technologies to limit the release of toxic compounds into the environment. Yang et al. (2019) investigated the effects of small-scale domestic waste incinerators in the emission of dioxins and dioxin-like compounds which has a major impact on the environment and human health.

14.3.2 Combustion

Combustion is an exothermic redox reaction of a substance in the presence of oxygen. Incomplete combustion process release PCDDs/PCDFs and related congeners to the surrounding. The generation of dioxins and furans from combustion sources is lower than the incineration sources. The main categories of combustion sources are cement kilns, firewood combustion, diesel vehicles, crematoria, coal-fired utilities and uncontrolled fires in landfill sites. Karstensen (2008) evaluated more than 2000 PCDDs/PCDFs from cement kiln and reported an emission level of 0.1 ng I TEQ/m^3 indicating the well-managed and operated modern cement kilns co-processing waste in Australia. Conesa et al. (2016) assessed the emission of persistent dioxin and furan pollutants using a long-term monitoring system from a cement plant. Further, an average of 0.23 ng I-TEQ (equivalent)/ton clinker of PCDDs/PCDFs was found to emit from the plant. Rivera-Austrui et al. (2014) demonstrated the installation of alternate fuels with clinker kiln stack significantly reduced the emission of harmful persistent organic compounds.

In India, total dioxin released to air from cement kilns, lime production, brick manufacturing units and glass production emits 83.64, 7.12, 50.54 and 0.03 g TEQ/a, respectively. Crematoria used for the purposes of burial found to form and release PCDDs/PCDFs from the cadavers containing chlorinated compounds and precursors. From crematoria, a total of 0.56 and 0.14 g TEQ/a dioxins released to air and residue, respectively, in India. The annual release of PCDDs/PCDFs from disposal/landfill in India to water was 1.22 g TEQ/a, to the product 70.16 g TEQ/a and to residue 3.44 g TEQ/a. Vehicle population in developing countries like India is increasing tremendously; the emission of PCDDs/PCDFs into the air by transportation is 9.57 g TEQ/a, especially for diesel engines. During the uncontrolled combustion processes like forest fires, open burning of wastes and accidental fires emitted 45.48 g TEQ/a in air and land in India.

14.3.3 Industrial

Processes in the chemical industry, pulp and paper industry, dry cleaning, brominated flame-retardants manufacturing, metallurgical processes, reactivation processes of granular carbon, etc. are some examples of industrial sources of dioxins and furans. These toxic chemicals are generated as a by-product during the manufacture of chlorinated pesticides, phenoxy herbicides and polyvinyl chloride synthesis. In pulp and paper industries, chemical bleaching of pulp with chlorine produced PCDDs/PCDFs and its congeners like 2,3,7,8-TCDD, 2,3,7,8-tetrachlorodibenzofuran (2,3,7,8-TCDF) and 1,2,7,8-tetrachlorodibenzofuran (1,2,7,8-TCDF). Wang et al. (2012) reported the average annual emission of concentration PCDDs/PCDFs from Chinese non-wood pulp and paper mills to be 27.59 g I-TEQ. Further, 2,3,7,8-TCDF highly formed during hypochlorite bleaching stage and removed during alkaline digestion, extraction and hydrogen peroxide bleaching. The recycling process of TV plastic materials resulted in the release of a higher amount (ten-fold increase) of dioxins/furans when compared with the original material (Ortuño et al. 2015).

In India, the major dioxin congener from pulp and paper mills was 1,3,6,8-TCDD found before effluent treatment, while 2,5-diphenyl furan congener was found in the secondary sludge of effluent treatment plant. Metallurgy industry is the major contributor to dioxin and furan emission in the atmosphere. Table 14.2 depicts the dioxin and furan emission from metallurgical industries in India. Rezaei et al. (2013) investigated and evaluated the release of dioxins and furans in the Iranian mining and ore processing industries and stated an annual release of 120 g TEQ/a of PCDDs/PCDFs.

Table 14.2 Dioxin and furan emission from metallurgical industries in India

	Release (g TEQ/a)		
Metallurgical industries	Air	Residue	Total
Iron ore sintering	260.06	0.16	260.22
Coke production	24.99	–	24.99
Iron and steel production	116.45	601.42	717.87
Foundries	27.63	1.3	28.93
Secondary copper production	32.12	404.68	436.8
Secondary aluminium production	17.5	200	217.5
Secondary lead production	4.53	2.8	7.33
Secondary zinc production	40.20	–	40.20
Brass and bronze production	0.138	–	0.138
Thermal non-ferrous metal production	16.06	–	16.06

14.3.4 Reservoir

Dioxins and furans are emitted into the atmosphere through redistribution from the accumulated sites such as vegetation, soils and landfills (Kulkarni et al. 2008). During natural calamities like volcanic eruptions, forest fires emit dioxins and furans accidentally into the environment. Over the past decades, the incidental release of dioxins and furans from the contaminated biota is increasing. In 2013, the Food Safety and Inspection Service (FSIS), an agency of the US Department of Agriculture (USDA) determined that the chicken imported to the United States from Chile was exposed to dioxins. In 2007, the European Commission detected the traces of dioxins (a contaminant of pentachlorophenol pesticide – no longer used) in guar gum from India. Pemberthy et al. (2016) showed and evaluated the presence of PCDDs/PCDFs and dioxin-like polychlorinated biphenyls in soybean, olive oil, fish oil, butter and shrimp used for human consumption in Colombia. Many such historical proofs manifest many sources of high-level release of dioxins in the food and threatened the food safety. Researchers reported the release of dioxins and furans by the microbial activity of chlorinated compounds under suitable environmental conditions. Composting, sewage decomposition and photolysis of chlorinated compounds are some typical examples of such sources. Ueno et al. (2005) elucidated the contamination of the marine environment with PCDDs/PCDFs from highly industrialized countries of Northeast Asia.

14.4 Dioxins/Furans in the Global Environment

Over the past few decades, several historical incidents have taken place which led to dioxin and furan contamination, especially during 1920–1970. Industrial activities involving organic/inorganic chloride are the prime source for the release of dioxins and furans to the environment. Dioxin/furan pollution is directly or indirectly linked to the food chain of humans, such as animal and poultry feed and dairy food products. One of the biggest food recalls in the world has happened in Ireland in 2008 because of chemical contamination in pork meat. The government of Ireland had to recall tons of pork and other pork-related products when it found out traces of dioxin over 200 times of the safe limit. Further, the food safety authorities traced back the incident to the contamination of animal feed with dioxins. In 1999, illegally disposed dioxin- and furan-contaminated industrial oils led to the contamination of poultry feed in Belgium. Subsequently, very high levels of dioxins and furans were found in poultry products.

Dioxin and furan contamination is observed in air, soil, water, plants and animals. The global government began to investigate the presence of dioxins and furans contamination in soil, air and water over the course of time to interrogate the severity of damage caused to the environment and humans. Taiwanese elementary school children were reported to be exposed to different persistent organic pollutants through classroom dust at lower than the threshold risk values (Gou et al. 2016). Bruce-Vanderpuije et al. (2019) reported about the epidemiological studies of

Table 14.3 Dioxin contamination levels in soil, water and air in different European countries

Sampling source	Region	Sampling year	PCDD/PCDF (pg TEQ/g)
Soil	Netherlands	1991	2.2–16
	Austria	1989–1993	1.6–14
	Germany	1992	1–5
	Belgium	1992	2
	France	1999	0.02–1
Water	Germany	1994	1–20
	Netherlands	1980–1990	8–21
Air	England	1991–1996	0–810
	Germany	1992	70–350
	Belgium	1993	86–129
	Netherlands	1991–1993	4–99
	Austria	1996	26–314

37 participants from Ghana who had no known accidental or occupational exposure to dioxins/furans but exhibited levels of ~5.3 pg WHO-TEQ/g. However, the data retraining to the levels of dioxins and furans are limited in environmental matrices. Table 14.3 represents the region-wise split up of dioxin and furan contamination in the soil, water and air.

14.4.1 Dioxin and Furan Contamination in Soil

One of the major sources of contamination of dioxins and furans is the soil and soil sediments. In soil, the contamination generally occurs through the deposition of atmospheric particulate contaminants. Interestingly, migration of dioxins and furans into the deep soil is very minimal. Further, ~90% of dioxins and furans are found within the shallow topsoil (10 cm). Similarly, the contaminants once settled in the soil will no longer be evaporated or lost due to decay. The soil contamination levels in Germany and Switzerland were estimated to be around 100 and 200 pg TEQ/g, respectively. Open dumping sites in developing countries like Cambodia, Philippines, India and Vietnam reported a release of 0.12–35 mg TEQ/year of dioxins and furans into the environment (Nguyen et al. 2003). The correlation of population density with soil dioxin-TEQs in Taiwan was reported by Hsu et al. (2018).

14.4.2 Dioxin and Furan Contamination in Water

Dioxins and furans are poorly soluble in water; however, it undergoes surface adsorption with the suspended solid particles which settle at the bottom of the stream substrate. Contamination in water is mainly due to the sewage sludge, paper and pulp industries, fertilizer industries, etc. Water contaminated with dioxins and furans is the primary source for bioaccumulation and biomagnification. Kumar et al. (2001) reported the presence of dioxins and furans in Ganges River dolphins and fish in India. Recently, surface sediments of rivers in Russia (the Neva River and the Eastern Gulf of Finland), Uruguay (Uruguay River) and Argentina (Gualeguaychú River) reported the presence of dioxins and furans (Metelkova et al. 2019; Matta 2018). The concentration levels of dioxins/furans in egg tissues of two species of salmon in Ontario lake are significantly higher than that of the levels determined by Canadian Council of Ministers of the Environment (Garner and Pagano 2019).

14.4.3 Dioxin and Furan Contamination in the Air

Incineration processes, combustion processes and several natural methods contribute to the dioxin and furan content in the atmosphere of all countries. Until 1990, the standard investigation was not carried out to estimate the toxic chemicals such as dioxins and furans in the atmosphere. Thus, it led to an outbreak of the maximum amount of chemicals in the atmosphere during that decade. Rahman et al. (2014) reported dioxin and furan contamination in ambient air at urban residential areas in Seoul, Korea. An industrial park in China (Tibet–Qinghai Plateau) reported being the source for dioxin/furan and dioxin-like compound pollution in the atmosphere (Hu et al. 2019).

14.5 Regulations

Foreseen, the dangerous effects of dioxins and furans, United States Environmental Protection Agency (USEPA) jumped into action to frame a strong regulatory framework called the "Dioxin Reassessment" and was reviewed by EPA's science advisory board. A strong regulation was framed to minimize the fatal impact of dioxins and furans in the environment as well as humans. The ultimate aim of the regulatory activities is to reduce the release of dioxins/furans into the environment from incinerator waste. Dioxins/furans are not produced in an intentional way for commercial activities or for any other industrial purpose, the key problem for developing guidelines and regulatory activities. Human exposure to dioxins/furans is through the food chain, and thus, regulatory actions must ensure ways to minimize the impact on the human food chain. Maximum Achievable Control Technology (MACT) was introduced by the USEPA in September 1999 for several contamination sources such as cement kilns, incinerators and combustion sources, to adopt this technology within their premises. Table 14.4 represents the Interim MACT Standards for Combustors in the United States (EPA 1999). POPs that are regulated through MACT were lead, cadmium, arsenic, beryllium, carbon monoxide, chromium, hydrocarbons and majorly dioxins and furans. Table 14.5 represents the international environmental standards for dioxins and furans from different sources.

Table 14.4 Interim MACT Standards for Combustors in the United States

Types of pollutants	MACT standards for different sources		
	Domestic and lightweight kilns	Cement kilns	Incinerators
Dioxins and furans (ng TEQ/dscm)	0.20	0.20 or 0.40	0.20
Hydrocarbons or carbon monoxide (ppmv dry)	–	10 or 100	–
Hydrocarbons (ppmv dry)	–	50	–
Hydrocarbons or carbon monoxide (ppmv dry)	20 or 100	20 or 100	10 or 100
Opacity (%)	–	20	–
Particulate matter (mg/dscm)	57	0.15	34
HCl/Cl_2 (ppmv dry)	600	86	21
Low volatile metals (μg/dscm)	110	54	97
Semivolatile metals (μg/dscm)	43	180	120
Mercury (μg/dscm)	120	120	45

Table 14.5 International environmental standards for dioxins and furans from different sources

Country	Sources of dioxins/furans	Standards for dioxin/furan emission
United States	Municipal solid waste incinerator (>35 Ton/day)	13 ng/m^3 (Total mass) about 0.1–0.3 ng I-TEQ/m^3
	Hazardous waste incinerator	0.2 ng I-TEQ/m^3
European Union	Municipal waste and hazardous waste incinerators	0.1 ng I-TEQ/m^3
Japan	Municipal solid waste incinerator (>4 Ton/h)	0.1 ng I-TEQ/m^3
	Municipal solid waste incinerator (2–4 Ton/h)	1.0 ng I-TEQ/m^3
	Municipal solid waste incinerator (<2 Ton/h)	5.0 ng I-TEQ/m^3
Hong Kong	Municipal solid waste incinerator	13 ng/m^3 (Total mass)
	Chemical waste treatment plants	0.1 ng I-TEQ/m^3
Canada	All incinerators	0.08 g I-TEQ/m^3.

14.6 Bioaccumulation of Dioxins/Furans

Human exposure to both dioxin and furan compounds is mainly through the food chain. The dairy products and aquatic seafood are the major sources of bioaccumulated products that happen to intrude the food chain of humans. The hydrophobic nature of dioxins and furans facilitates to form small particulate matter in both aquatic and other water resources. Dioxins and furans released as unintentional by-products that have an immediate effect on land and water thereby affect the ecosystem and human health.

14.6.1 Terrestrial Ecosystem

Out of all the sources, food exposure becomes the predominant source of toxic elements to humans in the terrestrial ecosystem. The main reason behind this scenario is that the concentration of abundant nutrients in each level of the food pyramid is different. Since both dioxins and furans are highly lipid-soluble, stable and less biodegradable, they reach the topmost part of the pyramid instantly. In terrestrial ecosystems, one of the main concerns is dairy products. Dioxins and furans emitted by incinerators, fly-ash machines, cement kilns, etc. are found to be less than 0.1–100 ng/m^3. Particles of this size are easily deposited on the plants, grass, weed and even soil because of certain climatic conditions such as precipitation and atmospheric currents. In vitro bioaccessibility of dioxins and furans from soil matrix within the human gastrointestinal tract ranged from 34.3% to 62.1% (Roberts et al. 2019). Cockerham and Young (1983) reported that the beach mice (*Peromyscus polionotus*) showed levels of 540–1300 ng/Kg TCDD in liver tissues. Young et al. (1976) studied the contamination levels in a reptile, *Cnemidophorus sexlineatus*, and estimated the levels of dioxins and furans to be around 360 ng/Kg in the visceral mass and 370 ng/Kg in the trunk of the reptile. Hryhorczuk et al. (1981) studied the contamination of dioxins and furans in grazing horse cattle and found levels over 165 ng/Kg in fat tissues and 57 ng/Kg in liver tissues. Cow grazing in an outside environment is directly exposed to such toxic chemicals contaminating their feed such as grass and hay. An amount of 1–50 pg TEQ/g (level of contamination) of dioxins/furans is directly ingested into the system of a higher organism. After ingestion, the gastrointestinal absorption of dioxins and furans by the animal takes place, thus resulting in a very low absorption rate. Generally, half-life times of dioxins and furans in cows are very low because lactation is the only route of excretion; moreover, the distribution of dioxins and furans throughout the body of cows is very high since it directly deposits on the adipose tissues of cows. Piskorska-Pliszczynska et al. (2017) reported that sugar beet pellets are the main source for dioxin and furans contamination in cow's milk. Dioxin and furan contamination has the ability to increase the chromosomal abnormalities and fragility in sheep (Perucatti et al. 2006), cow (Di Meo et al. 2010) and buffalo (Genualdo et al. 2012). Meat outlets in France revealed a detectable range of dioxins and furans to be around <1 pg TEQ/g fat. Similarly, in Austria, the dioxin and furan level goes up to 69 pg TEQ/g fat in cow's milk where the location is very close to a copper smelter. When compared with the bovine fat, the concentration of dioxins and furans in human fat is nearly 10 times higher. In vivo studies of dioxin and furan contamination on buffaloes reported 1,2,3,7,8-PCDD, 2,3,4,7,8-PCDF, 2,3,7,8-TCDD and 1,2,3,6,7,8-HCDD as the major contributors of TEF (Chirollo et al. 2018). Schuhmacher et al. (2019) compared the dioxin TEQs of NATO and WHO with that of the human breast milk TEQ which was significantly independent and moving in a decreased trend. Breast milk of humans has been observed to contain nearly 10–30 pg TEQ/g. From 2008 to 2011, dioxin and furan concentration in human breast milk was found to decrease in Canada (Rawn et al. 2017). Further, dioxin and furan contamination in human breast milk is higher in Europe than in North

Table 14.6 Dioxins and furans in human adipose tissue (ng/Kg wet tissue) in various countries in 1989

	Countries						
Congener	United States	Japan	Germany	Sweden	Canada	North Vietnam	South Vietnam
2,3,7,8-TCDD	7.2	9.0	150.0	3.0	6.4	<2	28.0
1,2,3,7,8-PCDD	11.1	15.0	19.2	10.0	10.0	<2	15.0
1,2,3,6,7,8-HCDD	96	70.0	77.0	15.0	81.0	11	100.0
1,2,3,7,8,9-HCDD	–	12.0	9.4	4.0	–	–	–
1,2,3,4,6,7,8-HCDD	164	77.0	56.0	97.0	135.0	28.0	178.0
Octa-CDD	707	230.0	267.0	414.0	830.0	104.0	1256.0
2,3,7,8-TCDF	–	9.0	0.9	3.9	–	–	–
1,2,3,7,8-PCDF	14.3	25.0	44.0	54.0	15.0	13.0	21.0
1,2,3,4,7,8-HCDF	–	15.0	10.0	6.0	–	–	–
1,2,3,6,7,8-HCDF	31.3	14.0	6.7	5.0	16.0	13.0	58.0
2,3,4,6,7,8-HCDF	–	8.0	3.8	2.0	–	–	–
1,2,3,4,6,7,8-HCDF	16.5	–	19.5	11.0	30.0	7.0	29.0
Octa-CDF	–	–	1.0	4.0	–	–	–

TCDD tetrachlorodibenzo-p-dioxin, *PCDD* polychlorinated dibenzo-p-dioxin, *HCDD* hexachlorodibenzo-p-dioxin, *CDD* chlorinated dibenzo-p-dioxin, *TCDF* tetrachlorodibenzo-p-furan, *PCDF* polychlorinated dibenzo-p-furan, *HCDF* hexachlorodibenzo-p-furan, *CDF*-chlorinated dibenzo-p-furan

America. Few other studies done between 1980 and 1990 show that the adults ingest around 150–300 pg/TEG/day on a median scale of around 2.3 pg TEQ/day, a maximum amount being 4 pg TEQ/Kg/day. The serum dioxin/furan levels among Michigan past and current workers (The Dow Chemical Company) had little or no impact of dioxin- and furan-contaminated sites; however, PCB 126 exhibited significant effect (Burns et al. 2008). Interestingly, a correlation with age and body fat with the workers exposed to dioxins and furans contamination was observed. There is very few information related to dioxin and furan contamination of human tissues mainly because of the complications and difficulties in sample collection and tracing back of source contamination. The ranges of contamination of both dioxins and furans are similar in developing as well as developed countries. The standard values that were recorded in several countries in 1989 in the human adipose tissue are listed in Table 14.6.

This includes several factors such as the number of toxicants present in the environment, bioaccumulation in the food chain, the bioavailability of such toxicants after absorption process metabolic changes that happen within humans. All these factors totally combine to form the bioaccumulation level in human tissues. Notably, once the toxicants get released into the environment, it directly attaches itself to the food sources, thereby entering the food chain of humans, an important mechanism of bioaccumulation in terrestrial ecosystems.

14.6.2 Aquatic Ecosystem

According to the USEPA, even a low concentration of dioxins and furans is highly lethal to aquatic organisms. It is estimated that the aquatic life accumulates such toxic elements in their body near the vicinity of dioxin- and furan-related industries. Edible parts of certain fish varieties such as catfish, yellow perches and suckerfish from Saginaw Bay, Michigan, United States, were estimated for the accumulation of dioxins and furans which was around 4.0–695 ng/Kg. It was reported that a certain type of fish variety called "bottom feeder fish" accumulated more dioxins and furans because it directly feeds on the adsorbed contaminants in water. Dioxins and furans were also targeted several species of crabs. A mean concentration level of 990 pg/g was observed in hepato-pancreas of crabs in Sweden which was due to the effluent discharge from the paper industry. One of the major congeners responsible for contamination of the aquatic ecosystem is later found to be 2,3,7,8– PCDD/PCDF. The acute and chronic levels of concentration are estimated to be 0.01 μg/L and 0.001 μg/L, respectively. Thompson et al. (1992) reported the elevated levels of dioxins and furans in shellfish and fish varieties in Australia. The overall predicted bioaccumulation factor in aquatic life is estimated to be around 3000–900,000, but the measured value was around 390–13,000. Assefa et al. (2019) investigated about the aquatic contamination of dioxins/furans in several species of Atlantic herring fishes using statistical modelling approach and concluded that the thermal sources were the biggest toxicants in herring fishes of the Baltic Sea. Biomagnification of dioxins/furans in lipid content of aquatic fish in China was found to have positive significant correlation on adverse health effects in humans who consume high-fat-content fish varieties (Zhou et al. 2019). Recently, Bartalini et al. (2019) reported inceptive immunosuppression of all sperm whales in the Mediterranean Sea that surpassed the threshold of 210 pg WHO-TEQ/g l.w.

14.7 Health Effects in Humans

High exposure to dioxin/furan compounds can result in various diseases of several effects depending on many factors. These factors include dosage of exposure, sex of the individual, duration of exposure and method of contact with the contaminants. Dioxins/furans enter into the human system by food chain through dairy products such as milk, meat, fish and eggs and are accumulated in the system for a very long

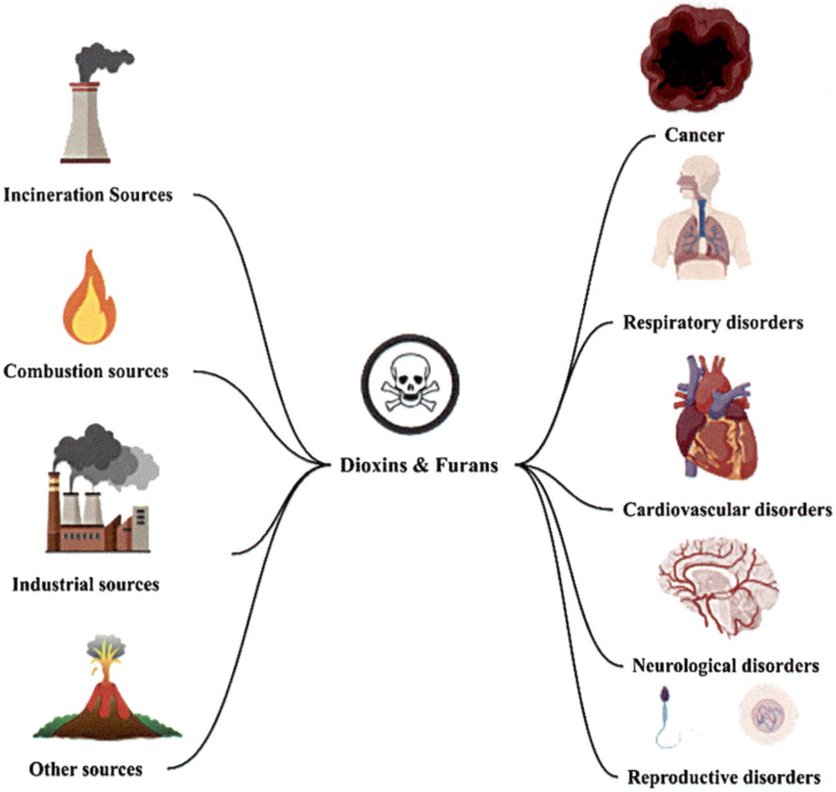

Fig. 14.2 Schematic representation of sources of dioxins/furans and its impact on humans

period. This subsequently leads to several complex diseases in humans in the form of neurodegenerative disorders, weakened immune system, kidney malfunction, skeletal deformities, liver abnormalities, heart disorders, reproductive impairments and sexual disorders. A health report from WHO suggests that more than 90% of the bioaccumulated dioxins/furans come from food ingestion. A schematic representation of the sources of dioxins and furans and possible disorders of such contamination is given in Fig. 14.2.

14.7.1 Cancer

Dioxins and furans are considered to be one among the deadliest carcinogens to humans according to a report from the WHO report. They are classified as level 1 carcinogen ever known to the history of mankind. 2,3,7,8-TCDD is the most studied congener of all compounds known to humans. The USEPA classifies this compound as the "strongest known carcinogen" and also as the "complete carcinogen" because even a very small dose can induce cancers. The main target organ for this compound is studied to be liver, whereas other tissues such as thyroid, lung and oral cavity were also found to be affected by cancer. The onset of cancer because of dioxins and furans is almost 50% dependent on the sex of the organism. In a standard hepatic carcinogenic study with mice and rats, male mice were most affected by cancer, and on the other hand, female rats were most affected by cancer. There are also several epidemiological case studies which reported the development of cancer because of dioxins and furans contamination. One such study was done with the population of Seveso, Italy, where the people were accidentally exposed to remarkable doses of dioxins and furans in the year 1976. These people were exposed to dioxins and furans nearly 100–1000 times compared with the general population. Several types of cancers such as myeloma, lymphoma, sarcoma and liver cancer were seen to be predominant among the groups of tested people of Seveso. In general population, zero or no cases were seen attributing to dioxins and furans exposure. There are several biochemical receptors in the human body that are majorly affected by dioxins and furans. These include glucose transporter proteins, vitamin C uptake, lipoprotein lipases, low-density lipoprotein receptor and insulin secretion receptor.

The mechanism of action of both dioxins and furans is that they directly change or modify the key protein kinase activity by activating the mitotic signals. This eventually modifies the nuclear transcription factors, thereby altering the primary response genes through receptor-mediated mechanisms. The main role of such nuclear transcription factors includes regulation of nutritional indexes and protection of the hormone-sensitive genes. Besides this, dioxins and furans act on several enzymes in the body such as cytochromes, estrogen receptors, interleukins and cytokines. This leads to cellular oxidative stress and blockage in growth factors. This ultimately leads to apoptosis which thereby results in the death of healthy cells. Estrogen receptor-positive cases of breast cancer were directly associated with the long-term airborne exposure of dioxins/furans (Danjou et al. 2019). Another mechanism by which dioxins and furans cause cancer effects in humans is by inducing both cell proliferation and cell differentiation which leads to hypoplasia, hyperplasia, metaplasia and neoplasia. All these forms of abnormalities are directly linked to several types of cancers including Hodgkin disease, non-Hodgkin disease and sarcoma. Effects of one of the most toxic forms of dioxin, TCDD, were studied on rodents, which showed the development of several forms of tumours in thyroid, skin, liver, lung, oral cavity and other tissues. It directly suppresses the immunity, indirectly causing the carcinogenic effects. Tests conducted on animals are more evident than that of humans that TCDD has the most carcinogenic effect among all congeners of dioxin and furan families.

Xu et al. (2016) provided detailed epidemiological evidence about the carcinogenic effects of TCDD on the human hematopoietic system through several types of cancers. The aryl hydrocarbon receptor mechanism of action by dioxins and furans is the most accepted mechanism by which dioxins and furans cause carcinogenic effects. Similarly, dioxins are known to upregulate the enzymes which regulate the metabolic activity of several drugs which leads to an imbalance in hormone secretion. Wimmerová et al. (2016) assessed the human risk associated with dioxin

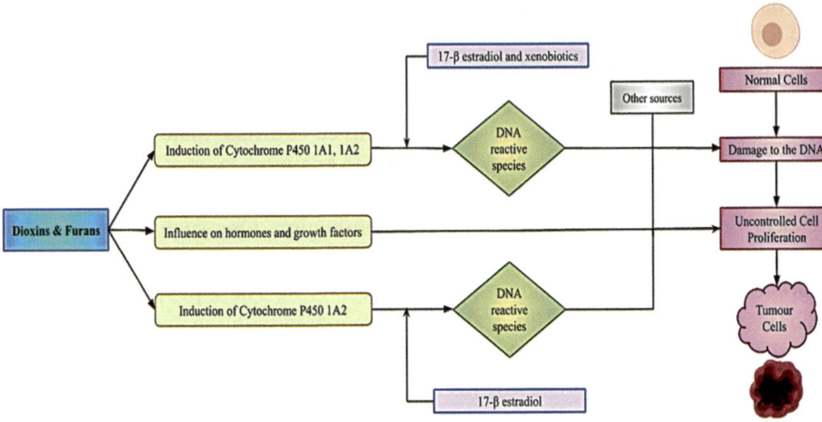

Fig. 14.3 Mechanism of tumour cell generation by dioxins/furans

exposure in terms of thyroid volume, FT4 serum level, cytochrome P450 1A1 and 1B1 expression and derived the relative effect potency in humans. The mechanism of tumour cell generation by toxicants such as dioxins and furans is represented diagrammatically in Fig. 14.3. Recently, Silveira et al. (2019) demonstrated the protective effect of resveratrol, chemoprotective activity, in rats exposed in utero to TCDD.

14.7.2 Respiratory Disorders

Several respiratory diseases are associated with the contamination of both dioxins and furans. Since the main source of contamination is through incineration and combustion of toxic materials, it is to be noted that the fly ash and smoke produced from such sources are the chief causal agents of respiratory ailments. Commonly seen respiratory problems of such smoke- and fly-ash-related sources are an allergy, asthma, nasal congestion, laboured breathing, wheezing, cough and itchy and watery nose. Since the toxic substances in the atmosphere directly reach the human body through respiratory air, the toxicants have the highest chance of reaching the lungs, thereby causing inflammation or damage to the lungs. The toxic substances that enter into the lungs have the capacity to create acute or chronic lung cancer. Kuratsune et al. (1971) studied the accidental exposure to PCB contaminated cooking oil during the late 1960s over 4000 people of Japanese and Taiwanese population. Several patients were reported to suffer from respiratory disorders such as damage in the nasal cavity and lungs along with several oil diseases. Not many studies have been performed on the respiratory system alone to estimate the effects of dioxin and furan contaminations.

14.7.3 Cardiovascular Disorders

Heart and its related organs are some among the main targets of toxic pollutants such as dioxins and furans. Several studies have been performed to estimate the damage it does to the cardiovascular system. Henríquez-Hernández et al. (2017) associated the persistence of cardiovascular diseases to the levels of dioxins/furans through contamination from various dietary sources of food. Kopf and Walker (2009) reported the reduced heart size, reduced blood flow to the heart and reduced contraction rate in the heart in fish model studies. Thinner ventricle walls, extensive cardiac dilation and reduced responsiveness to chronotropic stimuli were observed in the chicken embryo. Cardiocyte proliferation and cardiovascular apoptosis was commonly seen in all animal models exposed to higher levels of dioxins/furans.

Upregulation of several genes such as atrial natriuretic peptide (ANP), myosin heavy chain beta (MHC-β), and endothelin 1 (ET-1) resulted in heart failure as well as cardiac hypertrophy in mice models exposed to high doses of dioxins/furans (Fujisawa et al. 2019). Exposure to dioxins and furans in humans has also resulted in hypertension and cardiovascular diseases. Thackaberry et al. (2005) elaborated about the heart-related ailments in mammals using mice models. The disruption of aryl hydrocarbon receptors (AhR receptors) is the mechanism by which the toxicants cause human congenital heart disorder. The direct correlation between the amounts of exposure to body weight also plays an important role in the onset of several cardiovascular diseases. Studies have also shown that the human embryo is less sensitive to cardiovascular diseases which are induced by both dioxins and furans, but the developmental exposure will result in serious cardiovascular diseases.

14.7.4 Neurological Disorders

Several forms of dioxins and furans have known to cause multiple neurodegenerative disorders in humans as well as wildlife and such diseases, which have been studied over the past two decades. The direct target of dioxin compounds such as 2,3,7,8-TCDD decreases the effect of reactive oxygen species, thereby inhibiting the proliferation of nervous system cells. Because of its extensive usage as pesticides, hydraulic fluids, flame retardants, etc., dioxins and furans become easily accumulated in the food chain and biomagnified easily. During the early 1970s, the first study was carried out to understand the effects of dioxins and furans as a neurotoxic agent since there was acute and chronic exposure of chlorinated pesticides in the environment. After several studies, it is now established that dioxins and furans have adverse effects on attacking the central nervous system and to have neurobehavioral changes.

Fonnum and Mariussen (2009) reported the neurotoxic mechanisms (oxidative stress, calcium homeostasis disturbance and neurotransmitter transport) associated with environmental exposure of dioxins and furans. These chemicals directly target and induce the phospholipase A2 mobilization and induce protein kinase C, thereby creating oxidative stress to the nervous system (Fonnum and Mariussen 2009). They also change the astroglial intracellular Ca^{2+} and glutathione (GSH) levels, decrease glutamine synthetase immunoreactivity, and thereby, maintenance of the pH is lost. Dioxins and furans also are known to cause peripheral nerve damage by targeting the ROS system by suppressing the basal generation and inhibiting the proliferation of neuronal cells which is being investigated with human neuroblastoma cells. Being a lipophilic agent, TCDD directly accumulates in liver and fat tissues faster than that of the brain and central nervous system. This is in contrast with newborns and infants where the intoxication by TCDD is at a higher rate during lactation and gestation period. During the developmental stages of infant, serious neurotoxic diseases such as reduced neuronal calcium uptake, death of neurons, reduce neurotransmittance and cognitive defects are predominantly seen. This has been studied in extrahepatic tissues of mice and rats through the alteration of redox reaction balances. Even in this study, the ROS induction takes places which thereby leads to an increase in peroxidation reaction of lipid content and decrease in GSH content in several parts of tissues such as testis and liver. This correlates with the effect of TCDD on cell proliferation and reduction in the rate of uptake of thymidine to the cells.

Ames et al. (2019) reported about the association between exposure of dioxins/furans to the neuropsychological functioning of adolescent children and found no significant correlation between the two factors. Dioxins and furans entered into the body cannot be washed out easily by the human body. Women have two different methods of getting rid of it, one being the placental method and the other being lactation. But men, on the other hand, do not have any methods for getting rid of such toxicants, thus leading to bioaccumulation at a much faster rate than women.

14.7.5 Reproductive Disorders

Several animal models have been studied to report the effects of TCDD on both reproductive diseases and developmental studies. The whole dioxin and furan family of compounds has effects on gamete production, size of reproductive organs, number of spermatozoa, fertilization processes, number of ovules produced during the follicular phase of females, etc. The effect of fertility in males has been in fewer studies, but in females, the effect of TCDD has a major impact on fertility during the prenatal stages. The major mechanism by which the dioxin compounds such as TCDD acts on the reproductive and sexual system is that it targets the hypothalamic-pituitary axis, thereby inhibiting the functional signals of the system. Studies were carried out on rodents whose hypothalamus was affected which resulted in the decreased sexual activity. Several developmental delays during the differentiation of trophectoderm were also observed in such rodents. There was also a negative response in the action of several growth factors such as transforming growth factor-beta (TGF-ß) which is majorly responsible for the organogenesis process. The same effect has been seen in human fetal tissue cultures where the signs of organogenesis are absent or delayed. In a different study with the continuous exposure of TCDD on female rhesus monkeys, a serious reproductive disorder, endometriosis, was observed.

High exposure of TCDD can also result in continuous abortions in females which are observed even now in parts of Seveso, Italy. Several conditions based on the production of sperm in males such as oligospermia, asthenospermia and teratozoospermia are still observed in few populations in Seveso, Italy. Breastfeeding is one among the major mechanisms by which an offspring gets exposed to dioxins and furans. One such study was conducted by Finnish group researchers. The mothers were also affected by impaired fertility and endometriosis. Few researchers in the Netherlands and Germany also concluded similar results of that of the Finnish group of researchers. Few epidemiological studies were also conducted based on environmental exposure to the human fetus and found out that even without breastfeeding, the fetus was very vulnerable to both dioxins and furans from the external atmosphere. Gray et al. (1997a, b) and Mably et al. (1992a) reported the effects of TCDD and other such chemicals in the reproductive development of laboratory animals such as mice and rats. The animals showed signs of decreased sperm count; vaginal thread reduced accessory gland weight and delayed puberty.

Hale et al. (2019) assessed the contamination levels of dioxins/furans in American alligator embryos and concluded that the expression of several genes related to embryo–yolk mass ratio and embryo mass was associated with the dioxin/furan exposure. Different studies were performed at different dosage levels of exposure of TCDD and concluded the detection of several types of sexual deformities in both mice and rats. Hamm et al. (2000) reported about the lactation effects and deformities in rats were due to the exposure of about 2.0 g TEQ/kg. The decrease in sperm count in Holtzman rats was observed by Mably et al. (1992b) at a significantly lower dose of about 0.5 g/kg. In contrary, Faqi et al. (1998) reported the increased sperm production on the exposure of about 100 g PCB77/kg along with an increase in the size of testicles. Manikkam et al. (2012) demonstrated TCDD exposure on male Hsd:Sprague Dawley rats promoted transgenerational effects and sperm epimutations on male reproduction. Several reproductive disorders such as decreased fertility, lower daily sperm production, and low testosterone levels were found in Sprague Dawley rats which were exposed to higher doses of furans (Rehman et al. 2019). Foster et al. (2009) demonstrated utero exposure of TCDD induced changes in epididymal sperm; however, spermatogenesis was not affected. Pereg et al. (2001) reported the effects of intermediate reactive compounds that are metabolized from dioxin and furan contamination which rapidly causes secondary sexual deformities. Egeland et al. (1994) reported decreased testosterone and increased gonadotropin levels in several workers of the NIOSH cohort who were associated with the higher concentrations of dioxin compounds such as TCDD. In 1968, women were accidentally exposed to dioxin- and furan-contaminated rice oil in Japan led to reducing the birth weight and hyperpigmentation of Yusho disease (Tsukimori et al. 2012). During the time period 1977–1984, a test was conducted on 239 men and 296 women in Seveso, Italy, and reported that the sex ratio was changed to an excess of females over males. This corresponds to the TCDD contamination in that area during that period. This test was conducted to both parents who were exposed to high doses of TCDD during the exposure. There was also an increase in the probability of female births (p-value = 0.008) which was associated with the higher concentrations of TCDD in the serum samples of fathers. Though such reports have been made, biological explanation of such occurrences has not yet been made. Michalek et al. (1998) and Rogan et al. (1988) studied a similar type of pattern in other population where the dosage of TCDD exposure was much lower than in Seveso, Italy. A retrospective cohort study from Seveso Women's Health Study on TCDD exposure and reproductive health over past 30 years reported no association of serum TCDD with pregnancy, spontaneous abortion, gestation length and fetal growth (Wesselink et al. 2014). Besides this, there have not been major studies relating to the concentration of dioxin and furan exposure to the time of pregnancy.

14.8 Future Considerations and Conclusions

Dioxins and furans being defined as persistent, possessing characteristics of bioaccumulation, being highly stable and having a very high half-life, with a high range of atmospheric transport, are likely to produce several adverse effects on human health as well as the environment. Keeping in pace with the scientific advancements, it is possible to assess the risk associated with dioxin and furan incorporation into the environment. It is high time that the human associate the regulatory focus of emitted dioxins and furans along with the strategies to limit or stop emissions of such chemicals. A complete understanding and knowledge of formation and mechanism of action of dioxins and furans will help us a long way in battling effects that have on the environment as well as humans. Being identified as POPs, many countries including Hungary, Czech Republic, Ukraine, Slovakia, Slovenia and Croatia have already agreed to limit the total annual emissions of identified POPs, few among them being dioxins and furans. These countries have agreed to take effective measures under the circumstances of abnormal levels of POP emission with a set reference value. The year 1990 is set as a reference value year; the emission levels at any point of time should be less than that of the emissions that happened during the year 1990. Many of the Asian countries including India have taken an extensive measurement in reducing the emission levels of POPs. Such toxic and extremely harmful chemical compounds should be eliminated using existing advanced technologies to provide a healthy and better environment for the future generation.

14.8.1 Better Understanding and Identification of Sources

Over the past few decades, there have been several cases of dioxin and furan contamination into the environment. During the early stages, it was due to the incineration of chlorinated compounds. Later, this contamination was due to the construction of individual household incinerators which led to a rapid increase in the levels of dioxins and furans in the environment. Later during the same period, control of such constructions as well as metal processing of chlorine compounds have reduced much of the contamination. In view of such a matter, an analysis was performed by a European data agency reporting the creation of herbaria including the sedimentary samples from all the contaminated sites. This would be helpful in estimating the nature of sources over different periods of time and assessing them according to different environmental patterns. The agency also suggested that the government bodies would get access to contaminated samples of historical accidents. This led to the establishment of the Soil Monitoring Agency (SMA) to observe and maintain different sources of contamination of toxic chemicals or POPs including dioxins and furans.

14.8.2 Understanding the Mechanism of Action

Effective understanding of the transfer of dioxins and furans from soil to plants, soil to water and soil to air needs to be done carefully. Several agencies such as The French Institute for Agronomy Research (INRA), The French National Center for Scientific Research (CNRS) and Bureau for Geology and Mining Research (BRGM) have studied the effect of microorganisms in degrading dioxins and furans. Such areas of research should be appreciated and given more importance.

14.8.3 Development of New Techniques to Identify Contamination

Gas chromatography-mass spectroscopy (GC-MS) is a widely used assay method for reference of dioxins and furans. Though being a highly sensitive method, the presence of more than 210 different compounds and maximum variability between different laboratories, it has become extremely difficult to interpret the results with accuracy. The development of new biological and analytical assay methods for identification of dioxins and furans would facilitate the survey methods to be performed easily.

14.8.4 Development of Toxicological and Epidemiological Studies

The dose–response value of each compound of dioxins and furans needs to be understood and studied in order to understand the effects on humans and the environment. Several toxicokinetic and toxicodynamic methods are established by scientific advisory groups to estimate the cut-off limit of each compound. Similarly, epidemiological studies are necessary to assess the relation and distribution pattern of diseases related to dioxin and furan contamination.

With all the scientific advancements and technologies, it becomes easier to study the impact and its effects on humans and the environment. The most important effect is the poisoning of the food chain from the contaminated sites of the food supply. Similarly, animals and other organisms face the threat due to such harmful chemicals. Every effort should be made to minimize the exposure of such contaminants to the animals and the environment. Several ways of accomplishing this are building a modern scientific process to prevent the release of such contaminants to the outside environment and also biological methods, including the creation of hybrid organisms such as bacteria, fungi and yeast for the elimination of such toxic components. Thus, the development of high throughput technologies in the near future will certainly reduce the effects of such components in the environment.

Acknowledgements Authors thank Prof. M. Sivanandham, Secretary, SVEHT and SVCE Management for their support and encouragement.

References

Ames J, Warner M, Siracusa C, Signorini S, Brambilla P, Mocarelli P, Eskenazi B (2019) Prenatal dioxin exposure and neuropsychological functioning in the Seveso Second Generation Health Study. Int J Hyg Environ Health 222(3):425–433

Assefa A, Tysklind M, Bignert A, Josefsson S, Wiberg K (2019) Sources of polychlorinated dibenzo-p-dioxins and dibenzofurans to Baltic Sea herring. Chemosphere 218:493–500

Bartalini A, Muñoz-Arnanz J, Marsili L, Mazzariol S, Fossi MC, Jiménez B (2019) Evaluation of PCDD/Fs, dioxin-like PCBs and PBDEs in sperm whales from the Mediterranean Sea. Sci Total Environ 653:1417–1425

Boda H, Nghi TN, Nishijo M, Thao PN, Tai PT, Van Luong H, Anh TH, Morikawa Y, Nishino Y, Nishijo H (2018) Prenatal dioxin exposure estimated from dioxins in breast milk and sex hormone levels in umbilical cord blood in Vietnamese newborn infants. Sci Total Environ 615:1312–1318

Bruce-Vanderpuije P, Megson D, Jobst K, Jones GR, Reiner E, Sandau CD, Clarke E, Adu-Kumi S, Gardella JA Jr (2019) Background levels of dioxin-like polychlorinated biphenyls (dlPCBs), polychlorinated, polybrominated and mixed halogenated dibenzo-p-dioxins and dibenzofurans (PCDD/Fs, PBDD/Fs & PXDD/Fs) in sera of pregnant women in Accra, Ghana. Sci Total Environ 673:631–642

Burns CJ, Collins JJ, Budinsky RA, Bodner K, Wilken M, Rowlands JC, Martin GD, Carson ML (2008) Factors related to dioxin and furan body levels among Michigan workers. Environ Res 106(2):250–256

Chirollo C, Ceruso M, Pepe T, Vassallo A, Marrone R, Severino L, Anastasio A (2018) Levels and congeners distribution of dioxins, furans and dioxin-like PCBs in buffaloes adipose tissues sampled in vivo and milk. CyTA-J Food 16(1):1109–1114

Cockerham LG, Young AL (1983) Ultrastructural comparison of liver tissues from field and laboratory TCDD-exposed beach mice. In: Human and environmental risks of chlorinated dioxins and related compounds. Springer, Boston, pp 373–389

Conesa JA, Ortuño N, Abad E, Rivera-Austrui J (2016) Emissions of PCDD/Fs, PBDD/Fs, dioxin like-PCBs and PAHs from a cement plant using a long-term monitoring system. Sci Total Environ 571:435–443

Danjou AMN, Coudon T, Praud D, Lévêque E, Faure E, Salizzoni P, Le Romancer M, Severi G, Mancini FR, Leffondré K, Dossus L (2019) Long-term airborne dioxin exposure and breast cancer risk in a case-control study nested within the French E3N prospective cohort. Environ Int 124:236–248

Di Meo GP, Perucatti A, Genualdo V, Caputi-Jambrenghi A, Rasero R, Nebbia C, Iannuzzi L (2010) Chromosome fragility in dairy cows exposed to dioxins and dioxin-like PCBs. Mutagenesis 26(2):269–272

Egeland GM, Sweeney MH, Fingerhut MA, Wille KK, Schnorr TM, Halperin WE (1994) Total serum testosterone and gonadotropins in workers exposed to dioxin. Am J Epidemiol 139 (3):272–281
EPA US (1999) National air toxics program: the integrated urban strategy. Fed Regist 64:38705–38740
Faqi AS, Dalsenter PR, Merker HJ, Chahoud I (1998) Reproductive toxicity and tissue concentrations of low doses of 2, 3, 7, 8-tetrachlorodibenzo-p-dioxin in male offspring rats exposed throughout pregnancy and lactation. Toxicol Appl Pharmacol 150(2):383–392
Fonnum F, Mariussen E (2009) Mechanisms involved in the neurotoxic effects of environmental toxicants such as polychlorinated biphenyls and brominated flame retardants. J Neurochem 111 (6):1327–1347
Foster WG, Maharaj-Briceño S, Cyr DG (2009) Dioxin-induced changes in epididymal sperm count and spermatogenesis. Environ Health Perspect 118(4):458–464
Fujisawa N, Tohyama C, Yoshioka W (2019) Cardiotoxicity induced by 2, 3, 7, 8-tetrachlorodibenzo-p-dioxin exposure through lactation in mice. J Toxicol Sci 44 (7):505–513
Garner AJ, Pagano JJ (2019) Trends of polychlorinated dioxins, polychlorinated furans, and dioxin-like polychlorinated biphenyls in Chinook and Coho salmonid eggs from a Great Lakes tributary. Environ Pollut 247:1039–1045
Genualdo V, Perucatti A, Iannuzzi A, Di Meo GP, Spagnuolo SM, Caputi-Jambrenghi A, Coletta A, Vonghia G, Iannuzzi L (2012) Chromosome fragility in river buffalo cows exposed to dioxins. J Appl Genet 53(2):221–226
González N, Marquès M, Nadal M, Domingo JL (2018) Levels of PCDD/Fs in foodstuffs in Tarragona County (Catalonia, Spain): Spectacular decrease in the dietary intake of PCDD/Fs in the last 20 years. Food Chem Toxicol 121:109–114
Gou YY, Que DE, Chuang CY, Chao HR, Shy CG, Hsu YC, Lin CW, Chuang KP, Tsai CC, Tayo LL (2016) Dust levels of polybrominated diphenyl ethers (PBDEs) and polybrominated dibenzo-p-dioxins/furans (PBDD/Fs) in the Taiwanese elementary school classrooms: assessment of the risk to school-age children. Sci Total Environ 572:734–741
Gray LE Jr, Ostby JS, Kelce WR (1997a) A dose–response analysis of the reproductive effects of a single gestational dose of 2, 3, 7, 8-tetrachlorodibenzo-p-dioxin in male Long Evans Hooded rat offspring. Toxicol Appl Pharmacol 146(1):11–20
Gray LE Jr, Wolf C, Mann P, Ostby JS (1997b) In uteroexposure to low doses of 2, 3, 7, 8-tetrachlorodibenzo-p-dioxin alters reproductive development of female Long Evans hooded rat offspring. Toxicol Appl Pharmacol 146(2):237–244
Hale MD, Bertucci EM, Rainwater TR, Wilkinson PM, Parrott BB (2019) The impact of maternally derived dioxins on embryonic development and hepatic AHR signaling in a long-lived apex predator. Chemosphere 229:489–499
Hamm JT, Sparrow BR, Wolf D, Birnbaum LS (2000) In utero and lactational exposure to 2, 3, 7, 8-tetrachlorodibenzo-p-dioxin alters postnatal development of seminal vesicle epithelium. Toxicol Sci 54(2):424–430
Henríquez-Hernández LA, Luzardo OP, Zumbado M, Serra-Majem L, Valerón PF, Camacho M, Álvarez-Pérez J, Salas-Salvadó J, Boada LD (2017) Determinants of increasing serum POPs in a population at high risk for cardiovascular disease. Results from the PREDIMED-CANARIAS study. Environ Res 156:477–484
Hryhorczuk DO, Withrow WA, Hesse CS, Beasley VR (1981) A wire reclamation incinerator as a source of environmental contamination with tetrachlorodibenzo-p-dioxins and tetrachlorodibenzofurans. Arch Environ Health Int J 36(5):228–234
Hsu YC, Que DE, Gou YY, Tsou TC, Liu CK, Wang YL, Hou WC, Lin YH, Liu WY, Chao HR, Lee WJ (2018) National surveillance of 2, 3, 7, 8-substituted polychlorinated dibenzo-p-dioxins/furans in soil in Taiwan. Chemosphere 203:239–252
Hu J, Wu J, Xu C, Zha X, Hua Y, Yang L, Jin J (2019) Preliminary investigation of polychlorinated dibenzo-p-dioxin and dibenzofuran, polychlorinated naphthalene, and dioxin-like

polychlorinated biphenyl concentrations in ambient air in an industrial park at the northeastern edge of the Tibet–Qinghai Plateau, China. Sci Total Environ 648:935–942
Karstensen KH (2008) Formation, release and control of dioxins in cement kilns. Chemosphere 70 (4):543–560
Kopf PG, Walker MK (2009) Overview of developmental heart defects by dioxins, PCBs, and pesticides. J Environ Sci Health C 27(4):276–285
Kulkarni PS, Crespo JG, Afonso CA (2008) Dioxins sources and current remediation technologies- a review. Environ Int 34(1):139–153
Kuratsune M, Yoshimura T, Matsuzaka J, Yamaguchi A (1971) Yusho, a poisoning caused by rice oil contaminated with polychlorinated biphenyls. HSMHA Health Rep 86(12):1083
Li XD, Yan M, Chen T, Lu SY, Yan JH, Cen KF (2010) Levels of PCDD/Fs in soil in the vicinity of a medical waste incinerator in China: The temporal variation during 2007–2009. J Hazard Mater 179(1–3):783–789
Mably TA, Bjerke DL, Moore RW, Gendron-Fitzpatrick A, Peterson RE (1992a) In utero and lactational exposure of male rats to 2, 3, 7, 8-tetrachlorodibenzo-p-dioxin: 3. Effects on spermatogenesis and reproductive capability. Toxicol Appl Pharmacol 114(1):118–126
Mably TA, Moore RW, Goy RW, Peterson RE (1992b) In utero and lactational exposure of male rats to 2, 3, 7, 8-tetrachlorodibenzo-p-dioxin: 2. Effects on sexual behavior and the regulation of luteinizing hormone secretion in adulthood. Toxicol Appl Pharmacol 114(1):108–117
Manikkam M, Tracey R, Guerrero-Bosagna C, Skinner MK (2012) Dioxin (TCDD) induces epigenetic transgenerational inheritance of adult onset disease and sperm epimutations. PLoS One 7(9):e46249
Matta EJ (2018) Pollution in the Uruguay-Argentina border: critical data analysis from two rivers and the pulp mill effluent. Int J Environ Health 9(1):84–109
Metelkova L, Zhakovskaya Z, Kukhareva G, Rybalko A, Nikiforov V (2019) Occurrence of PCDD/ PCDFs, dioxin-like PCBs, and PBDEs in surface sediments from the Neva River and the Eastern Gulf of Finland (Russia). Environ Sci Pollut Res 26(8):7375–7389
Michalek JE, Rahe AJ, Boyle CA (1998) Paternal dioxin, preterm birth, intrauterine growth retardation, and infant death. Epidemiology 9:161–167
Mukerjee D (1998) Health impact of polychlorinated dibenzo-p-dioxins: a critical review. J Air Waste Manage Assoc 48(2):157–165
Nguyen HM, Tu BM, Watanabe M, Kunisue T, Monirith I, Tanabe S, Sakai S, Subramanian A, Sasikumar K, Pham HV, Bui CT (2003) Open dumping site in Asian developing countries: a potential source of polychlorinated dibenz-p-dioxins and polychlorinated dibenzofurans. Environ Sci Technol 37(8):1493–1502
Ortuño N, Lundstedt S, Lundin L (2015) Emissions of PBDD/Fs, PCDD/Fs and PBDEs from flame-retarded high-impact polystyrene under thermal stress. Chemosphere 123:64–70
Pemberthy D, Quintero A, Martrat MG, Parera J, Ábalos M, Abad E, Villa AL (2016) Polychlorinated dibenzo-p-dioxins, dibenzofurans and dioxin-like PCBs in commercialized food products from Colombia. Sci Total Environ 568:1185–1191
Pereg D, Tampal N, Espandiari P, Robertson LW (2001) Distribution and macromolecular binding of benzo [a] pyrene and two polychlorinated biphenyl congeners in female mice. Chem Biol Interact 137(3):243–258
Perucatti A, Di Meo GP, Albarella S, Ciotola F, Incarnato D, Jambrenghi AC, Peretti V, Vonghia G, Iannuzzi L (2006) Increased frequencies of both chromosome abnormalities and SCEs in two sheep flocks exposed to high dioxin levels during pasturage. Mutagenesis 21(1):67–75
Piskorska-Pliszczynska J, Maszewski S, Mikolajczyk S, Pajurek M, Strucinski P, Olszowy M (2017) Elimination of dioxins in milk by dairy cows after the long-term intake of contaminated sugar beet pellets. Food Addit Contam Part A 34(5):842–852
Psomopoulos CS, Bourka A, Themelis NJ (2009) Waste-to-energy: A review of the status and benefits in USA. Waste Manag 29(5):1718–1724

Quina MJ, Bordado JC, Quinta-Ferreira RM (2011) Air pollution control in municipal solid waste incinerators. The impact of air pollution on health, economy, environment and agricultural sources, pp 331–358

Rahman MM, Shon ZH, Ma CJ, Brown RJC, Pandey SK, Park CG, Bae IS, Sohn JR, Yoon HO, Kim KH (2014) Partitioning of dioxins (PCDDs/Fs) in ambient air at urban residential locations. Int J Environ Sci Technol 11(7):1897–1910

Rawn DF, Sadler AR, Casey VA, Breton F, Sun WF, Arbuckle TE, Fraser WD (2017) Dioxins/furans and PCBs in Canadian human milk: 2008–2011. Sci Total Environ 595:269–278

Rehman H, Ullah I, David M, Ullah A, Jahan S (2019) Neonatal exposure to furan alters the development of reproductive systems in adult male Sprague Dawley rats. Food Chem Toxicol 130:231–241

Ren Z, Zheng M (2009) Impacts of human activities on dioxins emissions at national scale. Chemosphere 76(6):853–859

Rezaei E, Farahani A, Buekens A, Chen T, Lu SY, Habibinejad M, Damercheli F, Moghadam SA, Gandomkar M, Bahmani A (2013) Dioxins and furans releases in Iranian mineral industries. Chemosphere 91(6):838–843

Rivera-Austrui J, Martínez K, Marco-Almagro L, Ábalos M, Abad E (2014) Long-term sampling of dioxin-like substances from a clinker kiln stack using alternative fuels. Sci Total Environ 485:528–533

Roberts SM, Lowney YW, Stuchal LD (2019) Bioaccessibility of polychlorinated dioxins and furans in soil from a Superfund site. Chemosphere 214:418–423

Rogan WJ, Gladen BC, Hung KL, Koong SL, Shih LY, Taylor JS, Wu YC, Yang D, Ragan NB, Hsu CC (1988) Congenital poisoning by polychlorinated biphenyls and their contaminants in Taiwan. Science 241(4863):334–336

Schuhmacher M, Mari M, Nadal M, Domingo JL (2019) Concentrations of dioxins and furans in breast milk of women living near a hazardous waste incinerator in Catalonia, Spain. Environ Int 125:334–341

Senthil Kumar K, Kannan K, Paramasivan ON, Shanmuga Sundaram VP, Nakanishi J, Masunaga S (2001) Polychlorinated dibenzo-p-dioxins, dibenzofurans, and polychlorinated biphenyls in human tissues, meat, fish, and wildlife samples from India. Environ Sci Technol 35 (17):3448–3455

Silveira LTR, de Mello Santos T, Camora LF, Pinho CF, Anselmo-Franci JA, Domeniconi RF, Justulin LA, Barbisan LF, Scarano WR (2019) Protective effect of resveratrol on urogenital sinus and prostate development in rats exposed in utero to TCDD (2, 3, 7, 8-tetrachlorodibenzo-p-dioxin). Reprod Toxicol 83:82–92

Thackaberry EA, Nunez BA, Ivnitski-Steele ID, Friggins M, Walker MK (2005) Effect of 2, 3, 7, 8-tetrachlorodibenzo-p-dioxin on murine heart development: alteration in fetal and postnatal cardiac growth, and postnatal cardiac chronotropy. Toxicol Sci 88(1):242–249

Thompson GB, Chapman JC, Richardson BJ (1992) Disposal of hazardous wastes in Australia: implications for marine pollution. Mar Pollut Bull 25(5–8):155–162

Tsukimori K, Uchi H, Mitoma C, Yasukawa F, Chiba T, Todaka T, Hirata T, Fukushima K, Wake N (2012) Maternal exposure to high levels of dioxins in relation to birth weight in women affected by Yusho disease. Environ Int 38(1):79–86

Ueno D, Watanabe M, Subramanian A, Tanaka H, Fillmann G, Lam PK, Zheng GJ, Muchtar M, Razak H, Prudente M, Chung KH (2005) Global pollution monitoring of polychlorinated dibenzo-p-dioxins (PCDDs), furans (PCDFs) and coplanar polychlorinated biphenyls (coplanar PCBs) using skipjack tuna as bioindicator. Environ Pollut 136(2):303–313

Van den Berg M, Birnbaum LS, Denison M, De Vito M, Farland W, Feeley M, Fiedler H, Hakansson H, Hanberg A, Haws L, Rose M (2006) The 2005 World Health Organization reevaluation of human and mammalian toxic equivalency factors for dioxins and dioxin-like compounds. Toxicol Sci 93(2):223–241

Wang LC, Hsi HC, Wang YF, Lin SL, Chang-Chien GP (2010) Distribution of polybrominated diphenyl ethers (PBDEs) and polybrominated dibenzo-p-dioxins and dibenzofurans (PBDD/Fs) in municipal solid waste incinerators. Environ Pollut 158(5):1595–1602

Wang X, Ni Y, Zhang H, Zhang X, Chen J (2012) Formation and emission of PCDD/Fs in Chinese non-wood pulp and paper mills. Environ Sci Technol 46(21):12234–12240

Wesselink A, Warner M, Samuels S, Parigi A, Brambilla P, Mocarelli P, Eskenazi B (2014) Maternal dioxin exposure and pregnancy outcomes over 30 years of follow-up in Seveso. Environ Int 63:143–148

Wimmerová S, van den Berg M, Chovancová J, Patayová H, Jusko TA, van Duursen MB, Murínová ĽP, Canton RF, van Ede KI, Trnovec T (2016) Relative effect potency estimates of dioxin-like activity for dioxins, furans, and dioxin-like PCBs in adults based on cytochrome P450 1A1 and 1B1 gene expression in blood. Environ Int 96:24–33

Xu J, Ye Y, Huang F, Chen H, Wu H, Huang J, Hu J, Xia D, Wu Y (2016) Association between dioxin and cancer incidence and mortality: a meta-analysis. Sci Rep 6:38012

Yang L, Liu G, Zhu Q, Zheng M (2019) Small-scale waste incinerators in rural China: Potential risks of dioxin and polychlorinated naphthalene emissions. Emerg Contam 5:31–34

Young AL, Thalken CE, Arnold EL, Cupello JM, Cockerham LG (1976) Fate of 2,3,7,8 tetrachlorodibenzo-p-dioxin (TCDD) in the environment: summary and decontamination recommendations (No. USAFA-TR-76-18). USAF Academy Colorado Springs

Zhou S, Zhu H, Huang S, Zhou J, Zhang S, Wang C (2019) Biomagnification and risk assessment of polychlorinated biphenyls in food web components from Zhoushan fishing ground, China. Mar Pollut Bull 142:613–619

The Role of Microorganisms in Remediation of Environmental Contaminants 15

Gaganpreet Kaur, Damanjeet Kaur, and Saurabh Gupta

Abstract

Environmental pollution by heavy metals, hydrocarbons, pesticides, dyes and other xenobiotic compounds is increasing day by day. Reclamation of these contaminants is of utmost importance for sustainable development. Various anthropogenic activities like industrial waste, mining, use of pesticides, deforestation and automobile emissions added a significant amount of these contaminants into the environment. Conventional remediation techniques such as physicochemical methods, electrochemical treatments and cationic and anionic exchange are relatively expensive and quite unsuccessful with minute concentration of pollutants. Besides these limitations, conventional methods also lead to the production of secondary pollutants. However, bioremediation employs the use of microbes as well as plants for the restoration of the environmental sites and water bodies. Exploitation of microbes such as bacteria, fungi and algae as remediation tools is emerging as an eco-friendly approach for the treatment of different pollutants and contaminants from air, soil and water. Comparing to conventional technologies, the microbe-based bioremediation is more economical and efficient in protecting the environment and minimizing the health hazards caused by pollutants. Nowadays, microbes have been also targeted using gene-editing tools such as genetic engineering, microarray technology and various omics approaches such as genomics, transcriptomics, metabolomics and proteomics for the regulation and optimization of the remediation process. Development in technology provides a new era of an integrated approach exploring the plant–microbe interaction and genetically engineered microbes for the restoration of

G. Kaur
Department of Biotechnology & Medical Sciences, Baba Farid Group of Institutions, Baba Farid College, Bathinda, Punjab, India

D. Kaur · S. Gupta (✉)
Department of Microbiology, Mata Gujri College, Fatehgarh Sahib, Punjab, India

R. Prasad (ed.), *Environmental Pollution and Remediation*, Environmental and Microbial Biotechnology, https://doi.org/10.1007/978-981-15-5499-5_15

polluted sites. This book chapter aims to summarize the different physiological, biochemical and molecular basis of remediation along with various advancements in remedial approaches for conserving an ecological balance by the use of microbes.

Keywords

Bioremediation · Ecosystem · Contaminants · Biodegradation · Genetic engineering

15.1 Introduction

Environmental contamination and deterioration caused by heavy metals, pesticides, petrochemicals, etc. has attracted increased attention worldwide. It is one of the most serious problems that humans are facing these days. The brisk expansion of population has led to the development of different technologies, industries, increased use of fertilizers and pesticides to cope with the demands for livelihood and elevated pressure on natural resources such as air, water and soil (Kumar et al. 2011; Shah 2014). A huge amount of waste containing agrochemicals, dyes, heavy metals and hydrocarbons has been generated by the industries and continuously dumped into the environment without any treatment. During the twentieth century, with rapid increase in industrialization, the discharge of solid and liquid contaminants in the environment has also reached critical levels, thus resulting in building up several pollution problems for the society (Shah 2014). The prominent sources of pollution in the surroundings are (Rao et al. 2010):

(i) Industrial activities such as mining of ores, processing of metals, petrochemical industry and chemical processing.
(ii) Waste discharged from electroplating, steel manufacturing, leather tanning, electronics and automobiles, pulp and paper industry, textile industry, etc.
(iii) Other activities include agricultural practices and atmospheric deposition.

These anthropogenic activities considerably alter the biogeochemical cycles, affecting the terrestrial as well as aquatic ecosystem, thus becoming a universal threat to the environment. The contaminants when released into the natural resources cause a detrimental effect on human, animals, flora, fauna and microorganisms. Apart from this, man compromise drinking water, food and air quality with greenhouse gases, nuclear wastes, pesticides, heavy metals and hydrocarbon pollution (Löffler and Edwards 2006; Rao et al. 2010; Azubuike et al. 2016). Human endeavours in the manufacturing and improvement of economic goods and services for fulfilling the needs of the population cannot be banned as they are mandatory for the survival of humans on the planet earth. As a substitute, we must look for cleaner processes, which lead to the formation of environmentally friendly things. We must concentrate on our strategy for the declination and elimination of the existing pollution in the environment (Ali 2010). The eradication of these wide ranges of contaminants (mentioned in the above points) from the environment is a sheer requirement to enhance the sustainability of our society (Adenipekun and Lawal 2012). The absolute key solution for pollution abatement is the use of "bioremediation". Among all available approaches, it is one of the most effective and inventive technologies, which uses microorganisms to deteriorate environmental pollutants causing pollution in the environment (Niti et al. 2013). Oil and pipeline spills in the 1970s to 1980s have triggered immense research on the degradation of petroleum hydrocarbons carried out by microbes and also demonstrated for the very first time that bioremediation is a powerful method for restoring surface soil affected with crude products of oil (Löffler and Edwards 2006). It is a multidisciplinary approach that centrally thrusts the field of microbiology (Niti et al. 2013). Microorganisms are the original recyclers present on mother earth to overcome challenges by converting toxic organic chemicals to harmless end products, i.e. CO_2 and H_2O (Joshi et al. 2010; Abatenh et al. 2017). Microorganisms are capable to survive in all places on the earth, because of their metabolism and ranges in all kinds of environments because of its varied nutritional capacity (Abatenh et al. 2017). Ever since, it has been discovered that the tiny organisms called microbes can biodegrade xenobiotics, especially pesticides in soil. Researchers are investigating the microbial diversity, which particularly belongs to polluted areas that can biodegrade a variety of contaminants (Joshi et al. 2010).

Remediation is a process of removal of contaminants from the environment using physical, chemical and biological means. The remediation approaches tackle the issues that cause potential damage from the "*origin*" of various pollutants (Lynch and Moffat 2005; Mehndiratta et al. 2013; Behera and Prasad 2020). Remediation is different from "restoration" and "reclamation" as it can tackle with the source that causes pollution, forbidding the "*pathway*" among the receiver and the source, removal of the susceptible receptor either on a permanent basis by the transition of an area in use or on temporary basis till pollution drops below a threshold level (Lynch and Moffat 2005). Different conventional remediation approaches such as membrane separation, filtration, ion exchange, reverse osmosis, electrochemical treatment and adsorption have been explored for the elimination of contaminants from the environment.

15.2 Conventional Methods of Remediation

Conventional methods (Fig. 15.1) used for remediation of waste or contaminants are discussed below:

15.2.1 Mechanical Separation

Mechanical separation is a physicochemical method in which contaminants in the soil are removed by gravitational separation (based on the differences in the density between fractions), cyclones (based on the Coriolis effect), sieving (separation based

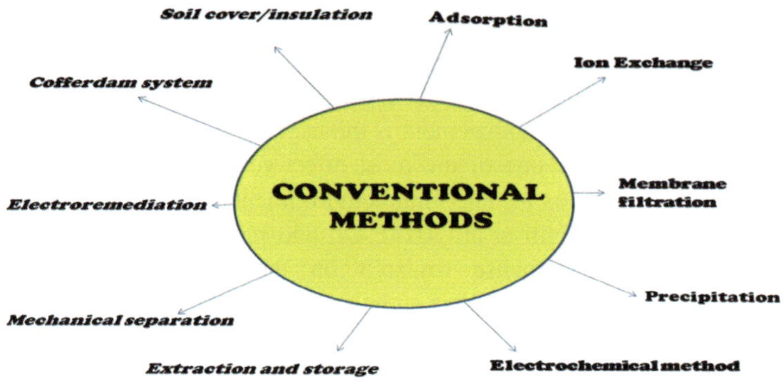

Fig. 15.1 Conventional methods of remediation which were used for treating contaminated sites

on the difference in particle grain size) or magnetic separation (based on magnetic induction). This method leads to a significant amount of volume reduction, but is not applicable for contaminants that are homogeneously distributed in soil (Das and Adholeya 2012).

15.2.2 Extraction and Storage

This process involves the separation of polluted soil cover, using a bulldozer or digger, followed by its storage in an accurate place for clean-up by using different technologies. The site where it is stored has to be covered to avoid erosion by wind or water (Das and Adholeya 2012).

15.2.3 Soil Cover/Insulation

It is a physical process that involves the covering of contaminated soil, which is composed of four layers: an upper protective layer, a drainage layer, a non-permeable layer, and a base. This is applied where enormous contamination is present underneath and separation as well as the removal of pollutants from the soil is not possible (Das and Adholeya 2012).

15.2.4 Adsorption

In this method, the transfer of the solute molecule is carried out at the interface present between two immiscible phases of the solvent that are in contact with each other. The elimination of hazardous particles present in effluents released by textile or paper and pulp industry is executed by the technique of adsorption on a suitable material or on granular activated carbon (Shah 2014).

15.2.5 Membrane Filtration

Membrane filtration technology includes reverse osmosis (RO) and electrodialysis. Reverse osmosis is greatly contributed towards the removal of salts and total dissolved solids from wastewater. Electrodialysis is efficient in removing the dissolved salts by the application of an electrical potential present across the water, which leads to shifting of positive and negative ions onto their corresponding electrodes through ion-permeable membranes (Shah 2014).

15.2.6 Electrochemical Method

This method prevents the formation of unwanted side products economically, but if suspended or colloidal solids are present in high concentration in the wastewater, they decelerate the electrochemical process. That is why such materials required to be removed efficiently before the process of electrochemical oxidation (Shah 2014).

However, physical as well as chemical approaches are generally utilized for eliminating organic compounds from the polluted sites, but these approaches have narrow efficiency as compared to removal by microorganisms (Haritash and Kaushik 2009; Das and Chandran 2011). These methods are quite ineffective when contaminants are present in low concentrations. Moreover, these techniques require high energy and are quite expensive (Fan and Krishnamurthy 1995). Compared with these approaches, bioremediation is emerging as an alternative strategy for the clean-up of polluted sites and water bodies through where remediation by other conventional technologies is not effective.

> Unlike a brute force approach, bioremediation can help make a very specific change to the system so that it heals. Frank Loeffler, PhD

The term bioremediation includes a wide variety of processes to control pollutants and their associated problems (Kandelbauer and Guebitz 2005). "Bio-remediate" means to utilize living organisms to remediate an environmental problem (Sasikumar and Papinazath 2003). Thus, the process that utilizes microbes or the biocatalysts produced by these to restore the surroundings modified by contaminants in the native state is known as "bioremediation" (Vidali and Moffat 2005; Salam and Das 2012; Prasad 2017, 2018). In broad terms, it is the scientific approach for removing contaminants from the environmental surroundings and maintaining the ecological balance by prevention of pollution in the future (Sasikumar and Papinazath 2003). The microbe-mediated remediation alters the harmful chemicals to less/non-toxic forms with less input of time, energy and chemicals (Haritash and Kaushik 2009). Furthermore, the microbe-based approaches are far more superior and efficient for the removal of various organic and inorganic contaminants from the polluted sites as compared to traditional approaches (Chowdhary et al. 2018; Thakare et al. 2021). Nanotechnology has also played a great role in remediation carried out by microbes. Increased bioavailability is the major challenge in bioremediation so it has also been combined with nanotechnology for the removal of contaminants from the soil (Mehndiratta et al. 2013; Prasad and Aranda 2018).

15.3 Bioremediation as a Sustainable Method for Remediation of Contaminants

Sustainable remediation approach has been described as a "treatment or a combination of treatments, which overall benefited humans and the surrounding living organisms (Jonsson and Haller 2014). The use of bioremediation seems to be the best alternative that involves microorganisms for resolving and eliminating contaminants by various treatment strategies (Agwu and Kalu 2012; Abatenh et al. 2017). Bioremediation is the sustainable, profitable and environmentally friendly approach for managing the polluted soils, sediments and groundwater (Banat et al. 2000; Kandelbauer and Guebitz 2005; Magan et al. 2010; Karigar and Rao 2011; Kumar et al. 2011; Nawaz et al. 2011; Niti et al. 2013; Fernando et al. 2018). Bioremediation utilizes the biological processes involving microbial metabolism and several enzymes to remediate the toxic effluents produced from sewage, domestic and industry into less toxic form (Kulkarni and Chaudhari 2007; Agwu and Kalu 2012; Gillespie and Philp 2013; Niti et al. 2013; Jonsson and Haller 2014; Alkorta et al. 2017; Shah 2017; Prasad and Aranda 2018). Bioremediation approaches focus on utilizing the metabolic capacity to attenuate the harmful effects of the contaminants through the following methods (Pandey et al. 2009): (i) detoxification of toxic compounds into less toxic form, (ii) complete degradation of contaminants and (iii) Immobilization of the contaminant.

Bioremediation can be operated both in situ and ex situ depending on the nature of contaminants and site conditions (Pandey et al. 2009; Agwu and Kalu 2012; Thakare et al. 2021):

15.3.1 In Situ Bioremediation

"*In situ* bioremediation" involves the remediation of contaminants under natural conditions of the environment by the use of the metabolic potential of microbes, without excavating or physical removal of contaminated soils or water from its original location (Sasikumar and Papinazath 2003; Kulkarni and Chaudhari 2007). Its cost-effectiveness and least disturbances due to no need of transport of pollutants from one site to another make it the most attractive alternative. Soil depth is the major restriction, thus rendering its use in a few cases (Niti et al. 2013; Shah 2017). Chemotaxis is a significant process in in situ bioremediation approaches, as microbes with their metabolic chemotactic abilities may shift into areas affected with pollutants. With the enhancement of chemotactic capabilities of microbial cells, in situ bioremediation will be a secure method in future to biodegrade the hazardous waste (Sasikumar and Papinazath 2003; Niti et al. 2013). In situ bioremediation has several advantages in comparison with conventional approaches; for instance, it is quite inexpensive and exhibits less deleterious effects on the surroundings along with limited production of secondary pollutants (Ateia et al. 2016).

15.3.2 Ex Situ Bioremediation

"*Ex situ* bioremediation" process involves the physical excavation of polluted soil or water from the environment and subsequent transport to other sites for treatment (Shah 2017). Ex situ bioremediation strategy is considered by interventions to biodegrade chemical contaminants (Kulkarni and Chaudhari 2007).

Both in situ and ex situ bioremediation approaches can be facilitated by different techniques (Fig. 15.2) such as bioreactors, biopiles, land farming, biocell treatment, composting, windrow, biostimulation, bioaugmentation, bioventing, biosparging, bioslurping and biosorption.

(a) *Bioreactor*: Bioreactor is basically a vessel that provides optimum conditions for catabolic activity of microbes for degradation of pollutants. A slurry (soil with water) or sludge is fed into the bioreactor along with microorganisms which change the pollutants into less toxic form by a sequence of metabolic reactions (Whiteley and Lee 2006; Agwu and Kalu 2012; Azubuike et al. 2016). Pretreatment of the polluted soil is required, or the pollutant can be stripped from the soil through vacuum extraction before placing it into the bioreactor (Das and Adholeya 2012). There are different modes in which bioreactor can be operated, i.e. batch, fed-batch, sequencing batch, continuous and multistage bioreactor (Azubuike et al. 2016).

(b) *Biopile*: This method of bioremediation involves the excavation of polluted soil above the ground. Nutrients and aeration are provided to increase the microbial activities, thus improving the bioremediation process (Azubuike et al. 2016). Biopiling provides favourable conditions for native aerobic microbes (Shah

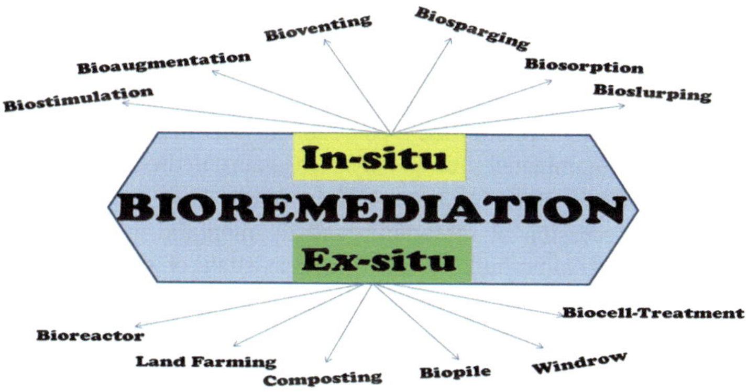

Fig. 15.2 Schematic representation types of bioremediation, upper portion showing in situ bioremediation and lower showing ex situ bioremediation

2017). This is an unrestrained microbial technology in which soils removed are mixed with soil amendments and placed on a treatment area, and bioremediation using forced aeration is carried out. The basic biopile system includes a treatment bed, an aeration system, an irrigation/nutrient system, and a leachate collection system. Factors such as moisture, heat, nutrients, oxygen and pH are controlled to enhance the rate of biodegradation (Shukla et al. 2010; Niti et al. 2013). Soil piles can be up to 20 feet high and can be covered with plastic to control the runoff, evaporation, and volatilization. Cost of maintenance and operation and the lack of power supply especially at remote which would enable uniform distribution of air in contaminated piled soil via air pump are some of the setbacks of biopiles (Azubuike et al. 2016).

(c) *Land Farming*: In this process, uncovered polluted soils are dispersed to form a thin layer on the surface of the ground. The metabolic activity of the indigenous microflora is stimulated via providing proper aeration, nutrients and moisture for the remediation of contaminants present in soil (Agwu and Kalu 2012; Niti et al. 2013). It is among one of the easiest bioremediation techniques because of its low-cost and less equipment requirement, which can be adopted owing to sufficient land availability (Das and Adholeya 2012; Azubuike et al. 2016). Due to the site of treatment, it is mostly considered as ex situ bioremediation, whereas in some cases, it is regarded as an in situ bioremediation technique (Azubuike et al. 2016). To enhance degradation, increased oxygen supply as well as mixing is done by ploughing, harrowing or milling at regular intervals (Das and Adholeya 2012). The area is also tilled to enhance aeration as well as to ensure that degradation and immobilization of contaminants within the upper and underlying soil layers occur (Whiteley and Lee 2006). Land farming is best suited for pesticide-contaminated soils, because of full-scale bioremediation technology in which the size and location of the spreading operation are decided based upon the application rate of pesticides to avoid concentrations that would be unsafe in soil, groundwater or crops (Niti et al. 2013).

(d) *Biocell Treatment*: This technique resembles farming of land except, the method in which the polluted soil has placed the form of a pile with alternative vent layers to enhance the air provision which is needed for the growth of bacteria (Agwu and Kalu 2012).

(e) *Composting*: It is a method in which a mixture of contaminated soil and organic materials is left into a pile of bed-type treatment with the microorganism to act on the highly contaminated sites, mostly used against hydrocarbon contaminants (Whiteley and Lee 2006; Agwu and Kalu 2012). This is a specific method involving a succession of mesophilic and thermophilic microorganisms and mixing with an organic bulking agent, such as straw or wood chips. The pile is aerated by forced aeration or pile turning. On the other hand, factors such as moisture content, pH and nutrient content are controlled (Whiteley and Lee 2006). The temperature is increased due to heat during the process of composting of organic compounds by microorganisms, and it may range from 55 °C to 65 °C. (Shukla et al. 2010; Das and Adholeya 2012; Niti et al. 2013).

Mechanical treatment is done by grinding, mixing and sieving, to remove non-degradable materials such as metals, plastics, glass and stones, which gives good conditions for the biological treatment of compostable materials (Das and Adholeya 2012). High substrate levels in composts can lead to co-metabolism of organic contaminants (Niti et al. 2013).

(f) *Windrow*: Soils with added amendments are layered in the form of long piles known as windrows. At first, the polluted soils are removed from the site, which is then screened for removing big rocks and waste debris. This soil is then shifted to a composting pad for providing containment and safety from extreme weather conditions. Modifications are used for bulking agents such as straw, alfalfa, manure, agricultural wastes or wood chips as well as a supplementary source of carbon (Niti et al. 2013). Bioremediation is further enhanced by preparing a thorough mixture of the windrow by periodic turning and a commercially available windrow turning machine (Niti et al. 2013; Azubuike et al. 2016). This periodical turning of contaminated soil, along with water addition, enhances aeration, homogeneous distribution of contaminants, nutrients and degradative activities of microbes, thus increasing the bioremediation rate (Azubuike et al. 2016). Factors such as humidity, pH, temperature and concentrations of explosives are monitored. The windrows are removed, and the compost is taken to the final disposal area, at the end of the composting period (Niti et al. 2013).

(g) *Biostimulation*: In this process, the naturally occurring microbial population is managed to monitor and provide an environment that optimizes the growth of microbes and their activity (Agwu and Kalu 2012). In this process, nutrients of the soil, minerals present in traces, acceptors or donors of electrons and several limiting factors as well as amendments are added to enhance the biotransformation of a broad range of pollutants occurring in the contaminated soil (Niti et al. 2013; Banik et al. 2014).

(h) *Bioaugmentation*: This approach includes the application of specific strains or microbial consortia either indigenous or exogenous to the polluted sites that target only specific domain of chemical compounds (Agwu and Kalu 2012; Das and Adholeya 2012; Shah 2017). Two points limit the usage of microbial cultures added in this method:
   (i) Exogenous cultures have to compete with a native population of microbes for the sustainability of useful microbes to complete the process.
   (ii) Native microbes are efficient degraders, and long-term exposure of these microbes to biodegradable waste is well managed by the native microflora (Das and Adholeya 2012).

Bioaugmentation is generally paired with the process of biostimulation by the addition of an optimal amount of water, nutrients and oxygen where the site is polluted; it increases the action of microbes added or enhancing co-metabolism (Niti et al. 2013).

(i) *Bioventing*: The method includes regulated airflow to cause stimulation by delivering O_2 to unsaturated zone to enhance the process of bioremediation, by uplifting the activities of native microorganisms. The amendments in bioventing are carried by the addition of nutrients such as nitrogen, phosphorus and moisture for an enhancement in bioremediation with the extreme goal to achieve the biological transformation of contaminated to a less toxic state (Niti et al. 2013; Azubuike et al. 2016; Shah 2017). Bioventing utilizes airflow at low rates along with only the required amount of oxygen which is necessary for biodegrading, minimizing the volatilization and release of pollutants to the environment. It is beneficial for simpler organic compounds which can be used for the contaminants present deep under the surface of soil and underground water (Shah 2017). This approach is very well-designed for soils of all types ranging from well-drained to coarse-textured (Niti et al. 2013). Bioventing enhances the in situ bioremediation of petroleum–oil–lubricants by providing O_2 to microbes that are present in the soil. In this system, oxygen is supplied directly by injecting air into the residual pollution of soil. A typical bioventing arrangement involves a well and a blower which provides aeration to the soil through well (Shukla et al. 2010).

(j) *Bioslurping*: This approach combines pumping increased by vacuum, soil vapour separation and followed by bioventing to accomplish underground water and soil bioremediation by indirect utilization of air for inducing contaminant biodegradation. It can also be utilized to degrade volatile as well as semi-volatile organic chemicals present in the polluted soils. In this process, excess of moisture in soil limits permeability of air and decreases the rate of transfer of oxygen which in turn reduces activities of microbes (Azubuike et al. 2016).

(k) *Biosparging*: This technique resembles bioventing because aeration is provided in the subsurface of soil to stimulate activities of microbes for promoting the removal of organic pollutants from polluted sites. The efficiency of biosparging is majorly dependent on two factors, namely, the permeability of soil and contaminant biodegradability (Azubuike et al. 2016). This process deals with the exploitation of native microbes to biodegrade organic contaminants of the soil. Biosparging technique is routinely utilized at the sites with mid-range molecular mass petroleum refining products such as diesel oil, while lightweight products of petroleum volatilize at a slow rate and are removed very quickly through sparging (Das and Adholeya 2012). Biosparging enhances the homogenizing in the saturated zone, thereby promoting its exposure between soil and underground water (Shah 2017).

(l) *Biosorption*: Biosorption is one of the most significant approaches among all methods from both the perspectives, i.e. ecological as well as practical for the removal of metals from aqueous solution by a combination of living or dead biomass through the usage of functionally active sites which includes carboxyl, imidazole, sulphydryl, amino phosphate, sulphate, thioether, phenol, carbonyl, amide and hydroxyl moieties (Whiteley and Lee 2006). The capability of microbes to combine with metals from its aqueous solution is known as biosorption. The components of a microorganism that bind to metals and are responsible for biosorption are called biosorbents. Biosorption of metals by the use of microbes is controlled by the parameters such as the nature of biosorbent undergoing the process, age of microbial culture and biomass from where it is originated, the nature of interactions that occur between metals with functional groups which are native to the cell wall of biomass, concentration of biosorbent present during the reaction, properties of metals to be removed and their concentrations present in solution and other factors: pH, temperature and the presence of other cations present during the whole process (Sen and Chakrabarti 2009). Bacteria are magnificent biosorbents because of the presence of their more surface-to-volume ratio along with a favourable number of chemosorption sites that are active, e.g. cell wall teichoic acid of bacteria (Banik et al. 2014).

15.4 Microbes Involved in Bioremediation

Microorganisms play a paramount role not only in controlling the biogeochemical cycles, for maintaining ecological balance, but also in maintaining a healthy environment for us, combating the plant diseases, and play a significant part in cleaning of atmospheric contaminants from the atmosphere (Malla et al. 2018). Microbes are the only organisms present in the environment which have the capability of exploitation of organic/inorganic compounds from the surrounding environment for their growth through many processes, for example, binding, oxidation, volatilization, and immobilization (Kulkarni and Chaudhari 2007; Malla et al. 2018). Microbes are empowered to inhabit different ecological niches where they can process extraordinary metabolic and physiological activities (Kulkarni and Chaudhari 2007). Microorganisms present in the environment such as bacteria, fungi, actinomycetes, protozoa, and algae carry out the process of recycling of organic compounds from where the new plant life can be raised and helps to transform organic waste (Fan and Krishnamurthy 1995). The microbes can be indigenous, i.e. belong to a particular contaminated space or can be isolated from anywhere else and then exposed to the contaminated sites (Shah 2014). Microorganisms are nature's recyclers because they have a vital metabolic potential for transformation of natural and synthetic compounds into energy sources and organic materials for their growth and to produce energy for their movement and respiration; therefore, they are considered as infallible (Fan and Krishnamurthy 1995; Rieger et al. 2002; Sasikumar and Papinazath 2003). Microbes break down organic chemicals into CO_2, H_2O and CH_4 (Ateia et al. 2016; Nandal et al. 2015). Microbes can substitute chemical or physical biodegradation processes by biological processes into a cheap and environmentally friendly way. Therefore, it is a promising, new environmental biotechnological technique as they can quickly respond and adapt to changes in the surrounding conditions (Sasikumar and Papinazath 2003; de Mora et al. 2005). Parameters such as pH, temperature, nutrient levels and aeration are essential to facilitate the metabolism of microbial cells for the biotransformation process carried out by them. When a microbe comes into proximity of complex organic matter, extracellular enzymes are liberated that helps in the remediation of materials with high molecular weight into soluble fractions, which are easily transferred by the cell wall for the process of assimilation. There are only particular species of microbes that have proved to be potential organisms that can convert specific substrates and can be grown under laboratory conditions as effective agents of bioremediation (Fan and Krishnamurthy 1995; Karigar and Rao 2011). The following are the microbes based on different categories discussed as per findings till now which are been utilized for the process of bioremediation.

15.4.1 Bacteria

Bacteria are the category of microbes which are easily obtained by the method which does not depend on additional clean up and are actively involved in biodegradation of organic pollutants (which serves as food to it) from polluted places (Haritash and Kaushik 2009; Agwu and Kalu 2012). Thus, this technique is a self-clearing process that ensures environmental sustainability. The use of bacteria in clearing up of soil contaminated with explosives, petroleum and hazardous wastes, sewage chlorinated solvents, pesticides and agricultural chemicals has been stated by many scientists (Agwu and Kalu 2012). Bacteria based on their oxygen uptake can be categorized into the following:

Aerobic Bacteria: Aerobic bacteria have been recognized for their degradative abilities that are isolated and primarily focus on Gram-negative bacterial strains that belong to the genera such as *Burkholderia*, *Ralstonia*, *Achromobacter*, *Alcaligenes*, *Comamonas*, *Rhodococcus*, *Pseudomonas*, *Sphingomonas* and *Mycobacterium*, which is usually reported for biodegradation of pesticides and hydrocarbons (alkanes as well as polyaromatic hydrocarbons, PAHs).

Anaerobic Bacteria: Aerobic bacteria are more often utilized as compared to anaerobic bacteria. Anaerobic bacteria are being used for the process of biodegradation of polychlorinated biphenyls (PCBs), which are subjected to reductive dehalogenation in the sediments of the river; dechlorination of solvents such as trichloroethylene (TCE) and chloroform is gaining the interest of the researchers (Kumar et al. 2011).

Apart from anaerobic bacteria, many other strains of bacteria are also potentially capable of oxidation of PCBs into simpler compounds such as *Pseudomonas*, *Burkholderia*, *Comamonas* and *Rhodococcus* sp. (Jing et al. 2018). Bacteria are not only the most effective mode for degradation of petroleum but also primary degraders of spilled oil (Das and Chandran 2011). Bacteria that are capable of bioremediation of petroleum products include *Bacillus* sp., *Aeromonas*, *Moraxella*, *Beijerinckia*, *Flavobacteria*, *Chrobacteria*, *Nocardia*, *Corynebacterium* sp., *Modococci*, *Streptomyces*, *Arthrobacter*, and *Aeromonas*. *Pseudomonas fluorescens*, *P. aeruginosa*, *B. subtilis*, *Alcaligenes* sp., *Acinetobacter lwoffi*, *Flavobacterium* sp., *Micrococcus roseus* were isolated from the contaminated stream that biodegrades the crude form of oil (Das and Chandran 2011; Joutey et al. 2013; Coelho et al. 2015). In areas where hydrocarbons cause contamination, heterotrophic genera of bacteria *Micrococcus*, *Mycobacterium*, *Vibrio*, *Pseudomonas*, *Corynebacterium*, *Arthrobacter*, *Achromobacter*, *Acinetobacter*, *Flavobacterium*, *Brevibacterium* and *Nocardia* are found to be present (Jonsson and Haller 2014). Among all of them, the best hydrocarbon-biodegrading bacteria was *Bacillus* (Joutey et al. 2013). The bacterial strains which are capable to often remediate aromatic hydrocarbons have been isolated so far, basically from the soil with bacteria, i.e. Gram-negative, belongs to the genus *Pseudomonas* (Coelho et al. 2015). Microbes play a significant role in biodegradation of heavy metals which are most volatile by the process of methylation. For instance, mercury (Hg II) easily undergoes methylation into gaseous methyl mercury by various bacterial species, for example, *Staphylococcus*, *P. aeruginosa, Brevibacterium iodonium*, *Bacillus pumilus*, and *Alcaligenes faecalis*. Bacteria are capable to leach higher concentrations of heavy metals such as arsenic, cobalt, copper, cadmium and zinc from polluted soils, for example, acidophilic iron bacteria like *Arthrobacter* spp., *Acidithiobacillus ferrooxidans*, *Pseudomonas veronii*, *Kocuria flava*, *Sporosarcina ginsengisoli*, *Bacillus cereus*, sulphur-oxidizing bacteria and *Burkholderia* spp. by the process of methylation and redox reactions (Mateos et al. 2006; Joutey et al. 2013; Coelho et al. 2015). Detoxification-based, especially arsenic (V), reduction has been best studied in *E. coli* (Mateos et al. 2006). Many catabolic plasmids had been discovered in the species such as *Alcaligenes*, *Arthrobacter*, *Flavobacterium*, *Actinobacter*, *Moraxella*, *Klebsiella* and *Pseudomonas* (Sayler and Ripp 2000). A wide range of bacteria metabolize pesticides such as genera *Alcaligenes*, *Flavobacterium*, *Pseudomonas* and *Rhodococcus* (Nawaz et al. 2011). Bacterial species that belong to genera *Bacillus*, *Cellulomonas*, *Acinetobacter*, *Pseudomonas*, *Rhodoseudomonas*, *Nitrobacter and Nitrosomonas* help in the process to mineralize the organic wastes (Panigrahi et al. 2005). Bacteria cells involved in the reduction of metals are *Geobacter metallireducens*, which has been recently added in the list of bacteria that eliminate radioactive waste of uranium, from water drainage during mining as well as from polluted underground waters (Sasikumar and Papinazath 2003).

15.4.2 Fungi

Fungi have been exploited by human beings for many applications such as degradation, or deterioration, of a broad range of contaminants by the method called mycodegradation or mycodeterioration (Leitão 2009). On the other hand, the main process known as *mycoremediation* utilizes fungus for the removal of chemical contaminants present in the soil. In this method, fungi uses certain biocatalysts and acids which are secreted naturally for the decomposition of hazardous chemicals into less or non-hazardous substances (Agwu and Kalu 2012; Prasad 2017, 2018; Prasad et al. 2021). Fungus, alone or in contribution with bacterial cells or plants, is an important component of biotechnology designed to biodegrade the polluted natural resources (Harms et al. 2011).

The application of fungus first came into attention for clean-up of polluted soil during the mid-1980s (Magan et al. 2010). The degradative ability to attack the wood of white-rot fungi was identified. White-rot fungi is a class of microorganisms that includes *Pleurotus ostreatus, Phanerochaete chrysosporium* and *Trametes versicolor* that produce efficient biocatalyst that can remediate a variety of organic pollutants, and the degradation activity is because of the lignin-degrading system of the fungi (Haritash and Kaushik 2009; Magan et al. 2010). It is also competent in decomposing dyes under aerobic conditions, and due to the production oxidoreductases, they are potent in degrading lignin and other aromatic compounds related to dyes (Ali 2010; Coelho et al. 2015). Lignin is a naturally occurring three-dimensional polymer present in woody plants and structurally complex; that is why it is resistant to microbial degradation (Magan et al. 2010). The fungal culture of *Phanerochaete chrysosporium* is well studied for its bioremediation capability, as well as *Nematoloma* spp. and *Pleurotus* spp. which can remediate a wide range of polycyclic aromatic hydrocarbons (PAHs) include anthracene, pyrene, di- and tri-benzoic acids, polychlorinated biphenyls (PCBs), 2,3,7,8-tetrachloro dibenzo-*p*-dioxin, ichlorodiphenyltrichloroethane (DDT), lindane, and chlordane. It is also been utilized as a model for the study of bioremediation of DDT, 2,3,7,8 tetrachloro-dibenzo-p-dioxin, lindane and benzopyrene as it helps in oxidation of organo-pollutants into CO_2 by the use of its enzyme system that degrades even under conditions when nitrogen is deficient (Nawaz et al. 2011; Salam and Das 2012). White-rot fungi, *Trametes pubescens*, was demonstrated as an efficient bioremediating agent for treating phenolic wastewaters (Ryan et al. 2005).

The phylum Basidiomycetes accounts for about 34% among all the species of fungi described, and it is inhabited in terrestrial environments only but very rare in aquatic habitat (Harms et al. 2011). The most popular edible mushroom belongs to class Basidiomycetes (*Agaricus bisporus*) is efficient in the accumulation of silver (Ag) as a demonstration given by Falandysz et al., and inspected Ag uptake from substrates which are enriched artificially. Weber and co-workers reported that *Boletus badius*is very effective in the accumulation of gold (Au) and arsenic (As), which is reserved in different mushroom parts, gold is accumulated in cap and stalk portion, while arsenic (As) gets collected in the hymenium region. Vetter investigated approximately 88 species of basidiomycetes fungus, with the accumulation of different heavy metals in certain *Agaricus* spp. and other species *Lepista nebularis*, *Flammulina velutipes*, *Macrolepiota rhacodes*, and *Clitocybe inversa.* A remarkable concentration of cadmium (Cd), i.e. 35 mg/kg was noted which accumulated in *Agaricus* sp. In comparison to it, the best Cd-hyperaccumulator plant *Thlaspi caerulescens* can concentrate over 100 mg/kg (Pletsch et al. 1999). Basidiomycetes species dwelling in soil and decomposes litter, i.e. *Stropharia* sp. and *Agrocybe* sp. popular to biodegrade organic contaminants. Although the process of mineralization of chemical substances present in the environment is not much common in brown-rot basidiomycetes especially *Gloeophyllum* spp. Decompose fluoroquinolone, chlorophenols, and antibiotics. The subphylum Pucciniomycotina of basidiomycete contains yeast mitospores belonging to the genus *Rhodotorula* that metabolizes cresols, crude oil containing compounds such as PAHs and RDX (Royal Demolition Explosive). The subphylum Mucoromycotina and the genera *Cunninghamella, Mucor* and *Rhizopus* include fungus that can degrade PAHs, pesticides, textile dyes, trinitrotoluene, etc. (Harms et al. 2011).

Fungi bear remarkable potency to bioremediate hydrocarbons with complex structure and long-chain length such as *Mucorales* and *Moniliales* show a higher rate of bioremediation ability (Fan and Krishnamurthy 1995). On the other hand, Fungal genera including *Neosartorya, Graphium, Aspergillus, Talaromyces, Cephalosporium,* and *Penicillium* as well as genera of yeast including *Yarrowia, Rhodotorula mucilaginosa, C. lipolytica, Pichi, Geotrichum* sp., *and Trichosporon mucoides* were isolated from soil polluted with petroleum and also found to be the effective organism for bioremediation of hydrocarbons in crude oil (Das and Chandran 2011). Alkane-utilizing yeast includes *C. lipolytica, C. tropicalis, Aureobasidion pullulans,* and *Rhodoturula rubra. Candida ernobi* and *Rhodotorula aurantiaca* have the ability to biodegrade diesel oil. *Candida methanosorbosa* is studied for biodegradation of aniline (Joutey et al. 2013). Another study says that species of the genera including *Cladophialophora* and *Exophiala* assimilate toluene. *Aspergillus* and *Penicillium* spp. biodegrade synthetic dyes, chlorophenols, pesticides, polycyclic aromatic hydrocarbons, aliphatic hydrocarbons and 2,4,6-trinitrotoluene. *Cordyceps, Fusarium* and *Pseudallescheria* spp. are reported for metabolizing polychlorinated dibenzo-*p*-dioxins(PCDDs). Mitospores of *Acremonium* spp. can remediate royal demolition explosive and polyaromatic hydrocarbons, whereas *Graphium* spp. has the ability to biodegrade methyl-*tert*-butyl ether, as well as *Phoma* spp. degrade pesticides, polyaromatic hydrocarbons and synthetic dyes. *Kluyveromyces, Candida, Pichia, Saccharomyces, Yarrowia,* and *Neurospora* are the potential bioremediaters of crude oil, trinitrotoluene, *n*-alkanes, *n*-alkylbenzenes, polyaromatic hydrocarbons and the endocrine-disrupting chemical (EDC) nonylphenol (Harms et al. 2011). Fungi used in bioremediation for the treatment of heavy metals include *Aspergillus versicolor, Penicillium canescens* and *Aspergillus fumigates* (Coelho et al. 2015). Thus, bioremediation approaches using fungus have noteworthy benefits (Nawaz et al. 2011).

15.4.3 Protozoa and Algae

Algae and protozoa are essential parts in the community of microbes in both aquatic as well as terrestrial ecosystems, which have been scantly reported for their role in biodegradation of hydrocarbons (Wasi et al. 2013). *Prototheca zopfi* is not only capable to utilize crude oil and a mixture of hydrocarbon substrate but also carries out a substantial amount of biodegradation of n-alkanes and isoalkanes as well as aromatic hydrocarbons (Das and Chandran 2011). Cerniglia et al. observed those nine cyanobacteria, five green algae, one red alga, one brown alga and two diatoms that are effective in oxidizing naphthalene (Joutey et al. 2013). Among the genus of green algae, *Chlorella* has been recognized as a potent species for biosorption of metals. Gale reported that the efficient removal of lead by blooms of *Chlorella, Cladophora, Rhizoclonium, Hydrodictyon, Spirogyra and Oscillatoria.* Scarano et al. stated that cadmium and lead phytochelatin complexes, formation occurred in a marine microalga when exposed to metal. Brown alga *Fucus serratus* was reported for the accumulation and transformation of arsenic into arsenic compounds (Wasi et al. 2013). Some research has shown that some of the species of fresh algae (e.g. *Scenedesmus platydiscus*, *S. quadricauda*, *Chlorella vulgaris* and *S. capricornutum*) have potential to uptake as well as degrade PAHs. Degradation of azo dyes as carbon (C) and nitrogen (N) sources by the usage of *C. vulgaris* and *C. Pyrenoidosa* was carried out, but degradation was based on the chemical structure of the dyes. *C. vulgaris*, *Lyngbyalagerlerimi, Nostoclincki, Oscillatoria rubescens, Elkatothrix viridis* and *Volvox aureus* were capable of decolorizing and removing methyl red, orange II, G-Red, basic cationic and basic fuchsin from the effluent waste. Some spp. of algae such as *Chlorella*, *Anabaena inacqualis*, *Westiellopsis prolifica*, *Stigeoclonium tenue* and *Synechoccus* sp. can accumulate and remove heavy metals, but the conditions of operation constrain the practical utilization of these organisms. The microalga *S. incrassatulus* was observed for removal of chromium (VI), cadmium (II) and copper (II) in continuous cultures. Green algae were also observed bioremediation of heavy metals; *C. Sorokiniana* for chromium (III) removal (Joutey et al. 2013). There are several possible hypotheses about the mechanism of protozoa speeding up the bioremediation process of organic pollutants, which generally includes the following six points:

1. The process to mineralize the nutrients that upgrades nutritional turnover.
2. Activation of bacteria that regulates the number of aged cells or excrete active components.
3. Grazing should be selective which decreases competition of resources and space for better growth of bacteria involved in degradation.
4. Physical hindrance can enhance overall O_2 content and the matter surface that is biodegraded.
5. Direct biodegradation can secrete specialized biocatalyst involved in remediation.
6. Sym-metabolism is the process that offers energy and carbon source for the degradation process carried out by bacteria (Joutey et al. 2013).

15.5 Microbial Enzymes Involved in Bioremediation

Enzymes act as a robust tool for sustaining pure environmental surroundings in many methods. Enzymes are being used for the motive of bioremediation in various industries, for example, oil, detergent, petroleum, agro-food, pulp and paper, textile, leather, animal feed and especially chemical- and biochemical-based industries. These biocatalysts are an efficient mode to maintain a clean environment by their effective use in waste management (Ahuja et al. 2004). Enzymes may present several advantages not only on traditional technologies but also on the bioremediation process carried out by microbes. Enzymes are used under extreme conditions are efficient by restricting the activity of microbes, not only when concentrations of pollutants are very low but also when microbial predators or antagonists are present

(Gianfreda and Bollag 2002). All the metabolic reactions in living organisms especially microorganisms are mediated by enzymes and the techniques that use enzymes are environmentally friendly processes, therefore such characteristics make them eco-friendly catalysts (Rao et al. 2010; Abatenh et al. 2017). All metabolic enzymes belong to the group of hydrolases, lyases, transferases, oxidoreductases, ligases and isomerases (Abatenh et al. 2017). Enzymes are biocatalysts that help in the facilitation of transformation of reactants into products by the provision of positive conditions to lessen the activation energy during reaction found universally in all living organisms (Ahuja et al. 2004; Karigar and Rao 2011). Enzymes are not expensive and catalysts that are friendly to nature with the capability of bioremediating the many harmful compounds that are not friendly to our environment as per studied present scenario of societal ecological standards (Rao et al. 2010; Karigar and Rao 2011).

Structurally an enzyme is a protein that consists of one polypeptide moiety, and the part of these biocatalysts that actively takes part in the catalytic processes is known as active sites. These biocatalysts have one or more groups that are necessary for their catalytic activity, associated with the active sites, either by covalent or by other bonds; this protein part in such enzymes is called apoenzyme. On the other hand, the non-protein part is generally called as the prosthetic group. Both apoenzyme and prosthetic groups combine to form holoenzyme (Karigar and Rao 2011). The action of enzymes may be intracellular, i.e. on the inner side of the cells from where they are originated or act extracellularly, i.e. in the absence of cells of their origination. If the enzyme is presently free, i.e. insoluble form, this type of catalysis is known as homogenous, or immobilized; on the other hand, if the enzyme is linked through a solid matrix, then catalysis is called as heterogeneous. The main producers of these catalysts are generally white-rot fungi, bacteria, and microbe-plant associations (Rao et al. 2010). The extracellular enzymes produced by fungi especially filamentous fungi is paramount because it is primarily utilized bioremediating or neutralizing toxic contaminants, for instance, polycyclic aromatic hydrocarbons (PAHs), which is a primary toxic pollutant that poses a potential risk to human health (Bonugli-Santos et al. 2010).

Enzyme-based degradation process usually occurs at a faster rate than abiotic oxidation reactions as well as organic chemical pollutants that moves into the metabolic pathway carried out by microbes present in the soil. The reason behind this enzyme-based reaction is the free energy potential which is being used by the microbes for degradation processes (Jonsson and Haller 2014). Table 15.1 shows various enzymes produced by various microorganisms and the substrate they act upon for degradation. The development of enzymes for the process of remediation usually done in three steps mentioned below:

(i) Isolation and identification of template enzyme from the natural resources.
(ii) Isolation of the genes that encode for a particular enzyme.
(iii) Improvement of enzymatic efficiency by improving enzyme properties in the laboratory (Sutherland et al. 2004).

Table 15.1 Various types of enzymes produced by microorganisms

Enzyme produced	Microorganism	Substrate act upon	References
Fungal laccase or p-diphenol: Dioxygen oxidoreductases include benzenediol: Oxygen oxidoreductases, ferroxidases, ascorbate oxidase	*Marasmius quercophilus*, (spent mushroom compost) *Pleurotus laccase, Pleurotus ostreatus, Trametes versicolor, Cerrena unicolor, Trametes villosa, Cerrena versicolor, Fusarium solani, Cladospora cladosporoides, Nematoloma frowardii, Rhus vernificiera, Pycnoporus coccineus, Coriolus (Trametes) versicolor, Agaricus biosporus*	Polyaromatic hydrocarbons (PAHs), aromatic amines, phenolic compounds, organophosphorus, polychlorinated biphenyls (PCBs), aminodinitrotoulenes, methoxyphenolic acids, metal complex phenolic acids, recalcitrant pollutants	Haritash and Kaushik (2009), Rao et al. (2014), Bonugli-Santos et al. (2010), Chandra and Chowdhary (2015), Rao et al. (2010), Duran and Esposito (2000), Mugdha and Usha (2012), Gianfreda and Rao (2004), Harms et al. (2011), Karigar and Rao (2011), Tanaka et al. (2000), Upadhyay et al. (2016), Fan and Krishnamurthy (1995) and Magan et al. (2010)
Bacterial laccases	*Azospirullium lipoferum, Bacillus subtilis, Brodetella campestris, Caulobacter crescentus, Escherichia coli, Mycobacterium tuberculosum, Pseudomonas syringae, Pseudomonas aeruginosa, Stenotrophomonas maltophilia*	Synthetic dyes	Chandra and Chowdhary (2015)
Tyrosinase or polyphenol or oxidase or catecholase		Monophenols, biphenols, carcinogenic phenols, aromatic amines	Duran and Esposito (2000), Harms et al. (2011) and Rao et al. (2014)
Peroxidases: 1. *Heme peroxidases*: Class I: Cytochrome C peroxidase, ascorbate peroxidase, catalase peroxidase Class II: Lignin peroxidase, mangsanese peroxidase, versatile peroxidase	*Phanerochaete chrysosporium, Coprinu scinerens, Arthromyces ramosus, Nematoloma frowardii* (white-rot fungi)	Dichlorodiphenyltrichloroethane (DDT), polychlorinated biphenyls (PCBs), polyaromatic hydrocarbons (PAHs), methoxy benzene, phenolic aromatic compounds, anilines, benzidines, heteroaromatic compounds (nylon)	Karigar and Rao (2011), Lalwani et al. (2014), Duran and Esposito (2000) and Gianfreda and Rao (2004)

Class III: Horseradish peroxidase			
2. *Nonheme peroxidases*:			
Thiol peroxidase, alkyl hydroperoxidase, nonheme haloperoxidasee, manganese catalase, NADH peroxidase			
Chloroperoxidase	*Caldariomyces fumago, Phanerochaete chrysosporium, Corprinopsis cinerea*	Phenolic compounds, chloride ions, phenols, synthetic dyes, polyaromatic hydrocarbons (PAHs)	Duran and Esposito (2000) and Harms et al. (2011)
Oxygenase		Halogenated organic compounds, aliphatic and aromatic compounds	Karigar and Rao (2011)
Cytochrome P 450 monooxygenase	*Cuninghamella elegans*	Phenol, catechol	Harms et al. (2011)
Manganese peroxidase	*Lentinula edodes, Nematoloma frowardii, Anthrocophyllum discolor*	Monoaromatic phenols, chlorophenols, aromatic dyes, pyrene, polyaromatic hydrocarbons (PAHs), aliphatic substances	Duran and Esposito (2000), Gianfreda and Rao (2004), Haritash and Kaushik(2009) and Rao et al. (2014)
Lignin peroxidase	*Nematolomafrowardii* (white-rot fungi)	Polyaromatic hydrocarbons (PAHs), pyrene, anthracene	Gianfreda and Rao (2004) and Rao et al. (2014)
Quinone reductase	White-rot fungi, brown-rot fungi	Quinones, catechol	Harms et al. (2011)
Cyanide dioxygenase	*Escherichia coli, Pseudomonas fluroscens*	Cyanide	Ebbs (2004)
Polylactic acid depolymerase	*Amycolatopsis* sp.	Polylactic acid	Gianfreda and Rao (2004)
Cellulase	*Asperigillus niger, Penicillium chyrsogenum, P. funicolosum, Bacillus subtilis, Trichoderma resei, Humicola* sp.	Solid municipal waste polyester	Gianfreda and Rao (2004) and Karigar and Rao (2011)
Xylanase	*Penicillium chyrsogenum, Aspergilus niger*	Lignin	Leitão (2009) and Rao et al. (2010)

(continued)

Table 15.1 (continued)

Enzyme produced	Microorganism	Substrate act upon	References
Esterase		Polyacrylates, polyurathanes, polyactides	Gianfreda and Rao (2004)
Phytase		Nylon, starch polymers	Gianfreda and Rao (2004)
Catechol deoxygenase	*Pseudomonas pseudoalcaligans*	Polychlorinated biphenyls (PCBs), chloroethenes, phenols	Duran and Esposito (2000)
Cyanide hydratase	*Fusarium lateritium, Cryptococcus* sp., *Candida guilleirmondii*	Nitriles, cyanide	Ebbs (2004)
Cyanidase or formamide cyanase	*Phanerochaete chrysosporium* (white-rot fungi), *Acremonium strictum, Thiobacillus thioparus*	Polyaromatic hydrocarbons (PAHs), phenolic compounds, anilinic compounds, azo dyes	Ebbs (2004)
Lipase	*Candida rugosa, Candida cylindracea, Phycomyces nitens*	Hydrocarbons, triglycerol fatty acids	Karigar and Rao (2011) and Tanaka et al. (2000)
Protease	*Penicillium chyrsogenum, Trametes versicolor*	Nylon	Gianfreda and Rao (2004)
Chitinase	*Actinobacteria*	Nylon	Gianfreda and Rao (2004)
Reductive dehalogenase	*Basidiomycetes* sp.	Chlorophenol	Harms et al. (2011)
Organophosphorus hydrolase		Organophosphates	Rao et al. (2014)
Bacterial hydrolase	*Archomobacter* sp., *Pseudomonas* sp., *Flavobacterium* sp., *Nocardia* sp., *Bacillus cereus*	Carbofuran carbaryl diazinon	Rao et al. (2010)
Nitrilase	*Nocardia* sp., *Rhodococcus* sp., *Fusarium solani, Aspergilus niger, Cryptococcus* sp., *Candida guilliermondii*	Aliphatic and alicyclic nitriles	Rao et al. (2010) and Ebbs (2004)

Enzymes that are particularly involved in bioremediation of hydrocarbon compounds are dehydrogenase, oxygenase and enzymes that can degrade lignin, which includes: fungal laccases and peroxidase including both lignin peroxidase and manganese peroxidase. These extracellular enzymes catalyze free radical formation by carrying out oxidation and rendering destabilization of bonds in a molecule (Haritash and Kaushik2009). A broad range of bacteria and fungi produces several extracellular enzymes including proteases, cellulases, amylases, xylanases, phosphatases, phytases and esterases. These biocatalysts not only are physiologically important for living organisms but also help to catalyze the hydrolysis of large molecules (Gianfreda and Rao 2004).

15.6 Molecular and Omnics Approaches Used for the Modification of Microbes to Increase Bioremediation

Applications of genetically engineered microbes (GEMs) in biodegradation have been in limelight to improvise the biodegradation of toxic wastes under in vitro conditions over past decades. The GEMs have the high biodegradative capacity; therefore, they are used in bioremediation process monitoring, strain monitoring, stress response, end-point analysis and toxicity assessment (Das and Chandran 2011; Singh et al. 2011; Joutey et al. 2013). A GEM is a microbe whose genetic material has been transformed by the use of genetic engineering techniques, which are collectively known as rDNA technology (Joutey et al. 2013). Recombinant DNA technology targets the utilization of various techniques including PCR, in situ hybridization, antisense RNA technique, enzyme-specific antibodies and site-directed mutagenesis, which allows betterment of the metabolic structure of microorganisms and helps in higher accumulation of heavy metals or decreases their harmful effect, by the cleaning of water bodies. The most commonly used method among these includes engineering techniques with single genes or operon system, construction of pathway and modification of the sequences of genes that already exist. Tests of DNA can show the presence of specific microbial potential and by using enzyme-specific immunoglobulins can predict the action of metabolic enzymes in the process of biodegradation of pollutants. DNA probes those targets only particular genetic sequences, i.e. the particular genes which are responsible for the bioremediating capability of a microbe, which can be used to identify and characterize a polluted site throughout the process of bioremediation. The genetically first two modified bacterial strains were *P. aeruginosa* (NRRL B-5472) and *P. putida* (NRRL B-5473); both of these contained genes for the biodegradation of naphthalene, salicylate and camphor. *Pseudomonas fluorescens* HK44, which is capable of degrading naphthalene, presents the first example of a GEM for purpose of bioremediation (Coelho et al. 2015). Enzymatic bioremediation improvised by molecular tools can be particularly suitable to meet the demands of a place where rapid remediation is necessary to eliminate pollutants. Molecular tools help us to increase expression levels by manipulations of not only physiochemical conditions but also at the genetic level of the organism, to enhance enzyme production in many different conditions (Peixoto et al. 2011). Currently, researchers can nick and insert the appropriate genes which are responsible for the manufacturing of particular enzymes that helps in the bioremediation of pollutants by GEMs is done through four activities/strategies mentioned below:

1. Alteration of enzymes for their specificity and affinity.
2. Pathways for construction as well as regulation.
3. Bioprocess development followed by the process of regulation and monitoring.
4. Bioaffinity bioreporting sensor applications for sensing of chemicals, reducing toxicity and analysis of endpoint (Abatenh et al. 2017).

Genes responsible for bioremediation of harmful contaminants such as toluene, chlorobenzene acids, pesticides that are halogenated and toxic wastes are reported. For every chemical, one individual plasmid is required because; a single plasmid can biodegrade all the harmful compounds of different types. The plasmids are placed under four subcategories:

1. OCT plasmid that biodegrades: decane, hexane and octane.
2. XYL plasmid which biodegrades toluene and xylene.
3. CAM plasmid that decomposes camphor.
4. NAH plasmid which biodegrades naphthalene (Joutey et al. 2013).

The bacteria named *Deinococcus radiodurans*is, a radioactive resistant organism, had been modified for the consumption of toluene as well as ionic highly radioactive mercury from nuclear waste (Prakash et al. 2013; Shah 2014). Recombinant polychlorinated biphenyl (PCB)-degrading microbes with improvised stability and survival ability in mixed populations of soil microbes is also been developed (Shah 2014). A recombinant strain of *E. coli* was created for biosorption ability five times enhanced for uranium radioactive isomer by changing the activity of particularly transporter genes, i.e., *nixA* of *Helicobacter pylori* and *merTP* of *Serratia marcescens*. The family of microbe, i.e. *Geobacteriaceae* has, also exhibited the ability for radioactive metal reduction (Prakash et al. 2013). Engineered *Deinococcus geothermalis* strains have been made for the biodegradation of the environment carrying a mixture of radioactive waste at high temperatures. A recombinant strain of *Acenitobacterbaumanii* was noticed to improve rates of biodegradation of crude oil-polluted sites (Coelho et al. 2015). Numerous strains of *Mycobacterium marinu, Sphingomonas desiccabilis, Ralstonia eutropha, P. putida, E. coli* and *B. idriensis* etc. have the genes that are integrated into their genomes, which help in biodegradation of hazardous metal compounds present in polluted surroundings. Further, the potential of the recombinant strain B3 of *Mesorhizobium huakuii* subsp. *Rengei* for the production of phytochelatins (PCs) and accumulation of cadmium Cd^{2+} is enhanced. These modified strains of bacteria are advantageous for the production of a newer plant–bacterium-mediated biodegradation system to remove heavy metals from rice crops when genetically engineered because strain B3 of *M. huakuii* sub spp. *Rengei* forms a symbiotic relationship with a nitrogen-fixing legume called *Astragalus sinicus*. It has been reported that a well-characterized model of a methanotrophic organism *Methylococcus capsulatus* is capable to biodegrade chromium Cr (VI) pollution over a broad range of concentration present in the pollutant (Singh et al. 2011). *Deinococcus radiodurans* is the organism resistant to radiations yet known which was engineered genetically to express an *E. coli* enzyme that can transform toxic ionic form of mercury to a less harmful form and a *Pseudomonas* spp. of an enzyme that helps in the breakdown of toluene (Panigrahi et al. 2005). In addition to this, recent advances of rDNA technologies have brought a revolution for formulating "suicidal genetically engineered microorganisms" to reduce anticipated dangers as well as to accomplish effective and safer clearing of highly polluted sites. The main concern is to build GEMs with an adequate degree of environmental certainty and protection (Joutey et al. 2013).

However, many modern reliable techniques of rDNA particularly genetic engineering, transcriptomics, proteomics, genomics, metabolomics and interactomics are promising modes to study the processes which are involved in the regulation of degradation ways of the pollutants (Kumavath and Deverapalli 2013; Malla et al. 2018). Apart from this, a wide range of genomic technologies such as biosensors and community profiling methods are accessible that can act "eco-genomics", to improve biodegradation on a wider scale (Gillespie and Philp 2013). The rate at which the pollutants are detoxified using GEMs depends on factors mentioned below such as:

(a) The actual composition of microbial communities present on site.
(b) Nature of the communities of microbes.
(c) The extent on which pollutant is present.
(d) Conditions of the environment during the process.

Thus, the process for optimizing the biodegradation process requires a combination of many complex variables, to foretell the fate of contaminants present in the environment. The utilization of omics approaches for the functional and taxonomic features of the community of microbes from polluted regions has led to the invention of a few newer bacteria which were not obtainable by the usage of conventional techniques. In recent times, higher-output omics tools are engaged to exploit the systemic biology of the consortium of microbes in different environments. However, implementing these biodegradation methods successfully needs elaborative knowledge and key insights of the factors which deal metabolism, growth, dynamics, structure and functions of the indigenous consortia of microbes present on the sites of pollution (Malla et al. 2018). Following are the methods involved in these molecular biology and omics approaches involved in the improvement of the field bioremediation:

15.6.1 Molecular Biology Tools in Advancement of Bioremediation of Pollutants

15.6.1.1 16S rRNA Technique and Bioremediation

Based on molecular techniques, denaturing gradient gel electrophoresis helps in separating 16S rRNA genes in polyacrylamide gel through a linear degree of temperature or of DNA denaturant which is formed by mixing of formamide and urea. The basic benefit of this approach, i.e. it generates an outline of the complete diversity of a community of the microbes by separation of a mixture of the populace of 16S rRNA gene products (Pandey et al. 2009). In some instances, one of the surprising applications of this approach in bioremediation is that finding; microbes that dominate biodegradation have closely resembled organisms that may be isolated and cultured from subsurface environments. Apart from this, one of the primary limitations of this approach is the knowledge about the phylogeny of the microbes associated with biodegradation of the pollutants does not surely predict various vital physiological features of the organism. For instance, microbes with 16S rRNA sequences of TCE-degrader *D. ethanogenes* may vary in chlorinated substances which can be degraded by it, and the prediction of compounds uncultured organism will decay might not be evident by analysing only 16S rRNA sequence of the microbe studied (Coelho et al. 2015).

15.6.2 Omnics-Based Advancements in Bioremediation of Pollutants

15.6.2.1 Metagenomics, Metatranscriptomics, Proteomics and Bioremediation

The genome of an organism composed of a complete set of genes that inherits information and converted to mRNA, followed by protein translation, whereas, Proteome is defined as the complete set of proteins, which include enzymes expressed in the living organism only under particulate environment conditions. The genomic sequence of many microbes is now accessible to be utilized for the organization of genome, while ecogenomics techniques could be utilized to identify polluted regions and monitor the process of bioremediation (Gillespie and Philp 2013). Application of genomics in the context of understanding microbes mediated biodegradation provides an aspect of the genes that are related to the susceptibility of microorganisms against harmful compounds in the soil not only biochemical-based factors but also molecular levels which are related to the mechanism (Dixit et al. 2015).

Metagenomics is considered by microbiologists as a very important and potent tool by which can be used to explain how genes are arranged in the microbes that are inhabiting in any kind environment and shows better degradation results in comparison to other approaches of bioremediation for the eradication of contaminants from the surrounding environment. However, metagenomics provides limited functions to explain the expression of the gene and its activity. Metagenomics, with a combination of rDNA techniques, has transformed the microbial world by focusing on the analysis of microbial diversity, their evolution as well as adaption. First, metagenomics has greatly enhanced the knowledge that how microbes form "bucket-brigades" for degrading xenobiotic compounds. It will help in the future to find out basic processes of microbes and mention how the composition of a community could be best complemented to enable degradation of a contaminant when quorum sensing between various species is compulsory, to occur through group process rather than by individual species. Metagenomics will provide an appropriate database that will, in turn, provide construction of huge stock of genes for the production of novel strains of microbes to be utilized for bioremediation purposes.

The term "proteomics" was introduced in 1995. Proteomic studies are particularly significant because it helps to observe the phenotype which is directly obtained by the action of proteins than genome sequence (Kumavath and Deverapalli 2013). Proteomics provides improved results when this technique is combined with metabolomics and transcriptomics. Profiling of protein for the treatment of polluted sites uses SDS–PAGE to characterize the microbes along with the ecology which is responsible for the process of bioremediation. While studies of ecology focused on the natural adaptations of the microorganisms in the surroundings they live, application of proteomics, the studies have given understanding into the systems used for adaptation, precisely under thermophilic conditions. Proteomics is also important in studying the physiological adaptations that microorganisms encounter during the process of biodegradation (Malla et al. 2018). The advancements in 2-D electrophoresis (2-DE) used in compartmental proteomics had been made by a technique based on multidimensional protein identification technology (MudPIT). Matrix-associated laser desorption/ionization time-of-flight MS (MALDI-TOF-MS) is then repeatedly used advancement of mass spectroscopy which has revolutionized environmental proteome technique to identify the protein of interest in few minutes, removed from 2-DE gels, by the generation of peptide mass fingerprinting for identification of protein of interest. Surface-enhanced laser-desorption-ionization MS (SELDI-TOF-MS) is the amalgamation of direct portions of sample loaded on a chip impregnated with MALDI-TOF-MS analysis. Several proteins that express differentially were analysed by the use of SELDI-TOF-MS in *Mytilus edulis* which was introduced to polyaromatic hydrocarbons and heavy metals (Kumavath and Deverapalli 2013). Though proteomics would continue to focus on functional transcriptomics, as the protein molecules are the leading players in an on-site enzymatic reaction by microbes, along with this every molecule of protein represents itself as the last product of gene expression. Complete profiling of protein provides not only knowledge about every single living organism, but also enlightens about the fate as well as the destination of a molecule of protein present in and out of the cell can be identified only with the help of transcriptomics and proteomics approaches both together (Coelho et al. 2015).

The subgroup of the genes that are transcribed in any of the given living organisms is called the transcriptome that presents a dynamic link among genome, proteome and cellular phenotype. DNA microarray is an efficient tool in transcriptomics that enables one to determine the expression level of every gene of mRNA of an organism (Kumavath and Deverapalli 2013). Metatranscriptomics studies may be used to determine the activity of expression of the functional genomes present in environmental samples for research which is about environment bioremediation. The rapid development of metatranscriptomics and metaproteomics has made it feasible to predict the functional activities of the consortium of microbes where anthropogenic activities have contaminated the sites by toxic compounds such as contamination by hydrocarbons and heavy metals, which can also strongly affect the diversity of microbes and its structural composition. But contamination varies that affects three main contents of the ecosystem, i.e. air, water and soil (Malla et al. 2018).

15.6.2.2 Metabolomics and Bioremediation

Metabolomics is an analysis of the metabolite profile of the cell under a given situation. It is one of the latest approaches to the "omics" tools which is a cutting-edge research and moving towards the study of cellular metabolites of microbes. Metagenomics, as compared to metabolomics, has already cemented its importance by playing a significant part to understand the biodiversity and functional viewpoint of the consortium of microbes. Applications of metabolomics approaches to certain samples of the environment have made it easy in developing models that may foretell activities of microbes occurring under different strategies used for bioremediation. Metabolomics also allows scientists a better understanding of the operations carried out by communities of microbes and the functional contributions they provide to the surroundings where they survive. When a microbial cell is subjected under the stress of the environment, it releases many primary and secondary metabolites which have been used by metabolomics to explore the functional role of these metabolites. Recently, various studies have been used in the analysis of microbial metabolome to investigate the bioremediation of contaminants produced by human activities (Malla et al. 2018).

Among these many operations are illustrations for the removing contaminants from the environment; under natural conditions, but these operations act as model systems for increasing the understanding of researchers towards the biology of microbes present within the ecosystem. Apart from the remarkable successes obtained in the laboratory and several failures in the environmental conditions. Lab-based research is not easy implicating into the field, more novel attempts are required to enhance the predictability of the results of these processes. The biology as a subject of bioremediation alone will not completely remove pollution, but also successful biodegradation needs the efforts of engineering also, to decrease the start-up time of operation to set quick and safer sensors that monitor the present situation which can lead to better microbial efficiency in the degradation process when required according to the need of environment (Malla et al. 2018). The major problem encountered in successful Remediation using GEMs is to provide hostile field conditions for the survival of modified organisms. Thus, under specific circumstances, GEMs as bioremediating agents can be warranted, despite the probability that these organisms might help in the welfare of the environment and efficacy in decontamination under real situations should be known (Gillespie and Philp

2013). The coming decade will belong to understand molecular processes and manipulation at the cellular level using bioinformatics and by more modifications and detailed analysis of various processes for the bioremediation of hazardous pollutants (Kumavath and Deverapalli 2013). Decontamination of contaminated sites by contaminants should be the first preference for authorities governing around the world. Bioremediation is a million-dollar project therefore cheap methods are being investigated constantly to promote sustainable living by using specified approaches (Wu et al. 2008).

15.7 Conclusion

To conclude, removal of toxic and hazardous chemical contaminants form the environment is of utmost concern, wherein bioremediation is a boon for the modern world of urbanization, and it has been accepted as a universal process for remediation of contaminated sites throughout the globe. It has been proved a less expensive and eco-friendly approach as compared to conventional methods used for remediation. Several microbes and microbial enzymes are exploited for this process and with the increasing industrialization researchers throughout the world trying to find out more alternatives by the production of genetically modified organisms to enhance the remediation process to cope with the demands of the future. Molecular tools have proved to be beneficial for bringing out innovations in this field. However, many omics approaches have been targeted for this purpose although they have been implied for bringing out modifications in the field of bioremediation, but there are still many constraints that pose the way towards progress in the improvement of microbial strains. Apart from these hurdles, bioremediation continues to be the most productive method of removal of pollutants and protecting earth.

References

Abatenh E, Gizaw B, Tsegaye Z et al (2017) Application of microorganisms in bioremediation-review. J Environ Microbiol 1(1):2–9

Adenipekun CO, Lawal R (2012) Uses of mushrooms in bioremediation: a review. Biotechnol Mol Biol Rev 7(3):62–68

Agwu A, Kalu AU (2012) Bioremediation and environmental sustainability in Nigeria. Int J Acad Res Prog Educ Dev 1(3):26–31

Ahuja SK, Ferreira GM, Moreira AR (2004) Utilization of enzymes for environmental applications. Crit Rev Biotechnol 24(2–3):125–154

Ali H (2010) Biodegradation of synthetic dyes – a review. Water Air Soil Pollut 213(1–4):251–273

Alkorta I, Epelde L, Garbisu C (2017) Environmental parameters altered by climate change affect the activity of soil microorganisms involved in bioremediation. FEMS Microbiol Lett 364(19)

Ateia M, Yoshimura C, Nasr M (2016) In-situ biological water treatment technologies for environmental remediation: a review. J Bioremed Biodegr 7(3):1–5

Azubuike CC, Chikere CB, Okpokwasili GC (2016) Bioremediation techniques–classification based on site of application: principles, advantages, limitations, and prospects. World J Microbiol Biotechnol 32(180):1–18

Banat IM, Makkar RS, Cameotra SS (2000) Potential commercial applications of microbial surfactants. Appl Microbiol Biotechnol 53(5):495–508

Banik S, Das K, Islam M, Salimullah M (2014) Recent advancements and challenges in microbial bioremediation of heavy metals contamination. JSM Biotechnol Biomed Eng 2(1):1035

Behera BK, Prasad R (2020) Environmental technology and sustainability. Elsevier. ISBN 9780128191033. https://www.elsevier.com/books/environmental-technology-and-sustainability/behera/978-0-12-819103-3

Bonugli-Santos RC, Durrant LR, Da Silva M et al (2010) Production of laccase, manganese peroxidase and lignin peroxidase by Brazilian marine-derived fungi. Enzyme Microb Technol 46(1):32–37

Chandra R, Chowdhary P (2015) Properties of bacterial laccases and their application in bioremediation of industrial wastes. Environ Sci Process Impacts 17(2):326–342

Chowdhary P, Hare V, Raj A (2018) Book review: environmental pollutants and their bioremediation approaches. Front Bioeng Biotechnol 6(193):1–2

Coelho LM, Rezende HC, Coelho LM et al (2015) Bioremediation of polluted waters using microorganisms. Adv Bioremed Wastewater Pollut Soil:1–22

Das M, Adholeya A (2012) Role of microorganisms in the remediation of contaminated soil. In: Microorganisms in environmental management. Springer, pp 81–111

Das N, Chandran P (2011) Microbial degradation of petroleum hydrocarbon contaminants: an overview. Biotechnol Res Int:1–13

de Mora AP, Ortega-Calvo JJ, Cabrera F, Madejón E (2005) Changes in enzyme activities and microbial biomass after "in situ" remediation of heavy metal-contaminated soil. Appl Soil Ecol 28(2):125–137

Dixit R, Malaviya D, Pandiyan K, Singh U et al (2015) Bioremediation of heavy metals from soil and aquatic environment: an overview of principles and criteria of fundamental processes. Sustainability 7(2):2189–2212

Duran N, Esposito E (2000) Potential applications of oxidative enzymes and phenoloxidase-like compounds in wastewater and soil treatment: a review. Appl Catal B 28(2):83–99

Ebbs S (2004) Biological degradation of cyanide compounds. Curr Opin Biotechnol 15(3):231–236

Fan CY, Krishnamurthy S (1995) Enzymes for enhancing bioremediation of petroleum-contaminated soils: a brief review. J Air Waste Manage Assoc 45(6):453–460

Fernando EY, Keshavarz T, Kyazze G (2018) The use of bioelectrochemical systems in environmental remediation of xenobiotics: a review. J Chem Technol Biotechnol

Gianfreda L, Bollag JM (2002) Isolated enzymes for the transformation and detoxification of organic pollutants, pp 495–538

Gianfreda L, Rao MA (2004) Potential of extracellular enzymes in the remediation of polluted soils: a review. Enzyme Microb Technol 35(4):339–354

Gillespie IM, Philp JC (2013) Bioremediation, an environmental remediation technology for the bioeconomy. Trends Biotechnol 31(6):329–332

Haritash AK, Kaushik CP (2009) Biodegradation aspects of polycyclic aromatic hydrocarbons (PAHs): a review. J Hazard Mater 169:1–15

Harms H, Schlosser D, Wick LY (2011) Untapped potential: exploiting fungi in bioremediation of hazardous chemicals. Nat Rev Microbiol 9(3):177

Jing R, Fusi S, Kjellerup BV (2018) Remediation of polychlorinated biphenyls in contaminated soils and sediments: state of knowledge and perspectives. Front Environ Sci 6(79):1–17

Jonsson A, Haller H (2014) Sustainability aspects of in-situ bioremediation of polluted soil in developing countries and remote regions 57–86

Joshi SM, Inamdar SA, Telke AA et al (2010) Exploring the potential of natural bacterial consortium to degrade mixture of dyes and textile effluent. Int Biodeterior Biodegrad 64 (7):622–628

Joutey NT, Bahafid W, Sayel H, El Ghachtouli N (2013) Biodegradation: involved microorganisms and genetically engineered microorganisms. In: Biodegradation-life of science, pp 289–320

Kandelbauer A, Guebitz GM (2005) Bioremediation for the decolorization of textile dyes—a review. In: Environmental chemistry, pp 269–288
Karigar CS, Rao SS (2011) Role of microbial enzymes in the bioremediation of pollutants: a review. Enzyme Res 2011:1–13
Kulkarni M, Chaudhari A (2007) Microbial remediation of nitro-aromatic compounds: an overview. J Environ Econ Manag 85(2):496–512
Kumar A, Bisht BS, Joshi VD, Dhewa T (2011) Review on bioremediation of polluted environment: a management tool. Int J Environ Sci 1(6):1079
Kumavath RN, Deverapalli P (2013) Scientific swift in bioremediation: an overview. Appl Bioremed Active Passive Approach:375–388
Lalwani G, Xing W, Sitharaman B (2014) Enzymatic degradation of oxidized and reduced graphene nanoribbons by lignin peroxidase. J Mater Chem B 2(37):6354–6362
Leitão AL (2009) Potential of Penicillium species in the bioremediation field. Int J Environ Res Public Health 6(4):1393–1417
Löffler FE, Edwards EA (2006) Harnessing microbial activities for environmental cleanup. Curr Opin Biotechnol 17(3):274–284
Lynch JM, Moffat AJ (2005) Bioremediation—prospects for the future application of innovative applied biological research. Ann Appl Biol 146:217–221
Magan N, Fragoeiro S, Bastos C (2010) Environmental factors and bioremediation of xenobiotics using white-rot fungi. Mycobiology 38(4):238–248
Malla MA, Dubey A, Yadav S et al (2018) Understanding and designing the strategies for the microbe-mediated remediation of environmental contaminants using omics approaches. Front Microbiol 9:1132):1–1132)18
Mateos LM, Ordóñez E, Letek M, Gil JA (2006) Corynebacterium glutamicum as a model bacterium for the bioremediation of arsenic. Int Microbiol 9(3):207–215
Mehndiratta P, Jain A, Srivastava S, Gupta N (2013) Environmental pollution and nanotechnology. Environ Pollut 2(2):49
Mugdha A, Usha M (2012) Enzymatic treatment of wastewater containing dyestuffs using different delivery systems. Sci Rev Chem Commun 2(1):31–40
Nandal M, Solanki P, Rastogi M, Hooda R (2015) Bioremediation: a sustainable tool for environmental management of oily sludge. Nat Environ Pollut Technol 14(1):181–190
Nawaz K, Hussain K, Choudary N et al (2011) Eco-friendly role of biodegradation against agricultural pesticide hazards. Afr J Microbiol Res 5(3):177–183
Niti C, Sunita S, Kamlesh K, Rakesh K (2013) Bioremediation: an emerging technology for remediation of pesticides. Res J Chem Environ 17(4):88–105
Pandey J, Chauhan A, Jain RK (2009) Integrative approaches for assessing the ecological sustainability of in situ bioremediation. FEMS Microbiol Rev 33(2):324–375
Panigrahi A, Mohapatra A, Panigrahi A (2005) Bioremediation an environmental friendly approach for sustainable aquaculture
Peixoto RS, Vermelho AB, Rosado AS (2011) Petroleum-degrading enzymes: bioremediation and new prospects. Enzyme Res 2011:1–7
Pletsch M, de Araujo BS, Charlwood BV (1999) Novel biotechnological approaches in environmental remediation research. Biotechnol Adv 17(8):679–687
Prakash D, Gabani P, Chandel AK et al (2013) Bioremediation: a genuine technology to remediate radionuclides from the environment. Microb Biotechnol 6(4):349–360
Prasad R (2017) Mycoremediation and environmental sustainability, vol 1. Springer International Publishing. ISBN 978-3-319-68957-9. https://link.springer.com/book/10.1007/978-3-319-68957-9
Prasad R (2018) Mycoremediation and environmental sustainability, vol 2. Springer International Publishing. ISBN 978-3-319-77386-5. https://www.springer.com/us/book/9783319773858
Prasad R, Aranda E (2018) Approaches in bioremediation: the new era of environmental microbiology and nanobiotechnology. Springer International Publishing (978-3-030-02369-0). https://www.springer.com/gp/book/9783030023683

Prasad R, Nayak SC, Kharwar RN, Dubey NK (2021) Mycoremediation and environmental sustainability, vol 3. Springer International Publishing. ISBN 978-3-030-54421-8. https://www.springer.com/gp/book/9783030544218
Rao MA, Scelza R, Scotti R, Gianfreda L (2010) Role of enzymes in the remediation of polluted environments. J Soil Sci Plant Nutr 10(3):333–353
Rao MA, Scelza R, Acevedo F et al (2014) Enzymes as useful tools for environmental purposes. Chemosphere 107:145–162
Rieger PG, Meier HM, Gerle M et al (2002) Xenobiotics in the environment: present and future strategies to obviate the problem of biological persistence. J Biotechnol 94(1):101–123
Ryan DR, Leukes WD, Burton SG (2005) Fungal bioremediation of phenolic wastewaters in an airlift reactor. Biotechnol Prog 21(4):1068–1074
Salam JA, Das N (2012) Remediation of lindane from the environment – an overview. Int J Adv Biol Res 2:9–15
Sasikumar CS, Papinazath T (2003) Environmental management: bioremediation of polluted environment. In: Proceedings of the third international conference on environment and health, pp 465–469
Sayler GS, Ripp S (2000) Field applications of genetically engineered microorganisms for bioremediation processes. Curr Opin Biotechnol 11(3):286–289
Sen R, Chakrabarti S (2009) Biotechnology–applications to environmental remediation in resource exploitation. Curr Sci:768–775
Shah MP (2014) Environmental bioremediation: a low-cost nature's natural biotechnology for environmental clean-up. J Pet Environ Biotechnol 5(4):1
Shah MP (2017) Prokaryotes: a promising agent in environmental bioremediation. Adv Recycl Waste Manag 2:e103
Shukla KP, Singh NK, Sharma S (2010) Bioremediation: developments, current practices, and perspectives. Genet Eng Biotechnol J 3:1–20
Singh JS, Abhilash PC, Singh HB et al (2011) Genetically engineered bacteria: an emerging tool for environmental remediation and future research perspectives. Gene 480(1–2):1–9
Sutherland TD, Horne I, Weir KM et al (2004) Enzymatic bioremediation: from enzyme discovery to applications. Clin Exp Pharmacol Physiol 31(11):817–821
Tanaka T, Yamada K, Tonosaki T et al (2000) Enzymatic degradation of alkylphenols, bisphenol A, synthetic estrogen and phthalic ester. Water Sci Tech 42(7–8):89–95
Thakare M, Sarma H, Datar S, Roy A, Pawar P, Gupta K, Pandit S, Prasad R (2021) Understanding the holistic approach to plant-microbe remediation technologies for removing heavy metals and radionuclides from soil. Curr Res Biotechnol. https://doi.org/10.1016/j.crbiot.2021.02.004
Upadhyay P, Shrivastava R, Agrawal PK (2016) Bioprospecting and biotechnological applications of fungal laccase. 3 Biotech 6(1):15
Vidali JM, Moffat AJ (2005) Bioremediation–prospects for the future application of innovative applied biological research. Ann Appl Biol 146(2):217–221
Wasi S, Tabrez S, Ahmad M (2013) Use of Pseudomonas spp. for the bioremediation of environmental pollutants: a review. Environ Monit Assess 185(10):8147–8155
Whiteley CG, Lee DJ (2006) Enzyme technology and biological remediation. Enzyme Microb Technol 38(3–4):291–316
Wu CH, Mulchandani A, Chen W (2008) Versatile microbial surface-display for environmental remediation and biofuels production. Trends Microbiol 16(4):181–188

Causes, Effects and Sustainable Approaches to Remediate Contaminated Soil

16

Meenu Gautam, Srishti Mishra, and Madhoolika Agrawal

Abstract

Growing population, increase in urbanization and escalating standards of living have contributed to substantial increases in both quantity and quality of generated wastes (51%, mining and metallurgical; 32%, agricultural; 13%, domestic and municipal; 3%, urban infrastructure and transportation; and rest 1%). Soil is the major sink for a wide range of pollutants, carried by discarded wastes which ultimately affect the terrestrial ecosystems. This chapter aims to describe the toxicological impacts of organic and inorganic pollutants on plants, animals and human beings. Besides, the study encapsulates different strategies to manage generated solid wastes and bioremediation of contaminated sites. Major soil pollutants include persistent organic compounds, volatile organic carbons, polycyclic aromatic hydrocarbons, polychlorinated biphenyls, pesticides, azo dyes, heavy metal(loid)s and non-metals. Pollutants in soil beyond their threshold levels may lead to adverse effects on plants such as alteration in plant community structure, crop yield loss, etc. Crop loss or nutritional loss in edible part of plants variably or invariably affects the countries' economy. Accumulation of contaminants in edible plant parts beyond FAO/WHO safe limits may lead to food chain contamination and affects the human health adversely. Several conventional technologies are available for remediation of contaminated sites, but phytoremediation (phytostabilization, phytofilteration, phytoextraction, phytodegradation and phytovolatilization) is the cost-effective and sustainable technique. Phytoremediation using naturally occurring hyperaccumulators, transgenic plants and organic/inorganic soil amendments and in conjugation with microbes have now been ascended as promising and highly efficient technologies in remediation of obstinate pollutants in soil. Thus, phytoremediation techniques

M. Gautam · S. Mishra · M. Agrawal (✉)
Laboratory of Air Pollution and Global Climate Change, Department of Botany, Institute of Science, Banaras Hindu University, Varanasi, India

R. Prasad (ed.), *Environmental Pollution and Remediation*, Environmental and Microbial Biotechnology, https://doi.org/10.1007/978-981-15-5499-5_16

need to be explored and require further advancement to expedite waste management at broader scale even under various environmental stress conditions.

Keywords

Environmental impacts · Organic/inorganic pollutants · Bioremediation · Soil · Sustainable management

16.1 Introduction

"Soil pollution" refers to the presence of chemicals in soil at a higher than normal concentration, is likely to cause adverse effects on non-targeted organisms and impedes the natural balance of the ecological system (The Food and Agriculture Organization (FAO) and Intergovernmental Technical Panel (ITPS) 2015). Although majority of the pollutants have anthropogenic origin, some occur naturally in the soil as minerals' components and can be toxic at high concentrations. Industrialization, urbanization, warfare activities, escalated standards of living and intensification of agriculture have left a legacy of polluted soils around the world (European Environment Agency (EEA) 2014; Bundschuh et al. 2012).

Accumulation of organic (persistent organic pollutants, POPs; volatile organic carbons, VOCs; polychlorinated biphenyls, PCBs; polycyclic aromatic hydrocarbons, PAHs; pesticides, etc.) and inorganic (metal(loid)s and non-metals) pollutants in soil is predominantly associated with the emissions from the rapidly expanding industries, disposal of metallic wastes, mine tailings, leaded paints and gasoline, agricultural application of pesticides and fertilizers (organic and inorganic), municipal waste generation, wastewater irrigation, coal combustion, spillage of petrochemicals and atmospheric depositions (Table 16.1). Contamination of soil is caused by either point or diffuse (non-point) sources:

16.1.1 Point-Source Pollution

Release of pollutants or contaminants to the soil by a specific or series of events within a particular area where one can easily identify the source and pollutants. Anthropogenic activities, viz. industries, mining sites, wastewater disposal, uncontrolled landfills, excessive application of agrochemicals, etc., are the main sources of point-source pollution.

16.1.2 Diffuse-Source Pollution

Spread of pollutants is over a wider area. Pollutants accumulate in soil by deposition and do not have a single or easily identified source. It occurs where emission,

Table 16.1 Major anthropogenic and natural sources of organic and inorganic pollutants in soil

No.	Types of pollutants	Major sources
Organic pollutants		
1.	Phenols	Distilleries, pulp and paper industries, coal mines, oil refineries, wood preservation, plants, pharmaceuticals, coke-oven batteries, herbicides, pesticides and their wastewaters
2.	Petroleum hydrocarbons	Refineries, industries and transportation
3.	Endocrine-disrupting chemicals	As plasticizers in industries, plastic resins' factories and polyurethane polymers manufacture
4.	Chlorinated phenols	Pulp and paper industries, tanneries, distilleries, dyes, paint manufacturing and pharmaceutical industries
5.	Pesticides	Industries, factories and agricultural applications
6.	Azo dyes	Textile, leather, paint, acrylic, cosmetics, plastics and pharmaceutical industries
7.	Melanoidins	Agro-based industries especially from cane molasses-based distilleries and fermentation industries
Inorganic pollutants		
1.	Aluminium	Mining and metallurgical industries and municipal, hospital and electronic wastes
2.	Iron	Metallurgical and mining, batteries, volcanic emissions and municipal and hospital wastes
3.	Zinc	Fertilizer, mining and metallurgical industries and municipal and hospital wastes
4.	Molybdenum	Combustion of fossil fuels and mining and metallurgical industries
5.	Magnesium	Mining, agriculture, fertilizers and municipal and hospital wastes
6.	Manganese	Municipal and agricultural wastes, combustion of fossil fuels and mining and mineral processings
7.	Cobalt	Wood preservatives and volcanic emissions
8.	Copper	Mining and metallurgical industries, municipal wastes incineration, carbon black production, electronics, wood preservatives and architecture
9.	Beryllium	Combustion of fossil fuels, electronics, municipal waste incineration, weathering of rocks, beryllium alloy and chemical industries
10.	Nickel	Metal electroplating and nickel mining industries, oil refineries, municipal wastes and combustion of fossil fuels
11.	Selenium	Incineration, coal, oil and mining, milling and metallurgical industries
12.	Lead	Mining and metallurgical industries, plastics, paints, pipes, batteries, gasoline and automobiles
13.	Cadmium	Fertilizers, plastics, pigments, oil refineries and mining and metallurgical industries
14.	Chromium	Tanneries, paints, pigments, fungicides and mining and metallurgical industries
15.	Mercury	Coal, vinyl chlorides, electrical batteries and thermometers

(continued)

Table 16.1 (continued)

No.	Types of pollutants	Major sources
16.	Barium	Dust control equipments and industrial controls
17.	Arsenic	Mines, smelters, oil refineries, pesticides, electrical waste, treated wood products, paints and herbicides
18.	Sodium, potassium and calcium	Municipal and agricultural wastes, fertilizers, volcanic emission, combustion of fossil fuels and weathering
19.	Nitrogen, sulphur and phosphorous	Fertilizers, volcanic emission, weathering, combustion of fossil fuels and municipal and agricultural wastes

Sources: Mishra et al. (2019), Yadav et al. (2017), Wuana and Felix (2011) and USEPA (2008)

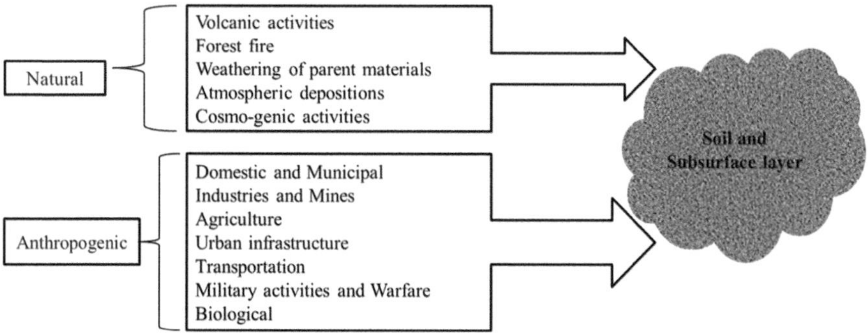

Fig. 16.1 Various natural and anthropogenic sources of soil pollution

transformation and dilution of pollutants takes place via air-soil-water systems before being transferred to soil (FAO and ITPS 2015).

16.2 Sources of Soil Pollution

16.2.1 Natural Sources

Events such as forest fires, volcanic eruptions and cosmogenic activities are the natural sources of soil pollution when many toxic elements are released into the environment (Fig. 16.1). These elements include metal(loid)s, dioxin-like compounds, PAHs and radionuclides (Deardorff et al. 2008). Dœlsch et al. (2006) reported high levels of PAHs and heavy metal(loid)s mainly mercury (Hg), copper (Cu), chromium (Cr), zinc (Zn) and nickel (Ni) in soil of Réunion, France, due to volcanic activity and weathering of the parent rocks. Arsenic (As) contamination is one of the global environmental issues where volcanic eruption (Ma et al. 2019) and weathering of minerals and ores (Mandaliev et al. 2013) occur. Li (2009) and Trendel et al. (1989) reported that PAHs in soil are of cosmogenic origin or due to diagenetic alteration of waxes in soil organic matter. Naturally occurring asbestos and radioactive gases in soil are mainly attributed to ultramafic rock specifically serpentine and amphibole (Bloise et al. 2016; Swartjes and Tromp 2008).

16.2.2 Anthropogenic Sources

Anthropogenic activities have been causing widespread environmental pollution to the land, air and water (Fig. 16.1). Amongst them mining and metallurgical industries followed by agriculture, municipal and urban infrastructure and transportation contribute significant proportions in wastes generation (Fig. 16.2).

16.2.2.1 Domestic and Municipal Wastes

According to the report by the World Bank, global municipal solid waste generation was estimated to be 1.3 billion tonnes (BT) per year (Hoornweg and Bhada-Tata 2012). Municipal wastes consist of many organic and inorganic pollutants such as heavy metal(loid)s, PAHs, VOCs, pharmaceutical compounds, personal care and their derivative products (Ghosh et al. 2014). Electronic waste contains valuable elements such as Cu, aluminium (Al), gold (Au) and many other hazardous substances (such as lead (Pb), Cd, Cr, brominated flame retardants and PCBs) (Fornalczyk et al. 2013). Use of pesticides such as dichlorodiphenyltrichloroethane (DDT), chlordane, etc. for the control of vector-borne diseases has led to soil pollution in urban and peri-urban areas (Mansouri et al. 2017). Table 16.1 shows various pollutants ensue from domestic and municipal sources. Biosolids from municipal wastewater treatment are the major sink for many organic and inorganic chemicals, and their land application can potentially introduce harmful contaminants into terrestrial environments (Haynes et al. 2009).

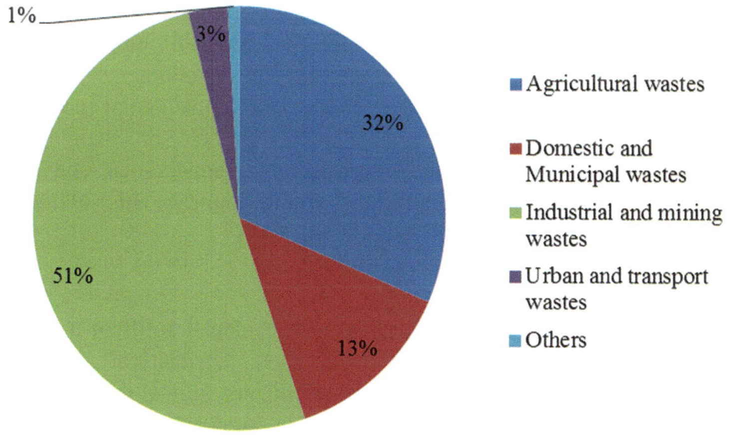

Fig. 16.2 Percentage contribution of various anthropogenic sectors in waste generations. (Source: Pappu et al. 2007)

16.2.2.2 Transportation and Urban Infrastructure

Widespread development of infrastructure such as housing, roads and railways has considerably contributed to environmental degradation by land consumption and soil sealing (Table 16.1). Transportation constitutes one of the main sources of soil pollution, perhaps not only because of the emissions from fuel combustion but also from petrol spills and the relevant activities as a whole (Mirsal 2008). Splashes generated by traffic during rainfall and run-off cause translocation of particles rich in heavy metals from the corrosion of vehicular parts and pavement abrasion (Venuti et al. 2016; Zhang et al. 2015a, b). Plastic, PAHs and rubber-derived compounds from urban infrastructure and transportation are also the sources of soil contamination (Kumar and Kothiyal 2016).

16.2.2.3 Industrial and Mining Wastes

According to the integrated pollution prevention and control (IPPC) directive, the European Union, potentially polluting activities associated with small- and large-scale industries can be grouped into the following six main categories (García-Pérez et al. 2007):

1. Energy industries
2. Production and processing of metal(loid)s
3. Mineral industry
4. Chemical industry
5. Waste management
6. Others (paper/board production, manufacture of fibres/textiles, tanneries, slaughterhouses, animal farm and manufacture of carbon or graphite)

Land in the vicinity of industries and factories are polluted by inappropriate storage of chemicals, spillages of raw materials, fuel ash, dusts, fires and refuses from the industrial activities (Alloway 2013). Small- and large-scale industries release huge amount of heavy metal(loid)s, gaseous pollutants, metal(loid)s, POPs, VOCs, radionuclides, etc., which in the environment persist for longer period of time even after the end of those activities (Table 16.1). These pollutants are dispersed by air and water to a larger distance, thereby contaminating residential and agricultural areas (Mileusnić et al. 2014). Salinization is another threat to soils mainly associated with the production of glass, rubber, pigment, ceramic, soap and detergent, processing of animal hide and metal(loid)s, leather tanning, chlor-alkali, textiles, oil/gas drilling and pharmaceuticals (Saha et al. 2017).

16.2.2.4 Agriculture

Sources of pollution in agricultural settings are accidental spills of hydrocarbons, utilization as fuels in machines, transportation, agricultural application of agrochemicals such as organic and inorganic fertilizers, animal manure, pesticides, weedicides and agricultural wastes (Table 16.1). Accidental spills of fuels and agrochemicals represent serious risks of soil pollution with POPs and heavy metal (loid)s (Osman et al. 2014). The fertilizers used for agriculture are rich sources of

Hg, Cd, Pb, Cu, Ni, Cu and natural radionuclides (^{238}U, ^{232}Th and ^{210}Po) (Kanter 2018; Stewart et al. 2005). Liu et al. (2015) and Cang (2004) found significant levels of heavy metal(loid)s in soil from livestock and poultry operations. In many countries, use of biomedical wastes as manure is also a chief source of soil pollution (Shankar and Shikha 2017).

16.2.2.5 Miscellaneous Sources

Nuclear testing laboratories and industries are the prime sources of radioactive substances (^{106}Rh, ^{131}I, ^{140}Ln, ^{144}Ce, ^{44}Ru, ^{106}Ru and ^{140}Ba) in the soil (Jadiyappa 2018). Long-term deposition of radionuclides in soil emits gamma radiation which is harmful for the health of soil living organisms (Jadiyappa 2018). The excreta of animals, birds and humans are also one of the contributory sources of soil pollution by biological agents (Clark 2014). In the developing countries, wastewater irrigation, wrong methods of agricultural practices and application of animal manures constitute serious soil pollution problems (Alloway 2013). Military and warfare activities accustom non-degradable weapons of destruction and chemicals (remains of ammunitions, landmines, leftover chemicals, radioactive and biological toxic agents) that persist in the affected soils for centuries after the end of the conflict (FAO and ITPS 2015).

16.3 Major Pollutants in Soil

Anthropogenic sources are the main drivers of wide range of organic as well as inorganic pollutants in the soil. Pollutants based on their chemical characteristics are categorized into organic and inorganic forms (Fig. 16.3).

16.3.1 Inorganic Pollutants

Small- and large-scale industries mainly associated with textiles, glass, rubber, chemicals, tanneries, pharmaceuticals, detergent and soap production are the major contributors of salts (mainly NaCl) produced in the world today (Fig. 16.3). Inorganic pollutants mainly include radionucleotides, non-metals, heavy metals and metalloids (Fig. 16.3). Heavy metal(loid)s refer to the group of metals and metalloids with high atomic density > 4.5 g cm^{-3} (Hawkes et al. 1997). Heavy metal(loid)s are further categorized into two main groups: (1) elements such as molybdenum (Mo), Ni, Zn, Cu, iron (Fe) and magnesium (Mg), which are essential to life and ecosystems but are noxious to plants and animals when their concentrations exceed certain threshold levels (Buchmann 2008), and (2) elements such as As, Cd, Hg, Cr and Pb, which do not pose any significant role in any biochemical processes, but their presence alters the soil quality and normal metabolic functioning in living beings (Edelstein and Ben-Hur 2018; Patinha et al. 2018).

Radionuclides in soil occur in three different forms, viz. primordial, cosmogenic and man-made radionuclides (United States Environmental Protection Agency

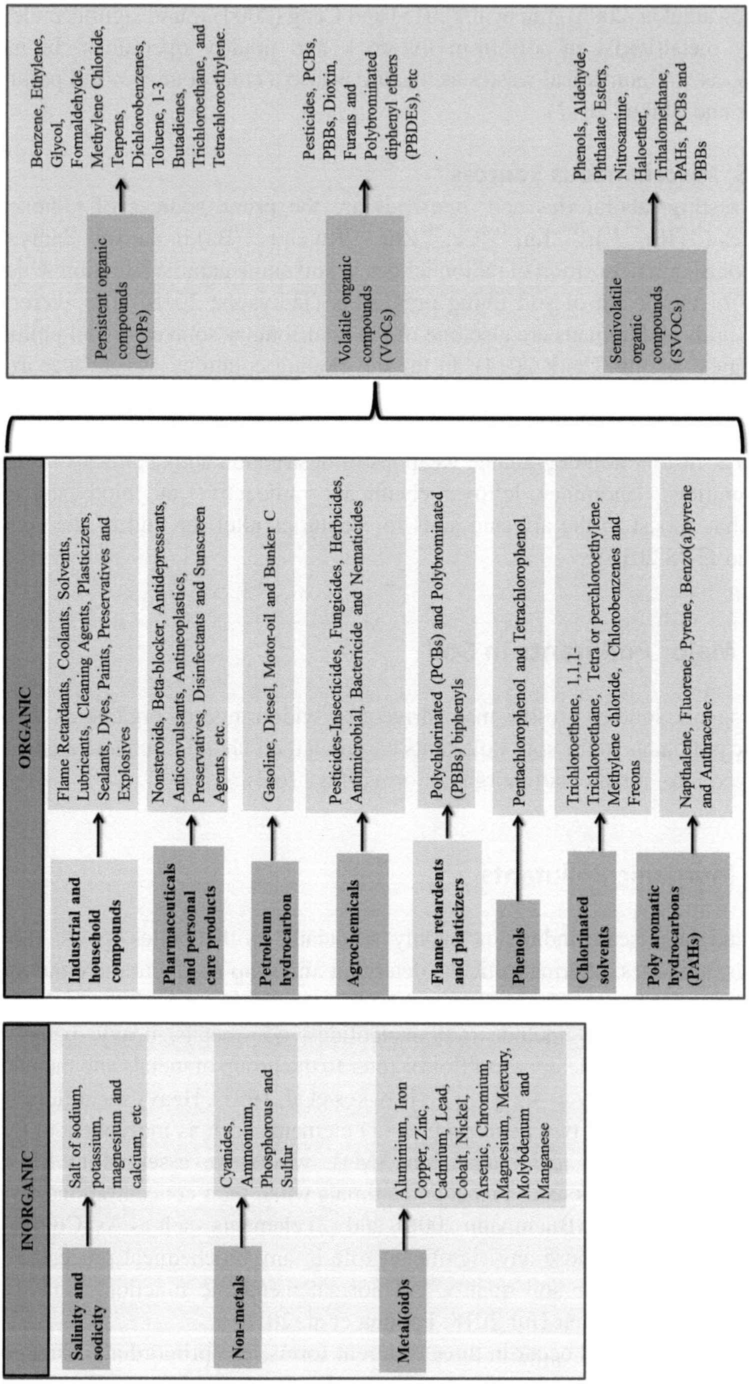

Fig. 16.3 Major organic and inorganic pollutants found in soil worldwide. (Sources: Rodríguez-Eugenio et al. 2018; Padmavathiamma et al. 2014)

(USEPA) 2006). Primordial radionuclides (^{235}U, ^{238}U, ^{232}Th and ^{40}K) are basically remain-over from the creation of the earth and have half-lives of hundreds millions of years. Primordial radionuclides end up in soil due to weathering of rocks. Cosmogenic radionuclides are produced by cosmic rays in the atmosphere and have long half-lives, while majority have shorter half-lives than the primordial radionuclides. Cosmogenic radionuclides include ^{14}C, ^{3}T and ^{7}Be. Anthropogenic activities such as nuclear testing and radiological events like the Chernobyl accident lead to deposition of radioactive particles in soil through air and water (Romanovskaia et al. 1998). Inappropriate disposal of radioactive substances also contributes to increasing contents of radionuclides in the soil (USEPA 2006).

16.3.2 Organic Pollutants

Organic pollutants include POPs, VOCs and semi-volatile (SVOCs) organic compounds (Fig. 16.3). POPs are recalcitrant towards degradation, highly toxic and are carcinogenic as well as mutagenic (Ashraf 2017). It includes pesticides, chlorinated solvents, industrial fluids and flame retardants (Bartrons et al. 2016). Short- and long-distance transportation of gaseous or particulate forms of POPs in the atmosphere is facilitated by air and water (Ashraf 2017). Besides, atmospheric dry and wet depositions constitute the main input of these compounds to the soil (Cousins et al. 1999). Indirect deposition of compounds in soil occurs through decomposition of litter fall from plants (Wania and Mclachlan 2001). Accumulation of POPs in soil horizons rich in organic matter may lead to their persistence for years (Masih and Taneja 2006). For an instance, it has been reported that 90% of United Kingdom's land are rich in POPs (Cousins and Jones 1998). There are about 7 key POPs, viz. PCBs, DDT, hexachlorocyclohexane (HCH), dieldrin, perfluorooctanesulfonate (PFOS), hexabromocyclododecane (HBCD) and endosulfan, which are restricted as per Stockholm Conference held in 2001 (Ashraf 2017). Organic pollutants such as gasoline and hydrocarbons as a result of fossil fuel burning and automobile exhaust are volatile in nature. World Wildlife Fund enlisted 67 different types of environmental hormones in 1997 that were considered harmful for the soil biota (Lyons 2005).

A myriad of organic pollutants such as biocides, pesticides, flame retardants, pharmaceuticals, surfactants, PAHs, PCBs, polychlorinated dibenzo-p-dioxins (PCDDs), polychlorinated dibenzofurans (PCDFs), etc. end up in the soil as a consequence of human activities (Rhind 2009). The most commonly found organic pollutants in soils are chlorinated compounds (e.g., PCDDs, PCBs and PCDFs), oil hydrocarbons (e.g., alkanes, alkenes and cycloalkanes), monomeric aromatic hydrocarbons (e.g., toluene, xylene, benzene and ethylbenzene), PAHs (e.g., chrysene, benzo(a)pyrene and fluoranthene), pesticides (e.g., alachlor, acetochlor, atrazine and bifenox), fungicides (e.g., penconazole, procymidone and metalaxyl), insecticides (e.g., endosulfan, heptachlor, captan, benomyl and endrin), pharmaceuticals (e.g., antibiotics, analgesics, nonsteroidal anti-inflammatories, antiparasitics and antimicrobials), hormones (e.g., oestrogens and androgens), sterols (e.g., dihydrocholesterol, cholesterol and coprostanol), flame retardants (e.g., polybrominated diethyl ethers, bisphenol-A and hexabromocyclododecanes), nitrosamines (e.g., nitrose dimethylamine, nitrose-di-n-propylamine, nitrose diethyl-amine and nitrosopyrrolidine) and their fate compounds (Fig. 16.3).

16.4 Factors Affecting Toxicity of Organic and Inorganic Pollutants in Soil

Even though wide range of pollutants is present, their reactivity and bioavailability in soil are controlled by many of their physico-chemical and biological properties (Kabata-Pendias and Pendias 2011; Kodešová et al. 2011).

16.4.1 Soil Texture and Mineralogy

Soil texture signifies the relative amounts of sand, silt and clay proportions in soil. High clay (particles size < 0.002 mm) fraction in soil has strong ability to bind positively charged ions due to their layered structure, large specific surface area, chemical and mechanical stability and high cation exchange capacity (Uddin 2017). On contrary, sandy soils have larger pore size and lower sorption capacity which lead to the movement of pollutants to ground and surface water (Uddin 2017). Thus, soils having higher amounts of clay and humus have high buffering and sorption capacity which, despite the increase in concentrations of contaminants, do not cause adverse biological effects (Różański et al. 2016). Presence of minerals such as layered silicates, oxides/hydroxides of Fe and Al, carbonates, sulphates, allophane and associated amorphous clays is inorganic, while humus is organic colloid, which has high cation adsorption capacity. Fijałkowski et al. (2012) showed the affinity of metal cations for clay minerals in a series $Cu^{2+}>Cd^{2+}>Fe^{2+}>Pb^{2+}> Ni^{2+}>Co^{2+}>Mn^{2+}>Zn^{2+}$.

16.4.2 The pH and Electrical Conductivity

16.4.2.1 Changes in Surface Charge

Change in pH and salinity of soil greatly influences the net negative charge on the charged colloids (clays, silicates, oxides/hydroxides of Fe and Al), their ion exchange capacities and the binding energies of their sorption sites (Proust 2015; Violante et al. 2010). Thus, increase in surface charge on inorganic and organic colloids lowers the availability of pollutants in soil.

16.4.2.2 Competition for Adsorption Sites

Under acidic condition of soil, more protons are available to the binding sites of clays, organic matter and oxides, thus enabling organic and inorganic species more available to biological organisms (Rampazzo et al. 2013). On the contrary, under high soil pH, the cations replace protons and get adsorbed to charged colloids so tightly that they are not readily bioavailable (Olaniran et al. 2013). Transition (Fe, Al, Zn, Cu, Cr, Ni, Mn, Pb and Co) as compared to alkaline earth cations has strong tendency to get adsorbed onto charge colloids and form inner-sphere complexes (Violante et al. 2010). Moreover, organic pollutants with low molecular weight behold less adsorption capacity onto silicates in soil (Lin et al. 2015).

16.4.2.3 Hydrolysis of Inorganic/Organic Species in Solution

Sorption capacity of soil increases with increasing soil pH and vice versa (Paulose et al. 2007). The lower the pH value, the more organic and inorganic elements can be found in solution. Soil pH catalyzes the hydrolysis reaction and subsequently influences degradation of pesticides, atrazine and inorganic salts (Zhang et al. 2013).

16.4.2.4 Dissolution of Inorganic/Organic Complexing Anions

High soil pH lowers high solubility of dissolve organic carbon and base cation concentrations, whereas low pH enhances their solubility in soil (Olaniran et al. 2013). Several studies have reported a positive correlation between pH and retention of Cd, Pb, Cu and Zn in soil (Deurer and Bottcher 2007).

16.4.3 Soil Organic Matter

Soil organic matter can reduce or increase the bioavailability of pollutants in soil through immobilization or mobilization by forming various insoluble or soluble complexes, respectively (Shrestha et al. 2019). A wide range of organic acids (formic, acetic, oxalic, succinic, malonic, maleic, citric, malic, lactic, fumaric acids and aconitic) acts as ligands for many cations in soil (Vranova et al. 2013; Fijałkowski et al. 2012). Generally, citric acid followed by malic>acetic>tartaric>oxalic acid is the most effective in terms of desorption of different metals (Zn, Cu, Hg, Pb, caesium (Cs) and Cd) in soil due to more carboxyl group to form stable ligand (Köchy et al. 2015).

16.4.4 Cation Exchange Capacity

Cation exchange capacity (CEC) of soils depends upon soil types, amounts and types of different colloids (Harter and Naidu 2001). Clayey soils have higher CEC value (30 cmol kg^{-1}) compared to sandy soils ($<$ 5 cmol kg^{-1}). Similarly, humus has very high CEC value compared to the inorganic clays (i.e., kaolinite). Thus, the greater the CEC value, more is the exchange sites on soil minerals for organic and inorganic species.

16.4.5 Oxidation-Reduction Potential

Oxidized and reduced soils have redox potential in the range of 400–700 and 250–300 mV, respectively (Pezeshki and DeLaune 2000). Redox potential plays a significant role in the reactivity of some soil oxides (Fe and manganese (Mn)) with organic and inorganic pollutants in soil (Alamgir 2016). It controls the predominant chemical speciation and sorption of metal(loid)s (As, Cr and selenium (Se)) in soil (Landner and Reuther 2004). Generally reducing condition favours decline in mobility of positively charged ions in soil (Gonsior et al. 1997).

16.5 Effects on Environment and Socioeconomic Segment

16.5.1 Effects on Soil

Organic and inorganic pollutants may cause alteration in soil pH, CEC and salinity and dispersion and/or flocculation of clays and adversely affect soil aggregation, mechanical strength and stability of soil (Zong and Lu 2019; Salem et al. 2017). Extremely high levels of nitrogen, phosphorous and potassium (NPK) in soil significantly lower the total porosity, water retention capacity and macroaggregate content and increase soil bulk density, plasticity index, coefficient of linear extensibility and tensile strength (Zong and Lu 2019). Soil pollution enables prodigious quantity of nitrogen to escape into the atmosphere through volatilization and denitrification (Fungo et al. 2019). Moreover, organic matter decomposition in soil emits sulphur dioxide and other associated compounds, instigating acid rain (Bricker and Rice 1993). Toxic elements, i.e. Cr, Cd and As, exhibit antagonistic behaviour with essential micronutrients in soil such as Zn, Cu, Mn, Mg and Fe for active binding sites of roots (Gautam et al. 2017). Various organic and inorganic pollutants inhibit the nitrification process and cause salinization of soil due to their highly saline properties. Soil pollution also causes loss of nutrients present in it, hindering plants ability to thrive therein, which consequently may result in soil erosion and disturbances in the balance of soil flora and fauna.

Soil biological properties play pivotal roles in affecting soil fertility and primary production through organic matter decomposition, nutrient cycling and aggregate formation (Brevik and Sauer 2015). Soil microbial communities and enzymatic activities are very sensitive to changes in soil properties accredited to pollution and thus are considered as indicators of soil fertility (Xu et al. 2015). Pollutants in soil interact with microorganisms and affect their normal metabolic functioning, thereby affecting the soil fertility and production (Saxena et al. 2015; Behera and Prasad 2020). Bastida et al. (2017) found that frequent use of agrochemicals adversely affects the water storage capacity, structure and function of microbial communities in soil.

16.5.2 Harmful Effects of Pollutants on Plants

Pollutants, specifically degradable, undergo physico-chemical and biological alterations before and after being deposited onto terrestrial ecosystems (Karthikeyan et al. 2004). Both organic and inorganic pollutants are available to plants either from soil or air (Fig. 16.4). Plants take up organic pollutants (pesticides, herbicides, weedicides, fertilizers and growth-promoting chemicals) mainly through leaf surfaces, roots and fruits which are then distributed within the plant either from cell to cell or through the vascular system (Rana and Rana 2015; Burken et al. 2005). Uptake of organic pollutants through roots is mediated by two pathways to the various parts of the plant via xylem vessels, i.e. symplastic and apoplastic (Fig. 16.4) (Kvesitadze et al. 2015). In former case, the tissue system is surrounded by plasmalemma and interconnected by plasmodesmata where movement of molecules within the conductive tissue occurs by mass flow and diffusion. In apoplastic pathway, cell wall and xylary elements form a continuous water-permeable column for short- and long-distance transport of solute by mass flow and diffusion. Certain chemicals are restricted to either apoplastic or symplastic pathways, while some are ambimobile (can follow both the domains efficiently). Uptake of organic molecule depends upon the molecular size, lipophilicity and dissociation constant (Schroder and Collins 2002). Generally, less lipophilic molecules take the apoplastic pathway, while more lipophilic one follow symplastic route (Karthikeyan et al. 2004). Moreover, pollutants with small molecular size can easily invade the membrane and vascular tissues of plants and vice versa (Kvesitadze et al. 2015).

The accessibility of inorganic pollutants (metal(loid)s and non-metals) in soil is an active process mainly governed by physico-chemical, biological and other environmental factors (Hajar et al. 2014). Moreover, phyto uptake and transportation of

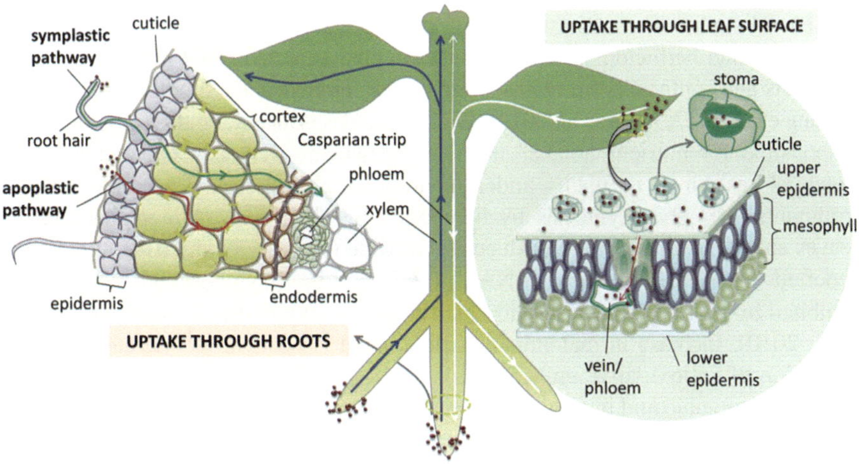

Fig. 16.4 Uptake of pollutants (brown dots) through leaves and roots over symplastic and apoplastic pathways (Source: Modified from Lv et al. 2019)

metal(loid)s and non-metals within the plant are principally reliant on the type of plant species, concentration of ions/molecules and their oxidation state (Tangahu et al. 2011). Intrinsic protein, proton pumps and transporters (IRT1, ZnT1, heavy metal ATPase-HMA2, and HMA4) facilitate the uptake and transportation of Zn, Cu, Cd, Pb, Ni and Fe to various parts of the plant (Martinoia 2018; Viehweger 2014). Contaminants translocate from roots to shoot by two regulatory mechanism, i.e. evaporation and transpiration (Tangahu et al. 2011). Plants are mainly of two types: accumulators and excluders. Accumulators continue to be present in spite of concerted pollutants in the shoots, whereas excluders confined pollutant phyto-uptake. Both accumulator and excluder manage to combat the high metal contents in their body parts through adapted defence strategies (Viehweger 2014).

Organic and inorganic pollutants in soil beyond National Oceanic and Atmospheric Administration (NOAA) soil quality guidelines may cause deleterious effects on plants (Buchmann 2008). They pose toxic effects on plant cell ultrastructure, biosynthesis, membrane stability and deoxyribonucleic acid (DNA) which consequently affect their metabolic, photosynthetic and reproductive processes. Organic and inorganic pollutants adversely affect the cellular ultrastructure such as distortion of the cell wall, leakage of the cytoplasm and alteration in the shape and size of chloroplasts from ovate to hexagonal (Xiong et al. 2017). Ramadass et al. (2015) found that organic pollutants, for example, bipyridylium and diphenylether, persuade cell membrane disruption and damage the plant tissues. Zhang et al. (2017) unravel the adverse effects of organic pollutants on the assemblage of micro-tubulins and associated proteins in the cell. Toxic effects of metal contents in plants may be direct and/or indirect. Some of the direct toxic effects associated with high metal contents include inhibition of cytoplasmic enzymes and damage to cell structures due to oxidative stress (Jadia and Fulekar 2009), whereas indirect toxic effect is composed of replacement of essential nutrients at cation exchange sites of plants (Taiz and Zeiger 2002).

Several studies have reported the impact of pollutants (toxic metal(loid)s, PCBs, antibiotics and herbicides) on plant cell biosynthesis, detrimental effects on photosynthesis and synthesis of proteins, amino acids, nucleic acids, lipids and hormones (Zhang et al. 2017; Asati et al. 2016). Photosynthesis is an essential process in plants, responsible for nutrient uptake, arbitrates growth as well as yield and provides potential resistance to plants under environmental stresses (Yang et al. 2010). Pollutants alter photosynthesis by hampering the formation of pigments causing leaves and stems to become translucent and white (Kaspary et al. 2014). It has been reported that ribonucleic acid (RNA), protein and lipid synthesis are significantly inhibited by prolonged exposure to hexazinone and chlorsulfuron treatments (Yang et al. 2010). Increase in malondialdehyde content is positively correlated with the inhibition of antioxidative enzyme activities in some plants under oxidative stress induced by organic and inorganic pollutants (Zhang et al. 2017; Gautam et al. 2016). Some studies have revealed that DNA damage can be caused by metals (Cd, Cr and Pb) and PAHs (Huang et al. 2013). A possible reason for this could be attributed to alteration in protein synthesis, enzyme activities and cell organelle dysfunction, which may result in inhibition of mitotic division in root tip cells (Van Dingenen

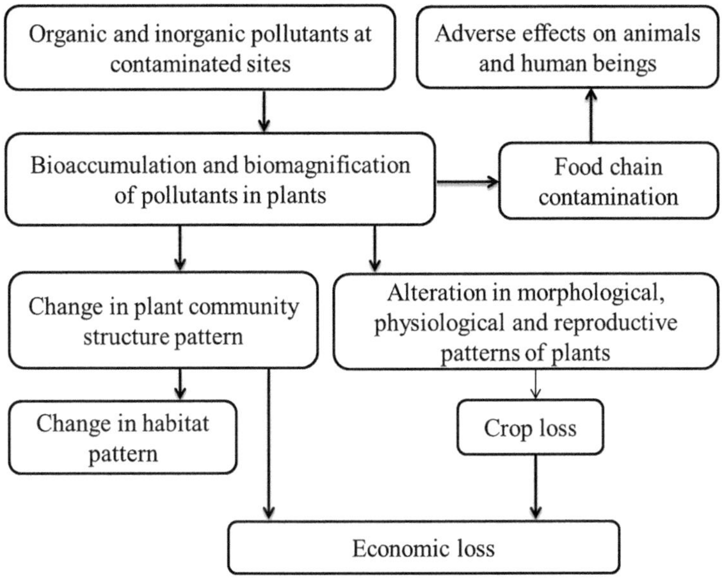

Fig. 16.5 A simplified diagram of the relationship between organic and inorganic pollutants in soil, change in plant community structure pattern, global climate change, crop loss, economic loss and health of animals and human beings

et al. 2016). Cell biosynthesis, membrane stability and synthesis of RNA, lipids, proteins and photosynthetic pigments, growth and yield are significantly affected in plants grown on contaminated sites (Zhang et al. 2017)

Habitat fragmentation or destruction, change in climatic condition, biogeochemical cycling and alteration in soil quality due to interference of wide range of persisting organic and inorganic pollutants (Fig. 16.5) are the major threat to plants diversity in an ecosystem (Krebs and Bach 2018; Bellard et al. 2012). Poorly developed soil structure, less available nutrients, readily available toxic constituents, poor soil biota and water-restricted conditions driven by soil pollution alter vegetation succession (Gautam et al. 2016). A number of studies have been reported on alteration in plant community structure due to accretion of pollutants in soil from various anthropogenic activities.

16.5.2.1 Changes in Plant Community Structure Pattern: Case Studies

Plant community structure study at abandoned red mud dumps of HINDALCO, Renukoot, showed less number of herbaceous, shrub and woody species when compared to forest site (Gautam et al. 2018). Relatively less number of species at red mud dumps was mainly attributed to high bulk density, salinity, alkalinity, exchangeable sodium percentage and toxic metal(loid)s (Cd, Cr, As and Pb) contents coupled with low porosity, moisture content, biological properties and available nutrients (Gautam et al. 2018). Species richness and Shannon-Wiener diversity index for all plant types were high at forest site compared to red mud dumps. The important value index (IVI) of sensitive species was low while that of tolerant species was higher and was accredited to altered soil properties.

According to Pandey et al. (2014), total number of plant species at Jharia coalfield (polluted site) was less when compared to the Central Institute of Mining and Fuel Research (reference site). At Jharia coalfield, total number of woody species was less than total number of herbaceous species. Multivariate analysis showed that number of woody species at study site was mainly governed by less sulphate and phosphorous contents in soil, whereas contaminated site with low total nitrogen and organic carbon contents were the prime factors governing the richness and IVI of herbaceous species.

A study by Vijayan (2011) showed that long-term spraying of endosulfan at eleven different panchayats in Kasargod district of Kerala reduced the plant biodiversity by 40–70% when compared to the site without endosulfan treatment. Besides, native species of the area such as *Hopea ponga, Cinnamomum malabatrum, Ixora polyantha, Premna serratifolia, Syzygium caryophyllatum* and *Embelia cheriyan-kottan* disappeared due to the adverse effects of organic pollutant on soil properties (Vijayan 2011).

Ensuing the rapid increase in global population, an estimate of 87% more food crops such as rice, wheat, soya bean, maize, etc. will be required to meet their demand for food by 2050 (Kromdijk and Long 2016). However, several restraints including abiotic and biotic stresses are likely to disrupt food security in near future, following the fact that increasing natural and anthropogenic activities are the major threats to food demand and security due to increasing severity of soil pollution on global scale (Riaz et al. 2019). Agricultural dependence on wastewater irrigation, chemical fertilizers, pesticides, fungicides, weedicides and rapid development of industries have increased the amount of toxic metal(loid)s and organic pollutants in cultivable land resulting in detrimental effects on soil-plant environment system (Bansiwal and Maheshvari 2018). In fact, persisting pollutants causes loss of soil productivity and also reduces crop yield. This challenge is adversely affecting the social and economic conditions of the world.

16.5.2.2 Crop Loss Due to Soil Pollution

Crop losses are the major threat to the wellbeing of rural families, economy of traders and governments as well as to food security worldwide (Avelino et al. 2015; Savary and Willocquet 2014). For instance, poor soil quality due to pollution-caused yield loss of annual crops such as rice ranged from 24 to 41% in Asia (Savary et al. 2000), potatoes from 5 to 96% in France (Rakotonindraina et al. 2012) and cotton up to 100% in Thailand (Castella et al. 2005). Yield loss of perennial crops such as apple and other stone fruits reached up to 5% in the Netherlands (Van Leeuwen et al. 2000), and for coffee, it ranged from 13 to 45% in Brazil (Barbosa et al. 2004). Crop losses due to pest attack, diseases, soil infertility and climate change for major food and cash crops (wheat, rice, maize, barley, soybeans, potatoes, coffee and cotton) were estimated between 20 and 40% at both country and regional levels in different continents (Oerke 2006; Cooke 2006). The phrase "losses between 20 and 40%" inadequately reflects the true costs of crop losses to farmers, fabrics, economic, environments, societies, public health and consumers (Savary et al. 2012). Crop losses owing to above-mentioned factors are direct and indirect:

Direct Losses

- Primary losses: Yield, quality, cost of control over losses, extra cost of harvesting, grading and replanting
- Secondary losses: Contamination of sowing and planting material, soil-borne diseases and weakening by premature defoliation of trees and perennials

Indirect Losses Agricultural farm, rural community, exporters, traders, wholesale retailers, consumers, government and environment.

Crop losses ultimately affect the financial structure of growers, distributors, wholesalers, transporters, retailers, food processors and others (Fig. 16.5). Economic loss basically is the reduction in economic benefits due to crop damage, the costs of labour, materials and inputs for the control of pests and diseases (Cerda 2017). An estimate of $1391 million has been incurred on the loss of crops and trees accredited to the use of pesticides (Pimentel and Burgess 2014).

16.5.3 Effects on Animals and Human Beings

Accumulation of organic and inorganic pollutants in edible or non-edible parts of crops and commercially important plants grown in contaminated soil may induce clinical disorders in biological systems (of animals and human beings) through food chain contamination (Li et al. 2018). Biological systems have no specific mechanism for the elimination of such pollutants from body parts; however, such xenobiotic after entering the body undergoes bioaccumulation and biomagnification (Tangahu et al. 2011). Organic and inorganic pollutants beyond their threshold levels for living beings (FAO/WHO 2001) are linked to a wide range of ailments (Table 16.2).

16.6 Advanced Technologies and Cost Incurred in Management of Wastes

According to the World Bank (2018), an estimate of 2.01 BT of total solid waste generation has been found in 2016 with a footprint of 0.74 kg per capita per day. Whereas in India, it ranged from 0.2 to 0.6 kg per capita per day with about 42 million tonnes (MT) of total solid wastes produced every year and likely to cross 260 MT in 2047 (World Bank 2018). Proper waste management has a lot of benefits such as reducing global warming and emission of noxious gases (methane, CH_4; carbon dioxide, CO_2; oxides of nitrogen, NO_x, and sulphur, SO_x; hydrogen peroxide, H_2O_2 and ammonia, NH_3), saving carbon footprint and maintaining the environment clean as well as pollution-free (Daniel and Perinaz 2012; Hoornweg and Bhada-Tata 2012). To manage the waste generated from agricultural, municipal, industrial and other sectors, several conventional and integrated advanced

Table 16.2 Harmful effects of excessive organic and inorganic pollutants in soil on human beings and animals

Types of pollutants	Harmful effects on human beings and animals	References
Organic pollutants		
Azo dyes	Skin irritation, nausea, vomiting, irritation in digestive tract and liver and kidney damage	Küçük and Liman (2018), Tadesse et al. (2017)
Phenols and chlorinated phenols	*Acute exposure*: Dryness of the mouth and throat, nausea, vomiting and diarrhoea *Inhalation and dermal contact*: Skin blisters and cardiovascular diseases *Chronic exposure*: Inhibit oxidative phosphorylation, damage mitochondrial structure, inhibition of circulatory system, methemoglobinemia, haemolytic anaemia, hypothermia, pulmonary oedema, arrhythmia, tachycardia, hypotension, central nervous system disorders, respiratory disease and heart failure *Ingestion*: Gastrointestinal damage, muscle tremors and death	Bharagava et al. (2020), Tadesse et al. (2017)
Endocrine-disrupting chemicals	Skin irritation, conjunctiva, mucous membranes of oral and nasal cavities, cryptorchidism, testicular lesions, prolongation of the oestrous cycle, hypospadia, obesity and anovulation	Bharagava et al. (2020), Sifakis et al. (2017)
Melanoidins	Severe toxic effects on fishes and other aquatic organisms	Bharagava et al. (2020)
Persistent organic pollutants	Allergies, hypersensitivity, damage to the central and peripheral nervous systems, neurobehavioural disorder, learning disabilities, endocrine system disorder, reproductive disorder, disruption of the immune system, mutagenicity and carcinogenicity	Ahmed et al. (2019), Bharagava et al. (2020), Alharbi et al. (2018)
Pesticides	Immune suppression, diminished intelligence, hormonal problems, reproductive abnormalities and cancer	Wang and Han (2019), Bharagava et al. (2020)
Petroleum hydrocarbons	*Dermal exposure*: Dermatitis, defatting injury and chemical burns *Inhalation*: Weakness, dementia, morbidity, mortality, central nervous system disorder, development of criminal/violent behaviour, memory and other cognitive deficits, cerebellar dysfunction, encephalopathy, metabolic acidosis and arrhythmia *Oral exposure*: Abdominal pain, irritation, vomiting and diarrhoea *Aspiration*: Fatal pneumonitis, coughing, wheezing, respiratory distress and hypoxia *Acute exposure*: Acidosis, dermatitis, pneumonitis, arrhythmia and encephalopathy	Bharagava et al. (2020), Varjani et al. (2018)

(continued)

Table 16.2 (continued)

Types of pollutants	Harmful effects on human beings and animals	References
Inorganic pollutants		
Aluminium	Aluminosis (pneumoconiosis followed by pulmonary fibrosis), neurotoxicity, Alzheimer's disease and breast cancer	Klotz et al. (2017)
Iron	Nausea, abdominal pain, seizure, cardiomyopathy, hepatic fibrosis, impotency, arthropathy, hereditary hemochromatosis, thalassemia, bone marrow failure and myelodysplastic syndrome	Zhang et al. (2015a, b)
Magnesium	Low blood pressure, nausea, diarrhoea, abdominal cramping and calcium deficiency	Institute of Medicine, Washington (1997)
Manganese	Low blood pressure, violent behaviour, hallucinations, schizophrenia, insomnia, muscle tremors, loss of appetite, apathy, dystonia, hypokinesia, lung disease, pneumonitis, impaired pulmonary and vascular function, improper foetus development and brain damage	Aschner et al. (2005)
Zinc	Dizziness, muscular cramps, vomiting, fatigue and renal damage	Yadav et al. (2017)
Copper	Stomach and intestine irritation, liver cirrhosis, brain and kidney damage and chronic anaemia	Yadav et al. (2017)
Cobalt	Diarrhoea, low blood pressure and paralysis	Yadav et al. (2017)
Nickel	Allergies; immunotoxic, neurotoxic, teratogenic, carcinogenic, genotoxic and mutagenic, lung cancer, infertility and hair loss	Yadav et al. (2017)
Barium	Muscle twitching, high blood pressure, respiratory failure, gastrointestinal dysfunction and cardiac arrhythmias	Yadav et al. (2017)
Cadmium	*Acute exposure*: Abdominal pain, gastrointestinal tract erosion, burning sensation, nausea, vomiting, salivation, muscle cramps, itai-itai disease, vertigo, shock, loss of consciousness, hepatic injury and coma *Chronic exposure*: Depression, DNA damage and cell death	Bharagava et al. (2020), Hassaan et al. (2016)
Chromium	Skin and nasal irritations, ulceration, eardrum perforation and lung cancer	Bharagava et al. (2020), Hassaan et al. (2016)
Arsenic	Melanosis, black foot disease, polyneuropathy, encephalopathy, disorder of cardiovascular and central nervous system, hemolysis, hepatomegaly, bone marrow depression and death	Jomova et al. (2011)
Lead	Headache, loss of memory, confusion, reduced consciousness, irritation, encephalopathy, acute psychosis and malfunctioning of kidneys, liver,	Bharagava et al. (2020), Hassaan et al. (2016)

(continued)

Table 16.2 (continued)

Types of pollutants	Harmful effects on human beings and animals	References
	endocrine and reproductive and central nervous systems	
Mercury	Loss of hearing, mental retardation, abnormal muscle tone, blindness, neurological deficits, dysarthria and developmental defects	Bhaargava et al. (2020)
Phosphorous	Chronic kidney disease, bone-related disorders, cardiovascular system disorder, cell damage, increased mortality, atherosclerosis and left ventricular hypertrophy	Komaba and Fukagawa (2016), Calvo et al. (2014)
Calcium	Irritability, headache, memory loss, lethargy, hypercalcemia, hypercalciuria, osteochondrosis, kidney stone, renal failure, coma and death	Whiting and Wood (1997)
Sodium	High blood pressure, hypertension and renal and cardiovascular diseases	Di Nicolantonio et al. (2017)
Potassium	Hypertension and cardiovascular disease	Adrogué and Madias (2014)
Sulphur	Eye irritation, chest pain, asthma and heart disease	Prasad (2014), Grimble (2006)
Nitrogen	Asthma, hyperthyroidism, methemoglobinemia, birth defects, insulin-dependent diabetes, central nervous system disorder, colon cancer, neural tube defects and digestive and respiratory failures	Davidson et al. (2012)

technologies are available on the basic principles of solid waste management (Agarwal et al. 2015):

1. 4Rs (refuse, reduce, reuse and recycle):
   - Refusal of buying anything which actually does not require
   - Reduction in the generation of garbage
   - Reutilizing the things to its maximum
   - Recycling the waste materials into useful forms wherever possible
2. Segregation of organic/biodegradable and inorganic/non-biodegradable wastes in separate container
3. Inculcation of different treatment techniques for different types of wastes to its nearest possible points

16.6.1 Conventional and Advanced Management Techniques

Composting It is simple and economically viable technique to manage agricultural and domestic (biodegradable) wastes using microorganisms and earthworm (Banerjee et al. 2019). Composts are very much rich in nutrients and are widely used as fertilizer in agriculture field and in horticulture. It maintains the soil health through increasing moisture-holding capacity and recycling nutrients into soil. However, composting of field emits methane and foul odour (Banerjee et al. 2019). Also, it may cause contamination of soil with toxic materials (organic and inorganic).

Anaerobic Digestion Composting of organic waste led to generation of biofuel (comprising 50–60% of methane) using anaerobes (Banerjee et al. 2019). The technique offers stabilization and disinfection of wastes like industrial sludge, farmland residue and animal slurries. The value-added part of the process is that the residue which is rich in nutrients and moisture can be used as fertilizer. Besides, energy and efficiency recovery of anaerobic digestion are better than composting.

Incineration Wastes are converted into ash under high temperature (980–2000 °C) with the emission of gaseous products (Banerjee et al. 2019). It is a stepwise process which leads to destruction of toxic material with recovery of energy from the wastes. Incineration reduces the volume of combustible waste to 80–90%, thereby facilitating the less requirement of land for its disposal (Banerjee et al. 2019). Additionally, the process is odourless, noise-free and environmentally safe.

Pyrolysis In pyrolysis, wastes are thermally degraded in absence of oxygen under the temperature ranged between 300 and 850 °C. Synthetic gas (CH_4, CO and H_2) and char (carbon and non-combustible materials) are the by-products of pyrolysis of waste material (Banerjee et al., 2019). Gases are further utilized for fuel, wax and tar production.

Gasification The process costs incentive and requires high power source. Gasification is a partial oxidation of wastes under insufficient oxygen condition under temperature >650 °C (Banerjee et al. 2019). In plasma gasification technology, high temperature (electric arc) is applied to the waste, thereby converting it into an inert residue (ash). Before thermal application, wastes are required to be dried and then segregated. During the process, gases produced as by-product comprise of H_2, CO and CH_4, which can be used as fuel. Unlike incineration, gasification does not emit any toxic gases like SOx and NOx because of insufficient oxygen. Non-combustible residual parts are disposed following proper handling.

Refuse-Derived Fuel (RDF) Alike gasification, RDF is obtained from partial oxidation of mixed municipal solid waste. Utilization of the fuel in combination with coal and other type of conventional fuel in industries could be a great input in minimizing the requirement of natural/synthetic fuel.

Landfilling This is the most common and ultimate way to manage all types of wastes (organic and inorganic) which do not require skilled employees and is a low-cost process. However, landfilling of wastes without proper pretreatment ranked lowest amongst all management techniques. Landfills are major source of greenhouse gas emissions and contamination of land and water with organic and inorganic pollutants.

Others Technologies Wastes from industrial and mining sectors are widely used through advanced engineered techniques in making construction materials, chemicals and other utilities (Gautam et al. 2016). Mining wastes such as Fe, Cu, Zn and Al tailings, coal washeries and overburden wastes are used as raw materials in the recovery of expensive minerals and manufacture of construction materials for embankments of roadways, railways, rivers, dams, bricks, concrete beams, tiles, lightweight aggregates, glasses, ceramics and fuel (Gayana and Chandar 2018; Indian Bureau of Mines 2002). Metallurgical wastes such as slags, red mud, fly ash and galvanizing residues are used in making cement, concretes, bricks, tiles, ceramics, blocks, polymers, composites, wood substitute products, paints, boards and in metal recovery (Matinde et al. 2018).

16.6.2 Bioremediation of Contaminated Sites

The engineered techniques in management of wastes from various sectors are although advanced and highly efficient but are still at their initial stages of development. Particularly in developing countries like India, these technologies do not offer a cost-effective option at present. The cost incurred in cleaning of contaminated sites in the USA alone is \$6–8 billion per year, with global costs in the range of \$25–50 billion (Tsao 2003). Conventional, mechanical or physico-chemical treatment techniques to manage polluted sites by excavation, soil washing, solidification/stabilization, electrokinetic remediation and soil incineration also suffer from limitations like cost ineffectiveness, intensive labour requirement and irreversible soil disturbances (Yeung 2010). Therefore, management of solid waste dumps and contaminated sites through biotechnological approaches is the only sustainable, cost-effective and environmentally benign options to safeguard the environment (Gautam et al. 2017; Meagher 2000). Proper management policies, suitable remedial strategies and sustainable utilization of resources without altering the natural ecosystem should be the prime aim of all researchers and decision-making bodies in order to combat soil pollution problems in a holistic way. The cost-effectiveness of bioremediation was reported by Blaylock et al. (1997), who were able to save 50–65% of cost, when bioremediation was used for the treatment of one acre of Pb-polluted soil compared with the use of conventional methods such as excavation and landfills.

Bioremediation is the use of biological agents (bacteria, fungi, plants and earthworm) to remove or neutralize harmful toxic substances by converting them to either less or non-toxic form in an eco-friendly manner for environmental safety (Mishra et al. 2019). Bioremediation is an effective means of mitigating both organic (hydrocarbons, halogenated organic compounds, pesticides, herbicides and other compounds) and inorganic (non-metals, metal(loid)s and radionuclides)

Table 16.3 Various techniques for remediation of organic and inorganic pollutants in soil

Types of contaminants	In situ	Ex situ
Organic pollutants		
Aromatic hydrocarbons and polycyclic aromatic hydrocarbons	Bioventing Microbial bioremediation Phytoremediation Thermal treatment	Biopiles Composting Land farming Bioreactors Thermal desorption Incineration
Petroleum	Microbial bioremediation Biosparging Slurping	Bioremediation Soil washing Thermal desorption
Chlorinated aliphatic hydrocarbons	Bioventing Microbial bioremediation Phytoremediation Slurping Thermal treatment	Biopiles Bioreactor Thermal desorption Incineration
Chlorinated and non-chlorinated phenols	Bioventing Microbial bioremediation Phytoremediation Slurping	Biopiles Composting Land farming Bioreactors Thermal desorption Incineration
Dioxins and furans	Thermal treatment Phytoremediation Microbial bioremediation	Soil stabilization and solidification
Inorganic pollutants		
Metal(loid)s and non-metals	Natural attenuation Phytoremediation Soil stabilization Thermal treatment (electrokinetics)	Chemical extraction and oxidation Soil washing, stabilization and solidification Solvent extraction
Hazardous chemicals	Microbial bioremediation Phytoremediation	Microbial bioremediation Soil washing Solvent extraction Thermal desorption

Sources: Lacatusu et al. (2017), Azubuike (2016), Petruzzeli et al. (2015), Agarry and Oghenejoboh (2015)

contaminants in soil by both in situ and ex situ remediation techniques (Wadgaonkar et al. 2019; Mosca Angelucci and Tomei 2016; Prasad 2017, 2018) (Table 16.3).

16.6.2.1 In Situ Bioremediation

In-situ bioremediation is accompanied by minimal interference to the environment at the contamination site. Besides, it incurs less cost when compared to conventional soil remediation techniques (ex situ bioremediation). In situ bioremediation techniques are further categorized into:

Intrinsic Bioremediation/Natural Attenuation Bioremediation relies on the natural environmental conditions and behaviour of soil microorganisms and plants that are indigenous. It occurs without human intervention other than monitoring. The techniques are widely used to treat hydrocarbons, dyes, chlorinated solvents and metal(loid)s at polluted sites (Roy et al. 2015; Frascari et al. 2015).

Enhanced In Situ Bioremediation Intrinsic bioremediation is enhanced by some additive agents for effective remediation of polluted sites. This includes bioventing, bioslurping, biosparging and phytoremediation (Azubuike et al. 2016).

a. *Bioventing*: It involves controlled stimulation of airflow by supplying oxygen to unsaturated zone (vadose) in order to increase microbial activities. Besides, soil amendments are used to supply moisture and nutrients for enhanced bioremediation of pollutants rendering them into a harmless form (Philp and Atlas 2005).
b. *Slurping*: The technique is the combination of vacuum-enhanced pumping, extraction of soil vapour and bioventing for indirect provision of oxygen for contaminant biodegradation (Gidarakos and Aivalioti 2007). It is effectively used in remediation of volatile and semi-volatile organic pollutants in soil (Table 16.3). The technique is not suitable for soil with low permeability.
c. *Biosparging*: Unlike bioventing, air is injected into the saturated zone of soil subsurface which causes skyward movement of VOCs into the unsaturated zone to stimulate bioremediation process. The effectiveness of the technique proportionally depends upon soil permeability and pollutant biodegradability (Philp and Atlas 2005).
d. *Phytoremediation*: The technique relies on the use of plants' interaction with physical, chemical and biological factors at polluted sites to mitigate the toxic effects of pollutants (Azubuike et al. 2016; Sarma et al. 2021).

16.6.2.2 Ex Situ Bioremediation

Ex situ bioremediation techniques implicate excavation of polluted soils from contaminated sites and its subsequent transportation to another site for treatment. The techniques are usually preferred after envisaging the type of pollutant, depth and degree of pollution, geology of the polluted site, geographical location and the cost of treatment (Philp and Atlas 2005). Following ex situ bioremediation techniques are:

a. *Biopile*: It is a bioremediation technology where excavated soils are amended with nutrient to form a compost piles and enclosed for further treatments. The entire setup composed of a treatment bed, aeration, irrigation and a leachate collection system. An aeration and irrigation system is buried under the soil to pass through air and nutrients. Moreover, pH, oxygen, heat and nutrients are controlled factors to enhance biodegradation. The biopile technique is mainly applicable to polluted sites having VOCs with low molecular weight (Whelan et al. 2015). In the process, airstream-containing VOCs when leave the soil are treated to remove or destroy before being discharged into the atmosphere.

b. *Windrows*: Microbial degradation of pollutants relies on periodic turning of piled polluted soil together with added water through increased aeration and uniform distribution of pollutants and nutrients (Barr et al. 2002). As compared to biopile, windrows exhibit higher efficiency towards removal of hydrocarbon in soil (Coulon et al. 2010). The use of windrow technique is more associated with the release of greenhouse gases due to reduced aeration in anaerobic zone within piled polluted soil (Hobson et al. 2005).
c. *Bioreactor*: It is a system mainly composed of batch, fed-batch, continuous and multistage. Controlled bioaugmentation, nutrient addition, increased pollutant bioavailability and mass transfer between pollutant and microbes facilitate efficient microbial degradation of pollutants in soil under specific set of conditions (temperature, pH, agitation, aeration rates, and substrate and inoculum concentrations). It is widely used in treatment of soil or water polluted with VOCs (Table 16.3) including toluene, benzene, ethylbenzene and xylene (Azubuike et al. 2016).
d. *Land farming*: It is the simplest technique owing to less equipment requirement and low cost. It is considered under both in situ and ex situ. In former case, bioremediation proceeds without excavation of soil when pollutants lie <1 m of the surface of the ground and vice versa (Nikolopoulou et al. 2013). Excavated polluted soils are treated with autochthonous microorganisms for aerobic biodegradation of pollutants (Silva-Castro et al. 2012), whereas in situ land farming is facilitated by ploughing, addition of nutrients and irrigation to stimulate biodegradation of pollutants using autochthonous bacteria.

16.6.2.3 Bioremediation Using Microbes, Plants and Their Association

Depending upon the types of biological organisms used, bioremediation is categorized into following:

a. *Microbial Bioremediation*

Microorganisms are omnipresent and considered to be the first evolved life forms on the earth. They are versatile and tolerant towards wide range of environmental conditions, i.e. from the small intestines of animals to frozen environments, hydrothermal vents, acidic lakes and bottoms of deep oceans (Seigle-Murandi et al. 1996). Table 4 illustrates certain microorganisms widely used for remediation of contaminants in soil and are resistant to adverse environmental conditions. Various factors such as pH, temperature, soil type and texture, nutrient amendments and oxygen significantly influence the microbial remediation process in soil (Vásquez-Murrieta et al. 2016; Sharma 2012). In order to thrive under extreme environmental conditions, microorganisms exhibit certain morphological adaptation at both cellular and colony levels (such as in shape, colour, texture, opacity, convexity, margin appearance, etc.) (Jeanson et al. 2015). Due to their adaptability, microorganisms are widely used for bioremediation of heavy metal(loid)s, hydrocarbons, polythenes, food wastes, greenhouse gases, etc. (Das 2014). A large number of microbial enzymes have been known in degradation and detoxification of organic and inorganic pollutants to safer or less toxic intermediates (Dash and Das 2012). Ligninolytic enzymes (e.g., manganese peroxidase, lignin peroxidase and laccase), monooxygenase, chrome reductase, azo-reductase and dioxygenase secreted by rhizospheric microbes have been found to be effectively used in bioremediation of various pollutants (Dash and Das 2012). Microbial bioremediation of organic compounds implicates either partial or complete mineralization by complex and genetically regulated physiological reactions (Joutey et al. 2013).

Metal(loid)s cannot be destroyed like organic contaminates, rather they are either removed or biotransformed to a stable form (Tangahu et al. 2011). Remediation of metal(loid)s includes biosorption, bioleaching, biomineralization, intracellular accumulation and enzyme-catalyzed transformation through redox reactions (Lloyd and Lovely 2001). Besides, microorganisms develop certain mechanisms such as uptake, adsorption, oxidation, reduction and methylation to protect them from metal toxicity (Igiri et al. 2018). Several metal-resistant bacteria (*Bacillus* sp., *Pseudomonas* sp., *Saccharomyces* sp., etc.) have been reported to thrive under metal-stressed condition by their accumulation and complexation into less toxic form (Díaz-Ramírez et al. 2008).

b. *Phytoremediation*

The term "phytoremediation" was initiated by the Environmental Protection Agency (EPA) in 1991 and was first addressed in open technical literature in 1993 by Cunningham and Berti. Phytoremediation, also referred to as botano-, green-, vegetative- or agro-remediation, is collectively used for green plant-based technologies for in situ and ex situ remediation of contaminated soil (Mahjoub 2013). It is publicly appealing remediation technology that utilizes plants, associated microbiota, various soil amendments and agronomic techniques to extract, contain, degrade, detoxify or immobilize organic (hydrocarbons, pesticides and chlorinated solvents) and inorganic contaminants in the soil (Helmisaari et al. 2007; Vyslouzilova et al. 2003). For developing countries, phytoremediation is an eco-friendly, aesthetically pleasing and cost-effective approach, but despite its potentiality, it is yet to become a commercially viable technology (Ghosh and Singh 2005; Sharma et al. 2021). Phytoremediation approach is further classified into various applications such as phytostabilization, rhizodegradation or phytodegradation, rhizofiltration or phytofiltration, phytoextraction and phytovolatilization (Fig. 16.6).

Phytostabilization The approach utilizes plant roots to confine the mobility and availability of contaminants in soil (Jadia and Fulekar 2009). Pollutants from soil are absorbed through roots and restricted in rhizospheric zone, rendering them harmless by preventing their leaching (Ekta and Modi 2018). Phytostabilization was found to be a most suitable technique to remediate Cu, Zn, Cd, Cr and As (Moosavi and Seghatoleslami 2013). Bacchetta et al. (2018) showed that *Helichrysum microphyllum* subsp. *tyrrhenicum* grown on mine wastes dumps of Sardinia in Europe can tolerate high concentration of Zn, Pb and Cd, thus appearing as a species

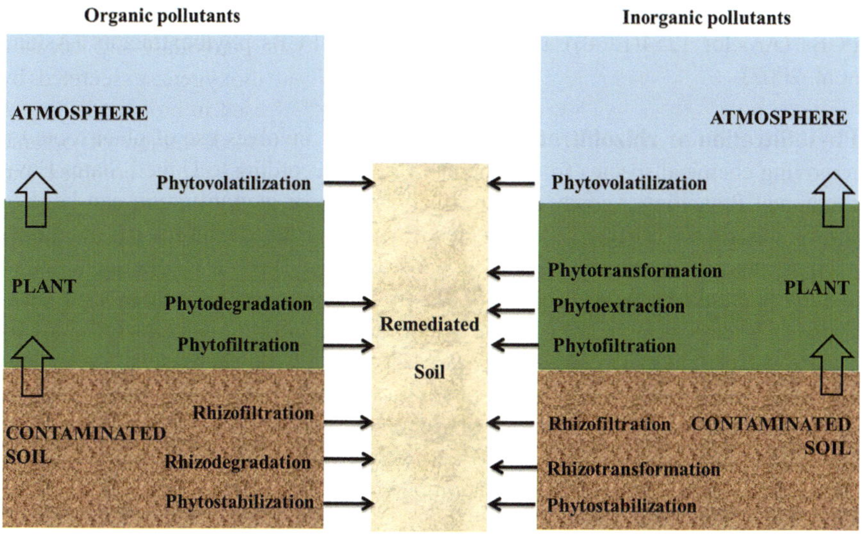

Fig. 16.6 Phytoremediation technologies for remediation of organic and inorganic pollutants from contaminated soil

suitable for phytostabilization. *Conocarpus erectus* grown in soil spiked with Ni (0, 50, 100 and 200 mg kg^{-1}), Pb (0, 600, 1200 and 2400 mg kg^{-1}), Cd (0, 20, 40 and 80 mg kg^{-1}) and Cr (0, 150, 300 and 600 mg kg^{-1}) has enormous potential for the phytostabilization of Cr, Ni and Cd in its roots (Tauqeer et al. 2019). Ferro et al. (1997) observed that alfalfa (*Medicago sativa*) did not cause the degradation of volatile organic constituents in soil treated with 40 and 660 μg kg^{-1} of ^{14}C-benzene, rather it enhanced its phytostabilization in soil.

Phytoextraction Plants absorb organic and inorganic chemical species from soil and translocate them to the aboveground harvestable parts to accumulate. In general, roots contain higher levels of contaminants than shoot despite the translocation is high which is primarily attributed to maximum accumulation capacity of the aboveground biomass (Suman et al. 2018; Wuana and Felix 2011). Malik and Ravindran (2018) reported that *Suaeda maritima* was found efficient in phytoextraction of metals (Zn, Cu, Cd, Cr, Mg, calcium (Ca) and potassium (K)) and chloride salts from soil contaminated with paper mill effluents. *Cymbopogon citratus* showed potentiality in extraction of Al, Zn, Ni, Pb, Cd, Pb, Cr and As, while *Chrysopogon zizanioides* was found efficient in extraction of Mn and Cu from soil mixed with sewage sludge at different red mud combinations (Gautam et al. 2017; Gautam and Agrawal 2017). White et al. (2003) in their study revealed that amongst 21 cultivars of *Cucurbita pepo* grown at Lockwood Farm of Hamden, CT, USA (contaminated with p, p′-DDE levels ranged from 200 to 1200 ng g^{-1}), cultivar ssp pepo had potentiality to phytoextract POPs from soil and translocate large quantities to aerial tissues. Moreover, plant species such as *C. pepo*, *Carex normalis* and *Festuca* *arundinacea* collected from the PCB storage site at Ontario (0.6–200 μg g^{-1} of total PCBs (Aroclor 1254/1260)) ascended as potential PCBs phytoextractors (Åslund et al. 2007).

Phytofiltration or rhizofiltration This technology involves use of plant roots for removing chemical species from aqueous wastes. According to United States Environmental Protection Agency (2000), rhizofiltration or phytofiltration can be used for Zn, Cu, Ni, Cd, Pb and Cr which are principally retained within the roots. Plants such as mustard, sunflower, spinach, rye, tobacco and corn have been identified as suitable candidates for infiltration of Pb from water (Mukhopadhyay and Maiti 2010). Abhilash et al. (2009) found maximum Cd content in roots followed by leaves and peduncle of *Limnocharis flava* grown in Cd-contaminated water, thus showing the efficacy of *L. flava* in phytofiltration of Cd (>93%). *Micranthemum umbrosum* was found effective in removing organic As species such as monomethylarsenic acid (CH_5AsO_3) and dimethylarsenic acid ($C_2H_7AsO_2$) from oxic environment through phytofiltration process (Islam et al. 2017).

Phytovolatilization The technique uses green plants to extract contaminants such as As, Hg, Se and volatile organic compounds from polluted soils, transform them into volatile forms and transpire them into the atmosphere from their leaves or stem (Limmer and Burken 2016; Karami and Shamsuddin 2010). According to Sakakibara et al. (2010), *Pteris vittata* was found efficient in remediating about 90% As from soil containing 37% of arsenite and 63% of arsenate. Similarly, *Lepidium latifolium*, *Artemisia douglasiana*, *Caulanthus* sp., *Fragaria vesca* and *Eucalyptus globulus* were grown in soil contaminated with mercury (450–1605 mg kg^{-1}, where *Caulanthus* sp. showed a higher proficiency in removing mercury from soil (emission rate 92.6 ng m^{-2} h^{-1}) as compared to other plant species (Wang et al. 2012). Limmer and Burken (2016) highlighted the phytovolatilization efficiency of several plants, i.e. *Eucalyptus* sp., *Populus* sp., *Salix* sp. and Pine, in remediating volatile organic compounds in soil.

Phytodegradation or phytotransformation Plants and its associated microorganisms are involved in uptake, metabolization and degradation of pollutants especially organic compounds (Ekta and Modi 2018; Thakare et al. 2021). Phytoconversion of Cr (VI) (more toxic) to Cr (III) (less toxic) form using halophytes was reported by Cacador and Duarte (2015). Some plants are successfully used to decontaminate polluted soil and sludge dumps using root exudates (Ekta and Modi 2018). Newman and Reynolds (2004) enlisted the plants, i.e., *Leucaena* sp., *Populus* sp., *Brassica* sp., *Helianthus* sp., *Secale cereale*, *Cucurbita* sp., *Arabidopsis* sp., *Bruguiera* sp., *Kandelia* sp., *Nicotiana tabacum*, *Sorghum* sp., etc. commonly used in phytodegradation of organic pollutants (such as pesticides, PAHs, PCBs, chlorinated compounds, gasoline additives and de-icing agents) in soil.

Typically, four main strategies currently exist to augment phytoremediation of organic as well as inorganic pollutants from soils: (1) use of natural phytoaccumulators, (2) enhancement of phytoremediation process by soil amendments, (3) microbe-assisted phytoremediation and (4) genetic alterations in plants used for remediation purposes (McGrath et al. 2002).

Use of Natural Phytoaccumulators There are about 400 plant species belonging to 45 plant families known to be effectively used in remediation of polluted soil (Ekta and Modi 2018). Some of the families are Asteraceae, Brassicaceae, Convolvulaceae, Euphorbiaceae, Fabaceae, Lamiaceae, Poaceae and Scrophulariaceae (Gautam and Agrawal, 2019; Ekta and Modi 2018). Plants such as *Ludwigia* sp., *Dracaena* sp., *Phragmites australis*, *Rhizophora mangle*, *Sparganium* sp., *Aegiceras corniculatum*, *Cannabis sativa, Arrhenatherum elatius*, *Arabidopsis halleri*, *Brassica* sp., *Corrigiola telephiifolia, Raphanus sativus*, *Thlaspi caerulescens*, *Alyssum* sp., *Arabidopsis* sp., *Eichhornia crassipes*, *Salix* sp., *Euphorbia* sp., *Helianthus annuus*, *Pteris vittata*, *Jatropha curcas*, *Populus* sp., *Spartina maritima*, *Potamogeton pectinatus*, *Ricinus communis*, *Trifolium alexandrinum*, *Zea mays*, *Spinacia oleracea*, *Verbascum speciosum*, *Vetiveria* sp., *Ambrosia artemisiifolia*, *Lycopersicon esculentum*, etc. have very high bioaccumulation potential for organic pollutants (POPs, PAHs and PCBs) and heavy metals (Cd, Zn, As, Cu, Fe, Pb, Hg, Mn, Mg, Cr and Ni) in soil (Table 16.4).

Brassica juncea cultivar Pusa Bold grown in soil amended with cow dung manure exhibited high potential towards extraction of Cu, Fe, Cd, Zn, Mn, Cu, Pb and Cr when compared to the cultivar Kranti (Gautam and Agrawal 2019). A study conducted by Gautam and Agrawal (2019) on the herbaceous community at abandoned red mud dumps showed that dominant species, viz. *Brachiaria mutica*, *Cynodon dactylon*, *Dactyloctenium aegyptium*, *Digitaria ischaemum*, *Digitaria longiflora*, *Eragrostis cynosuroides*, *Launaea asplenifolia*, *Parthenium hysterophorus*, *Sporobolus diander* and *Stylosanthes scabra* (with high IVI), exhibited high metal accumulation and tolerance capabilities. Plant species were recommended to be used in sustainable phytomanagement of abandoned red mud dumps.

Enhancement of Phytoremediation Using Soil Amendments Bioavailability of pollutants in soil can be controlled by the amalgamation of either organic or inorganic amendments. Organic amendments include animal manure, sewage sludge, biochar, vegetative dry dust, bacteria and plant-growth promoting rhizobacteria (PGPR), litter waste, woodchips, rice husks, straw, etc. (Gautam et al. 2017; Wiszniewska et al. 2016). Inorganic amendments include bauxite residue, fly ash, ethylenediaminetetraacetic acid (EDTA), oxides of iron and aluminium, silicon (Si), inorganic fertilizers, liming agents (calcium carbonate, $CaCO_3$; and calcium oxide, CaO) and sulphur-containing compounds (hydrogen sulphide, H_2S) (Vu and Gowripalan 2018; Gautam and Agrawal 2017). Shrestha et al. (2019) showed reduced phyto-uptake of metals (Zn, Cd, Pb, Co and Ni) by switchgrass (*Panicum virgatum*) grown in soil amended with vermicompost, thermophilic compost and coconut coir. In a study conducted by Gautam and Agrawal (2017), sewage sludge addition to soil (1:2 (w/w)) enhanced the uptake of Fe, Zn, Cu, Ni, Cr, Pb, Cd

Table 16.4 Plants and microorganisms widely used in remediation of organic and inorganic pollutants from soil

Types of biological organisms	Organic and inorganic pollutants	References
Microorganisms		
Bacillus sp., *Bordetella* sp. and *Pseudomonas* sp.	Co, Zn, Cd and Ni	Das and Dash (2014)
Staphylococcus aureus and Methylococcus capsulatus	Cr	Bawa and Omairi (2017); Das and Dash (2014)
Penicillium chrysogenum and *Pseudomonas putida*	Monocyclic aromatic hydrocarbons	Abatenh et al. (2017)
Pseudomonas sp., *Achromobacter* sp., *Flavobacterium* sp., *Acinetobacter* sp., *Bacillus* sp., *Alcaligenes odorans*, *Arthrobacter* sp., *Citrobacter koseri* and *Serratia ficaria*	Crude oil, petrol and diesel	Abatenh et al. (2017)
Pseudomonas alcaligenes, *Pseudomonas mendocina* and *P. putida*, *Pseudomonas veronii*, *Achromobacter* sp., *Flavobacterium* sp., *Acinetobacter* sp., *Coprinellus radians*, *Candida viswanathii* and *Bacillus licheniformis*	PAHs	Abatenh et al. (2017), Mardani et al. (2016)
Myrothecium roridum, *Pycnoporus sanguineus*, *Phanerochaete chrysosporium*, *Trametes trogii*, *Exiguobacterium* sp., *Bacillus cereus* and *Acinetobacter baumannii*	Industrial dyes	Abatenh et al. (2017)
P. chrysosporium	PCBs	Elangovan et al. (2019)
Aspergillus niger, *Aspergillus fumigatus*, *Fusarium solani*, *Penicillium funiculosum*, *Tyromyces palustris*, *Gloeophyllum trabeum*, *Trametes versicolor*, *Acinetobacter* sp., *Pseudomonas* sp., *Ralstonia* sp. and *Microbacterium* sp.	Aliphatic and aromatic hydrocarbon	Abatenh et al. (2017)
A. odorans, *Bacillus subtilis*, *Corynebacterium propinquum* and *Pseudomonas aeruginosa*	Phenol	Abatenh et al. (2017), Hasan and Jabeen. (2015)
Saccharomyces cerevisiae, *Cunninghamella elegans* and *Escherichia coli*	As	Abatenh et al. (2017), Su et al. (2009)
Pseudomonas fluorescens, *P. aeruginosa* and *Aeromonas* sp.	Cu	Abatenh et al. (2017)
P. fluorescens, *P. aeruginosa*, *Lysinibacillus sphaericus*, *Aerococcus* sp., *Azotobacter* sp. and *Rhodopseudomonas palustris*	Pb	Abatenh et al. (2017), Ashraf (2017)
P. fluorescens, *P. aeruginosa*, *B. subtilis*, *B. cereus*, *Bordetella* sp., *Microbacterium* sp. and *Pseudomonas* sp.	Zn	Abatenh et al. (2017), Ashraf et al. (2017)

(continued)

Table 16.4 (continued)

Types of biological organisms	Organic and inorganic pollutants	References
Aerococcus sp., *R. palustris*, *Bacillus* sp., *Bordetella* sp. and *P. aeruginosa*	Cd	Abatenh et al. (2017), Chellaiah (2018)
P. fluorescens, *P. aeruginosa*, *Geobacter* sp., *Leptospirillum ferrooxidans*, *Acidithiobacillus thiooxidans*, *Acidithiobacillus ferrooxidans* and *A. niger*	Fe	Abatenh et al. (2017), Nguyen and Lee (2015)
Lysinibacillus sphaericus	Co	Abatenh et al. (2017)
P. aeruginosa, *Aeromonas* sp., *B. subtilis*, *B. cereus*, *Bordetella* sp. and *Pseudomonas* sp.	Ni	Abatenh et al. (2017), Ashraf (2017)
S. cerevisiae	Mg	Joutey et al. (2013)
Vibrio alginolyticus, *Brochothrix thermosphacta* and *Moraxella urethralis*	Al	Kurniawan et al. (2018), Titah et al. (2019)
P. fluorescens and *P. aeruginosa*	Mn	Abatenh et al. (2017)
Plants		
Ludwigia octovalvis and *Dracaena reflexa*	Crude oil, petrol and diesel	Almansoory et al. (2015), Agamuthu and Dadrasnia (2013)
Phragmites australis	PAHs	Di Gregorio et al. (2015)
Populus sp. and *Arabidopsis thaliana*	Silver nanoparticles	Wang et al. (2013)
Rhizophora mangle	Total petroleum hydrocarbon	Moreira et al. (2013)
Sparganium sp. and *Aegiceras corniculatum*	PCBs and PBBs	Chen et al. (2015), Gregorio et al. (2015)
Luffa acutangula	Anthracene and fluoranthene	Somtrakoon et al. (2014)
Spartina maritime, *Arabidopsis* sp. and *Corrigiola telephiifolia*	As	Azubuike et al. (2016), Rezania et al. (2016)
S. maritime, *Eichhorina crassipes*, *Cannabis sativa*, *Haumaniastrum katangense* and *Vetiveria zizanioides*	Cu	Gautam and Agrawal (2017), Azubuike et al. (2016)
Spartina maritime, *Plectranthus amboinicus*, *Carex pendula*, *Sorghum halepense* and *Betula occidentalis*, *Helianthus annuus*, *Brassica nigra*, *Medicago sativa* and *Cymbopogon citratus*	Pb	Gautam et al. (2017), Azubuike et al. (2016)
S. maritime, *E. crassipes*, *C. citratus* and *Nicotiana glauca*	Zn	Gautam et al. (2017), Azubuike et al. (2016), Rezania et al. (2016)
E. crassipes, *C. sativa*, *Thlaspi caerulescens*, *Solanum photeinocarpum*, *Rorippa globosa*, *Arabidopsis* sp. and *C. citratus*	Cd	Gautam et al. (2017), Azubuike et al. (2016), Rezania et al. (2016)
E. crassipes and *C. citratus*	Cr	Gautam et al. (2017), Azubuike et al. (2016)

(continued)

Table 16.4 (continued)

Types of biological organisms	Organic and inorganic pollutants	References
Limnocharis flava and *Elodea canadensis*	Fe	Rezania et al. (2016)
E. canadensis	Co	Rezania et al. (2016)
Amaranthus paniculatus, *Nicotiana glauca* and *Alyssum markgrafii*	Ni	Azubuike et al. (2016), Rezania et al. (2016)
C. citratus	Al	Gautam et al. (2017)
C. citratus,	Mg	Gautam et al. (2017)
Vetiveria zizanioides	Mn	Gautam and Agrawal (2017)

PBBs polybrominated biphenyls, *PCBs* polychlorinated biphenyls, *PAHs* polycyclic aromatic hydrocarbons, *Al* aluminium, *Zn* zinc, *Cu* copper, *Mg* magnesium, *Mn* manganese, *Fe* iron, *Co* cobalt, *Ni* nickel, *Cr* chromium, *Cd* cadmium, *As* arsenic, *Pb* lead

and As by *C. zizanioides*; however, their phyto-uptake was decreased significantly with increase in red mud addition (5, 10 and 15% (w/w)) to sludge-amended soil. Similarly, increased phytoavailability of heavy metal(loid)s (Al, Fe, Zn, Co, Cu, Ni, Mn, Mg, Cd, Cr, Pb and As) to lemongrass due to biowaste amendments (cow dung manure and sewage sludge) in soil was remarkably reduced by red mud treatments which was mainly accredited to increased soil pH and oxides of iron and aluminium (Gautam et al. 2017).

Highest accumulation of Pb in roots and shoot was attained by *Eucalyptus camaldulensis* cultivated in Pb-contaminated soil inoculated with *Alcaligenes eutrophus* when compared to other soil amendments (EDTA, compost, Hoagland solution and their mixture). Crude oil-polluted soil of Akala-Olu, Nigeria, characterized by the presence of ten non-carcinogenic and six carcinogenic PAHs was efficiently remediated by *Fimbristylis littoralis* (87%), *Hevea brasiliensis* (92%), *C. citratus* (85%) and *Vigna subterranea* (89%) grown in soil amended with organic as compared to inorganic amendments (Nwaichi et al. 2015). Plants used for phytoremediation of PCBs in contaminated soil include *Lespedeza cuneate*, *Phalaris arundinacea*, *M. sativa*, *Lathyrus sylvestris*, *Salix alaxensis*, *Sparganium* sp. and *Picea glauca* (Jing et al. 2018). Tu et al. (2011) validated a significant reduction in PCB content of soil by 31.4 and 78.4% after the first and second years of study using *M. sativa*.

Microbe-Assisted Phytoremediation The effectiveness of bioremediation could be enhanced using integrated approach of plants and microbes (Qi et al. 2019; Hussain et al. 2018). Microbial community in the rhizospheric region is stimulated by the release of plant rhizodeposits, thus establishing a gradient of interactions between both (Tabassum et al., 2017). The plant-microbe associations influence the pH and composition of root exudates, thereby affecting the bioavailability and phyto-uptake of pollutants in the soil (Sarwar et al. 2017). Non-symbiotic and symbiotic relationships between plants and microbes are making them a single candidate for bioremediation of contaminated soil. Microbes specifically plant growth-promoting rhizobacteria (PGPR) enhance plants' growth and tolerance towards varying environmental stresses by different action mechanisms (solubilizing phosphates, producing phytohormones or fixing nitrogen) or by altering the plant metabolism (increasing the absorption of water and minerals) that consequently increases root development, enzymatic activities of the plant, support growth and development of other microorganisms beneficial to plants and suppress plant pathogens (Jacoby et al. 2017).

Genetic Alterations in Plants for Effective Remediation Many organic and inorganic pollutants are recalcitrant to phytoremediation (Doty 2008). There are many genes involved in metabolism, uptake, translocation and sequestration of pollutants. Thus, the most feasible method for enhancing the effectiveness of phytoremediation is to overexpress those genes in transgenic plants through genetic engineering (Cherian and Oliveira 2005). Depending upon the strategy, transgenic plants can be genetically modified for enhanced accumulation of contaminants in different parts of the plant. There are several available reports on bioremediation of pollutants through transgenic plants (Agnihotri and Seth 2019; Shah and Pathak 2019). Transgenic plants such as *Brassica juncea*, *Arabidopsis thaliana* and *Thlaspi caerulescens* have been reported to effectively remove the heavy metals (Cd, Cr, Pb, Zn and As) from soil (Agnihotri and Seth 2019). Genetically engineered plant (*Arabidopsis thaliana*) with two expressed bacterial genes (one gene converts arsenate into arsenite and the second binds to the arsenite) have been reported to absorb arsenic efficiently from polluted soil and store in the vacuoles (Finnegan and Chen 2012). Phytoremediation of organic pollutants is significantly improved with transgenic plants (Hannink et al. 2002). For an instance, explosives such as 2,4,6-trinitrotoluene (TNT), hexahydro-1,3,5-trinitro-1,3,5-triazine (RDX) and glycerol trinitrate (GTN) are phytotoxic, and phytoremediation of such pollutants through non-transgenic plants is relentlessly hindered (Van Aken 2009). But expression of bacterial genes (nitroreductase, cytochrome P450 and pentaerythritol tetranitrate reductase) specifically involved in degradation of TNT, RDX and GTN, respectively, enables the plants, viz. *Nicotiana tabacum*, *Arabidopsis thaliana* and *Populus* sp., to be more tolerant of the pollutant in phytoremediation.

16.7 Future Prospects

A large area of land has been impacted by the organic and inorganic pollutants due to anthropogenic activities which have become a major environmental issue and health concern worldwide. The analysis for pollutants in soil would be useful in reassuring the soil quality, food and health security of mankind. Several advanced technologies are available to manage waste and contaminated sites, but there is growing interest in in situ and ex situ remediation techniques using green plants, microbes and microbe-assisted phytoremediation. Plants with high biomass production should be accentuated and enhanced through genetic engineering for effective remediation of pollutants from the environment. Similarly, genetic modification of microorganisms used in bioremediation to compete with indigenous microbial population is essential for the successful bioremediation. Bioremediation approach would showcase sustainable and revitalizing strategies to escalate the options for reutilization of contaminated sites, thus impacting the economic gains of the country. Besides, it offers a build-up of stronger communities through partnerships amongst organizations and individuals keeping the socio-economic and environmental concerns in the centre of the table.

16.8 Conclusions

Contamination of soil with organic and inorganic pollutants is the major cause for alteration in plant community structure pattern and crop loss worldwide. Such pollutants beyond threshold levels for animals may lead to several toxic effects through food chain contamination. Remediation of contaminated sites requires attention towards inculcation and development of sustainable approach because although several advanced technologies exist, they are very costly, labour-intensive and also not environmentally benign options. Therefore, utilization of microorganisms, plants and their integrated approach could make a better tactics to degrade xenobiotic compounds. However, phytoremediation using naturally growing hyperaccumulators, different soil amendments, microbial assistance and transgenic plants are the most sustainable and viable approaches to remediate the extensively large areas of polluted land without causing any environmental harm. Thus, there is the need of upcoming era to focus on the research initiatives on the exploration of different bioremediation techniques in stress environment and to unravel the mechanisms involved. Such studies would be helpful in identifying efficient microbe and plant species for effectual bioremediation of contaminated sites. Nonetheless, bioremediation also provides economic, efficient and sustainable remediation technology to manage contaminated sites at global level.

Acknowledgements Authors are thankful to the head and the coordinator, CAS in Botany, coordinators, ISLS, DST-FIST, UGC-UPE, DST-PURSE, Institute of Science, Banaras Hindu University and Varanasi for basic facilities. Meenu Gautam is thankful to the Council of Scientific and Industrial Research (File No. 09/013(0857)/2018-EMR-I), New Delhi, for research associateship.

References

Abatenh E, Gizaw B, Tsegaye Z et al (2017) Application of microorganisms in bioremediation-review. J Environ Microbiol 1(1):02–09

Abhilash PC, Pandey VC, Srivastava P et al (2009) Phytofiltration of cadmium from water by Limnocharis flava (L.) Buchenau grown in free-floating culture system. J Hazard Mater 170:791–797

Adrogué HJ, Madias NE (2014) The impact of sodium and potassium on hypertension risk. Sem Nephrol 34:257–272

Agamuthu P, Dadrasnia A (2013) Dynamics phytoremediation of Zn and diesel fuel in co-contaminated soil using biowastes. J Bioremed Biodeg 4(2):1–5
Agarry SE, Oghenejoboh KM (2015) Enhanced aerobic biodegradation of naphthalene in soil: kinetic modelling and half-life study. Int J Environ Biorem Biodegrad 3:48–53
Agarwal R, Chaudhary M, Singh J (2015) Waste management initiatives in India for human well-being. Eur Sci J 2015:105–127
Agnihotri A, Seth CS (2019) Transgenic Brassicaceae: a promising approach for phytoremediation of heavy metals. In: Prasad MNV (ed) Transgenic plant technology for remediation of toxic metals and metalloids. Academic Press/Elsevier, London, pp 239–255
Ahmed KEM, Frøysa HG, Karlsen OA et al (2019) Effects of defined mixtures of POPs and endocrine disruptors on the steroid metabolome of the human H295R adrenocortical cell line. Chemosphere 218:328–339
Alamgir M (2016) The effects of soil properties to the extent of soil contamination with metals. In: Hasegawa H, Rahman IMM, Rahman MA (eds) Environmental remediation technologies for metal-contaminated soils. Springer, Tokyo, pp 1–19
Alharbi OM, Khattab RA, Ali I (2018) Health and environmental effects of persistent organic pollutants. J Mol Liq 263:442–453
Alloway BJ (2013) Bioavailability of elements in soil. In: Selinus O (ed) Essentials of medical geology. Springer, Dordrecht, pp 351–373
Almansoory AF, Hasan HA, Idris M et al (2015) Potential application of a biosurfactant in phytoremediation technology for treatment of gasoline-contaminated soil. Ecol Eng 84:113–120
Asati A, Pichhode M, Nikhil K (2016) Effect of heavy metals on plants: an overview. Int J Appl Innov Eng Manag 5:2319–4847
Aschner M, Erikson KM, Dorman DC (2005) Manganese dosimetry: species differences and implications for neurotoxicity. Crit Rev Toxicol 35:1–32
Ashraf MA (2017) Persistent organic pollutants (POPs): a global issue, a global challenge. Environ Sci Pollut Res 24:4223–4227
Åslund MLW, Zeeb BA, Rutter A et al (2007) In situ phytoextraction of polychlorinated biphenyl-(PCB) contaminated soil. Sci Total Environ 374:1–12
Avelino J, Cristancho M, Georgiou S et al (2015) The coffee rust crises in Colombia and Central America (2008–2013): impacts, plausible causes and proposed solutions. Food Sec 7:303–321
Azubuike CC, Chikere CB, Okpokwasili GC (2016) Bioremediation techniques–classification based on site of application: principles, advantages, limitations and prospects. World J Microb Biotechnol 32:180
Bacchetta G, Boi ME, Cappai G et al (2018) Metal tolerance capability of *Helichrysum microphyllum* Cambess. subsp. Tyrrhenicum Bacch., Brullo & Giusso: a candidate for phytostabilization in abandoned mine sites. Bull Environ Contam Toxicol 101:758–765
Banerjee P, Hazra A, Ghosh P et al (2019) Solid waste management in India: a brief review. In: Ghosh SK (ed) Waste management and resource efficiency. Springer, Singapore, pp 1027–1049
Bansiwal K, Maheshvari RP (2018) Soil pollution & its solutions. Int J Eng Res Gen Sci 7:550–553
Barbosa DHSG, Vieira HD, Souza RM et al (2004) Field estimates of coffee yield losses and damage threshold by *Meloidogyne exigua*. Nematol Bras 28:49–54
Barr D, Finnamore JR, Bardos RP, Weeks JM, Nathanail CP (2002) Biological methods for assessment and remediation of contaminated land: case studies. Construction Industry Research and Information Association, London. https://www.brebookshop.com/samples/139819.pdf. Assessed on 01/09/2019
Bartrons M, Catalan J, Penuelas J (2016) Spatial and temporal trends of organic pollutants in vegetation from remote and rural areas. Sci Rep 6:25446
Bastida F, Torres IF, Romero-Trigueros C (2017) Combined effects of reduced irrigation and water quality on the soil microbial community of a citrus orchard under semi-arid conditions. Soil Biol Biochem 104:226–237
Bawa S, Omairi T (2017) Bioremediation of higher amounts of chromium (VI) by the methane oxidising bacteria Methylococcus capsulatus. EC Microbiol 12:2–7

Behera BK, Prasad R (2020) Strategies for soil management. In: Behera BK, Prasad R (eds) Environmental technology and sustainability. Elsevier, pp 143–167

Bellard C, Bertelsmeier C, Leadley P et al (2012) Impacts of climate change on the future of biodiversity. Ecol Lett 15:365–377

Bharagava RN, Saxena G, Mulla SI (2020) Introduction to industrial wastes containing organic and inorganic pollutants and bioremediation approaches for environmental management. In: Saxena G, Bharagava R (eds) Bioremediation of industrial waste for environmental safety. Springer, Singapore, pp 1–18

Blaylock MJ, Salt DE, Dushenkov S et al (1997) Enhanced accumulation of Pb in Indian mustard by soil-applied chelating agents. Environ Sci Technol 31:860–865

Bloise A, Punturo R, Catalano M, Miriello D, Cirrincione R (2016) Naturally occurring asbestos (NOA) in rock and soil and relation with human activities: the monitoring example of selected sites in Calabria (southern Italy). Ital J Geosci 135:268–279

Brevik EC, Sauer TJ (2015) The past, present, and future of soils and human health studies. Soil 1:35–46

Bricker OP, Rice KC (1993) Acid rain. Annu Rev Earth Planet Sci 21:151–174

Buchmann MF (2008) NOAA Screening Quick Reference Tables, NOAA OR & R Report 08-1. Office of Response and Restoration Division, National Oceanic and Atmospheric Administration 34, Seattle, WA

Bundschuh J, Litter MI, Parvez F et al (2012) One century of arsenic exposure in Latin America: a review of history and occurrence from 14 countries. Sci Total Environ 429:2–35

Burken JG, Ma X, Struckhoff GC et al (2005) Volatile organic compound fate in phytoremediation applications: natural and engineered systems. Z Naturforsch C 60:208–215

Caçador I, Duarte B (2015) Chromium phyto-transformation in salt marshes: The role of halophytes. In: Ansari A, Gill S, Gill R, Lanza G, Newman L (eds) Phytoremediation. Springer, Cham, pp 211–217

Calvo MS, Moshfegh AJ, Tucker KL (2014) Assessing the health impact of phosphorus in the food supply: issues and considerations. Adv Nutr 5:104–113

Cang L (2004) Heavy metals pollution in poultry and livestock feeds and manures under intensive farming in Jiangsu Province, China. J Environ Sci 16:371–374

Castella JC, Dollon K, Savary S (2005) Path coefficient analysis to assess yield losses due to a multiple pest complex in cotton in Thailand. Int J Trop Insect Sci 25:39–49

Cerda R (2017) Assessment of yield and economic losses caused by pests and diseases in a range of management strategies and production situations in coffee agroecosystems. PhD thesis. Montpellier SupAgro, Montpellier, Fr. https://agritrop.cirad.fr/583804/1/ TESIS_PhD_RCERDA _vf.pdf. Assessed on 13/07/2019

Chellaiah ER (2018) Cadmium (heavy metals) bioremediation by *Pseudomonas aeruginosa*: a minireview. Appl Water Sci 8:154

Chen J, Zhou HC, Wang C, Zhu CQ, Tam NF-Y (2015) Short-term enhancement effect of nitrogen addition on microbial degradation and plant uptake of polybrominated diphenyl ethers (PBDEs) in contaminated mangrove soil. J Hazard Mater 300:84–92

Cherian S, Oliveira MM (2005) Transgenic plants in phytoremediation: recent advances and new possibilities. Environ Sci Technol 39:9377–9390

Clark L (2014) Disease risks posed by wild birds associated with agricultural landscapes. In: Matthews KR, Sapers GM, Gerba CP (eds) The produce contamination problem. Academic Press, London, pp 139–165

Cooke BM (2006) Disease assessment and yield loss. In: Karl R, Matthews, Sapers GM, Gerba CP (eds) The epidemiology of plant diseases, 2nd edn. Springer, Dordrecht, pp 43–80

Coulon F, Al Awadi M, Cowie W et al (2010) When is a soil remediated? Comparison of biopiled and windrowed soils contaminated with bunker-fuel in a full-scale trial. Environ Pollut 158:3032–3040

Cousins IT, Jones KC (1998) Air–soil exchange of semi-volatile organic compounds (SOCs) in the UK. Environ Pollut 102:105–118

Cousins IT, Beck AJ, Jones KC (1999) A review of the process involved in the exchange of semi-volatile organic compounds (SVOC) across the air-soil interface. Sci Total Environ 228:5–24
Cunningham SD, Berti WR (1993) Remediation of contaminated soils with green plants: an overview. In Vitro Cell Dev Biol 29:207–212
Daniel H, Perinaz BT (2012) What a waste: a global review of solid waste management. Urban development series knowledge paper. The World Bank, Washigton
Das S (ed) (2014) Microbial biodegradation and bioremediation, 1st edn. Elsevier, pp 1–642
Das S, Dash HR (2014) Microbial bioremediation: a potential tool for restoration of contaminated areas. In: Das S (ed) Microbial biodegradation and bioremediation. Elsevier, London, pp 1–21
Dash HR, Das S (2012) Bioremediation of mercury and the importance of bacterial mergenes. Int Biodeter Biodegr 75:207–213
Davidson E, David M, Galloway J (2012) Excess nitrogen in the U.S. environment: trends, risks, and solutions. Issues in Ecol 15:1–16
Deardorff T, Karch N, Holm S (2008) Dioxin levels in ash and soil generated in Southern California fires. Organohalogen Compd 70:2284–2288
Deurer M, Böttcher J (2007) Evaluation of models to upscale the small scale variability of Cd sorption in a case study. Geoderma 137:269–278
Díaz-Ramírez IJ, Escalante-Espinosa E, Favela-Torres E et al (2008) Design of bacterial defined mixed cultures for biodegradation of specific crude oil fractions, using population dynamics analysis by DGGE. Int Biodeterior Biodegrad 62:21–30
Di-Nicolantonio JJ, Mehta V, O'Keefe JH (2017) Is salt a culprit or an innocent bystander in hypertension? A hypothesis challenging the ancient paradigm. Am J Med 130:893–899
Dœlsch E, Van de Kerchove V, Saint Macary H (2006) Heavy metal content in soils of Réunion (Indian Ocean). Geoderma 134:119–134
Doty SL (2008) Enhancing phytoremediation through the use of transgenics and endophytes. New Phytol 179:318–333
Edelstein M, Ben-Hur M (2018) Heavy metals and metalloids: sources, risks and strategies to reduce their accumulation in horticultural crops. Sci Hortic 234:431–444
EEA (2014) Progress in management of contaminated sites. European Environment Agency. https://www.eea.europa.eu/data-and-maps/indicators/progress-in-management-of-contaminated-sites/progress-in-management ofcontaminated 1. Assessed on 03/08/2019
Ekta P, Modi NR (2018) A review of phytoremediation. J Pharmacogn Phytochem 7:1485–1489
Elangovan S, Pandian SBS, Geetha SJ et al (2019) Polychlorinated Biphenyls (PCBs): environmental fate, challenges and bioremediation. In: Arora PK (ed) Microbial metabolism of xenobiotic compounds. Springer, Singapore, pp 165–188
FAO & ITPS (2015) Status of the World's Soil Resources (SWSR) – main report. Food and Agriculture Organization of the United Nations and Intergovernmental Technical Panel on Soils, Rome, Italy. http://www.fao.org/3/a-i5199e.pdf. Assessed on 22/07/2019
FAO/WHO (Codex Alimentarius Commission) (2001) Food additives and contaminants. Joint FAO/WHO Food Standards Program, ALINORM 01/12A:1-289
Ferro A, Kennedy J, Doucette W, Nelson S, Jauregui G, Mc Farland B, Bugbee B (1997) Fate of benzene in soils planted with alfalfa: uptake, volatilization, and degradation. In: Kruger EL, Anderson TA, Coats JR (eds) Phytoremediation of soil and water contaminants. ACS Symposium Series, Washington, DC, pp 664, 223–237
Fijałkowski K, Kacprzak M, Grobelak A et al (2012) The influence of selected soil parameters on the mobility of heavy metals in soils. Inżynieria I Ochrona Środowiska 15:81–92
Finnegan P, Chen W (2012) Arsenic toxicity: the effects on plant metabolism. Front Physiol 3:182
Fornalczyk A, Willner J, Francuz K et al (2013) E-waste as a source of valuable metals. Arch Mater Sci Eng 63:87–92
Frascari D, Zanaroli G, Danko AS (2015) In situ aerobic cometabolism of chlorinated solvents: a review. J Hazard Mater 283:382–399
Fungo B, Lehmann J, Kalbitz K et al (2019) Ammonia and nitrous oxide emissions from a field Ultisol amended with tithonia green manure, urea, and biochar. Biol Fertil Soils 55:135–148

García-Pérez J, Boldo E, Ramis R et al (2007) Description of industrial pollution in Spain. BMC Public Health 7:40

Gautam M, Agrawal M (2017) Phytoremediation of metals using vetiver (*Chrysopogon zizanioides* (L.) Roberty) grown under different levels of red mud in sludge amended soil. J Geochem Explor 182:218–227

Gautam M, Agrawal M (2019) Identification of metal tolerant plant species for sustainable phytomanagement of abandoned red mud dumps. J Appl Geochem 104:83–92

Gautam M, Pandey D, Agrawal SB, Agrawal M (2016) Metals from mining and metallurgical industries and their toxicological impacts on plants. In: Singh A, Prasad SM, Singh RP (eds) Plant responses to Xenobiotics. Springer, Singapore, pp 231–272

Gautam M, Pandey D, Agrawal M (2017) Phytoremediation of metals using lemongrass (*Cymbopogon citratus* (DC) Stapf.) grown under different levels of red mud in soil amended with biowastes. Int J Phytorem 19:555–562

Gautam M, Pandey B, Agrawal M (2018) Identification of indicator species at abandoned red mud dumps in comparison to residential and forest sites, accredited to soil properties. Ecol Indic 88:88–102

Gayana BC, Chandar KR (2018) Sustainable use of mine waste and tailings with suitable admixture as aggregates in concrete pavements-a review. Adv Concr Constr 6:221–243

Ghosh M, Singh SP (2005) A review on phytoremediation of heavy metals and utilization of it's by products. Asian J Energy Environ 6:18

Ghosh PS, Das MT, Thakur IS (2014) In vitro toxicity evaluation of organic extract of landfill soil and its detoxification by indigenous pyrene-degrading *Bacillus* sp. ISTPY1. Int Biodeterior Biodegrad 90:145–151

Gidarakos E, Aivalioti M (2007) Large scale and long term application of bioslurping: the case of a Greek petroleum refinery site. J Hazard Mater 149:574–581

Gonsior SJ, Sorci JJ, Zoellner MJ et al (1997) The effects of EDTA on metal solubilization in river sediment/water systems. J Environ Qual 26:957–966

Gregorio S, Giorgetti L, Castiglione MR et al (2015) Phytoremediation for improving the quality of effluents from a conventional tannery wastewater treatment plant. Int J Environ Sci Technol 12:1387–1400

Grimble RF (2006) The effects of sulfur amino acid intake on immune function in humans. J Nutri 136(6):1660S–1665S

Hajar EWI, Sulaiman AZB, Sakinah AM (2014) Assessment of heavy metals tolerance in leaves, stems and flowers of *Stevia rebaudiana* plant. Procedia Environ Sci 20:386–393

Hannink NK, Rosser SJ, Bruce NC (2002) Phytoremediation of explosives. Critic Rev Plant Sci 21:511–538

Harter RD, Naidu R (2001) An assessment of environmental and solution parameter impact on trace-metal sorption by soils. Soil Sci Soc Am J 65:597–612

Hasan SA, Jabeen S (2015) Degradation kinetics and pathway of phenol by Pseudomonas and Bacillus species. Biotechnol Biotechnol Equip 29:45–53

Hassaan MA, El Nemr A, Madkour FF (2016) Environmental assessment of heavy metal pollution and human health risk. Am J Water Sci Eng 2:14–19

Hawkes JC, Pyatt DG, White IMS (1997) Using Ellenberg indicator values to assess soil quality in British forests from ground vegetation: a pilot study. J Appl Ecol:375–387

Haynes RJ, Murtaza G, Naidu R (2009) Inorganic and organic constituents and contaminants of biosolids: implications for land application. Adv Agron 104:165–267

Helmisaari HS, Salemaa M, Derome J (2007) Remediation of heavy metal–contaminated forest soil using recycled organic matter and native woody plants. J Environ Qual 36:1145–1153

Hobson AM, Frederickson J, Dise NB (2005) CH_4 and N_2O from mechanically turned windrow and vermicomposting systems following in-vessel pre-treatment. Waste Manag 25:345–352

Hoornweg D, Bhada-Tata P (2012) What a waste: a global review of solid waste management. Urban development series; knowledge papers no. 15. World Bank, Washington, DC. https://openknowledge.worldbank.org/handle/10986 /17388. Assessed on 11/09/2019

Huang HB, Chen GW, Wang CJ et al (2013) Exposure to heavy metals and polycyclic aromatic hydrocarbons and DNA damage in Taiwanese traffic conductors. Cancer Epidemiol Pre Biomark 22:102–108

Hussain F, Hussain I, Khan AHA et al (2018) Combined application of biochar, compost, and bacterial consortia with Italian ryegrass enhanced phytoremediation of petroleum hydrocarbon contaminated soil. Environ Exp Bot 153:80–88

Igiri BE, Okoduwa SI, Idoko GO et al (2018) Toxicity and bioremediation of heavy metals contaminated ecosystem from tannery wastewater: a review. J Toxicol 2018(2568038):1–16

Indian Bureau of Mines (2002) Indian minerals year book. Indian Bureau of Mines Press, Nagpur (1947–1991)

Institute of Medicine (1997) Dietary reference intakes for calcium, phosphorus, magnesium, vitamin D and fluoride. Standing Committee on the Scientific Evaluation of Dietary Reference Intakes, National Academies Press

Islam MS, Sikder MT, Kurasaki M (2017) Potential of *Micranthemum umbrosum* for phytofiltration of organic arsenic species from oxic water environment. Int J Environ Sci Technol 14:285–290

Jacoby R, Peukert M, Succurro A et al (2017) The role of soil microorganisms in plant mineral nutrition-current knowledge and future directions. Front Plant Sci 8:1617

Jadia CD, Fulekar MH (2009) Phytoremediation of heavy metals: recent techniques. Afr J Biotechnol 8(6):921–928

Jadiyappa S (2018) Radioisotope: applications, effects and occupational protection. In: Rahman ROA, Saleh HEM (eds) Principles and applications in nuclear engineering: radiation effects, thermal hydraulics, radionuclide migration in the environment. Intech Open, New Delhi, pp 19–47

Jeanson S, Floury J, Gagnaire V et al (2015) Bacterial colonies in solid media and foods: a review on their growth and interactions with the micro-environment. Front Microbiol 6:1284

Jing R, Fusi S, Kjellerup BV (2018) Remediation of polychlorinated biphenyls (PCBs) in contaminated soils and sediment: state of knowledge and perspectives. Front Environ Sci 6:79

Jomova K, Jenisova Z, Feszterova M et al (2011) Arsenic: toxicity, oxidative stress and human disease. J Appl Toxicol 31:95–107

Joutey NT, Bahafid W, Sayel et al (2013) Biodegradation: involved microorganisms and genetically engineered microorganisms. Biodegrad Life Sci:289–320

Kabata-Pendias A, Pendias H (2011) Trace elements in soils and plants. CRC, Boca Raton

Kanter DR (2018) Nitrogen pollution: a key building block for addressing climate change. Clim Change 147:11–21

Karami A, Shamsuddin ZH (2010) Phytoremediation of heavy metals with several efficiency enhancer methods. Afr J Biotechnol 9:3689–3698

Karthikeyan R, Davis LC, Erickson LE et al (2004) Potential for plant-based remediation of pesticide-contaminated soil and water using nontarget plants such as trees, shrubs, and grasses. Crit Rev Plant Sci 23:91–101

Kaspary TE, Lamego FP, Cutti L et al (2014) Determination of photosynthetic pigments in fleabane biotypes susceptible and resistant to the herbicide glyphosate. Planta Daninha 32:417–426

Klotz K, Weistenhöfer W, Neff F et al (2017) The health effects of aluminum exposure. Dtsch Arztebl Int 114:653

Köchy M, Hiederer R, Freibauer A (2015) Global distribution of soil organic carbon–Part 1: masses and frequency distributions of SOC stocks for the tropics, permafrost regions, wetlands and the world. Soil 1:351–365

Kodešová R, Kočárek M, Kodeš V et al (2011) Pesticide adsorption in relation to soil properties and soil type distribution in regional scale. J Hazard Mater 186:540–550

Komaba H, Fukagawa M (2016) Phosphate-a poison for humans? Kidney Int 90:753–763

Krebs J, Bach S (2018) Permaculture-scientific evidence of principles for the agroecological design of farming systems. Sustainability 10:3218

Kromdijk J, Long SP (2016) One crop breeding cycle from starvation? How engineering crop photosynthesis for rising CO_2 and temperature could be one important route to alleviation. Proc R Soc B Biol Sci 283:20152578

Küçük D, Liman R (2018) Cytogenetic and genotoxic effects of 2-chlorophenol on *Allium cepa* L. root meristem cells. Environ Sci Pollut Res 25:36117–36123

Kumar V, Kothiyal NC, Saruchi (2016) Analysis of polycyclic aromatic hydrocarbon, toxic equivalency factor and related carcinogenic potencies in roadside soil within a developing city of Northern India. Polycycl Aromat Compd 36:506-526

Kurniawan SB, Purwanti IF, Titah HS (2018) The Effect of pH and aluminium to bacteria isolated from aluminium recycling industry. J Ecol Eng 19(3):154–161

Kvesitadze G, Khatisashvili G, Sadunishvili T et al (2015) Plants for remediation: Uptake, translocation and transformation of organic pollutants. In: Öztürk M, Ashraf M, Aksoy A, Ahmad M, Hakeem K (eds) Plants, pollutants and remediation. Springer, Dordrecht, pp 241–308

Lacatusu AR, Lacatusu R, Dumitru M et al (2017) Decontamination of a petroleum hydrocarbons polluted soil by different bioremediation strategies. Ann Univ Craiova-Agric Montanol Cadastre Ser 46:326–334

Landner L, Reuther R (2004) A critical review of current knowledge on fluxes, speciation, bioavailability and risk for adverse effects of copper, chromium, nickel and zinc. In: Landner L, Reuther R (eds) Metals in society and in the environment. Springer, Cham, pp 1–339

Li A (2009) PAHs in comets: an overview. In: Käufl HU, Sterken C (eds) Deep impact as a world observatory event: synergies in space, time, and wavelength. Springer, Berlin/Heidelberg, pp 161–175

Li G, Sun GX, Ren Y, Luo XS et al (2018) Urban soil and human health: a review. Eur J Soil Sci 69:196–215

Limmer M, Burken J (2016) Phytovolatilization of organic contaminants. Environ Sci Technol 50:6632–6643

Lin DR, Hu LJ, Xing BS et al (2015) Mechanisms of competitive adsorption organic pollutants on hexylene-bridged polysilsesquioxane. Materials 8:5806–5817

Liu X, Zhang W, Hu Y et al (2015) Arsenic pollution of agricultural soils by concentrated animal feeding operations (CAFOs). Chemosphere 119:273–281

Lv J, Christie P, Zhang S (2019) Uptake, translocation, and transformation of metal-based nanoparticles in plants: recent advances and methodological challenges. Environ Sci Nano 6(1):41–59

Lloyd JR, Lovely DR (2001) Microbial detoxification of metals and radionuclides. Curr Opin Biotechnol 12:248–253

Lyons G (2005) Policy requirements for protecting wildlife from endocrine disruptors. Environ Health Perspect 114:142–146

Ma Q, Han L, Zhang J et al (2019) Environmental risk assessment of metals in the volcanic soil of Changbai mountain. Int J Environ Res Pub Health 16:2047

Mahjoub B (2013) Plants for soil remediation. In: Gaspard S, Ncibi MC (eds) Biomass for sustainable applications: pollution remediation and energy. RSC Publication, Cambridge, pp 25–106

Malik ZH, Ravindran KC (2018) Biochemical tolerance of *Suaeda maritima* L. (Dumort) as a potential species for phytoextracting heavy metal and salt in paper mill effluent contaminated soil. J Drug Deliv Therapeu 8:241–245

Mandaliev PN, Mikutta C, Barmettler K et al (2013) Arsenic species formed from arsenopyrite weathering along a contamination gradient in circumneutral river floodplain soils. Environ Sci Technol 48:208–217

Mansouri A, Cregut M, Abbes C et al (2017) The environmental issues of DDT pollution and bioremediation: a multidisciplinary review. Appl Biochem Biotechnol 181:309–339

Mardani G, Mahvi AH, Hashemzadeh-Chaleshtori M et al (2016) Degradation of phenanthrene and pyrene using genetically engineered dioxygenase producing Pseudomonas putida in soil. Genetika 48:837–858

Martinoia E (2018) Vacuolar transporters–Companions on a longtime journey. Plant Physiol 176:1384–1407

Masih A, Taneja A (2006) Polycyclic aromatic hydrocarbons (PAHs) concentrations and related carcinogenic potencies in soil at a semi-arid region of India. Chemosphere 65:449–456
Matinde E, Simate GS, Ndlovu S (2018) Mining and metallurgical wastes: a review of recycling and re-use practices. J South Afr Inst Min Metall 118:825–844
McGrath SP, Zhao FJ, Lombi E (2002) Phytoremediation of metals, metalloids and radionuclides. Adv Agron 75:1–56
Meagher RB (2000) Phytoremediation of toxic elemental and organic pollutants. Curr Opin Plant Biol 3:153–162
Mileusnić M, Mapani BS, Kamona AF et al (2014) Assessment of agricultural soil contamination by potentially toxic metals dispersed from improperly disposed tailings, Kombat mine, Namibia. J Geochem Explor 144:409–420
Mirsal IA (2008) Sources of soil pollution. In: Mirsal IA (ed) Soil Pollution: Origin, monitoring & remediation, 2nd edn. Springer, Dillenburg/Berlin, pp 137–173
Mishra S, Bharagava RN, More N et al (2019) Heavy metal contamination: an alarming threat to environment and human health. In: Sobti R, Arora N, Kothari R (eds) Environmental biotechnology: for sustainable future. Springer, Singapore, pp 103–125
Moosavi SG, Seghatoleslami MJ (2013) Phytoremediation: a review. Adv Agric Biol 1:5–11
Moreira IT, Oliveira OM, Triguis JA (2013) Phytoremediation in mangrove sediments impacted by persistent total petroleum hydrocarbons (TPH's) using *Avicennia schaueriana*. Mar Pollut Bull 67:130–136
Mosca Angelucci D, Tomei MC (2016) Ex situ bioremediation of chlorophenol contaminated soil: comparison of slurry and solid-phase bioreactors with the two-step polymer extraction and bioregeneration process. J Chem Technol Biotechnol 91:1577–1584
Mukhopadhyay S, Maiti SK (2010) Phytoremediation of metal mine waste. Appl Ecol Environ Res 8:207–222
Newman LA, Reynolds CM (2004) Phytodegradation of organic compounds. Curr Opin Biotechnol 15:225–230
Nguyen VK, Lee JU (2015) Effect of sulfur concentration on microbial removal of arsenic and heavy metals from mine tailings using mixed culture of *Acidithiobacillus* spp. J Geochem Explor 148:241–248
Nikolopoulou M, Pasadakis N, Norf H, Kalogerakis N (2013) Enhanced ex situ bioremediation of crude oil contaminated beach sand by supplementation with nutrients and rhamnolipids. Mar Pollut Bull 77:37–44
Nwaichi EO, Frac M, Nwoha PA et al (2015) Enhanced phytoremediation of crude oil-polluted soil by four plant species: effect of inorganic and organic bioaugumentation. Int J Phytorem 17:1253–1261
Oerke EC (2006) Crop losses to pests. J Agric Sci 144:31–43
Olaniran A, Balgobind A, Pillay B (2013) Bioavailability of heavy metals in soil: impact on microbial biodegradation of organic compounds and possible improvement strategies. Int J Mol Sci 14:10197–10228
Osman NI, Chapple CR, Abrams P et al (2014) Detrusor underactivity and the underactive bladder: a new clinical entity? A review of current terminology, definitions, epidemiology, aetiology and diagnosis. Eur Urol 65:389–398
Padmavathiamma PK, Ahmed M, Rahman HA (2014) Phytoremediation-a sustainable approach for contaminant remediation in arid and semi-arid regions–a review. Emir J Food Agric:757–772
Pandey B, Agrawal M, Singh S (2014) Coal mining activities change plant community structure due to air pollution and soil degradation. Ecotoxicology 23:1474–1483
Pappu A, Saxena M, Asolekar SR (2007) Solid wastes generation in India and their recycling potential in building materials. Build Environ 42:2311–2320
Patinha C, Armienta A, Argyraki A, Durães N (2018) Inorganic pollutants in soils. In: Duarte AC, Cachada A, Rocha-Santo T (eds) Soil pollution. Academic Press/Elsevier, Portugal, pp 127–159
Paulose B, Datta SP, Rattan RK, Chhonkar PK (2007) Effect of amendments on the extractability, retention and plant uptake of metals on a sewage irrigated soil. Environ Pollut 146:19–24

Petruzzelli G, Pedron F, Rosellini I et al (2015) In: Ansari AA, Gill SS, Gill R, Lanza GR, Newman L (eds) Phytoremediation. Springer, pp 31–43

Pezeshki SR, DeLaune RD (2000) Effects of soil oxidation-reduction conditions on internal oxygen transport, root aeration, and growth of wetland plants. In: Holland Marjorie M, Warren Melvin L, Stanturf John A (eds) Proceedings of a conference on sustainability of wetlands and water resources. Oxford, Mississippi, pp 139–145

Philp JC, Atlas RM (2005) Bioremediation of contaminated soils and aquifers. In: Atlas RM, Philp JC (eds) Bioremediation: applied microbial solutions for real-world environmental cleanup. American Society for Microbiology (ASM) Press, Washington

Pimentel D, Burgess M (2014) Environmental and economic costs of the application of pesticides primarily in the United States. In: Pimentel D, Peshin R (eds) Integrated pest management. Springer, Dordrecht, pp 47–71

Prasad R (2014) Major sulphur compounds in plants and their role in human nutrition and health–An overview. Proc Indian Natl Sci Acad 80:1045–1054

Prasad R (2017) Mycoremediation and environmental sustainability, vol 1. Springer International Publishing. ISBN 978-3-319-68957-9. https://link.springer.com/book/10.1007/978-3-319-68957-9

Prasad R (2018) Mycoremediation and environmental sustainability, vol 2. Springer International Publishing. ISBN 978-3-319-77386-5. https://www.springer.com/us/book/9783319773858

Proust D (2015) Sorption and distribution of Zn in a sludge-amended soil: influence of the soil clay mineralogy. J Soils Sediments 15:607–622

Qi X, Hao X, Chen X et al (2019) Integrated phytoremediation system for uranium-contaminated soils by adding a plant growth promoting bacterial mixture and mowing grass. J Soils Sediments 19:1799–1808

Rakotonindraina T, Chauvin JE, Pellé R et al (2012) Modeling of yield losses caused by potato late blight on eight cultivars with different levels of resistance to *Phytophthora infestans*. Plant Dis 96:935–942

Ramadass K, Megharaj M, Venkateswarlu K et al (2015) Ecological implications of motor oil pollution: earthworm survival and soil health. Soil Biol Biochem 85:72–81

Rampazzo N, Todorovic GR, Mentler A et al (2013) Adsorption of glyphosate and amino methyl phosphonic acid in soils. Int Agrophy 27:203–209

Rana SS, Rana MC (2015) Advances in Weed Management. Department of Agronomy, College of Agriculture, CSK Himachal Pradesh Krishi

Rezania S, Taib SM, Din MFM et al (2016) Comprehensive review on phytotechnology: heavy metals removal by diverse aquatic plants species from wastewater. J Hazard Mater 318:587–599

Rhind SM (2009) Anthropogenic pollutants: a threat to ecosystem sustainability? Philos Trans R Soc B Biol Sci 364:3391–3401

Riaz M, Arif MS, Ashraf MA, Mahmood R et al (2019) A comprehensive review on rice responses and tolerance to salt stress. In: Hasanuzzaman M, Fujita M, et al (eds) Advances in rice research for abiotic stress tolerance. Woodhead Publishing, pp 133–158

Rodríguez-Eugenio N, McLaughlin M, Pennock D (2018) Soil pollution: a hidden reality. Food and Agriculture Organization of United Nation, Rome, pp 1–142

Romanovskaia VA, Sokolov IG, Rokitko PV et al (1998) Ecological consequences of radioactive pollution for soil bacteria within the 10-km region around the Chernobyl Atomic Energy Station. Mikrobiologiia 67:274–280

Roy M, Giri AK, Dutta S, Mukherjee P (2015) Integrated phytobial remediation for sustainable management of arsenic in soil and water. Environ Int 75:180–198

Różański SŁ, Castejón JMP, Fernández GG (2016) Bioavailability and mobility of mercury in selected soil profiles. Environ Earth Sci 75:1065

Saha JK, Selladurai R, Coumar MV et al (2017) Major inorganic pollutants affecting soil and crop quality. In: Soil pollution – an emerging threat to agriculture. environmental chemistry for a sustainable world, vol 10. Springer, Singapore, pp 75–104

Sakakibara M, Watanabe A, Inoue M et al (2010) Phytoextraction and phytovolatilization of arsenic from As-contaminated soils by *Pteris vittata*. In: Proceedings of the annual international conference on soils, sediments, water and energy, vol 12, article 26. https://scholarworks.

umass.edu/soilsproceedings/vol12/iss1/26?utm_source=scholarworks.umass.edu%2Fsoilsproceedings%2Fvol12%2Fiss1%2F26&utm_medium=PDF&utm_campaign=PDFCoverPages. Assessed on 14/08/2019

Salem HM, Abdel-Salam A, Abdel-Salam MA et al (2017) Soil xenobiotics and their phyto-chemical remediation. In: Hashmi M, Kumar V, Varma A (eds) Xenobiotics in the soil Environment, Soil biology, vol 49. Springer, Cham, pp 267–280

Sarma H, Forid N, Prasad R, Prasad MNV, Ma LQ, Rinklebe J (2021) Enhancing phytoremediation of hazardous metal(loid)s using genome engineering CRISPR–Cas9 technology. J Hazard Mater. https://doi.org/10.1016/j.jhazmat.2021.125493

Sarwar N, Imran M, Shaheen MR et al (2017) Phytoremediation strategies for soils contaminated with heavy metals: modifications and future perspectives. Chemosphere 171:710–721

Savary S, Willocquet L (2014) Simulation modeling in botanical epidemiology and crop loss analysis. Plant Health Instr:1–173

Savary S, Willocquet L, Elazegui FA (2000) Rice pest constraints in tropical Asia: quantification of yield losses due to rice pests in a range of production situations. Plant Dis 84:357–369

Savary S, Ficke A, Aubertot JN, Hollier C (2012) Crop losses due to diseases and their implications for global food production losses and food security. Food Security 4:519–537

Saxena G, Marzinelli EM, Naing NN et al (2015) Ecogenomics reveals metals and land-use pressures on microbial communities in the waterways of a megacity. Environ Sci Technol 49:1462

Schröder P, Collins C (2002) Conjugating enzymes involved in xenobiotic metabolism of organic xenobiotics in plants. Int J Phytorem 4:247–265

Seigle-Murandi F, Guiraud P, Croize J et al (1996) Bacteria are omnipresent on *Phanerochaete chrysosporium* Burdsall. Appl Environ Microbiol 62:2477–2481

Shah K, Pathak L (2019) Transgenic energy plants for phytoremediation of toxic metals and metalloids. In: Narasimha M, Prasad V (eds) Transgenic plant technology for remediation of toxic metals and metalloids. Academic Press, London, pp 319–340

Shankar S, Shikha (2017) Management and remediation of problem soils, solid waste and soil pollution. In: Singh R (ed) Principles and applications of environmental biotechnology for a sustainable future, Applied environmental science and engineering for a sustainable future. Springer, Singapore, pp 143–171

Sharma S (2012) Bioremediation: features, strategies and applications. Asian J Pharm Life Sci 2 (2):202–213

Shrestha P, Bellitürk K, Görres JH (2019) Phytoremediation of heavy metal-contaminated soil by switchgrass: a comparative study utilizing different composts and coir fiber on pollution remediation, plant productivity and nutrient leaching. Int J Environ Res Public Health 16:1261

Sifakis S, Androutsopoulos VP, Tsatsakis AM et al (2017) Human exposure to endocrine disrupting chemicals: effects on the male and female reproductive systems. Environ Toxicol Pharmacol 51:56–70

Silva-Castro GA, Uad I, Go´nzalez-Lo´pez J et al (2012) Application of selected microbial consortia combined with inorganic and oleophilic fertilizers to recuperate oil-polluted soil using land farming technology. Clean Technol Environ Policy 14:719–726

Solid Waste Management, World Bank (2018). https://www.worldbank.org/en/topic/urbandevelopment/brief/solid-waste-management. Assessed on 12/08/2019

Somtrakoon K, Chouychai W, Lee H (2014) Phytoremediation of anthracene-and fluoranthene-contaminated soil by *Luffa acutangula*. Maejo Int J Sci Technol 8:221

Stewart WM, Dibb DW, Johnston AE (2005) The contribution of commercial fertilizer nutrients to food production. Agronomy J 97:1–6

Su YJ, Lin JQ, Lin JQ, Hao DH (2009) Bioaccumulation of arsenic in recombinant Escherichia coli expressing human metallothionein. Biotechnol Bioprocess Eng14(5):565–570

Suman J, Uhlik O, Viktorova J, Macek T (2018) Phytoextraction of heavy metals: a promising tool for clean-up of polluted environment? Front Plant Sci 9:1476

Swartjes FA, Tromp PC (2008) A tiered approach for the assessment of the human health risks of asbestos in soils. Soil Sediment Contam 17:137–149

Tabassum B, Khan A, Tariq M et al (2017) Bottlenecks in commercialization and future prospects of PGPR. App Soil Ecol 121:102–117
Tadesse GL, Guya TK, Walabu M (2017) Impacts of tannery effluent on environments and human health: a review article. Adv Life Sci Technol 54:10
Taiz L, Zeiger E (2002) Plant physiology. Sinauer Associates, Sunderland
Tangahu BV, Abdullah S, Rozaimah S et al (2011) A review on heavy metals (As, Pb, and Hg) uptake by plants through phytoremediation. Int J Chem Eng 2011(939161):1–31
Tauqeer HM, Hussain S, Abbas F et al (2019) The potential of an energy crop "*Conocarpus erectus*" for lead phytoextraction and phytostabilization of chromium, nickel, and cadmium: An excellent option for the management of multi-metal contaminated soils. Ecotox Environ Saf 173:273–284
Thakare M, Sarma H, Datar S, Roy A, Pawar P, Gupta K, Pandit S, Prasad R (2021) Understanding the holistic approach to plant-microbe remediation technologies for removing heavy metals and radionuclides from soil. Curr Res Biotechnol. https://doi.org/10.1016/j.crbiot.2021.02.004
Titah HS, Purwanti IF, Tangahu BV et al (2019) Kinetics of aluminium removal by locally isolated Brochothrix thermosphacta and Vibrio alginolyticus. J Environ Manag 238:194–200
Trendel JM, Lohmann F, Kintzinger JP et al (1989) Identification of des-A-triterpenoid hydrocarbons occurring in surface sediments. Tetrahedron 45:4457–4470
Tsao DT (2003) Overview of phytotechnologies. In: Tsao DT (ed) Phytoremediation. Advances in biochemical engineering/biotechnology, vol 78. Springer, Berlin/Heidelberg, pp 1–50
Tu C, Teng Y, Luo YM, Sun XH, Deng SP, Li ZG, Liu WX, Xu ZH (2011) PCB removal, soil enzyme activities, and microbial community structures during the phytoremediation by alfalfa in field soils. J Soils Sediments 11:649–656
Uddin MK (2017) A review on the adsorption of heavy metals by clay minerals, with special focus on the past decade. Chem Eng J 308:438–462
United States Environmental Protection Agency (2006) Radionuclides in soil RADTOWN, USA. Office of Radiation and Indoor Air (6608J). http://large.stanford.edu/courses/2014/ph241/eller1/docs/soil.pdf. Assessed on 02/07/2019
US EPA (2008) Green remediation incorporating: sustainable environmental practices into remediation of contaminated sites. Office of Solid Waste and Emergency Response, Washington, DC
USEPA (2000) Electrokinetic and phytoremediation in situ treatment of metal-contaminated soil: stateof- the practice. Draft for Final review. EPA/542/R-00/XXX. United States Environmental Protection Agency, Office of Solid Waste and Emergency Response Technology Innovation Office, Washington, DC
Van Aken B (2009) Transgenic plants for enhanced phytoremediation of toxic explosives. Curr Opinion Biotechnol 20:231–236
Van Dingenen J, Blomme J, Gonzalez N et al (2016) Plants grow with a little help from their organelle friends. J Exp Bot 66(22):6267–6281
Van Leeuwen GCM, Stein A, Holb I et al (2000) Yield loss in apple caused by *Monilinia fructigena* (Aderh. & Ruhl.) Honey, and spatio-temporal dynamics of disease development. Eur J Plant Pathol 106:519–528
Varjani SJ, Joshi RR, Kumar PS et al (2018) Polycyclic aromatic hydrocarbons from petroleum oil industry activities: effect on human health and their biodegradation. In: Varjani S, Gnansounou E, Gurunathan B, Pant D, Zakaria Z (eds) Waste bioremediation. energy, environment, and sustainability. Springer, Singapore, pp 185–199
Vásquez Murrieta MS, Hernández Hernández OJ, Cruz Maya JA et al (2016) Approaches for removal of PAHs in soils: Bioaugmentation, biostimulation and bioattenuation. In: Soil contamination-current consequences and further solutions. InTech, Rijeka, pp 329–342
Venuti A, Alfonsi L, Cavallo A (2016) Anthropogenic pollutants on top soils along a section of the Salaria state road, central Italy. Ann Geophys
Viehweger K (2014) How plants cope with heavy metals. Bot Stud 55:35
Vijayan VS (2011) Report on monitoring of endosulfan residues in the 11 panchayaths of Kasaragod district, Kerala, Kerala State Council for Science, Technology and Environment

Sasthra Bhavan, Thiruvananthapuram. http_cdn.cseindia.org_attachments_0.03602700 _1498929209_endo_report1%2 0(1).pdf. Assessed on 22/07/2019
Violante A, Cozzolino V, Perelomov L et al (2010) Mobility and bioavailability of heavy metals and metalloids in soil environments. J Soil Sci Plant Nutr 10:268–292
Vranova V, Rejsek K, Formanek P (2013) Aliphatic, cyclic, and aromatic organic acids, vitamins, and carbohydrates in soil: a review. Sci World J:1–15
Vu TH, Gowripalan N (2018) Mechanisms of heavy metal immobilisation using geopolymerisation techniques–a review. J Adv Concrete Technol 16:124–135
Vyslouzilova M, Tlustos P, Száková J et al (2003) As, Cd, Pb and Zn uptake by *Salix* spp. clones grown in soils enriched by high loads of these elements. Plant Soil Environ 49:191–196
Wadgaonkar SL, Ferraro A, Nancharaiah YV et al (2019) In situ and ex situ bioremediation of seleniferous soils from northwestern India. J Soils Sediments 19:762–773
Wang J, Han R (2019) Removal of pesticide on food by electrolyzed water. In: Ding T, Oh DH, Liu D (eds) Electrolyzed water in food: fundamentals and applications. Springer, Singapore, pp 39–65
Wang J, Feng X, Anderson CWN, Xing et al (2012) Remediation of mercury contaminated sites – a review. J Hazard Mater 221:1–18
Wang J, Koo Y, Alexander A et al (2013) Phytostimulation of poplars and Arabidopsis exposed to sliver nanoparticles and Ag? at sublethal concentrations. Environ Sci Technol 47:5442–5449
Wania F, McLachlan MS (2001) Estimating the influence of forests on the overall fate of semivolatile organic compounds using a multimedia fate model. Environ Sci Technol 35:582–590
Whelan MJ, Coulon F, Hince G, Rayner J et al (2015) Fate and transport of petroleum hydrocarbons in engineered biopiles in Polar Regions. Chemosphere 131:232–240
White JC, Wang X, Gent MP et al (2003) Subspecies-level variation in the phytoextraction of weathered p, p ‘-DDE by *Cucurbita pepo*. Environ Sci Technol 37:4368–4373
Whiting SJ, Wood RJ (1997) Adverse effects of high-calcium diets in humans. Nutr Rev 55:1–9
Wiszniewska A, Hanus-Fajerska E, Muszyńska E et al (2016) Natural organic amendments for improved phytoremediation of polluted soils: a review of recent progress. Pedosphere 26:1–12
Wuana RA, Felix EO (2011) Heavy metals in contaminated soils: a review of sources, chemistry, risks and best available strategies for remediation. Int Scholar Res Netw 2011(402647):1–20
Xiong D, Huang J, Peng S et al (2017) A few enlarged chloroplasts are less efficient in photosynthesis than a large population of small chloroplasts in *Arabidopsis thaliana*. Sci Rep 7:5782
Xu C, Peng C, Sun L et al (2015) Distinctive effects of TiO_2 and CuO nanoparticles on soil microbes and their community structures in flooded paddy soil. Soil Biol Biochem 86:24–33
Yadav A, Chowdhary P, Kaithwas G et al (2017) Toxic metals in environment, threats on ecosystem and bioremediation approaches. In: S. Das & Singh (eds.) Handbook of metal-microbe interactions and bioremediation. Boca Raton: CRC Press/Taylor & Francis Group, p 813
Yang X, Guschina IA, Hurst S et al (2010) The action of herbicides on fatty acid biosynthesis and elongation in barley and cucumber. Pest Manag Sci 66:794–800
Yeung AT (2010) Remediation technologies for contaminated sites. In: Chen Y, Zhan L, Tang X (eds) Advances in environmental geotechnics. Springer, Berlin/Heidelberg, pp 328–369
Zhang X, Wang H, He L et al (2013) Using biochar for remediation of soils contaminated with heavy metals and organic pollutants. Environ Sci Pollut Res 20:8472–8483
Zhang Y, Zhai W, Zhao M et al (2015a) Effects of iron overload on the bone marrow microenvironment in mice. PloS one 10:e0120219
Zhang Y, Zhai W, Zhao M, Li D et al (2015b) Effects of iron overload on the bone marrow microenvironment in mice. PloS one 10:1–17
Zhang C, Yao F-E-N-G, Liu YW et al (2017) Uptake and translocation of organic pollutants in plants: a review. J Integer Agric 16:1659–1668
Zong Y, Lu S (2019) Does long-term inorganic and organic fertilization affect soil structural and mechanical physical quality of paddy soil? Archiv Agro Soil Sci https://doi.org/10.1080/03650340.2019.1630823. Assessed on 05/08/2019

GPSR Compliance

The European Union's (EU) General Product Safety Regulation (GPSR) is a set of rules that requires consumer products to be safe and our obligations to ensure this.

If you have any concerns about our products, you can contact us on ProductSafety@springernature.com

In case Publisher is established outside the EU, the EU authorized representative is:

Springer Nature Customer Service Center GmbH
Europaplatz 3
69115 Heidelberg, Germany

Batch number: 10755222

Printed by Printforce, the Netherlands